IET CONTROL, ROBOTICS AND SENSORS SERIES 09

Embedded Mechatronics System Design for Uncertain Environments

Other volumes in this series:

Volume 8 **A History of Control Engineering, 1800–1930** S. Bennett
Volume 18 **Applied Control Theory, 2nd Edition** J.R. Leigh
Volume 20 **Design of Modern Control Systems** D.J. Bell, P.A. Cook and N. Munro (Editors)
Volume 28 **Robots and Automated Manufacture** J. Billingsley (Editor)
Volume 33 **Temperature Measurement and Control** J.R. Leigh
Volume 34 **Singular Perturbation Methodology in Control Systems** D.S. Naidu
Volume 35 **Implementation of Self-tuning Controllers** K. Warwick (Editor)
Volume 37 **Industrial Digital Control Systems, 2nd Edition** K. Warwick and D. Rees (Editors)
Volume 39 **Continuous Time Controller Design** R. Balasubramanian
Volume 40 **Deterministic Control of Uncertain Systems** A.S.I. Zinober (Editor)
Volume 41 **Computer Control of Real-time Processes** S. Bennett and G.S. Virk (Editors)
Volume 42 **Digital Signal Processing: Principles, devices and applications** N.B. Jones and J.D.McK. Watson (Editors)
Volume 44 **Knowledge-based Systems for Industrial Control** J. McGhee, M.J. Grimble and A. Mowforth (Editors)
Volume 47 **A History of Control Engineering, 1930–1956** S. Bennett
Volume 49 **Polynomial Methods in Optimal Control and Filtering** K.J. Hunt (Editor)
Volume 50 **Programming Industrial Control Systems Using IEC 1131-3** R.W. Lewis
Volume 51 **Advanced Robotics and Intelligent Machines** J.O. Gray and D.G. Caldwell (Editors)
Volume 52 **Adaptive Prediction and Predictive Control** P.P. Kanjilal
Volume 53 **Neural Network Applications in Control** G.W. Irwin, K. Warwick and K.J. Hunt (Editors)
Volume 54 **Control Engineering Solutions: A practical approach** P. Albertos, R. Strietzel and N. Mort (Editors)
Volume 55 **Genetic Algorithms in Engineering Systems** A.M.S. Zalzala and P.J. Fleming (Editors)
Volume 56 **Symbolic Methods in Control System Analysis and Design** N. Munro (Editor)
Volume 57 **Flight Control Systems** R.W. Pratt (Editor)
Volume 58 **Power-plant Control and Instrumentation: The control of boilers and HRSG systems** D. Lindsley
Volume 59 **Modelling Control Systems Using IEC 61499** R. Lewis
Volume 60 **People in Control: Human factors in control room design** J. Noyes and M. Bransby (Editors)
Volume 61 **Nonlinear Predictive Control: Theory and practice** B. Kouvaritakis and M. Cannon (Editors)
Volume 62 **Active Sound and Vibration Control** M.O. Tokhi and S.M. Veres
Volume 63 **Stepping Motors, 4th Edition** P.P. Acarnley
Volume 64 **Control Theory, 2nd Edition** J.R. Leigh
Volume 65 **Modelling and Parameter Estimation of Dynamic Systems** J.R. Raol, G. Girija and J. Singh
Volume 66 **Variable Structure Systems: From principles to implementation** A. Sabanovic, L. Fridman and S. Spurgeon (Editors)
Volume 67 **Motion Vision: Design of compact motion sensing solution for autonomous systems** J. Kolodko and L. Vlacic
Volume 68 **Flexible Robot Manipulators: Modelling, simulation and control** M.O. Tokhi and A.K.M. Azad (Editors)
Volume 69 **Advances in Unmanned Marine Vehicles** G. Roberts and R. Sutton (Editors)
Volume 70 **Intelligent Control Systems Using Computational Intelligence Techniques** A. Ruano (Editor)
Volume 71 **Advances in Cognitive Systems** S. Nefti and J. Gray (Editors)
Volume 72 **Control Theory: A guided tour, 3rd Edition** J. R. Leigh
Volume 73 **Adaptive Sampling with Mobile WSN** K. Sreenath, M.F. Mysorewala, D.O. Popa and F.L. Lewis
Volume 74 **Eigenstructure Control Algorithms: Applications to aircraft/rotorcraft handling qualities design** S. Srinathkumar
Volume 75 **Advanced Control for Constrained Processes and Systems** F. Garelli, R.J. Mantz and H. De Battista
Volume 76 **Developments in Control Theory towards Glocal Control** L. Qiu, J. Chen, T. Iwasaki and H. Fujioka (Editors)
Volume 77 **Further Advances in Unmanned Marine Vehicles** G.N. Roberts and R. Sutton (Editors)
Volume 78 **Frequency-Domain Control Design for High-Performance Systems** J. O'Brien
Volume 80 **Control-oriented Modelling and Identification: Theory and practice** M. Lovera (Editor)
Volume 81 **Optimal Adaptive Control and Differential Games by Reinforcement Learning Principles** D. Vrabie, K. Vamvoudakis and F. Lewis
Volume 83 **Robust and Adaptive Model Predictive Control of Nonlinear Systems** M. Guay, V. Adetola and D. DeHaan
Volume 84 **Nonlinear and Adaptive Control Systems** Z. Ding
Volume 86 **Modeling and Control of Flexible Robot Manipulators, 2nd edition** M. O. Tokhi and A.K.M. Azad
Volume 88 **Distributed Control and Filtering for Industrial Systems** M. Mahmoud
Volume 89 **Control-based Operating System Design** A. Leva *et al.*
Volume 90 **Application of Dimensional Analysis in Systems Modelling and Control Design** P. Balaguer
Volume 91 **An Introduction to Fractional Control** D. Valério and J. Costa
Volume 92 **Handbook of Vehicle Suspension Control Systems** H. Liu, H. Gao and P. Li
Volume 93 **Design and Development of Multi-Lane Smart Electromechanical Actuators** F.Y. Annaz
Volume 94 **Analysis and Design of Reset Control Systems** Y.Guo, L. Xie and Y. Wang
Volume 95 **Modelling Control Systems Using IEC 61499, 2nd Edition** R. Lewis & A. Zoitl
Volume 96 **Cyber-Physical System Design with Sensor Networking Technologies** S. Zeadally and N. Jabeur (Editors)
Volume 99 **Practical Robotics and Mechatronics: Marine, Space and Medical Applications** I. Yamamoto
Volume 100 **Organic Sensors: Materials and Applications** E Garcia-Breijo and P Cosseddu (Editors)
Volume 102 **Recent Trends in Sliding Mode Control** L. Fridman JP. Barbot and F. Plestan (Editors)
Volume 104 **Control of Mechatronic Systems** L. Guvenc, B.A Guvenc, B. Demirel, M.T Emirler
Volume 105 **Mechatronic Hands: Prosthetic and Robotic Design** P.H. Chappell
Volume 107 **Solved Problems in Dynamical Systems and Control** D. Valério, J. T Machado, A. M. Lopes and A. M. Galhano
Volume 108 **Wearable Exoskeleton Systems: Design, Control and Applications** S. Bai, G.S. Virk and T.G.Sugar
Volume 111 **The Inverted Pendulum in Control Theory and Robotics: From Theory to New Innovations** O. Boubaker and R. Iriarte (Editors)
Volume 112 **RFID Protocol Design, Optimization, and Security for the Internet of Things** A. X. Liu, M. Shahzad, X Liu and K. Li
Volume 113 **Design of Embedded Robust Control Systems Using MATLAB® / Simulink®** P.H. Petkov, T.N. Slavov and J.K. Kralev
Volume 114 **Signal Processing and Machine Learning for Brain-Machine Interfaces** T. Tanaka and M. Arvaneh (Editor)
Volume 119 **Swarm Intelligence Volumes** 1 – 3 Y. Tan (Editor)
Volume 121 **Integrated Fault Diagnosis and Control Design of Linear Complex Systems** M. Davoodi, N. Meskin and K.Khorasani

Embedded Mechatronics System Design for Uncertain Environments

Linux®-based, Rasbpian®, ARDUINO® and MATLAB® xPC Target Approaches

Cheng Siong Chin

The Institution of Engineering and Technology

Published by The Institution of Engineering and Technology, London, United Kingdom

The Institution of Engineering and Technology is registered as a Charity in England & Wales (no. 211014) and Scotland (no. SC038698).

First published 2018

The Institution of Engineering and Technology
Michael Faraday House
Six Hills Way, Stevenage
Herts, SG1 2AY, United Kingdom

www.theiet.org

British Library Cataloguing in Publication Data
A catalogue record for this product is available from the British Library

ISBN 978-1-78561-322-7 (hardback)
ISBN 978-1-78561-323-4 (PDF)

Typeset in India by MPS Limited

Contents

Foreword xi
Acknowledgments xiii
Outlines xv

1 Introduction 1
1.1 Introduction to embedded system 1
1.2 Example of embedded system using Athena III PC104 2
1.3 Example of embedded systems using ARDUINO® 3
1.4 Example of embedded system using Raspberry Pi 5
1.5 Example of embedded system using PIC 7
1.6 Motivations 8
1.7 Systematic design approach for prototyping embedded systems 10
References 15

2 Linux®-based embedded system design 17
2.1 Linux® operating system 17
2.2 Building Linux® for embedded systems 20
2.3 Program layouts in Linux® 25
2.4 System design and architecture 27
2.4.1 Main process design 30
2.4.2 Sensor process design 33
2.4.3 Sensor fusion thread design 37
2.4.4 Control process design 39
2.4.5 Actuator driver design 39
2.4.6 Network communication thread design 40
2.5 Testing of components for control systems 42
2.5.1 Inertial measurement unit 42
2.5.2 DVL sensor unit 47
2.5.3 Image video unit 50
2.5.4 Depth sensor unit 56
2.6 Kalman filter 59
2.7 Graphical user interface 59
References 67

3 Modeling and simulation of embedded underwater vehicle system 69
3.1 Introduction 69
3.2 Overview of remotely operated underwater vehicle 69

3.3 Dynamics modeling of remotely operated underwater vehicle 71
3.3.1 Hydrodynamic damping model 73
3.3.2 Hydrodynamic-added mass model 79
3.4 Validation of experimental results 84
3.4.1 Heave model identification 84
3.4.2 Yaw model identification 88
3.5 Simulation of remotely operated underwater vehicle model 92
3.6 Simulating external disturbance for remotely operated underwater vehicle model 97
3.7 Launch and recovery process model 101
3.8 Control systems design 102
3.8.1 Sliding-mode control 104
3.8.2 Proposed fuzzy-based genetic algorithm for SMC 105
3.8.3 Proportional-integral-derivative 110
3.9 Remotely operated underwater vehicle sea trial 117
References 119

4 xPC-Target embedded system design 121
4.1 Introduction 121
4.2 Overview of hardware interfacings for simulations testing 123
4.3 Hardware interfacings 124
4.4 Hardware-in-the-loop testing using xPC-Target 134
4.4.1 Create xPC-Target real-time kernel using desktop PC as target PC 136
4.4.2 Create xPC-Target real-time kernel using Athena II-PC104 as target PC 139
4.5 Creating xPC-Target Simulink® block diagrams 142
4.6 Using RS232, analog, and digital I/O in xPC-Target 151
4.7 Infrared sensor model 158
4.8 Incremental encoder model 159
4.9 Identification of a servo DC motor 164
4.10 PID speed control of servo DC motor 168
4.11 Sliding-model speed control of servo DC motor 169
4.12 Linear quadratic regulator 171
4.13 Digital speed control of servo DC motor 175
4.14 Case study: marine robotic vehicle with uncertainties using xPC-Target system 177
4.14.1 System design and architecture 178
4.14.2 Underwater robotic vehicle dynamic model 181
4.14.3 Steady-state thruster's dynamics 183
4.14.4 Underwater robotic vehicle—horizontal subsystem model 188
4.14.5 Controller design 199
4.14.6 Implementation and testing 202
References 207

5 PIC embedded system design **211**
5.1 Overview of MPLAB IDE 211
5.2 Intelligent vacuum robot system design 212
5.2.1 System design and architecture 212
5.2.2 Programming and system implementation 216
5.2.3 Testing 234
5.3 Remote temperature-sensing system design for patients 235
5.3.1 System design and architecture 236
5.3.2 Programming and system implementation 240
5.3.3 Testing 241
5.4 Wall-climbing robot system design 243
5.4.1 System design and architecture 245
5.4.2 Programming and system implementation 253
5.4.3 Testing 259
5.5 Magnetic conveyor system design 260
5.5.1 System design and architecture 261
5.5.2 Programming and system implementation 274
5.5.3 Testing 284
References 287

6 ARDUINO® embedded system design **291**
6.1 Remotely operated vehicle system design 291
6.1.1 System design and architecture 291
6.1.2 Programming and system implementation 303
6.1.3 Testing 307
6.2 Smart control of marine-tracked vehicle for surveillance 313
6.2.1 System design and architecture 314
6.2.2 Programming and system implementation 320
6.2.3 Testing 325
6.3 A sloth bear-inspired pole-climbing robot 327
6.3.1 System design and architecture 327
6.3.2 Programming and system implementation 335
6.3.3 Testing 343
References 344

7 Raspberry Pi-embedded system design **345**
7.1 Fouling detection system 345
7.1.1 System design and architecture 347
7.1.2 Programming and system implementation 367
7.1.3 Testing 381
7.2 Multi-hop microprocessor-based prototype system for remote vibration and image monitoring 387
7.2.1 System design and architecture 388
7.2.2 Programming and system implementation 394
7.2.3 Testing 395

7.3 Face recognition system 399
7.3.1 System design and architecture 399
7.3.2 Programming and system implementation 399
7.3.3 GUI using PyQt 419
7.3.4 Testing 421
References 438

Index **441**

Foreword

Rapid prototyping of an embedded control system design is mainly constrained by the system configurations, specifications, and the operating system. Book publications that communicate how to create such an embedded system from initial concepts (mechanical and electrical), modeling and simulation (mathematical relationships), and graphical-user interface (software) to actual implementation (mechatronics system testing) are not available. Hence, this makes the control system design difficult to realize when dealing with physical mechatronic systems design during design stage.

There are many mechatronic systems of interest in the current literature and industrial applications. The application on a marine vehicle such as an underwater robotic vehicle (URV) reflects the multidisciplinary that unifies mathematical aspects from the various and separate engineering disciplines. URVs are used for a wide range of applications, such as exploring the extreme depths of our ocean, harvesting resources ranging from food to oil, monitoring our environment, supporting national defense operations, transporting goods, and supporting recreational activities. Most of the systems operate in an extreme environment with challenges such as great hydrostatic pressures, strong forces, harmful interaction with electronics, and high attenuation of electromagnetic signals. The human–machine interaction and real-world applications of the systems have complicated the embedded system design process. There is also a strong demand for sensor platforms for monitoring and development in method to enable the development of smarter and more efficient systems in both prototyping stage as well as in actual implementation. In addition, other land robotic vehicles for vacuum cleaning and climbing robot for inspection facing similar problems are demonstrated in this book. Applications on magnetic levitation for conveyor system, temperature monitoring, face recognition and vision-based inspection using camera and vibration sensor working in uncertain environment are shown.

In order to realize a more efficient and sustainable embedded system design for control system design, the low-cost microcontrollers were often used to construct a motion control system for a typical robotic system. Although the position control was easily realized, more advanced control applications are above the ability of the cards, such as the implementation of different control algorithms for speed and position control in underwater, information flow, and real-time requirement, etc. The above-mentioned control systems consist of ADCs, DACs, SBCs, or DSPs which are interfaced to workstations or host PCs. Most often, the real-time operating systems are needed, such as QNX, Unix, or RT-Linux®. In these systems, much work is needed to deal with the interface codes, drivers, and communications among the hardware. Although it is more cost-effective and computation time-efficient especially for

Linux®-based operating system, it often requires much expertise on computer engineering and low-level programming than control system design. As a result, the time spent on computer programming increases with significant time allocated for control algorithm design decreases.

Hence, in this book, the xPC-Target™ by MathWorks for prototyping, simulation, testing, and deploying real-time systems using standard PC hardware is proposed. The xPC Target can be seen in the wide range of applications from automotive system design, process control, vibration control, control engineering education to the robotic systems. However, the conventional-based design using Linux®-based operating system and the use of PIC-Microchip™, ARDUINO®, and Raspberry Pi™ are also documented.

This book has been prepared for the needs of those who seek an application on developing embedded system easy for control purpose. The choice of material is guided by the basic objective of making an engineer or student capable of dealing with embedded system design. The book can be used as a textbook, reference book, laboratory exercise for a first course in Linux®-based embedded system, xPC-Target, PIC-Microchip, ARDUINO®, and Raspberry Pi. For more information about MATLAB® and xPC-Target, I encourage you to use the help, manual, and online resources in the respective packages provided.

The contents of the book are organized into seven main chapters with program codes. The main chapters include sections on Linux®-based system design, model-based simulation, and xPC-Target system design. In these chapters, it contains applications on various real-life systems such as the underwater robotics vehicle, vacuum cleaning, climbing robot for inspection, magnetic levitation for conveyor system, temperature monitoring, face recognition and vision-based inspection using camera, and vibration sensor. As it is not possible to cover every single subject in each chapter, the chapters therefore emphasize on certain topics in the embedded systems design process. The software adopted throughout the chapters are Ubuntu/Fedora™, Python™, Qt™, MATLAB®/Simulink®, and other open-source software development kits from PIC-Microchip, ARDUINO®, and Raspberry Pi.

- Chapter 1 gives an introduction to *embedded system design*.
- Chapter 2 discusses the embedded system design software and configurations using *Linux®-based operating system* and *system architecture design*. It also provides the implementation and testing of Linux®-based system on an underwater vehicle.
- Chapter 3 provides the *modeling and simulation of embedded system* on the underwater vehicle.
- Chapter 4 illustrates embedded system prototyping and testing using *xPC-Target*.
- Chapter 5 presents the use of microcontroller such as *PIC-Microchip* in real applications.
- Chapter 6 shows the use of microcontroller such as *ARDUINO®* in real applications.
- Chapter 7 demonstrates the use of microcontroller such as *Raspberry Pi* in real applications.

Acknowledgments

Finally, I wish to acknowledge my great debt to all those who have contributed to the development of embedded system designs into the exciting field it is today and specifically to the considerable help and education I have received from final-year project students, industry partners, and my colleagues. In particular, I would first like to thank my late master thesis advisor Professor Neil Munro, who had made control engineering alive by using computer-aided software. I had benefited from his knowledge in the robust control and the computer-aided control engineering. I would also like to express my sincere gratitude to Dr. Micheal Wai Shing Lau for his guidance and assistance during the tough time in Nanyang Technological University. Also, I thank my former Research Assistant Mr. Weipeng Lin and project students who contributed to the embedded system work and testing using Ubuntu and Fedora for all these years. In addition, I would also like to thank Prof. Tuncer Ören from the School of Information Technology and Engineering in University of Ottawa, Canada, for his motivation and advice. Finally, I would like to thank the editorial staff and reviewers in providing comments to the text.

I would like to express my gratitude to my family. In particular, I would like to thank Irene (my wife), Giselle (late elder daughter), Millicent (my second daughter), Tessalyn (third daughter), and Alyssa (youngest daughter) for their support and understanding, without which the undertaking would not have been possible. I would like to share my joy and results with my strong and kind daughter, Giselle. Although she passed away at the age of four after a few consecutive open-heart surgeries, without her present and altitude toward life, many things would not have happened.

Associate Professor Cheng Siong Chin
PhD, MSc, BEng, Dip. Mech
FHEA, FIMarEST, SMIEEE, Eur Ing, MIET, CEng,
Newcastle University in Singapore

Outlines

This book provides how to create an embedded system from initial concepts to modeling and simulation, programming to the actual implementation. Rapid prototyping of an embedded control system design is constrained by the system configurations, specifications, and the operating system. There are many mechatronic systems of interest in the current literature and industry applications. The application on a vehicle such as an underwater robotic vehicle (URV) and land robots reflects the multidisciplinary that unifies mathematical aspects from the various and separate engineering disciplines. Applications on magnetic levitation for conveyor system, temperature monitoring, face recognition and vision-based inspection using camera and vibration sensor working in uncertain environment are also shown. Most of the systems operate in an extreme environment with challenges such as great hydrostatic pressures, model uncertainties, strong forces, harmful interaction with electronics, and high attenuation of electromagnetic signals. The human–machine interaction and real-world applications of the systems have complicated the embedded system design process.

In order to realize a more efficient and sustainable embedded system design for control system design, the low-cost microcontrollers were often used to construct a motion embedded control system. In these systems, much work is needed to deal with the interface codes, drivers, and communications among the hardware. Although it is more cost-effective and computation time-efficient especially for Linux®-based operating system, it often requires much expertise on computer engineering and low-level programming than control system design. As a result, the time spent on computer programming increases with significant time allocated for control algorithm design decreases. Hence, in this book, the xPC-Target by MathWorks for prototyping, simulation, testing, and deploying real-time systems using standard PC hardware is proposed for ease of programming. However, the conventional-based design using Linux®-based operating system and the use of PIC-Microchip™, ARDUINO®, and Raspberry Pi™ are also documented for completeness. In summary, the proposed embedded systems design process can be summarized into system design and architecture, software programming, system implementation, and testing. The software adopted throughout the chapters are Ubuntu/Fedora, Python™, Qt™, MATLAB®/Simulink®, and other open-source software development kits from PIC-Microchip, ARDUINO®, and Raspberry Pi.

Chapter 1
Introduction

1.1 Introduction to embedded system

In today's competitive markets, engineers face an ongoing challenge to produce complex engineering systems with a high level of performance, reliability, and value at a low price. This demand is brought about by the advances in microprocessor technologies (small and cheap) that make what was state of the art yesterday fade in comparison with what might be possible tomorrow. Hence, a mechatronics engineer is the one who views a system as a whole and offers optimum embedded solution problems. To perform correctly, contemporary systems and products rely on harmonious interactions between mechanical systems, sensors, actuators, and computers to realize with embedded control systems as shown in Figure 1.1.

Therefore, a mechatronics engineer should transcend barriers that existed in the past between various engineering disciplines and acquire the necessary skills and expertise that enable him/her to:

- Select and integrate mechanical components and drives to the microprocessor.
- Select sensors and implement appropriate interfacing circuits.
- Develop a model of the processes involved.
- Design, simulate, and implement appropriate control systems.
- Use software and hardware to build the mechatronic systems.

Besides, the plant dynamics is not known and is highly nonlinear with uncertainties due to varying loads and environment. Hence, it is essential to understand the use of embedded control system to overcome the system or plant working in an uncertain environment.

There exist a few definitions of what is an embedded system. In this book, an embedded system is a compact computer system used to accomplish a few specific tasks. It is usually embedded into a completed system that includes physical mechanical and electrical parts. The embedded system is tasked to work on preprogrammed jobs. However, with artificial intelligence, it can be intelligent to handle different scenarios. Typically, the embedded systems are produced in a small form factor with inputs and outputs (for digital inputs/outputs, analog inputs/outputs, high-definition multimedia interface (HDMI) port, universal serial bus (USB) port, and other port configurations) for broad applications. The human operator has direct

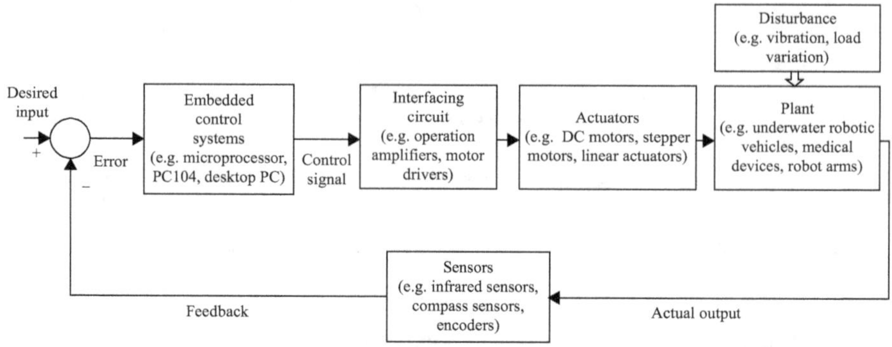

Figure 1.1 Example of the typical embedded control system

interaction with the system through debugging interfaces, firmware, supporting software, and network to obtain the performance of the embedded system. It enables the output responses or messages to be monitored. Sometimes, tailored inputs can be used to obtain insight into the behavior of the systems.

The PIC™ microcontroller, ARDUINO®, and Raspberry Pi™ microcontroller are the common types of microcontrollers used in the embedded system. For example, Raspberry Pi is widely used as a low-cost small single-board computer (SBC) within the embedded system. Also, other SBCs that have several embedded form-factors like PC/104 and PC/104-Plus are commonly used. They are Athena III PC104, Athena III PC/104, Helios PC/104, Helix PC/104, Rhodeus PC/104 SBC, Aries PC/104-Plus, Atlas PCI/104-Express, and Venus 3.5 Inch Core i7/i5 from Diamond Systems Corporation. The users can further optimize the power consumption and performance based on the applications and engineering cost factor. The subsequent paragraph will show the example of embedded systems used in different applications. They are briefly described in this section to give the readers the types of embedded systems design and challenges faced in these applications.

1.2 Example of embedded system using Athena III PC104

The ROV [1–4] in Figure 1.1 is used to dock and recover a smaller underwater vehicle during the sea mission. The vehicle is installed with six electric brushless thrusters. The following sensor and actuators are connected to an SBC using Diamond Systems ATHENA III PC104 with additional data acquisition (DAQ) card operating via Ubuntu operating software in Figure 1.2. There are six thrusters to move the vehicle in designated directions.

In Figure 1.3, the host PC with a graphical user interface was developed by open-source platform using Python-based integrated development environment (IDE) to communicate with the remotely operated underwater vehicle via an Ethernet connection. The remaining components are housed in an aluminum enclosure. The vehicle has an umbilical cable linked to the surface vessel for communication, and

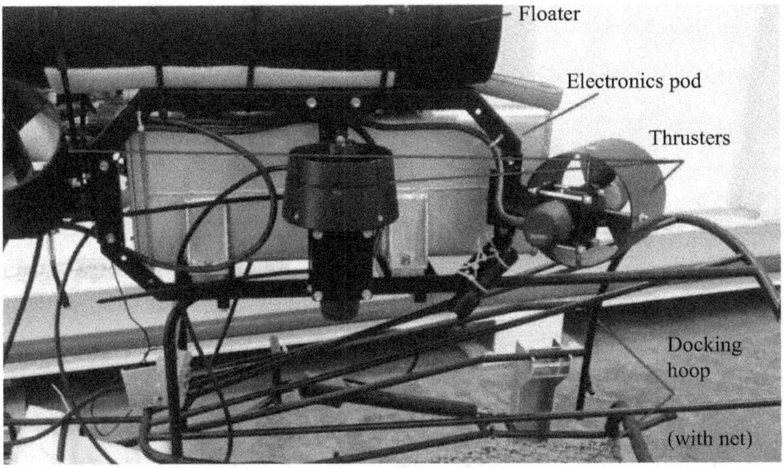

Figure 1.2 Remotely operated underwater vehicle prototype with docking hoop near to the sea surface [2]

at the same time to provide the power (230 V AC power). The cage with a net (see Figure 1.2) attached to the bottom of the vehicle is opened and closed by a pair of linear actuators to hold the recovered vehicle in position. For example, the main components used in the embedded system are as follows.

(a) NavQuest Doppler velocity log (DVL) to measure the velocity of the remotely operated underwater vehicle.
(b) Druck pressure sensor to measure the depth of the remotely operated underwater vehicle.
(c) XSENS inertia measurement unit (IMU) to measure the positions and angular positions of the remotely operated underwater vehicle.
(d) Two USB cameras for underwater image capturing.
(e) USB Thrustmaster Joystick for remote control.
(f) Six DC Tecnadyne brushless electric thrusters connected to an ISO-6 thruster driver card to control the speed of the six propellers.
(g) Two ultramotion linear actuators with step drive to lower and raise the docking hoop.

1.3 Example of embedded systems using ARDUINO®

Another low-cost remotely operated underwater vehicle using microcontroller from ARDUINO® is shown in Figure 1.4. The motions of the vehicle include moving in the direction of the *X*- (forward), *Z*- (heave), and the rotation around the *Z*-axes (yaw). There is a webcam camera that faced front for navigation and inspection purpose. There are two horizontal thrusters and two vertical thrusters to maneuver the vehicle in the shallow water.

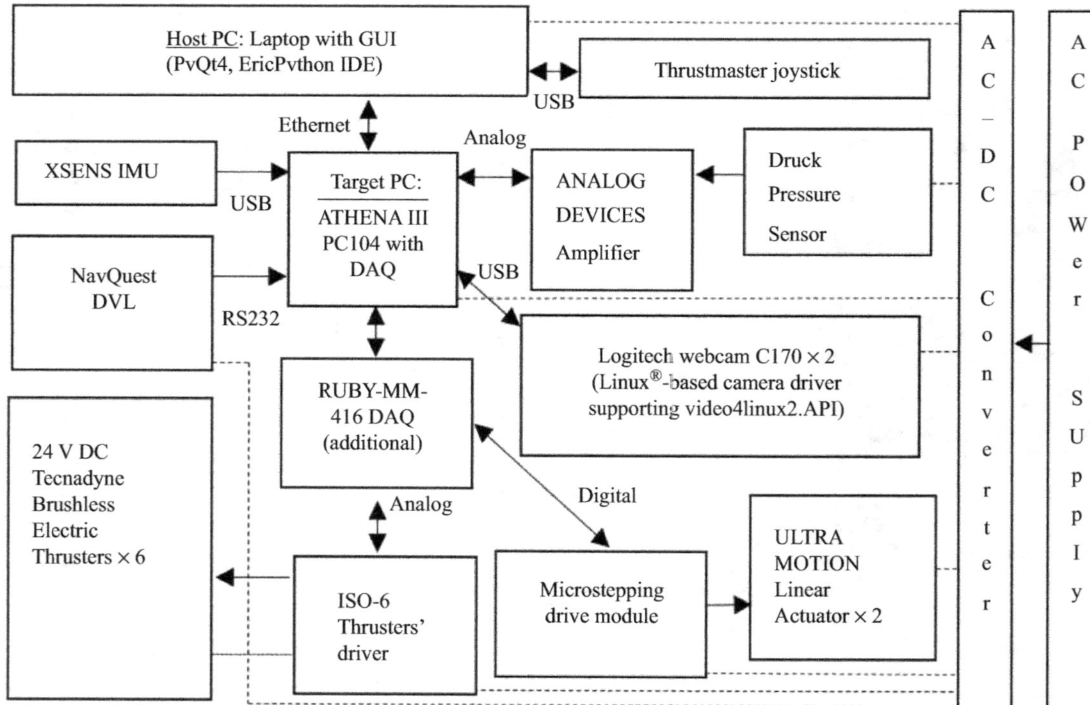

Figure 1.3 System architecture design for a remotely operated underwater vehicle

Figure 1.4 Basic components of the low-cost remotely operated underwater vehicle

ARDUINO® Yun™ is needed to communicate with the remote control (iPhone) and laptop to control the thrusters and the webcam. ARDUINO® Yun has a microcontroller called ATmega32U4 microcontroller and a microprocessor named Atheros AR9331, running Linux® and the OpenWrt wireless stack. The Yun works with Wi-Fi. The Wi-Fi-enabled remotely operated underwater vehicle can be controlled by an iPhone application, "Yun buddy," in shallow water. ARDUINO® Yun receives and analyzes commands from the "Yun Buddy" through the router and controls the trajectories of the vehicle accordingly. As shown in Figure 1.5, four motors are connected to the ARDUINO® Yun. The communication link to establish communication between the ARDUINO® Yun, laptop, and iPhone via wireless local area network (WLAN) can be set up. The "Yun Buddy" interface displays a list of possible commands which allow the available directional controls for the vehicle. The USB webcam is attached to the ARDUINO® Yun to achieve real-time video feedback. The ARDUINO® Yun and the computer will connect to the router's network. Secure Shell (SSH) is used on Ubuntu Linux® to set up a connection between the ARDUINO® Yun and the webcam.

1.4 Example of embedded system using Raspberry Pi

The next example of an embedded system is an image recognition of biofouling [5–7] on ship's hull using Raspberry Pi (see Figure 1.6). The printed fouling images were used to simulate the presence of the fouling on the ship's hull. The left-hand side shows the microprocessor connected via mobile connection (situated on board the "ship") into cloud storage and to the local machine on the right side. The host device ("onshore") is used for fouling recognition and analysis. The microprocessor (i.e., Raspberry Pi 3 Model B) operates on Raspbian (see Figure 1.7), and the local machine uses Ubuntu operating system. The results obtained from the Raspberry Pi camera module can be displayed via HDMI cable on the host screen.

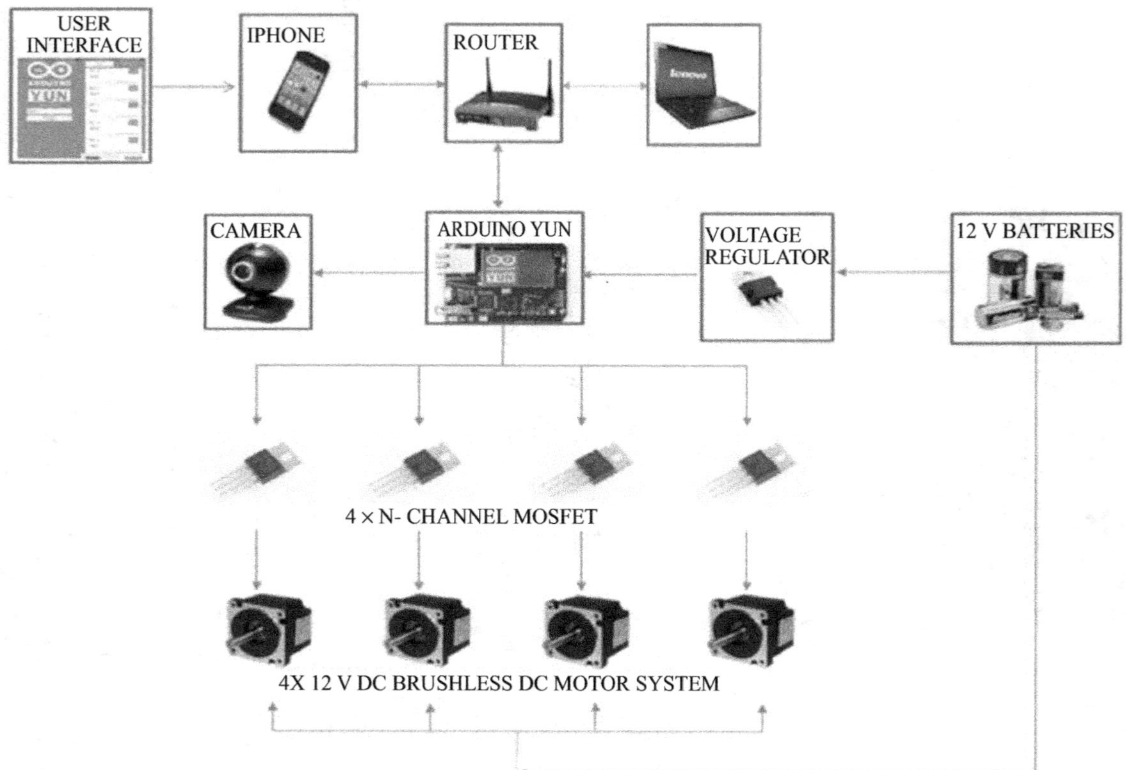

Figure 1.5 System architecture design of a low-cost remotely operated underwater vehicle

Figure 1.6 Overview of fouling image recognition system setup that includes a microprocessor connected via mobile connection (left) and a host device (right) in the laboratory [5]

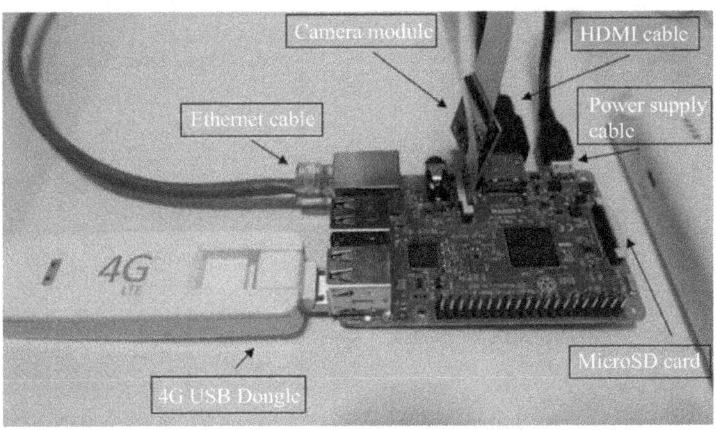

Figure 1.7 Raspberry Pi configuration for fouling image recognition system [5]

1.5 Example of embedded system using PIC

The last example is the vacuum cleaning robot [8,9] using PIC18F4520 from PIC-Microchip™ (see Figure 1.8). The PICkit 2 Development Programmer/Debugger (PG164120) can control the hardware via C language programming software. With the Microchip's powerful MPLAB- IDE, the PICkit 2 enables in-circuit debugging. The code can be easily downloaded to the microcontroller using the MPLAB-IDE environment. The main components used in the embedded system can be seen in Figure 1.9. The digital input/output ports that consist of 16 pins are used by 2-IR Sensors, 7 LED, 4 WAY DIP Switch, 1 Buzzer. Two servo motors and one servo

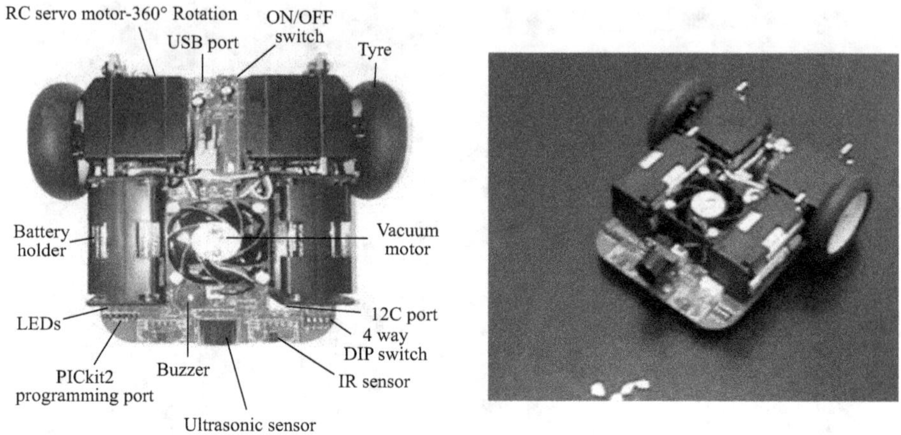

Figure 1.8 Two-wheel drive vacuum robot using PIC18F4520 [8,9]

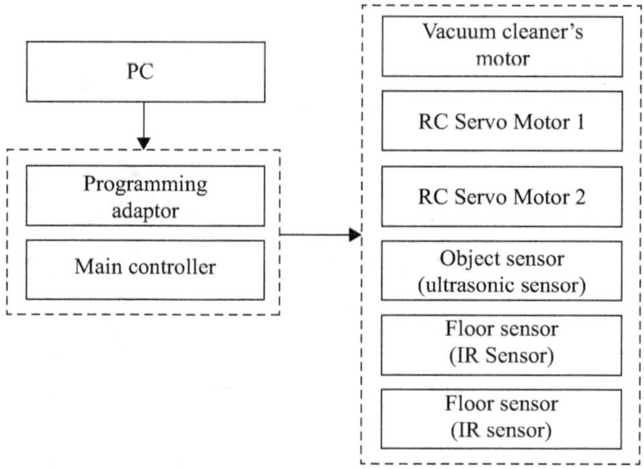

Figure 1.9 System architecture design of 2-wheel drive vacuum robot [8,9]

motor are used for driving the two wheels and providing the suction, respectively. The analog input port is used for an ultrasonic sensor to detect the object in front of the robot. The two infrared sensors are used to detect the presence of the table's surface to prevent the robot from falling.

1.6 Motivations

As shown in the examples mentioned above, the use of various microcontrollers embedded into the systems operate in an uncertain environment. They are used in underwater, ship, or land subjected to unpredicted external disturbances caused by the wind, wave, underwater current, temperature, pressure, and other unforeseen parameters. Moreover, the interdisciplinary of the system involved in computing, communication network, mechanical, electrical and electronics design to achieve commonly shared goals require coordination. The problems faced in the embedded mechatronics system design operating under the uncertain environment are summarized as follows.

- Numerical modeling for sufficient control purpose.
- Interdisciplinary nature requires systematic design approach.

First, the numerical modeling of such vehicle is quite challenging due to its nonstreamlined body and inaccuracy that may result from the numerical modeling using the computer-aided software. Numerical modeling and control techniques are essential in many engineering applications. This technique has been evolved and played an essential role in industry and institutions for the past few decades. However, in the control systems prototyping stage, it is constrained mainly by model complexity, embedded system configurations, and interface testing. There is no standard method to design embedded systems due to different operating systems used, interfacing requirements, sensor, and actuators specifications. The need to propose a systematic approach for Linux®-based and Windows-based embedded systems is essential. For a start, we need to examine the types of system available to devise the systematic approach.

Second, there is lack of information and detail on how the embedded system can be used and configured systematically. For example, the Linux® has a graphical interface in common to the Window-based embedded system. However, Linux® is different from other operating systems as it is open-source software that is free and readily customizable as compared to the Window-based system. However, it sometimes quite difficult to use and configure the devices if users do not have much knowledge. In the applications like constructing a system that needs much time and effort in testing, the writing of C/C++ programming language can be difficult when the embedded system uses the Linux®-based operating system.

To implement the program more accessible, new software and hardware are necessary for the real-time operating environment. This book covers a real-time control system using xPC Target™ (https://www.mathworks.com/tagteam/37937_xpc_target_selecting_hardware_guide.pdf) for underwater robotic system control. By utilizing the software package MATLAB®, Simulink®, Real-Time Workshop™, xPC Target, and a C/C++ compiler, the I/O boards are interfaced between the Simulink® blocks and the embedded system in the manner of hardware-in-the-loop simulation. It provides a convenient way to implement a real-time underwater robotic control system that does not require any low-level language programming.

The PIC-Microchip (https://www.microchip.com/about-us/company-information/about) is a provider of microcontroller that is headquartered in Arizona, USA. A development programmer/compiler is used to control the hardware via

C language programming software. It is a low-cost development tool with an easy-to-use interface for the programming and debugging Microchip's Flash families of the microcontrollers. The full-featured Windows® programming interface supports baseline, midrange, families of 8-bit, 16-bit, 32-bit microcontrollers, and many Microchip Serial EEPROM products. To facilitate the prototype testing, analysis, and improvements, the completed program in C code can be implemented by downloading the program into the embedded processor.

The open-source platform named ARDUINO® (https://www.arduino.cc/en/Guide/Introduction) was created in Ivrea Interaction Design Institute as an easy tool for fast prototyping for students without a background in electronics and programming. It consists of both a physical programmable circuit board (often referred to as a microcontroller) and a piece of software, or IDE that runs on your computer, used to write and upload computer code to the physical board. The ARDUINO® platform has become quite popular. Unlike most previous programmable circuit boards, the ARDUINO® does not need a separate piece of hardware (called a programmer) to load new code onto the board. The user can just use a USB cable. Additionally, the ARDUINO® IDE uses a simplified version of C++, making it easier to program. Finally, ARDUINO® provides a standard form factor that breaks out the functions of the microcontroller into a more accessible package.

On the other hand, the Raspberry Pi (https://www.raspberrypi.org/about) is a SBC designed in the United Kingdom by the Raspberry Pi Foundation. As the Raspberry Pi runs on a Linux® Operating System called Raspbian, it is not commonly used by many users. Hence, the hosting SSH connection will enable convenient usage of Raspberry Pi. The programming script is programmed and stored in the Python IDE. The camera is commonly used for image and video recording. It is connected via flex cable connector for saving the files in the Raspberry Pi.

Although systems arising in the applications and embedded systems are entirely different, they have at least one common area such as several interacting subsystems, components, and people. Interactions among the parts of a complex system may give rise to unpredictable behaviors that can have desirable or undesired impacts. The design step should provide valuable tools and technologies to understand, predict, and evaluate the behavior of complex systems for developing and evaluating methods to steer the system toward more desirable states to enable a more efficient, robust, and cost-in-mind engineered complex systems that are penetrating to modern societies.

1.7 Systematic design approach for prototyping embedded systems

As mentioned earlier, the problems such as the system uncertainties and poor measurement caused by the external disturbances and numerical modeling errors could exist. There should exist a unifying approach to handle the prototyping process under such scenarios. The proposed steps can be summarized into mainly system design and architecture, software programming, system implementation, and testing. These steps are used in the subsequent chapters for the readers to follow systematically.

1. **Understanding requirements and specifications for applications:** It involves examine essential operational characteristics, network communication, system configurations, external working environment, and problems facing;
2. **Formulate control objectives**: It involves defining the control parameters;
3. **System design and architecture:** It involves select physical hardware availability, costs, knowledge requirements, functional characteristics, identify additional component interactions, types of energy storage system, constraints or limitations, and control methodology;
4. **Software programming:**. It includes selecting physical software by considering the software functionality, availability, knowledge of software principle, functional characteristics, constraints or limitations and programming, numerical modeling of embedded system using computer-aided software;
5. **System implementation:** It involves implementating the open-loop (at the first instance to test the subsystems) or closed-loop system (the next step to validate the functionality of the closed-loop system) via hardware–software-in-the-loop simulations (typcially in the laboratory or controlled environment);
6. **Testing:** In an external or actual environment, it includes examining the design functionality to meet objectives, validating design in a different environment to test for robust performance, and finetuning design for improvement before finalizing the embedded system design.

As seen in the proposed steps, the applications have to be defined. The problems and the challenges of the systems need to be identified. It includes visiting the prototype available in the markets or Internet and searching for useful publications to gain insight into the problems related to the applications. In addition, it requires gaining more information from the end-user (or even vendors who supplied the subcomponents) to understand the specifications of the components required.

After formulating the control objectives, the control variables have to be decided. The control parameters need to be measurable or at least estimated from the given controllable and observable system. If not, there is a need to perform simulation after the hardware and software are selected. The number of control variables are usually more than the one that leads to a multivariable system with many states. Subsequently, the selection of hardware is essential as it determines whether the additional numerical model of the individual subsystem is needed. For example, some devices that generate ON/OFF function do not need dynamic modeling. The sensor needs to be calibrated, and the communication ports such as RS232, USB, and others need to be considered. Hence, the types of hardware used are essential. Undoubtedly the cost is another factor that decides the selection of hardware. The types of energy storage system will need to be chosen. For example, lithium iron phosphate battery or $LiFePO_4$ batteries have become popular due to their high specific power, high specific energy density, long cycle life, low self-discharge rate, and high discharge voltage for a variety of applications. However, the balance between the specifications, cost, and availability in a given project time frame should be considered.

Another consideration is the software. It can be grouped into software used for operating system within the microcontroller or SBCs and numerical model simulation. The operating system should be easy to use or learn from given online resources or manual. The software for modeling the dynamic model of the systems is usually performed in MATLAB® and Simulink®. The numerical condition of the model is crucial as it determines the accuracy of the simulated results. In fact, most of the time, the simulated and experimental results do not match. However, it provides an insight into the behavior of the dynamic systems or a model that can be used for sufficient control purpose. Other computer-aided design software such as SolidWorks™, AutoCAD™, and Pro/ENGINEER™ are used to obtain the physical model representation and parameters such as mass, moment of inertia, and required tolerances for fabrication and numerical model simulation.

The controller is then designed to control the system in such a way that it meets the control objectives. For example, a camera is used to capture and recognize the captured images. If there is a need to position the camera in a specific location, it is evident that a controller is required to position the camera. After the embedded system is selected and designed, the selected hardware and software will require testing. The hardware with software-in-the-loop testing will provide a mean to simulate the system in real time. The primary objectives are to test the functionalities of the hardware, software interfacing, and codes that are downloaded to the target or embedded system. The control system design needs to ensure it can perform satisfactorily, i.e., it can control the parameters to the desired values. The next step is to validate the system in the designated working environment. The embedded system should be tested in a different environment or perturbed with external disturbances to show that the system is robust with the current settings (i.e., enclosure design, onboard power supply, user input devices, and graphic user interface). Fine tuning of the control system in the embedded system will be performed if there exist some issues such as an error in control parameters and inconsistent results at different test occasions. Sometimes, it requires modifications in both hardware and software. The final step is to package the prototype into a presentable product for demonstration and possible commercialization. It requires some efforts in improving the product esthetics and rearranging the layout of the electronics and mechanical parts.

In summary, the following guidelines can be adopted in the embedded system design.

1. Application classification
   (a) Application domain
   (b) Type of device/microcontroller to be interconnected
   (c) Type of interconnection: Coupled ☐ Embedded ☐
   (d) Technological readiness level (according to TRL levels—1 to 9)

2. Description of specifications
3. Type of operating system

Type of operating system	Yes/No
Linux®	☐
MATLAB®	☐
Microsoft Windows	☐
UNIX	☐
Mac OS X	☐
Other(s):	☐

4. Programming Language and Standards

Technologies	Yes/No
OPC-UA	☐
JSON	☐
MATLAB®-based	☐
.NET	☐
MODBUS	☐
JAVA	☐
X10	☐
C / C++	☐
MTConnect	☐

Standards	Yes/No
IEC 61131-3	☐
IEC 61850	☐
IEC 61499	☐
IEC 62264	☐
Smart Message Language (SML)	☐
IEC 62714	☐
ISO/IEC9506	☐
Others	☐

5. Sketch of the interface/system design and architecture
6. Type of input/output

Analog I/O variables: read ☐, write ☐, other:
A number of analogs I/O:
Digital I/O variables: read ☐, write ☐, other:
A number of digital I/O:
Other communication ports: RS232 ☐, ICP ☐, USB ☐, SCSI ☐, HDMI ☐, other:
Handling programs: Start ☐ , Stop ☐, upload ☐, select ☐, getStatus ☐, other:
Handling events: event notification ☐, subscription ☐, other:
Other(s):

7. Implementation and testing

Type of tests	Yes/No
Site testing	☐
Laboratory testing	☐
Hardware-in-the-loop testing	☐
Software-in-the-loop simulation/validation	☐
Other(s):	☐

8. What not to follow in the implementation of the embedded system
9. Type of applications that may benefit from the embedded systems

Type of application	Yes/No	Score {1–5}
Supervision	☐	
Monitoring	☐	
Distributed DAQ	☐	
Real-time control	☐	
Simulation	☐	
High-level control	☐	
Other(s):	☐	

However, it is essential to evaluate the following performance of the embedded system design after implementation. Note that “high,” “moderate,” “low,” and “not used” can be applied to evaluate each performance parameter as shown.

1. Current usability level in the application domain
2. Current level of readiness
3. Current technological integrity level
4. Current level of automation
5. Current availability of technologies
6. Current economic benefit
7. Current potential of the embedded system
8. Current robustness degree

In summary, it is not easy to start the system design from scratch without a good reference and guides. This book will provide guidance on how to perform rapid system prototyping to realize the embedded system for specific applications operating in an uncertain environment. Note that, it is not possible to show all applications and scenarios in the book. Nevertheless, this book will serve a good starting point for an individual who wants to obtain an idea on how to prototype an embedded system.

References

[1] Chin, CS, Lin, WP, and Lin, JY. Experimental Validation of Open-Frame ROV model for Virtual Reality Simulation and Control. **Journal of Marine Science and Technology** 2018, 23(2), 267–287.

[2] Lin, WP, Chin, CS, Looi, LCW, Lim, JJ, and the, EME. Robust Design of Docking Hoop for Recovery of Autonomous Underwater Vehicle with Experimental Results. **Robotics**. Guest Editors: Prof. Thor I. Fossen and Prof. Ingrid Schjølberg, NTNU,Trondheim, Norway 2015, 4(4), 492–515.

[3] Lin, WP, Chin, CS, and Mesbahi, E. Remote Robust Control and Simulation of Robot for Search and Rescue Mission in Water. **MTS/IEEE OCEANS** 2014, Taipei, Taiwan.

[4] Lin, WP and Chin, CS. Remote Underwater Dual Cameras Video Image Acquisition System using Linux Based Embedded PC104. **MTS/IEEE OCEANS** 2014, Taipei, Taiwan.

[5] Chin, C, Si, JT, Clare, AS, and Ma, MD. Intelligent Image Recognition System for Marine Fouling using Softmax Transfer Learning and Deep Convolutional Neural Networks. **Complexity** 2017, 5730419, 9. https://doi.org/10.1155/2017/5730419.

[6] Guo, JY, Chin, CS, Maode, MA, and Clare, AS. Intelligent Fouling Detection System using Haar-like Cascade Classifier with Neural Networks. **Advances in Computer Communication and Computational Sciences**. Springer, 2018. DOI: 10.1007/978-981-13-0344-9_34

[7] Guo, JY, Chin, CS, Clare, AS, and Maode, MA. Interactive Vision-based Intelligent System for Active Macfouling and Microfouling Detection on Hull. **MTS/IEEE OCEANS** 2016, Shanghai, China.

[8] Chin, C. Application of an Intelligent Table Top Vacuum Robot Cleaner in Mechatronics System Design Education. **Journal of Robotics and Mechatronics** 2011, 23(5), 645–657.

[9] Chin, CS and Yue, KM. Vertical Stream Curricula Integration of Problem-Based Learning (PBL) using Autonomous Vacuum Robot in Mechatronics Course. **European Journal of Engineering Education** 2011, 36(5), 485–504.

Chapter 2
Linux®-based embedded system design

2.1 Linux® operating system

This section provides a brief overall view of Linux® operating system. A useful introductory notes on Linux® can be found in http://www.tldp.org/LDP/intro-linux/intro-linux.pdf. Linux® is an operating system (OS) or a kernel distributed under an open-source license for free. It has libraries, compilers, debugging, and development tools. Various companies such as RedHat and Mandriva provide packaged Linux® distributions with user-friendly interfaces developed by the community (see Figure 2.1). A user can begin the required applications without the need to type any character at the terminal.

The Linux® system is based on open-source GNU tools that provide a standard approach to use the system. Most distributions come with a complete set of development tools offer precompiled packages for the programmers to install. A few popular GNU software is as follows.

- GCC: The GNU C Compiler
- Bash: The GNU shell
- Gnome: the GNU desktop environment
- GDB: The GNU Debugger

The Linux® kernel is not part of the GNU project but it consists large part of development and utilities tools that are not Linux®-specific. Such a system can be called a GNU/Linux® system. The kernel is important in the system. It controls the communication between the peripherals and the underlying hardware. The kernel also ensures that processes are begun and stopped at exact time. The kernel is a huge topic that contains large amounts of information to discuss in detail. Note that most of the things in Linux® system are considered as a file except for the process files.

- Special files: the mechanism used for input and output.
- Directories: files that are lists of other files.
- Links: a system to make a file or directory visible in multiple parts of the system's file tree.
- Sockets: a special file type, similar to Transmission Control Protocol (TCP)/IP sockets, providing interprocess networking protected by the file system's access control.

Figure 2.1 Example of packaged Linux® distributions from Ubuntu 16.04.2 LTS

Figure 2.2 Example of file system in Ubuntu 16.04.2 LTS

- Named pipes: these act more or less like sockets and form a way for processes to communicate with each other without using network socket semantics.

The important details of the file system after the command typed in terminal: "ls –l" are shown in Figure 2.2. The "Owner" named root is the user (i.e., mcschin1) owning the file. The "Group" is the user group assigned to. The "Permission string: -rwxr-xr—"is the file permissions for the user, the group members, and others are explained in Figure 2.3.

The standard root partition consists most basic commands (see Table 2.1), system configuration files and server programs, some temporary space, system

```
- rwx r-x r--
| \ / \ / \ /
|  v   v   v
|  |   |   \__ Other permissions: [r] 'r'ead access granted
|  |   |                          [-] 'w'rite access denied
|  |   |                          [-] e'x'ecuting denied
|  |   \__ Group permissions: [r] 'r'ead access granted
|  |                          [-] 'w'rite access denied
|  |                          [x] e'x'ecuting granted
|  \__ Owner permissions: [r] 'r'ead access granted
|                         [w] 'w'rite access granted
|                         [x] e'x'ecuting granted
\__ File type: [-] normal file
               (alternatives: 'd'irectory, sym'l'ink,
               'c'haracter device, 'b'lock device, ...)
```

Figure 2.3 File permissions for user, group members, and others (see https://askubuntu.com/questions/700258/how-to-check-if-running-as-root-not-asking-for-sudo)

Table 2.1 Basic Linux® commands

Commands	Descriptions
ls	displays a list of files in the current working directory, like the dir command in DOS
cd	change directories
passwd	change the password for the current user
file	display file type of file with name filename
cat	throws content of textfile on the screen
pwd	display present working directory
exit	leave this session
man	read man pages on command
info	read Info pages on command

libraries, and the home directory of the administrative user in data partition. It contains all the data to begin and run the system. Swap space is only accessible for the system hidden from user during normal operation. The kernel is on a different partition as it is the most essential file of the system. It is a good idea to visit the root directory. The kernel and accompanying data files can be found in */boot* partition. It follows a set pattern as shown.

- partition for user programs (*/usr*)
- partition for third party and extra software (*/opt*)
- partition consisting user data (*/home*)
- partition to keep temporary data (*/var*)

The shell is an advanced method of communicating with the system. The shell allows the user to handle a system easily for task automation. Most configuration files are stored in the /etc directory, and the content can be viewed using the cat command.

Devices of a computer are presented to the system as an entry in the /dev directory. The advantage of the way of handling devices is that neither the system nor the user needs to know the specification of devices. As a security measure, these files are usually kept in separate parts from the main system files. It helps to set stricter permissions where necessary. Many of these files require more permissions like /var/tmp that needs to be editable for all users. The subdirectories can contain more essential data such as configuration files and programs.

2.2 Building Linux® for embedded systems

The Athena-II/Hercules-II SBC Linux® 2.6.23 is used to demonstrate the installation of the Linux® (www.diamondsystems.com). The software development kit (SDK) from a vendor should provide all the tools required for creating Linux® 2.6.23 based or newer platform images for the SBC. The OS image contains programs for demonstrating and validating the hardware capabilities. Functions exercised by the bootable Linux® OS image include:

- Ethernet support with local-area network installed
- DHCP, SSH, SCP, FTP
- PS/2 mouse and keyboard
- USB 1.0-2.0 ports
- Serial ports RS-232/422/485
- IDE storage devices
- Diamond Universal Driver 6.0x for Linux® 2.6.23
- DAQ demo programs (for AV model boards)

Before installing the Linux®-based OS, the development computer must meet the following hardware requirements as seen on the main website: http://www.diamondsystems.com/products/linuxdevkit

- 1 GHz 32-bit (x86) or 64-bit (amd64) processor
- 1 GB of system memory (32-bit), 2 GB of system memory (64-bit)
- 7 GB free hard drive space for complete installation
- One or more of the following media drives: DVD-ROM drive and USB 2.0 port
- Software must be running Linux® flavor OS

For example, the steps to compile Universal Driver Kernel Module (named DSCUD) are as follows. Note that the user to choose either Ubuntu 16.04.2 or Fedora 26 as the OS. Most commands in these steps are written for Fedora 11. However, a corresponding command can be found for the latest version of Fedora and Ubuntu.

1. Installing Fedora (see https://fedoraproject.org/wiki/Installing_Fedora_11_Leonidas) or Ubuntu OS (see https://tutorials.ubuntu.com/tutorial/tutorial-install-ubuntu-desktop#0). Briefly, it should cover the following steps.
   (a) Download Fedora/Ubuntu
   (b) Check if the computer will boot from the USB

   (c) Make BIOS changes using F2 or Delete key on the keyboard if necessary
   (d) Try Fedora/Ubuntu before you install it
   (e) Create bootable USB
   (f) Install Fedora/Ubuntu
   (g) Create a username, password, and computer name. You will log in with this user id after the installation is complete.

2. Essential tools for compilation. The OS does not come with gcc/g++ compiler. So, it can be installed by running the following command terminal:

```
$ yum install gcc

$ yum install ncurses-devel

$ yum install gcc-c++
```

   Note that the requirements for compilation are as follows.
   (a) Linux® kernel source (2.4.x.x to 2.6.x.x)
   (b) The Fedora 11 live OS is built using 2.6.29.4 kernel. So, download the same kernel source code. You can refer to the Fedora version history at https://en.wikipedia.org/wiki/Fedora_version_history#Fedora_11
   (c) kernel module source code

3. Download the Kernel Source. If the embedded system is using a standard version of the Linux® kernel, you can download the source from the www.kernel.org website. If it is a patched or custom kernel, you will have to get the kernel source from the organization that provided the custom kernel. Extract the kernel source on the development system so that it is available in /usr/src/Linux®. Below is an example of how this is normally done.

```
$ tar xvzf kernel-source-2.6.x.tar.gz

$ mv linux®-2.6.x /usr/src/linux®-2.6.x
```

4. Prepare the Kernel Source tree. The kernel source is not ready to be used for building kernel modules until it is configured. In some kernel configurations, you must also begin the kernel compile process for it to include all the code required.
   (a) Edit Make file under /usr/src/linux-2.6.x/
   (b) Add the following line in the field "EXTRAVERSION"=.4-167.fc11.i586
   (c) This will vary based on the running kernel. The version can be found in terminal by running following command terminal

```
$ uname-r
```

   It will display like 2.6.9.4-167.fc11.i586, then the extraversion will be ".4-167.fc11.i586". The above procedure is to avoid driver version magic error while loading in the running kernel. The first step is to configure the kernel. Run this command.

```
$ cd /usr/src/linux

$ make menuconfig
```

You will be presented with a menu that lists many options. You should select the correct CPU type "Processor Type and Features." Then go to "Loadable Module Support" and make sure that "Enable loadable module support" is enabled. Exit the menu making sure it saves the kernel configuration. Run this command to finish build the kernel source.

```
$ make
```

This kernel source is now ready to be used to build kernel modules.

5. Building dscudkp.ko module

```
Run ./install.sh
```

This installer compiles a kernel module which supports interrupt-driven DAQ features of Diamond Systems products. If you are not using the interrupt-driven features of the board, you do not need to run this installer. Online documentation for this driver is available at http://docs.diamond-systems.com/DSCUD/. Hit [CONTROL-C] to exit this installer or [ENTER] to continue.

```
Step One: Locate Kernel Source Code
------------------------------------------
The installer will now scan your system looking for Linux® kernel source code. The
/usr/src directory and the /lib/modules directory will be scanned.
Select a Linux® kernel version below that you will run on your TARGET system. Type
the number next to the selection and hit ENTER.
0 ) Kernel 2.6.24.1 (/lib/modules/2.6.24.1-DMP/build)
1 ) Kernel 2.6.27.9 (/lib/modules/2.6.27.9-DMP/build)
2 ) Kernel 2.6.24.1 (/usr/src/linux-2.6.29.4)
3 ) Kernel 2.6.24.7 (/usr/src/linux-2.6.24.7)
4 ) Kernel 2.6.27.9 (/usr/src/linux-2.6.27.9)
For example: Enter '2'

Step Two: Compile Kernel Module
-----------------------------------------
The installer is now ready to compile the Linux® kernel module. If errors occur, see the
README file in this directory as well as the online DSCUD documentation for help.
This kernel module will only load under the exact Linux® kernel version which you have
installed in the directory /usr/src/linux-2.6.29.4.
```

```
--> Compiling kernel module for your system <--
rm-f dscudkp.ko dscudkp.o dscudkp.mod.*
make-C /usr/src/linux-2.6.24.1 SUBDIRS=/usr/local/DSCUD modules
make[1]: Entering directory `/usr/src/linux-2.6.24.1'
CC [M] /usr/local/DSCUD/dscudkp.o
Building modules, stage 2.
MODPOST 1 modules
CC /usr/local/DSCUD/dscudkp.mod.o
LD [M] /usr/local/DSCUD/dscudkp.ko
make[1]: Leaving directory `/usr/src/linux-2.6.24.1'
--> Installing module dscudkp.o in /lib/modules/misc <--
mkdir-p /lib/modules/misc
cp dscudkp.ko /lib/modules/misc/


Step Three: Final Instructions

----------------------------

The dscudkp kernel module has been installed in /lib/modules/misc/. You must copy this
file to the same location on your target system. The load.sh script will load the kernel
module so that it can be used by the driver. You must run this script each time the Linux®
system boots. See the README file for help with this. Driver installation has
completed.
```

In the example above, the dscudkp.ko was compiled to be used on the system that has Linux®-2.6.29.4. If the user has tried all the steps in development PC, copy /lib/modules/misc/dcudkp.ko, and /usr/local/DSCUD-7.00 directory to target board and all the files must be placed in the same location. Once copying the files to the target system is done, change directory to DSCUD-7.00 and run the command to load the driver

```
$ cd /usr/local/DSCUD-7.00/

$ ./load.sh
```

Now, installing and setting up DSCUD support on our target system is done.

6. Compiling Test/Demo application. The driver is a library which you link to your application. Here is an example.

```
$ gcc myapp.c-o myapp-I/usr/local/DSCUD-7.00 -L/usr/local/DSCUD-7.00 –ldscud-
7.00 –lm -lpthread
```

7. Execution of compiled application

   Become a root user and run the executable file which was generated in the previous step as follows

```
$./myapp
```

Now, the list of components on Image is as follows.

- Intel 82574 gigabit Ethernet driver which is located at /lib/modules/2.6.29.4-
- 167.fc11.i586/kernel/Driver/net/e1000e/ path in the image
- VGA Display driver for the integrated graphics controller in Tunnel Creek
- USB 2.0/1.1 device driver for 4 USB ports
- ALC262 Audio driver
- SSH server for remote desktop
- HTTP server
- FTP server
- Minicom serial port application.
- Athena III Demo applications which are located at /home/dscguest/Desktop and /home/Diamond/Desktop/ Path
- DSCUD-7.00 library located at /usr/local/DSCUD-7.00/ path

8. Intel Ethernet driver compilation.

   First, install kernel headers. Install the kernel-devel package and it will vary based on a running kernel. Run the following command

```
$ rpm –i kernel-devel-2.6.29.4-167.fc11.i586
```

It will install the kernel headers in /usr/src/kernel/ directory and make ensure that headers are installed in the above-mentioned path.

Second, compile Ethernet driver. Extract Ethernet tar file e100e-2.1.4.tar.gz as follows

```
$ tar –xvf e100e-2.1.4.tar.gz
```

It will be extracted in current directory and change directory inside Ethernet driver source as follows.

```
$ cd e100e-2.1.4/src/
```

Run make command, and it will compile the driver, then generate e1000e.ko driver file in the same directory.

```
$ make
```

Third, load Ethernet driver. Become root user and Use insmod command to load the driver as follows

```
$ insmod e1000e.ko
```

If it displays e1000e.ko file exists, remove the older driver and insert the compiled latest driver as follows.

```
$ rmmod e1000e.ko

$ insmod e1000e.ko
```

Finally, load Ethernet driver at boot time. Always user will be expected to load all the drivers at boot time itself. So, load the Ethernet driver at boot time as follows. The Fedora kernel already has e1000e.ko at /lib/modules/2.6.29.4-167.fc11.i586/kernel/Driver/net/e1000e/e1000e.ko

Replace the above driver with our working compiled driver. It will load the driver at boot time itself. Note that there may exist some differences in the results. However, it is good to visit some of the resources on the internet related to the common error and the solutions to resolve them.

- https://www.linuxquestions.org/questions/
- https://ubuntuforums.org/
- https://stackoverflow.com/questions/
- https://fedoraforum.org/
- http://www.linuxforums.org/forum/

2.3 Program layouts in Linux®

This section provides the software architecture design for the control system realization. A brief description includes the real-time embedded OS Linux® environment for sensor and fusion module, control system design, actuator driver module, network communication module, and lastly graphical user interface (GUI) design module. Multiple tasks (i.e., objects containing operations to be performed) approach will be used in the programming. This is realized by using multithreading techniques. It allows multiple signals from multiple sensors to be acquired simultaneously, and real-time network communication between host and target system to be achieved.

A process may be divided into some independent units known as threads. A thread is a dispatchable unit of work. The process is a collection of one or more threads and associated system resources. The difference between a process and thread is shown in Figure 2.4. A process-based user thread performs the same processing as the real-time operating system (RTOS)-based interrupt service routines (ISRs) would. It uses the call to queue a message for receipt by another local thread or by another process.

The development process (see Figure 2.5) is based on Linux® OS. The C program will be edited using VI editor as shown in the host PC. It is compiled using GCC compiler. The remote debugger (using remote GDB Debugger) is used for the remotely debugging purpose on the host PC. Debugging is something that cannot be avoided. We have to debug a section of code. GDB is effective when it is

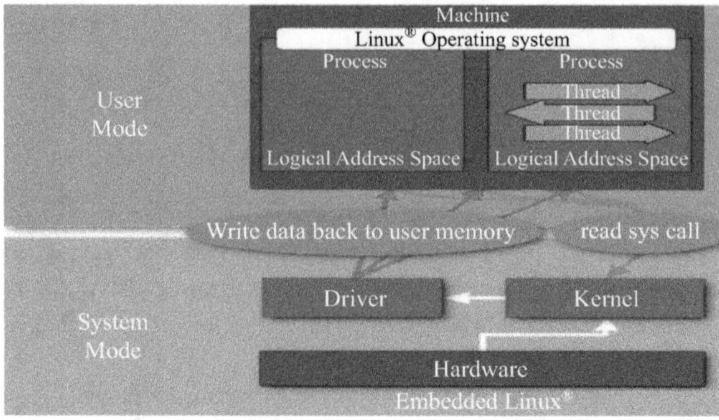

Figure 2.4 Linux® process-based threads [1]

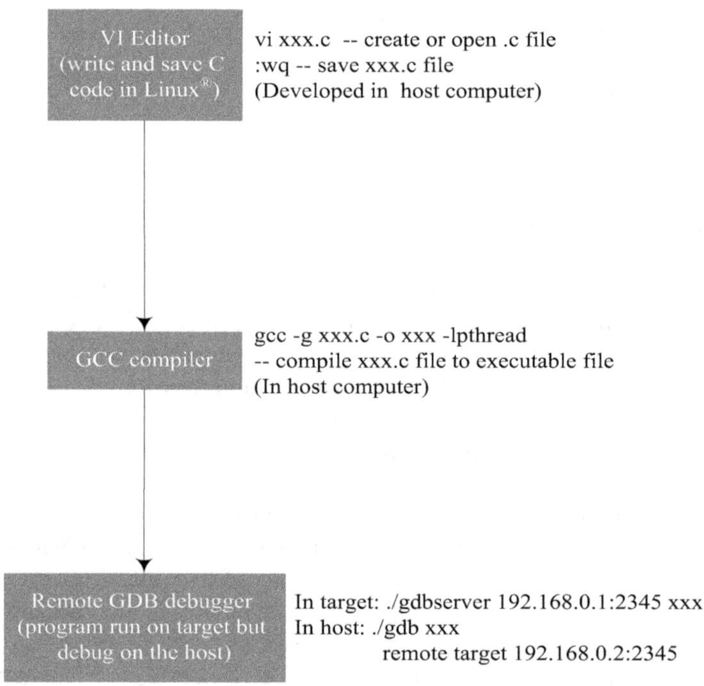

Figure 2.5 Development process uses a Linux® and C program [1]

debugging a program. In GCC compiler, it is accomplished using the -g command line argument. The -ggdb switch can be used which includes debugging symbols which are specific to gdb.

As shown in Figure 2.6, the main program that consists of all the subprograms is named as "Main.c". Most problems require a computer program to solve some

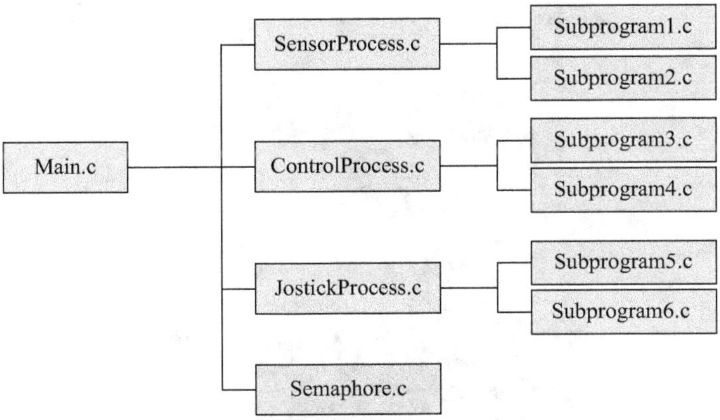

Figure 2.6 C-program files on main and subprograms [1]

complex works in the same program. Using subprograms allow us to deal with bite-size pieces of a problem individually. In this case, the subprograms are namely: SensorProcess, ControlProcess, JoystickProcess, and Semaphore. To implement specific functions in the main and subprograms, we first write the main program that references all of the subprograms in the desired order and then start writing the subprograms.

2.4 System design and architecture

The programming architecture and the C program files are needed for the control system design of a typical remotely operated underwater vehicle prototype. The details shown in this section onwards can be found in the following references [1–3]. The prototype in Figure 2.7 (and see Figure 1.1) has an open-frame design and is 1.5 m long, 1 m wide, and 1.7 m high. It has a dry weight of around 135 kg and a working depth of at least 100 m. It is designed for launch and recovery of a smaller vehicle. The vehicle is installed with six electric brushless thrusters. The roll and pitch motion are not controlled as the vehicle's metacentric height is quite large to give good static stability.

The host PC with a GUI is developed by open-source platform using Python-based IDE as the IDE to communicate with the vehicle via an Ethernet connection. The remaining components are housed in an aluminum enclosure with openings for subsea electrical connectors. The vehicle has an umbilical cable linked to the surface vessel for communication, and at the same time to provide the power (230 V AC power) for the vehicle. It needs to position itself at a predetermined target near the aft of the vessel as shown in Figure 2.8. During the mission, it will be launched from the surface vessel. The operator on boards the vessel and moves the vehicle remotely from the vessel to the vicinity of the desired target point (see position 1 in Figure 2.8). At this point, the vehicle will maneuver autonomously to the final desired depth and heading angle (see position 2 in Figure 2.8) to launch the smaller

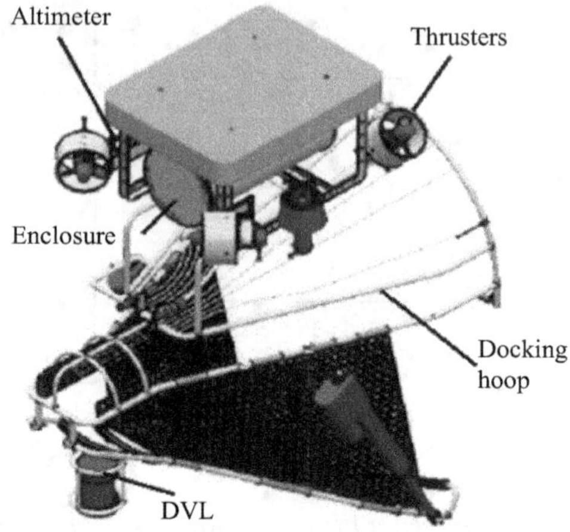

Figure 2.7 Remotely operated underwater vehicle design [1]

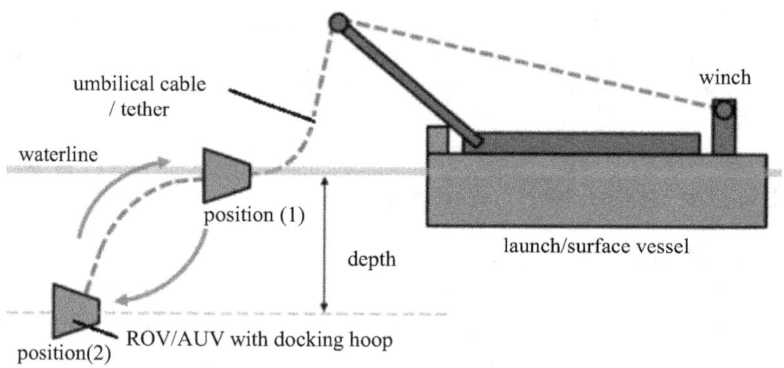

Figure 2.8 Schematic of remotely operated underwater vehicle launching (position 1) and recovery (position 2)

vehicle, and station-keeping at the same location to recover it after the mission by the launch vessel.

The following sensor and actuators are connected to a target PC using Diamond Systems ATHENA PC104 where Figure 1.2 shows the proposed hardware and software architecture. The hardware connection of the entire robotics vehicle control system includes target system (i.e., Athena PC104 board) and host computer/laptop connecting via Ethernet cable, altitude/position sensor, depth sensor, DVL sensor, camera, and six thrusters. The joystick is connected to the host PC. In the program, a layer is often used to describe the different levels at which we can place a program. In this program, we can stack or group the following layers to differentiate different functions used in the control systems design. There are five layers as shown by different colors in Figure 2.9.

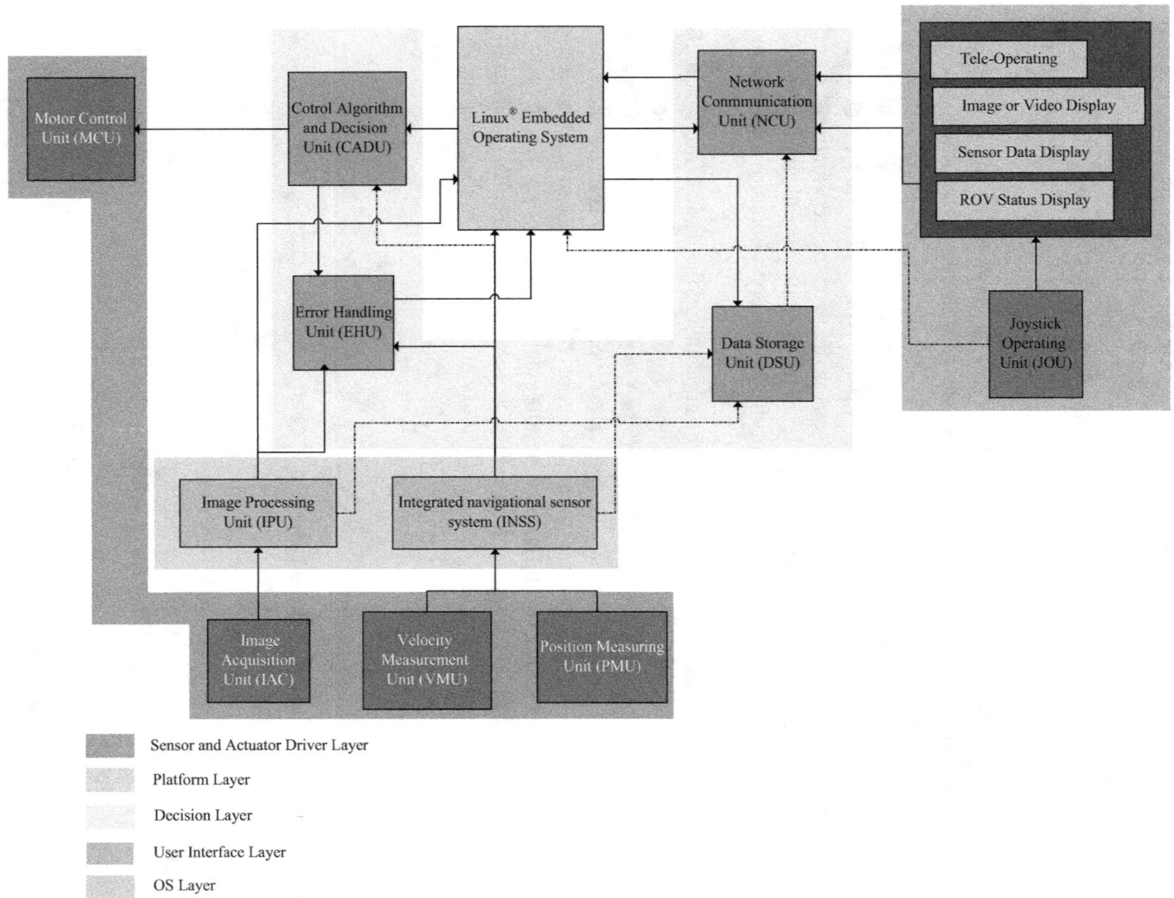

Figure 2.9 Software layers communication in a control system [1]

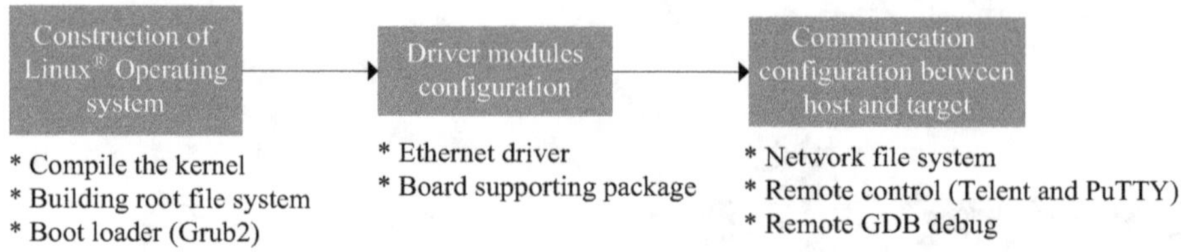

Figure 2.10 Development process for building OS layer [1]

- **Sensor and actuator driver layer**: The driver layer handles the low-level driver functions required to operate the sensor and actuator, including sensor DAQ and motor driver.
- **Platform layer:** It contains code that processes the data which collected form the sensors. It serves as the bridge for the driver layer and higher level algorithm layer to convert the low-level raw sensor data for the applications.
- **Decision layer:** This layer contains high-level control algorithms, data storage, error handling, and Ethernet socket unit for remotely operated underwater vehicle.
- **OS layer**: Embedded Linux® system is configured using the basic Linux® kernel (2.6.39.4), and a brief process of building Linux® OS layer is shown in Figure 2.10. Briefly, the two most important responsibilities of the kernel are process management and file management. However, the kernel is responsible for many other things. One aspect is input/output (I/O) management, which is essentially the accessing of all the peripheral devices.
- **User interface layer**: The user interface layer provides useful interaction between the host and remotely operated underwater vehicle. On the other hand, the user can send a command to control the vehicle easily, and allow vivid information to be displayed on the laptop's screen.

2.4.1 Main process design

Figure 2.11 shows the overall data flow in a control system design. There exist a main and other few processes which use the producer-and-consumer relationship to achieve interprocess communication as shown in Figure 2.12.

The main process only runs once when it boots up. It is responsible for parameter initialization as well as creating and booting two child processes (i.e., SensorProcess and ControlProcess), which continues until the program gets terminated. The primary purpose of the producer-and-consumer relationship is to process the input data before sending it to the next process.

- Producer (SensorProcess): Sensors data collection and sensor data fusion;
- Consumer (ControlProcess): Control algorithm, actuator driver program, and network communication between the host computer and target system.

SensorProcess generates the processed sensor data and video stream into share memory while the control process receives and uses them from the shared memory.

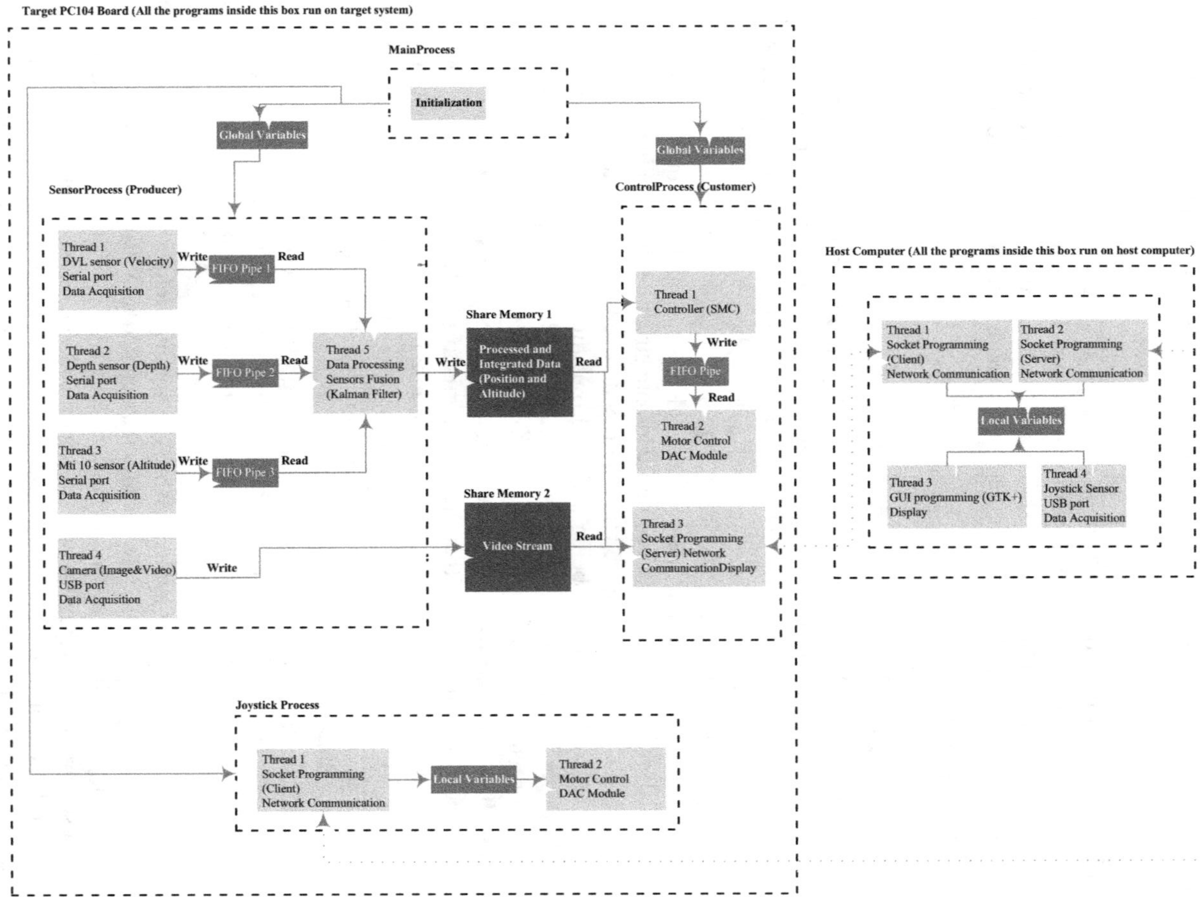

Figure 2.11 Overall data flowchart of the control system [1]

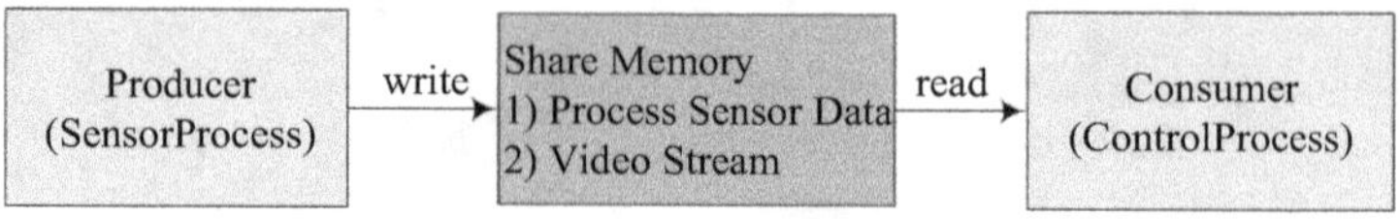

Figure 2.12 Producer-and-consumer concept (example on FIFO Pipe 1) [1]

The video stream collected from camera sensor is only for display in the host computer and does not involve in the control algorithm. It has another independent share memory to store video streams and separate from other sensor data which need further processing. A few specific designs of share memory are as follows:

- Build share memory –shmid();
- Mapping address –shmat();
- Cancel mapping –shmdt();

For protecting the data stored in share memory, producer and consumer process cannot write or read from the same location in the shared memory at the same time. Here, the synchronization and mutual exclusion between the sensor process and control process are important. We used a semaphore to ensure data safety. Semaphores are used to protect critical regions of code or data structures. Remember that each provides access to a critical piece of data. It would be very dangerous to allow one process to alter a critical data structure that is being used by another process. One way to achieve this would be to use "locking system" around the critical piece of data that is being accessed, but this is a simplistic approach that would degrade the system. A Linux® semaphore data structure contains the following information:

Count: This field keeps track of the count of processes wishing to use this resource. A positive value means that the resource is available. A negative or zero value means that processes are waiting for it. An initial value of 1 means that one and only one process at a time can use this resource. When processes want this resource, they decrement the count, and when they have finished with this resource, they increment the count.

Waking: This is the count of processes waiting for this resource which is also the number of process waiting to be awakened when this resource becomes free.

Wait queue: When processes are waiting for this resource, they are put in this wait queue.

Lock: A buzz lock used when accessing the waking field.

Semaphore is equipped with two operations, denoted as V (input signal) and P (output signal). The value of the semaphore is either 0 or 1. Suppose the initial count for a semaphore is 1, the first process to come along will see that the count is positive and decrement it by 1, making it 0. The process now "owns" the critical piece of code or resource that is being protected by the semaphore. When the process leaves the critical region, it increments the semaphore's count. The most optimal case is where there are no other processes contending for ownership of the critical region. If another process wishes to enter the critical region while it is owned by a process, it too will decrement the count. As the count is now negative (−1), the process cannot enter the critical region. Instead, it must wait until the owning process exits it. Linux® makes the waiting process sleep until the owning process wakes it on exiting the critical region. The waiting process adds itself to the semaphore's wait queue and sits in a loop checking the value of the waking field and calling the scheduler until waking is nonzero.

The owner of the critical region increments the semaphore's count, and if it is less than or equal to zero, then there are processes sleeping, waiting for this resource. In the optimal case, the semaphore's count would have been returned to its initial value of 1, and no further work would be necessary. The owning process increments the waking counter and wakes up the process sleeping in the semaphores wait queue. When the waiting process wakes up, the waking counter is now 1, and it knows that it may now enter the critical region. It decrements the waking counter, returning it to a value of zero, and continues. All access to the waking field of the semaphore is protected by a buzz lock using the semaphore's lock.

The working principles of the main process program (see Figure 2.13) are as follows:

- Create two share memories named shared memory 1 and share memory 2 with shmid() function. Share memory 1 is responsible for sensor fusion data communication between SensorProcess and ControlProcess, while share memory 2 is used for video streams transmission.
- Initialize semaphore named SEM_ID for preventing data which stored in share memory 1 from being read-write by SensorProcress and ControlProcess simultaneously.
- Use IF-ELSE and FORK() function to create three children processes called Sensor process, Control process, and JoystickProcess; the former one is responsible for sensor DAQ and sensor data fusion, the latter process is used for control decision, actuator driver, as well as network communication.
- Map the process to share memory 1 with shmat() function, and then call the SensorProcess.c file to cycle through in SensorProcess. Meantime, V operation of SEM_ID guarantees the safety of the share memory data.
- In ControlProcess that is similar to SensorProcess, it performs the mapping work first, and then conducts the P operation to ensure the OS runs Sensor-Process. Otherwise, Control process will stop and wait for Sensor process. Next, it creates two threads, namely ControlProcess.c to undertake control operation and socket.c to transmit sensor data from target-to-host for displaying and data manipulation. To close the PV operation, V operation of semaphore has to be used after each operation.

2.4.2 Sensor process design

The purpose of sensor process is to collect data from multiple sensors and use the sensor fusion module to integrate the sensor data using Kalman filtering on the position and altitude information. Under Linux® environment, all devices are denoted by a special file. To gather data from multiple devices, it uses multiple file

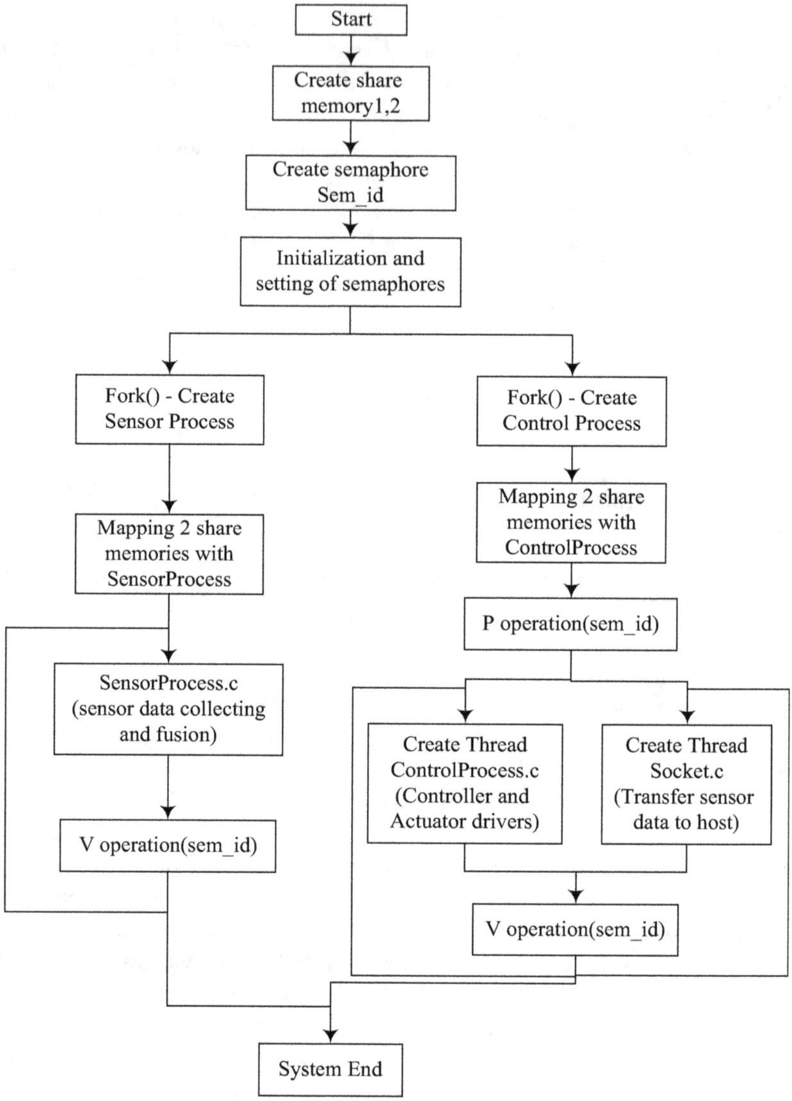

Figure 2.13 Main process program flowchart [1]

descriptors approach. It has the following sequence for deployment: multiprocess, select() function, poll() function, nonblocking I/O model with polling, asynchronous I/O, and multithreads. Multiprocess consumes relatively more resources when using select() or poll() function. The remotely operated underwater vehicle system needs to capture four channel serial data from altitude sensor, depth sensor, DVL sensor, and camera. Also, it needs to create another thread to deal with sensor data, namely sensor fusion module. Hence, the sensor process has a total of five threads running in parallel.

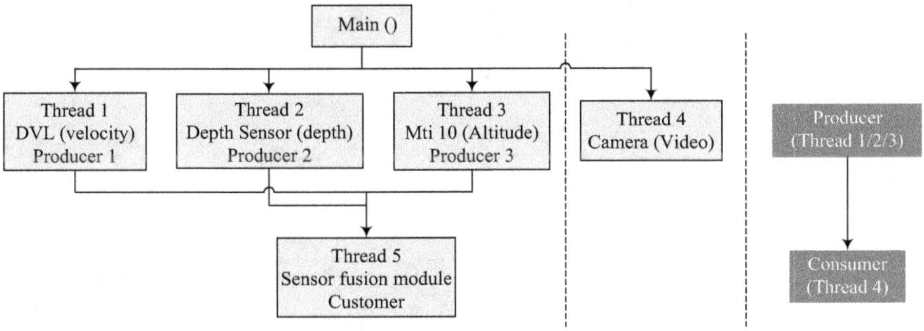

Figure 2.14 Program structure of sensor process [1]

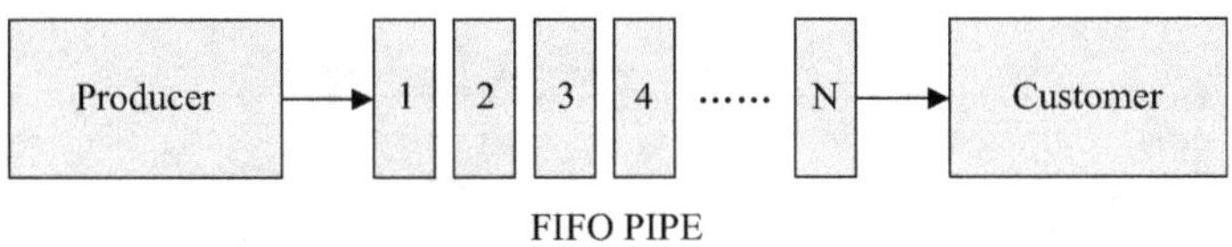

Figure 2.15 Producer–consumer model with FIFO PIPE [1]

As observed in Figure 2.14, the three threads refer to DVL sensor, depth sensor, and MTI 10 altitude sensor, respectively. These three threads have similar software structure, namely, to read data using serial ports. Figure 2.15 shows the first-in-first-out (FIFO) pipe model. A pipe is a mechanism for interprocess communication; data written to the pipe by one process can be read by another process. The data is handled in the FIFO order. A pipe needs to be open at both ends simultaneously. If one read from a pipe that does not have any processes writing, it will return end-of-file. Writing to a pipe or FIFO that does not have a reading process is considered as an error condition; it generates a SIGPIPE signal. The pipes do not allow file positioning. Both the reading and writing operations have to be done sequentially; reading from the beginning of the file and writing at the end.

The details in reading the serial port sensor (using a pipe) are as follows:

- Open serial port;
- Setting parameters of serial port—parity, baud rate, stop bit, and data bit;
- Read data from serial port—read();
- Close serial port.

For thread 4, it is used to collect camera data. It will read the image streams from USB ports device and write them into share memory 2 for subsequent display.

As shown in Figure 2.16, threads 1 to 3 serve as the producer, and thread 5 becomes the consumer. The producer–consumer model has its buffer pipe. It can be

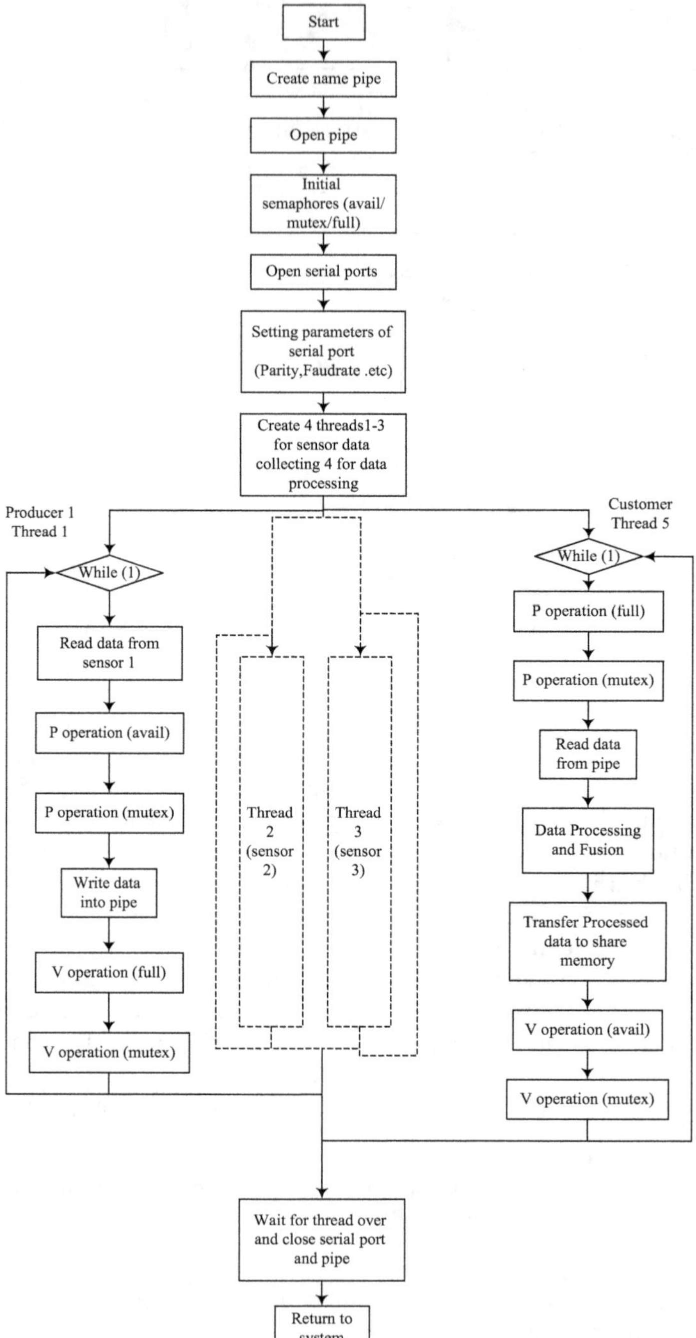

Figure 2.16 Sensor process program flowchart [1]

developed in parallel without interference with other threads. FIFO has a maximum size subjected to the following conditions:

- The consumer must wait for the producer to execute if the queue is empty.
- The producer must wait for the consumer to consume if the queue is full.

Here, we use semaphore method to guarantee these two conditions. The semaphore solution to the producer–consumer problem tracks the state of the queue with two semaphores. First, one is named "avail" while another is "full." The detail operations of the two semaphore are similar to the P/V operations as mentioned above. To resolve the mutual exclusion issue of multithread, the binary semaphore named "mutex" was used.

Figure 2.16 shows the flow of the SensorProcess. The program uses a time-sharing scheduling algorithm in Linux®. It will execute each thread, avoiding preemption among the threads. With this scheduling algorithm, there is no significant delay in communicating with the sensors. It can guarantee the acquisition of each channel with minimal data loss.

2.4.3 Sensor fusion thread design

The Kalman filter is used in some multisensor systems when it is necessary to fuse dynamic low-level redundant data in real time. The filter uses the statistical characteristics of a measurement model to recursively determine estimates for the fused data that are optimal in a statistical sense. If the system can be described with a linear model and both the system and sensor error can be modeled as white Gaussian noise, the Kalman filter will provide unique statistically optimal estimates for the fused data.

An extended Kalman filter (EKF), in a mathematically optimal manner, utilizes a wide variety of navigation sensors for aiding the inertial navigation system (INS). The Kalman filter is based on an error-state model which provides a much higher total navigation performance than would be obtained from the independent navigation sensors. The DVL-aided INS is capable of handling submerged autonomous operation for some time. In Figure 2.17, the core DVL-aided INS system consists of the IMU and the navigation equations, the EKF, and the following aiding sensors: DVL, compass (optional), and depth/pressure sensor. Depending on position accuracy requirements, the navigation system must get some position measurement updates. For submerged position updates, the system needs underwater transponders for position updates as shown in Figure 2.17.

Note that the remotely operated underwater vehicle model is highly nonlinear, we need to design an EKF for the sensor estimation. A linearized model of the vehicle at different operating conditions is required before the estimation. The linear equations of motion are obtained by linearization of the nonlinear equations about a time-varying reference trajectory or an equilibrium point using Taylor series. The details will not be covered in this chapter. However, the following EKF code based on the linearized matrices at the equilibrium condition is shown in Figure 2.18.

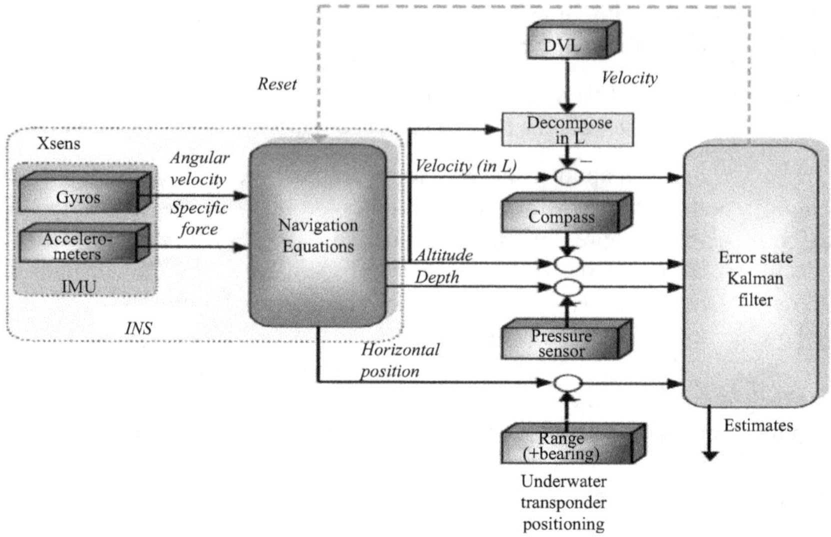

Figure 2.17 Integrated INS structure (Cited: https://www.scubadivingchicago.us/underwater-vehicles/integrated-inertial-navigation-system-structure.html)

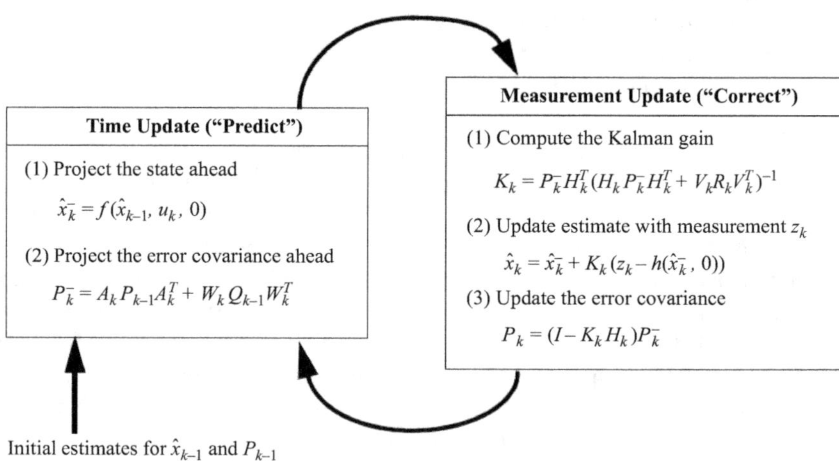

Figure 2.18 Extended Kalman filter [1, references therein] process with associated parameters

The parameters used in EKF are given as follows:

x: m state vector
A: $m \times m$ state transition matrix (linearized matrix)
B: $m \times p$ input transition matrix (linearized vector)

u: p input vector
W: Jacobian matrix of partial derivatives of with respect to process noise vector w
V: Jacobian matrix of partial derivatives of with respect to measurement noise vector v
K: Kalman filter gain
P: State covariance matrix
Q: $m \times n$ process noise covariance matrix
R: $m \times n$ measurement noise covariance matrix

2.4.4 Control process design

The control process consists of three main tasks. There are namely:

- design a controller (using sliding mode control algorithm);
- implement thrusters' driver;
- establish network communication between the host computer and target system.

With multithread technology, each thread is responsible for each task. Thread 1, thread 2, and thread 3 are used for controller design, actuator driver, and network communication, respectively. As shown in Figure 2.19, threads 1 and 2 constitute the single feedback control system. The numerical model of the vehicle is used to compute the control input or voltage to the actuators or thrusters to move the vehicle to the desired position. The sensor data will be stored in the memory that will be used to compute the error (e) to the controller.

2.4.5 Actuator driver design

After obtaining the control signal from the control algorithm, the control signal will drive the six thrusters (attached to the two DAQ cards) via the motor driver. In this case, Thread 2 is used for the actuator driver program. The analog control signal comes from the analog output ports of the DAQ card that requires a driver function to drive the thrusters. The flowchart of the program is shown in Figure 2.20.

This program describes the DAC operation. Some output, speed, and repeat cycle mode are required in the function, and then the D/A values are output via an

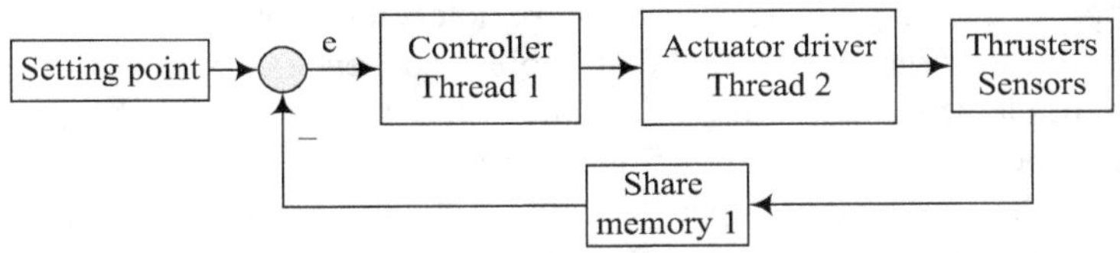

Figure 2.19 Block diagram of control process design for remotely operated underwater vehicle [1]

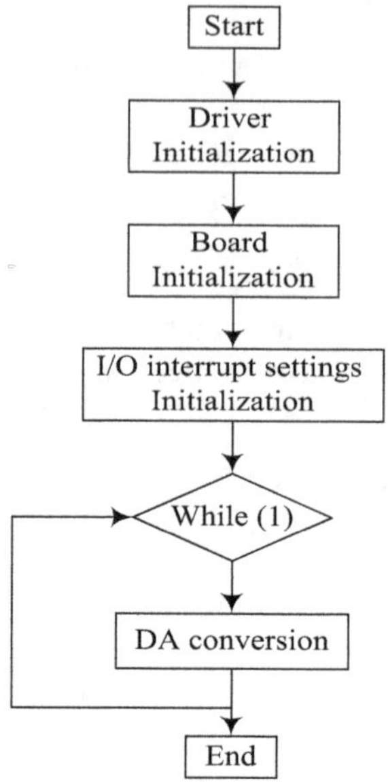

Figure 2.20 DAC program (thread 2) flowchart [1]

interrupt routine in memory. This generates a continuous waveform for each of the D/A channels. The steps are shown as follows:

- Initialize the driver, using the driver version for validation;
- Set the board type, base address to 0×280, interrupt level 5, initialize, and register the board with the driver;
- Set the number of conversions (must be a multiple of the FIFO depth), conversion rate (must be less than 100 kHz), the cycle flag, the internal clock flag, the low channel and the high channel (must be between 0 and 3), external gate enable flag, internal clock gate flag, and channel alignment;
- Enable channels (any of 0–3), the output codes, and write the output codes to the selected output channels.

2.4.6 Network communication thread design

All the fusion sensor data and video streams need to be transferred from target to host for manipulation. This module is responsible for communication between the host computer and target system (on the remotely operated underwater vehicle). The host link to target is via Ethernet cable, and hence, we have to use socket method to implement the network communication. A network socket is an endpoint of an interprocess communication flow across a computer network. As shown in Figure 2.21, the server is on target PC while the client works on the host PC.

For the server side, the initialization parameter includes setting socket address, port number, and TCP/UDP (User Datagram Protocol) communication mode. The following is the sequence for the socket server–client.

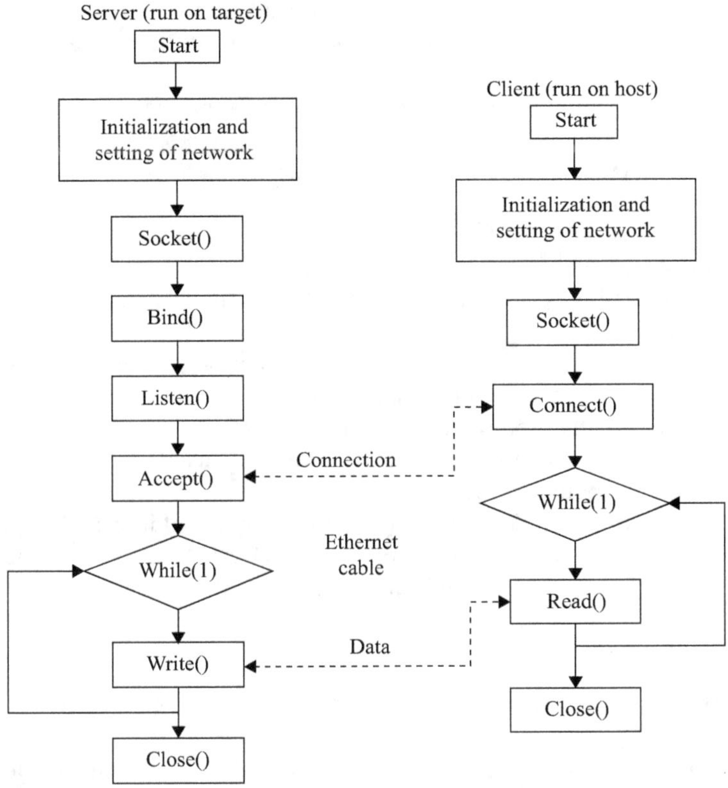

Figure 2.21 Socket server–client program (Thread 3) flowchart [1]

- Create a socket with the socket() system call;
- Bind the socket to an address using the bind() system call. For a server socket on the Internet, an address consists of a port number on the host machine;
- Listen for connections with the listen() system call;
- Accept a connection with the accept() system call. This call typically blocks until a client connects to the server;
- Send and receive data using write() system calls.

The following are the socket-layer functions used by the user process to send or receive packets and to do other socket operations.

- Socket(2) creates a socket, connect(2) connects a socket to a remote socket address, the bind(2) function binds a socket to a local socket address, listen(2) tells the socket that new connections shall be accepted, and accept(2) is used to get a new socket with a new incoming connection. Socketpair(2) returns two anonymous connected sockets (only implemented for a few local families like AF_UNIX);
- Send(2), send to(2), and sending(2) send data over a socket, and recv(2), recvfrom(2), and recvmsg(2) receive data from a socket. Poll (2) and select(2)

wait for arriving data or a readiness to send data. Also, the standard I/O operations like write(2), writev(2), sendfile(2), read(2), and readv(2) can be used to read and write data;

- Getsockname(2) returns the local socket address and getpeername(2) returns the remote socket address. Getsockopt(2) and setsockopt(2) are used to set or get socket layer or protocol options. Ioctl(2) can be used to set or read some other options;
- Close (2) is used to close a socket. Shutdown (2) closes parts of a full-duplex socket connection;
- Seeking, or calling pread(2) or pwrite(2) with a nonzero position is not supported on sockets;
- It is possible to do nonblocking I/O on sockets by setting the O_NONBLOCK flag on a socket file descriptor using fcntl(2). Then all operations that would block will (usually) return to EAGAIN (operation should be retried later); connect(2) will return EINPROGRESS error. The user can then wait for various events via poll(2) or select(2).
- An alternative to poll(2) and select(2) is to let the kernel inform the application about events via a SIGIO signal. For that, the O_ASYNC flag must be set on a socket file descriptor via fcntl(2) and a valid signal handler for SIGIO must be installed via sigaction(2).

2.5 Testing of components for control systems

It is important to test the components or devices used for the control systems after creating the software architecture for control purpose. Before moving on to the numerical modeling of the system or plant, the following components are used.

- Inertial measurement unit;
- DVL sensor unit;
- Image video unit;
- Depth sensor unit.

Note that there are few approaches to test the sensor. However, due to the space limitations, the following approaches are documented.

2.5.1 Inertial measurement unit

The Xsens MTi-30 AHRS sensor (see Figure 2.22) is often used for motion measurements. The MTi is 57 mm×42 mm×24 mm (Length × Width × Height) and weighs 55 g. The static accuracy of the sensor is 0.5°. The data are transmitted through a USB to serial cable. Real-time data (both raw and calibrated) can be obtained via low-level communication commands. The maximum sample rate is 2 kHz. The detailed specifications are as shown in Table 2.2.

Through integration on an acceleration signal $a(t)$, we can get the velocity signal $v(t)$ and position signal $x(t)$ if the initial position and initial velocity are known. After the first integration, the initial velocity should be added to the result, as the

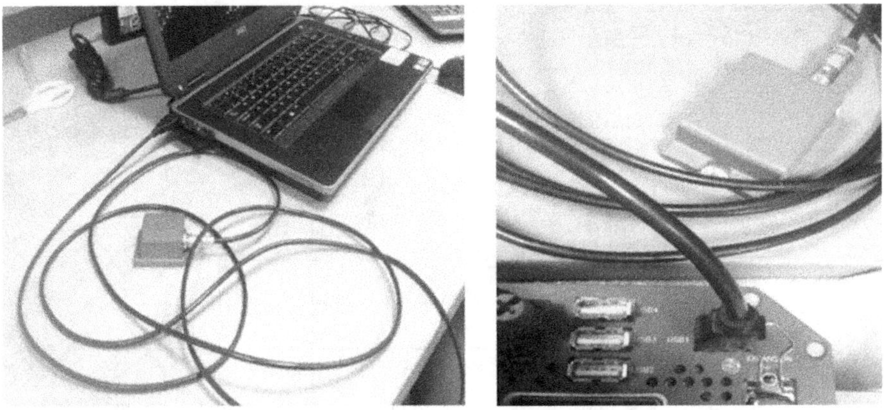

Figure 2.22 Specifications of Mti-30 AHRS connection to laptop (left) and Athena PC104 (right) [1]

Table 2.2 Specifications of MTi-30 AHRS (Cited: https://www.xsens.com/wp-content/uploads/2013/11/MTi_usermanual.pdf)

Input voltage	4.5–34 V or 3 V3	Latency	< 2 ms
Typical power consumption	480–570 mW	Interfaces	RS232/RS485/422/UART/ USB (no converters)
IP rating	IP 67 (encased)	Standard full range gyro	450 °/s
Temperature	−40 to 85 °C	Standard full range acc	50 m/s^2
Vibration and shock	MIL STD-202; 2,000 g	In-run bias stability gyro	18 °/h
Sampling frequency	10 kHz	Bandwidth gyro	415 Hz
Output frequency	Up to 2 kHz	Bandwidth acc	375 Hz

initial position should be added after the second integration. These operations are illustrated by the following equations:

$$v(t) = v(t_0) + \int_{t_0}^{t} a(\tau)\mathrm{d}\tau \tag{2.1}$$

where t_0 is the initial time, and $v(t_0)$ is the initial velocity. To get the position signal from velocity, the following relationship is used:

$$x(t) = x(t_0) + \int_{t_0}^{t} v(\tau)\mathrm{d}\tau \tag{2.2}$$

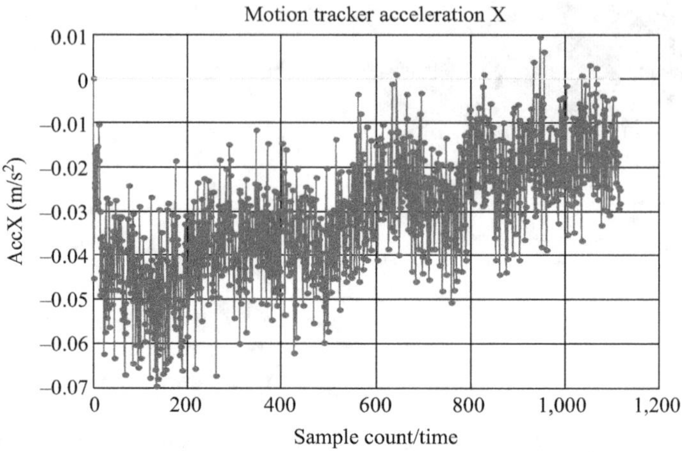

Figure 2.23 An acceleration signal of motion tracker in the X-direction (stationary) [1]

Therefore, for a double integration to be performed on acceleration, the two initial position and velocity should be known to avoid integration errors. Here, the initial velocity is zero, and initial position is at the original point. We used double integrating acceleration data to determine the Cartesian coordinates. The acceleration of remotely operated underwater vehicle will be measured with an accelerometer via the motion tracker XSENS MTi-30 ARHS. Unfortunately, the accelerometer has a drift phenomenon which was caused by a small DC bias in the acceleration signal. Ideally, there is no DC bias from the accelerometer for the measurement of vibration. This drift can cause significant integration errors. If the acceleration signal was integrated without any form of filtering, the output could become infinite over time.

Figure 2.23 is an example of an exaggerated acceleration signal that has a slight negative DC bias. Figure 2.24 illustrates an acceleration signal after the double integration. Two methods are developed here to eliminate the random bias and drift to improve the accuracy of integrating results. To solve the problem of the drifting, a high-pass filter can be used to remove the DC component of the acceleration signal. The frequency response of the filter has a very low cutoff frequency as compared to the bandwidth of the signal. By performing filtering operating before the integrating, drift errors can be eliminated.

The proposed filter uses the Fast Fourier Transform (FFT) to remove low-frequency content near DC. The FFT filter process is as follows. First, the acceleration signal is taken, and the low-frequency component coefficients are modified. It is followed by the inverse FFT to obtain the filtered signal. In Ribeiro's paper [4], a lower frequency coefficient (below about 0.7 Hz, which is the cutoff frequency of the filter) to a value equal to the cutoff frequency coefficient to attenuate the DC component was suggested. The low-frequency coefficients are

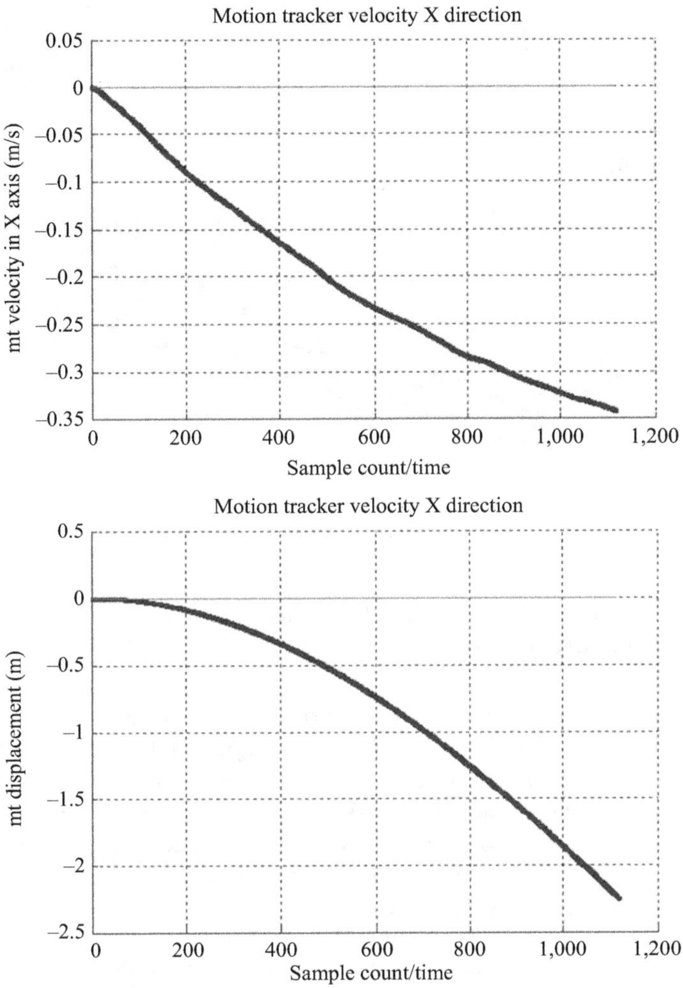

Figure 2.24 Display of integration errors due to drift and bias for velocity (top) and displacement (bottom)

located at the beginning and end of the FFT sequence. The method was tested based on the experiments. The results are shown in Figures 2.25 and 2.26.

For a stationary signal, the double integration process with FFT filtering works very well. It overcomes the drift and bias of the sensor. The result is accurate from the start of the time interval. The difference of the values of the velocity and position at time equals zero, and the actual initial condition values are negligible. This implies that initial velocity and position can be calculated accurately without direct measurement. Unfortunately, as shown in Figure 2.27, the shape of the acceleration signal has changed after the FFT filter. It implies that some motion signals were missing.

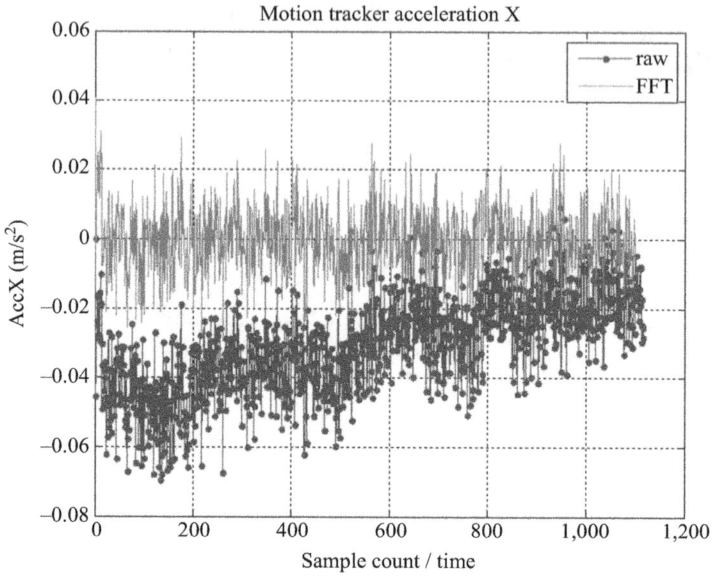

Figure 2.25 Comparison of raw and FFT-filtered acceleration signal (stationary) [1]

Hence, an alternative method was proposed to remove the drift and bias error of acceleration. The accelerometer bias has a significant impact on the overall performance of an IMU as the output could lead to errors in the calculated position. Here, a dead-reckoning method was used to identify whether the remotely operated underwater vehicle moves or remains as stationary. The detailed procedures are stated in Figure 2.28. First, an acceleration signal will be collected. Second, the mean value of the signal average (acc) will be computed. If the average (acc) is between −0.03 and 0.03 without points larger than 0.15, then the signal is considered as signal bias. Otherwise, it will be identified as motion (i.e., the vehicle is moving). When the remotely operated underwater vehicle is moving, the Butterworth high-pass filter is used to remove the drift and bias.

The dead-reckoning filter is illustrated in the following experiment. The motion tracker was mounted on the vehicle, and then using the joystick to control the remotely operated underwater vehicle along a straight line in the X-direction for a distance of around 2 m in open loop (without any control). At the same time, the velocity of the remotely operated underwater vehicle was captured for comparison with the integration of acceleration of motion tracker. As shown in Figure 2.29, the acceleration profile after dead-reckoning is quite similar to the raw data. Figure 2.30 shows that the velocity profile of the motion tracker (after integration) is quite similar to the velocity from the DVL.

The same measurement method was deployed for the Y-direction, and the results are shown in Figures 2.31 and 2.32. The results show that the signals have the same signature as the original data and indicate the correct trajectory.

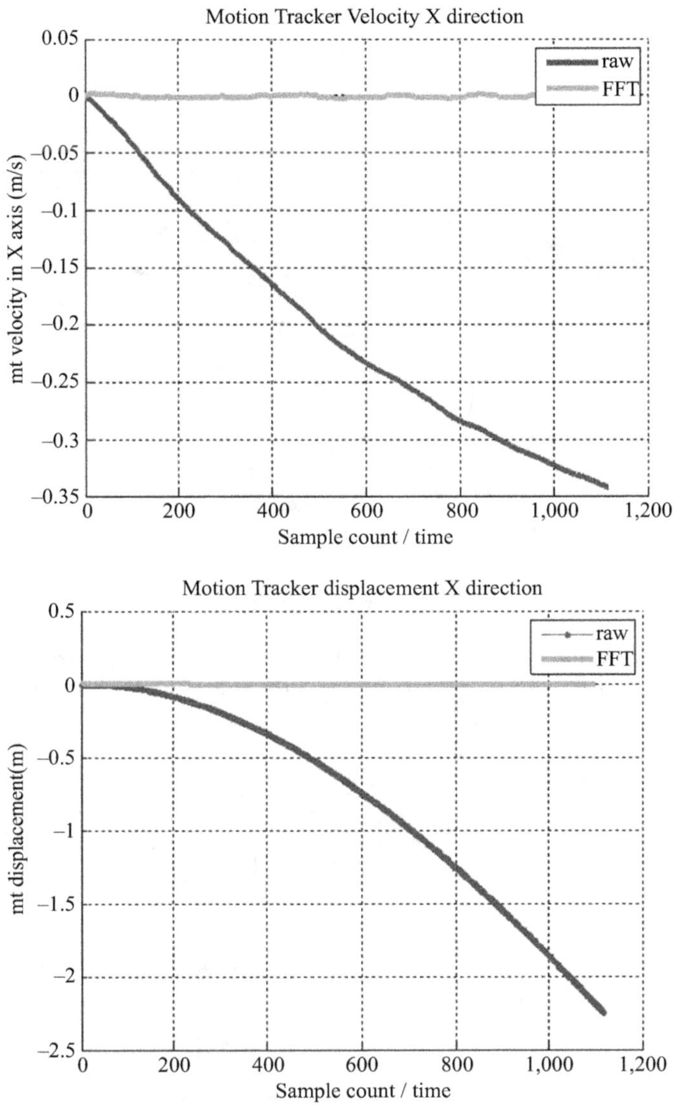

Figure 2.26 Comparison of raw and FFT-filtered integral velocity (top) and displacement (bottom) [1]

2.5.2 DVL sensor unit

NavQuest 600 is the world's smallest DVL by LinkQuest Inc (see: http://www.link-quest.com/html/intro_nq.htm). It is used for precision navigation and positioning of underwater vehicles. It has small size, light weight, capability for fast and accurate velocity output, significantly longer range, and smaller minimum altitude as

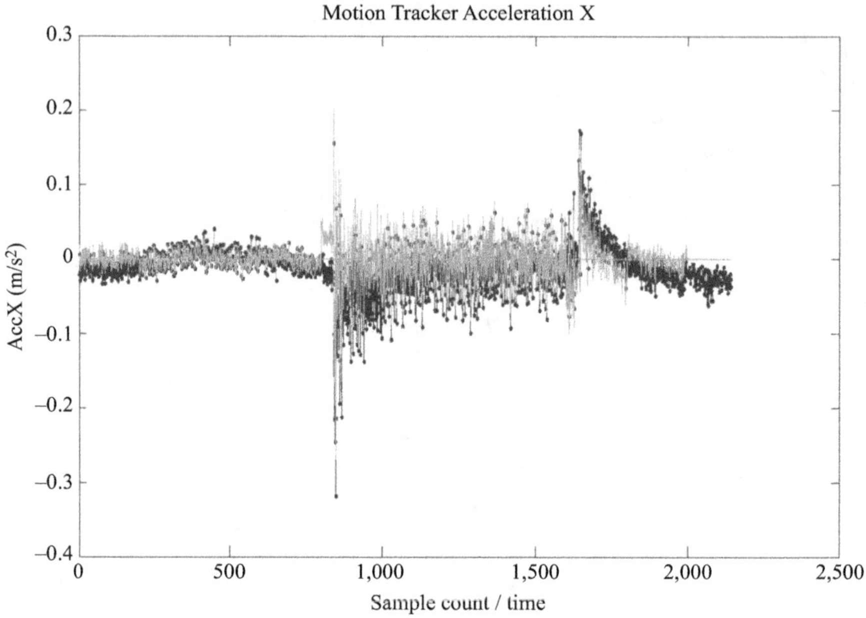

Figure 2.27 Example of an acceleration signal before and after FFT filter (Dark color: original signal; light color: FFT filter signal) [1]

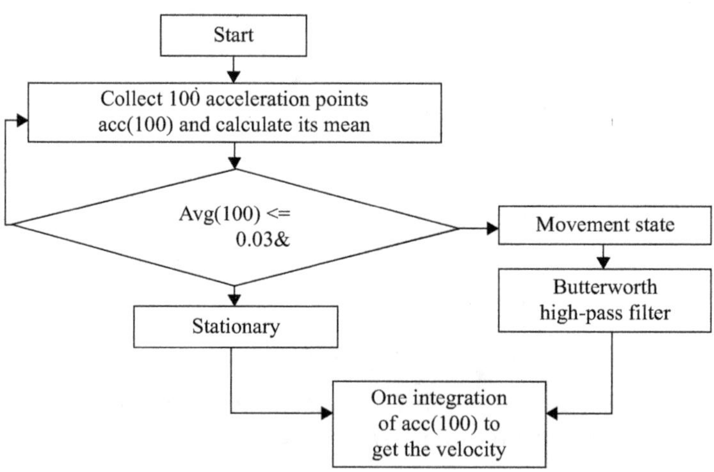

Figure 2.28 Flowchart of dead-reckoning filter [1]

compared to other products. The specifications for the DVL NavQuest 600 Micro Sensor are listed as follows.

- Maximum altitude (Range): up to 110 m;
- Minimum altitude: 0.3 m;

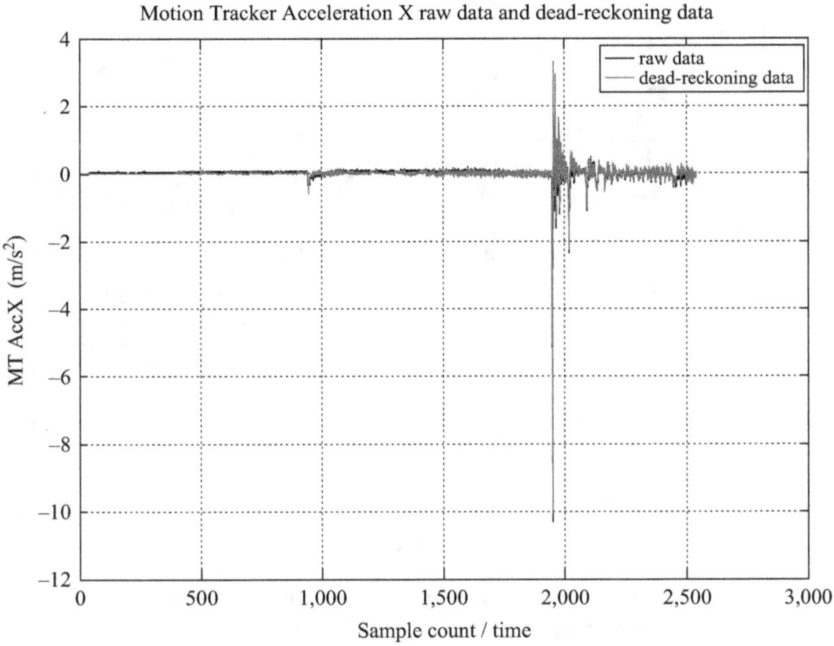

Figure 2.29 Comparison of raw and dead-reckoning filter acceleration signal in the X-direction [1]

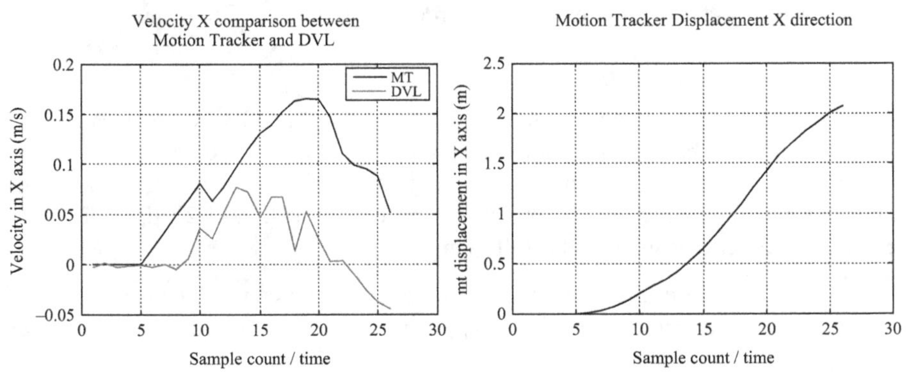

Figure 2.30 Velocity (comparison between DVL and motion tracker) and position signal after dead-reckoning [1]

- Accuracy: 1% ± 1 mm/s;
- Accuracy (Model P): 0.2% ± 1 mm/s;
- Operating frequency: 600 kHz;
- Depth: up to 6,000 m.

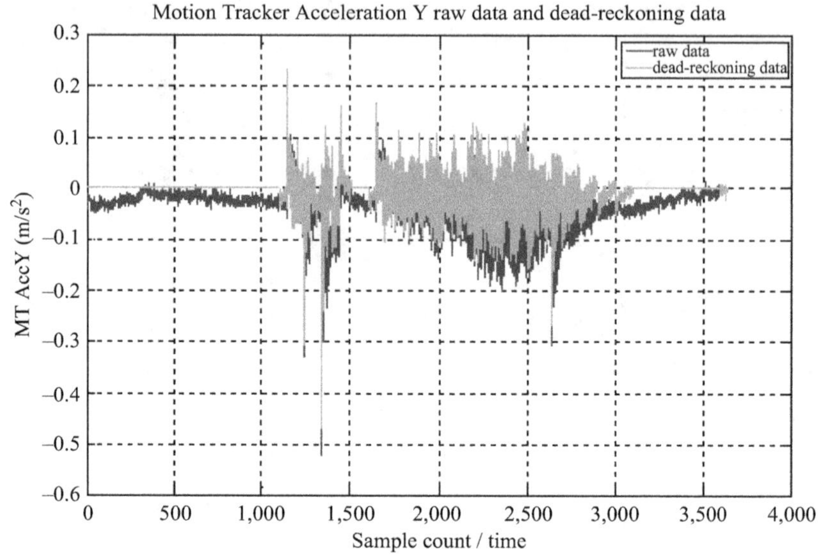

Figure 2.31 Comparison of raw and dead-reckoning filter acceleration signal in the Y-direction [1]

The ASCII commands were obtained from LinkQuest Inc. We will use the functions and retrieve the data via the Linux®-based environment. The plots of the DVL's velocities are shown in Figure 2.33, where the velocities are all zero (since it was not moving). This sanity check shows that the DVL is working. Note that it is important to test the DVL in water and not in the air in order not to damage the device.

2.5.3 Image video unit

The acquisition program for the vision in water using the USB Logitech Webcams C170 cameras can be realized using application functions and data structures provided by Video4Linux2. Most information in this section is obtained from the paper published in IEEE/MTS OCEAN 2014 Conference [5]. V4L2 supports many devices and functions such as the following:

- The video capture interface grabs video data from a camera device.
- The video output interface allows applications to drive peripherals which can improve video images.

The camera driver (see: http://www.ideasonboard.org/uvc/) used is Linux®-uvc. It is an open-source driver for USB video devices. It can support Video4Linux2 API. The UVC specification covers many devices which support video streaming for both video input and output.

After the image (or video) is captured, a multiplatform toolkit for creating GUIs called GTK+ will be used. The acquisition program realized the USB camera (i.e., USB Logitech Webcams C170 cameras) image DAQ using the driver functions and data structures provided by Video4Linux2 as shown in Figure 2.35.

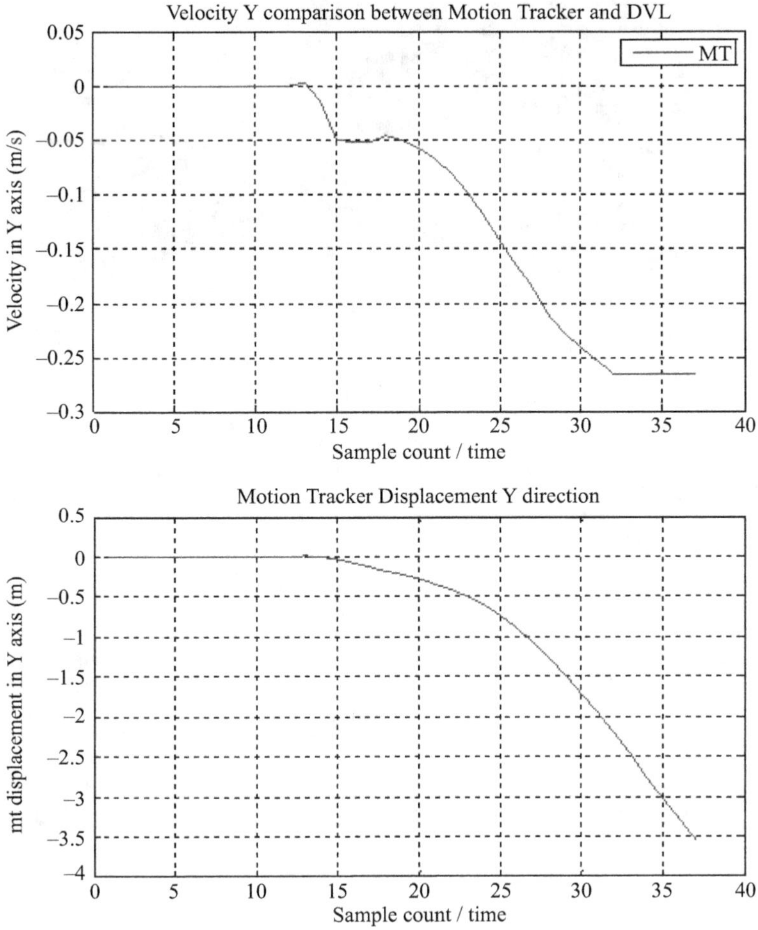

Figure 2.32 Velocity and position signal in the Y-direction after dead-reckoning [1]

Under the Linux® OS, all peripherals are treated as a special file called a device file. System calls and a variety of libraries offer interfaces between the kernel and application programs directly or indirectly, while the driver is the interface between kernel and peripherals, it completes the initialization and release equipment, and documents in various operations end-processing functions. A USB webcam driver module used is Linuxuvc which can support Video4Linux2 API. First, we need to compile the linuxuvc and Video4Linux2 files in the host computer and get the videodev.ko and videouvc.ko. Then install the two modules into the target system using insmod commands. The following shows the video image capture program using the functions and data structure which are defined by the interfaces application programming. The detailed flowchart is shown in Figure 2.36.

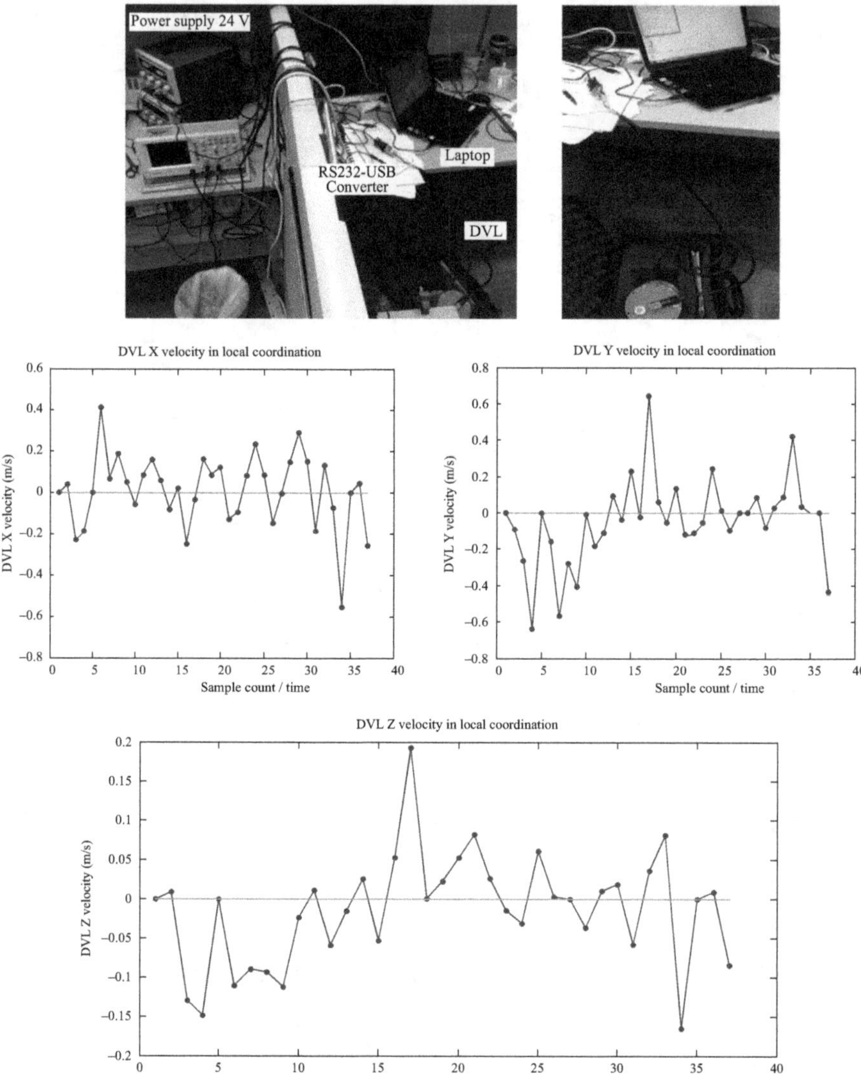

Figure 2.33 DVL-captured screen/testing and DVL's velocity signal in the X-, Y-, and Z-directions [1]

The steps used for the video image capture program are described (see Figure 2.34) as follows:

- Open the camera device and get the corresponding file descriptor called fd, to operate the device;
- Gather the device attributes, set the image format in Video4Linux2 API, using ioctl function to operate the device;

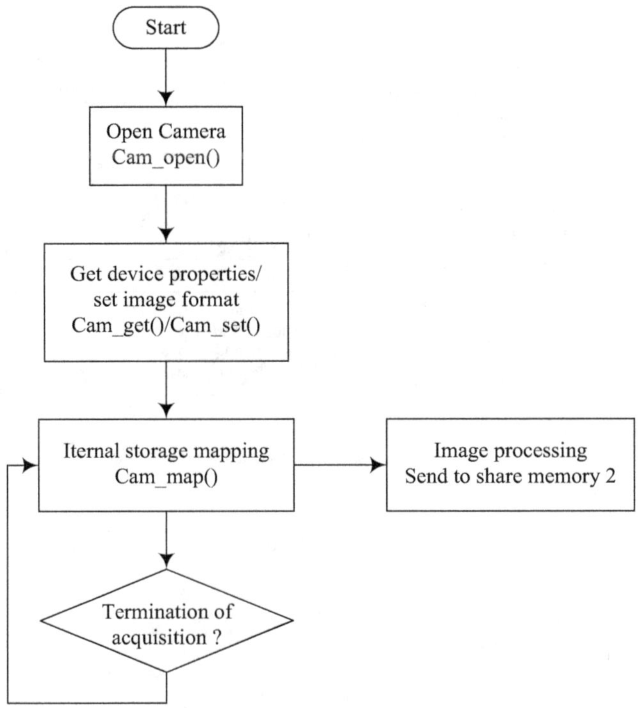

Figure 2.34 Image acquisition program flowchart [1,5]

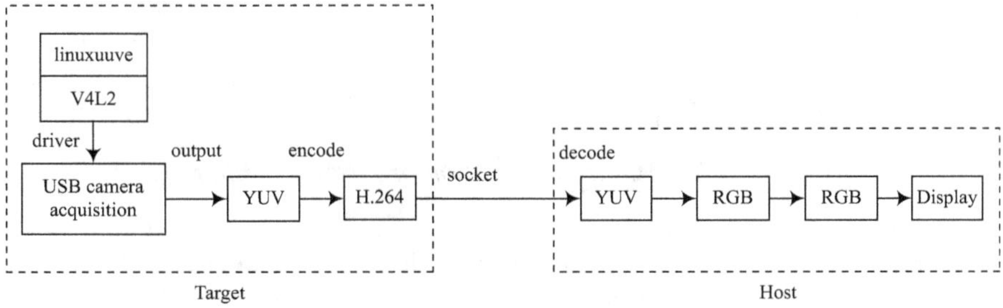

Figure 2.35 Process of camera capture system [1,5]

- Apply Ioctl(fd, VIDIOC_QUERYCAP, &cap) to Query device capabilities, Ioctl(fd, VIDIOC_S_FMT, &cap) to set the image format and Ioctl(fd, VIDIOC_S_PARM, &setfps) to set the video format;
- Map the internal memory using mmap (NULL, buf.length, PROT_READ| PROT_WRITE, MAP_SHARED, fd, buf.m.offset) map the work between user space and kernel space;

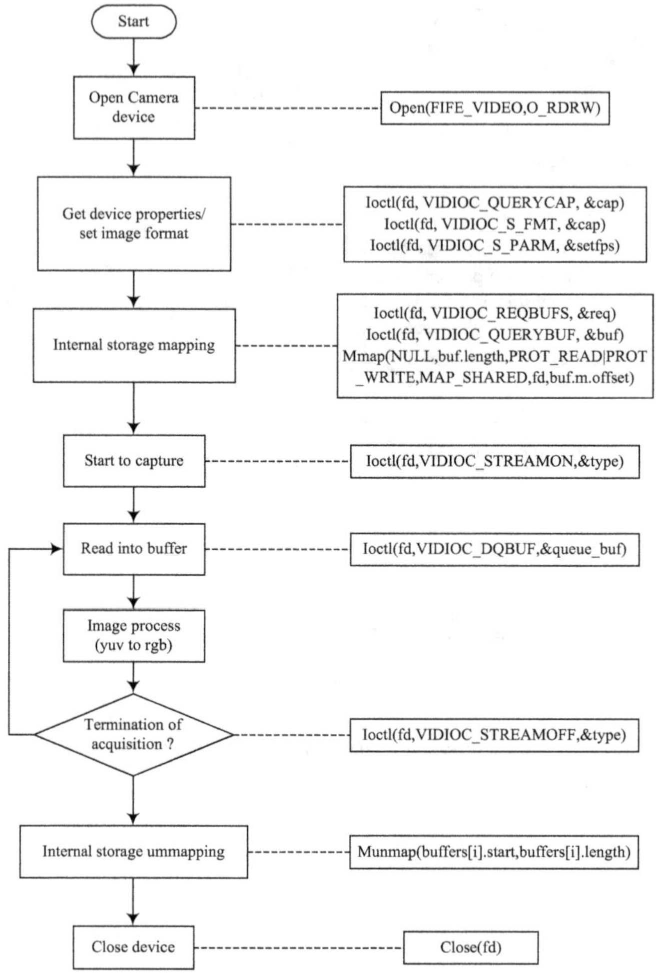

Figure 2.36 Video image acquisition program flowchart [1,5]

- Start to capture one frame image with ioctl(fd,VIDIOC_STREAMON,&type) function and read the image raw data using ioctl(fd,VIDIOC_STREAMON, &type);
- Transfer image file to the host computer for display using the Ethernet cable, but the DAQ format from the camera is YUV, it takes up too much space and is not suitable for transmission. Hence, we should do the image process here.
- Refresh the frame buffer to capture the next frame using Using ioctl(fd, VIDIOC_STREAMOFF, &type) functions;
- Release the memory and close device.

The hardware architecture of the camera system is shown in Figure 2.37. The underwater video image acquisition system was designed using client–server model

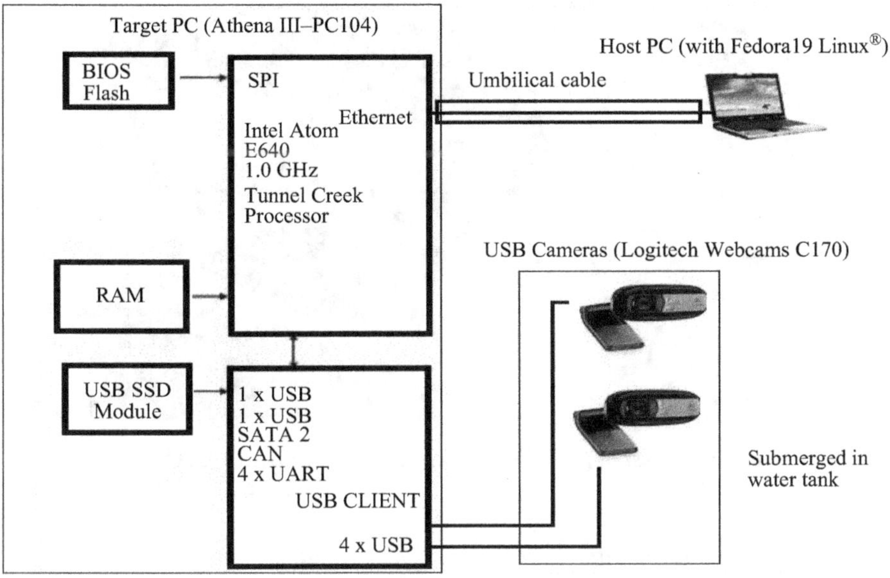

Figure 2.37 Hardware architecture of the camera system [1,5]

to transmit two-channel video image data. It includes two sections: target board and a host computer connected by an Ethernet cable (inside an umbilical cable). The target uses Athena which has a stable CPU reset chip and kernel power supply chip to ensure the system stability. To capture a high-quality video image from different angles in underwater, two USB Logitech Webcams C170 cameras were used for video acquisition. The cameras support USB Video Class Linux® device driver (UVC) driver and YUV422 data format. The YUV422 data format shares U and V values between two pixels. For example, Pixel Number = 0 has pixel values of U0Y0V0, and Pixel Number = 1 has pixel values of U0Y1V0.

Linux® OS operates on the host computer. The bootloader Grub2, compiled kernel, root file system, and the dynamic modules were saved into the 8G flash disk located in the Athena III board. The host PC can communicate with the development board using NFS via Ethernet cable. Target system board's IP address is 192.168.0.2, while host computer's IP address is set as 192.168.0.1 to ensure they are on the same subnet. During the test, the Linux® system startup time is approximately 12.20 s. The video image acquisition programs run on the target machine after it was successfully compiled in the host computer. The dual camera's video image acquisition system was tested in a 150 L water tank in a laboratory environment. The test setup of the entire system is shown in Figure 2.38. The two cameras are placed in the water tank to capture the images.

To obtain a clear and high-quality image, apart from hardware factors, the main point is to ensure that every image frame data and context is not lost in the translation during the transmission (i.e., encoding and decoding) program. We captured one frame data (resolution: 320×240) as an example and save it into a file

Figure 2.38 Test setup of the video image acquisition system [1,5]

Table 2.3 Single frame data (resolution 320×240) file size in different phases of the transmission process [1,5]

File format	YUV422	YUV420	H264 (target machine)	H264 (host computer)	RGB24
Size (byte)	153,600 bytes	115,200 bytes	16,052 bytes	16,052 bytes	460,800 bytes

in every format conversion phase and then verify the size of the image as well as a comparison of image quality by using a pYUV reader and VLC media player.

In Table 2.3, the server and client side have the same size of .h264 file; we can verify that there is no data loss during the UDP socket transmission via Ethernet cable. Converting YUV420 frame data to an H264 file using X264 encoder, the compression ratio is 16,052/115,200 = 13.93%, which significantly saves the network bandwidth, reduces time delay and ensures the real-time transmission performance.

The remote underwater dual cameras video image acquisition system using Athena PC104 platform loaded with embedded Linux® was tested. To realize the system, Linux® OS was transplanted with UVC driver, V4L2, X264, and FFMPEG. As shown in the test results, the proposed setup was able to perform multiple channel camera image acquisition, data compression, and transmission without much interference with each other and data lost. Hence, this proposed system setup is suitable for underwater multichannel signal acquisition applications.

2.5.4 Depth sensor unit

PDCR1830 Pressure Sensor (see: http://www.eurotronbenelux.nl/products/druk-sensoren/ptx-pdcr-1830-series-depth-and-level-pressure-sensors.html) is the latest generation of fully submersible high-performance sensors for the measurement of

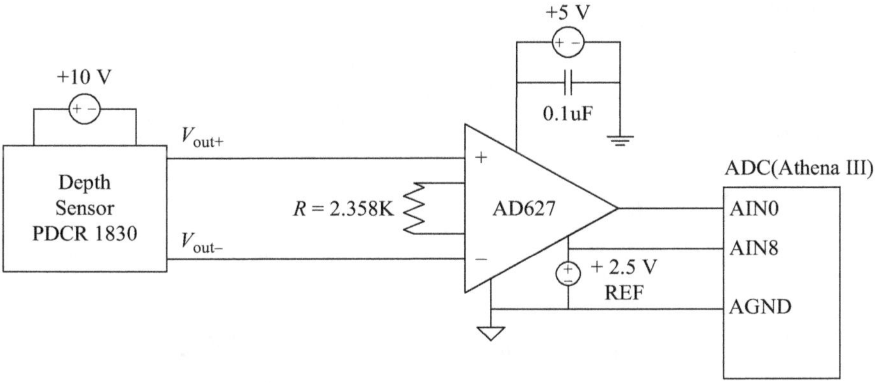

Figure 2.39 Amplifier circuit connection circuit diagram [1]

hydrostatic liquid levels. The micromachined silicon element is sealed within a titanium pressure module assembly, fully isolated from the pressure media. The pressure sensor was tested in a water tank to test the functionality. The following pressure specifications are shown.

- Pressure sensor type: PDCR_1830;
- Accuracy: ±0.1% FSR (full-scale range);
- Pressure range: 50 mV (for 2.5 to 5 psig (i.e., 3.5 m));
- Cable length: 1 m (3 feet);
- Wires layout: 6-wire configuration.

The full-scale output of the depth sensor is quite small voltage (0–100 mV) and the analog to digital convertor of Athena III board is between 0 and 10 V. To obtain more accurate digital data, the output of the depth sensor should be conditioned before analog processing in the Target. Here, we used AD627 to amplify the analog signal of the depth sensor.

- Define the gain of amplifier circuit: Gain $= 5 + (200\ \text{k}\Omega/R_{\text{G}})$;
- Ideal gain = 10 V/100 mV = 100, here we used $R_G = 2.358\text{k}\Omega$, Gain = 90.

Figure 2.39 shows the AD627 configured to amplify the output signal of the depth sensor. The circuit works in single supply mode (+5 V). The REF pin is set to 2.5 V.

The test environment and results are shown below. The calibrated relationship between the pressure and water column height is shown in Figure 2.40. As observed, the result is not linear across all water depths.

- Test depth: 1 m maximum;
- Test H_2O temperature: 23 °C;
- Test excitation voltage: 10 V at 5 mA maximum;
- Sensor's sensitivity data (from GE-Druck's Calibration Test Report).

Computed sensor's sensitivity: 0.653 mV/V/psig (See the underlined in Table 2.4) for water column not exceeding 10.55 m.

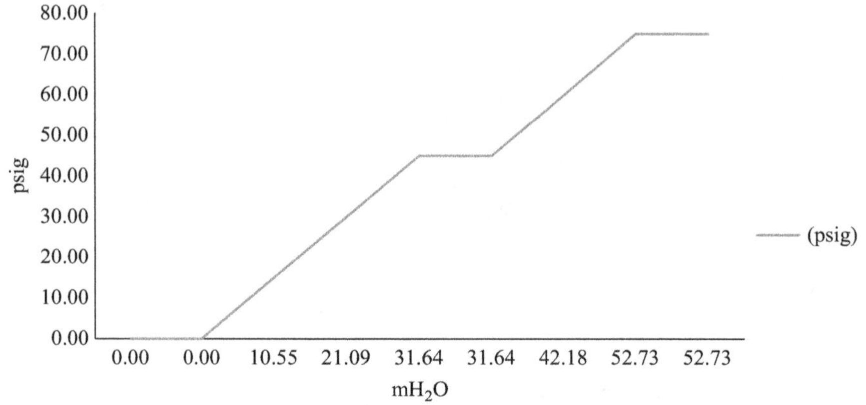

Figure 2.40 Calibrated results of PCR1830 sensor (from GE-Druck's Calibration Test Report) [1]

Table 2.4 Sensitivity of PCR1830 pressure sensor at different mH_2O [1]

Full-range output/FRO	(mV)	98.00	98.00	98.00	98.00	98.00	98.00	98.00
Pressure	(mH_2Og)	10.55	21.09	31.64	31.64	42.18	52.73	52.73
Pressure	(psig)	15.00	30.00	45.00	45.00	60.00	75.00	75.00
% deviation FRO	(mV)	−0.03	−0.06	−0.05	−0.06	−0.01	0.05	0.05
Deviation	(mV)	−2.94	−5.88	−4.90	−5.88	−0.98	4.90	4.90
Sensitivity	mV/psig	6.53	3.27	2.18	2.18	1.63	1.31	1.31
Excitation voltage input	V	10.00	10.00	10.00	10.00	10.00	10.00	10.00
Sensitivity	mV/V/psig	0.653	0.327	0.218	0.218	0.163	0.131	0.131

Table 2.5 Multiplier (1/ Sensitivity) of PCR1830 pressure sensor at different mH_2O [1]

Pressure	(mH_2Og)	10.55	21.09	31.64	31.64	42.18	52.73	52.73
Multiplier (M) = 1/sensitivity	psig/mV/V	1.531	3.061	4.592	4.592	6.122	7.653	7.653
	mH_2O/mV/V	1.076	2.152	3.228	3.228	4.305	5.381	5.381

Multiplier (= 1/Sensitivity) is computed for different water column heights as seen in Table 2.5. Hence, we have different values of the multiplier for different water heights. It is considered in the C-program.

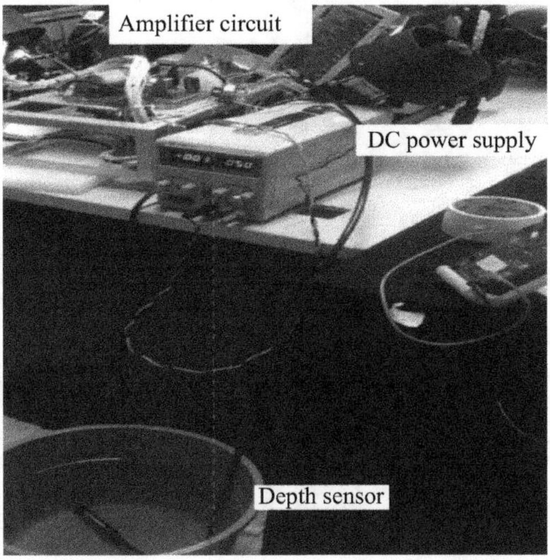

Figure 2.41 Experimental setup for the pressure sensor [1]

The setup for the test is shown in Figure 2.41. The sensor was submerged in the polyethylene water tank (0.5 m diameter × 1.035 m height) with a maximum storage size of 193 L.

2.6 Kalman filter

To improve the precision of remotely operated underwater vehicle navigation, a loosely integrated IMU/DVL navigation method is proposed. The two subsystems measure and calculate the velocity of the vehicle, respectively in Figure 2.42. Then the velocity data are fused using indirect error Kalman filter. This method does not need to do any modification of IMU or DVL, and it gives quite a good performance.

In this proposed system, IMU subsystem is working on a high frequency while the Kalman filter is running at a low DVL output frequency. As a result, the estimated velocity can be obtained. The IMU and DVL were fixed in the remotely operated underwater vehicle. The acceleration of IMU and velocity of DVL were recorded. Figures 2.43–2.46 show the sensor data fusion results in the *X*- and *Y*-directions. We can see that, after the Kalman filter, the errors of velocities become smaller.

2.7 Graphical user interface

A GUI is a type of interface that allows users to interact with electronic sensors, command inputs, and actuators through visual indicators and graphical icons such as secondary notation, as opposed to text-based interfaces, text navigation, or typed command labels. The GUI is seen as an essential requirement for enhanced operator dexterity. Telepresence systems have for some time been investigated as a

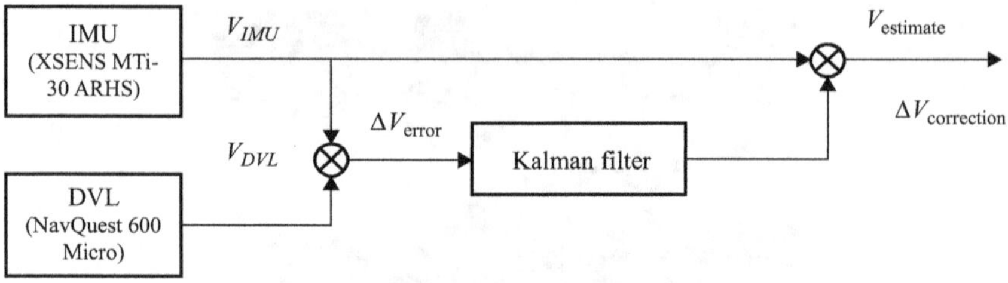

Figure 2.42 Block diagram of Kalman filtering [1]

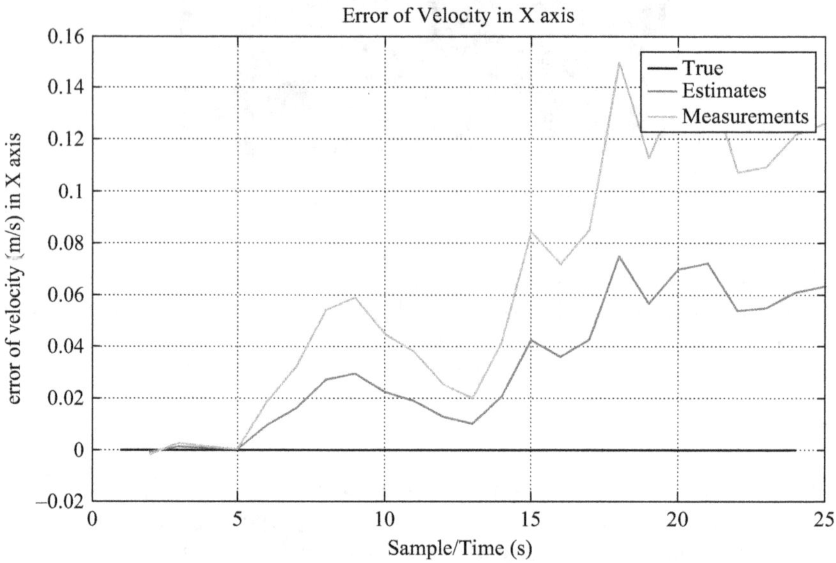

Figure 2.43 Velocity error in the X-direction [1]

means of realizing highly dexterous and intuitive systems for remotely operated systems. The AUV Docking Hoop GUI control system panel consists of two primary functions: send control commands to control the thrusters manually and automatically; display sensors and thrusters' status in real time. The actual layout of the GUI control panel is shown in Figure 2.47.

Creating the GUI applications on Linux® can be performed in different ways. Using the simplest and the most functional programming languages and libraries under the Linux® desktop using the Qt library with Python programming language which called "PyQt" was chosen, when Python was compared with other programming languages (Perl, Python, Rexx, Tcl). In terms of run time and memory consumption, they often turn out better than Java and not much worse than C or C++. Hence, Python programming language was chosen. On the other hand, Qt is

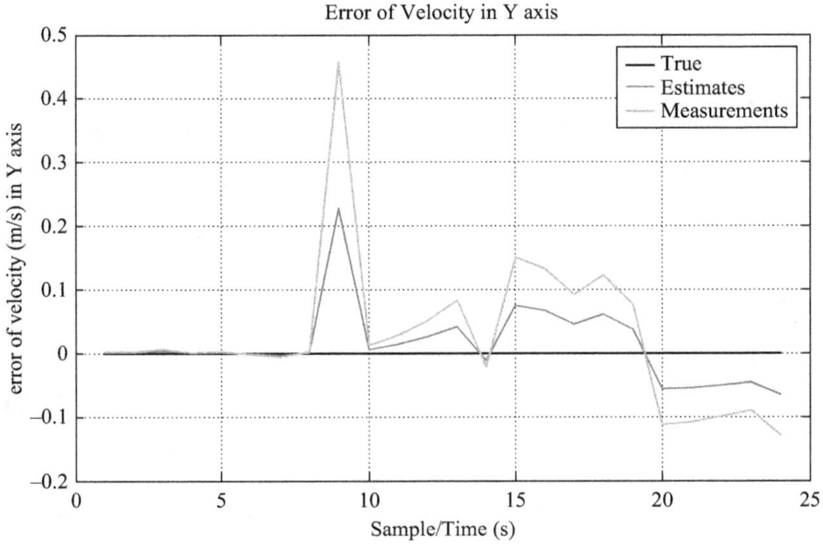

Figure 2.44 Velocity error in the Y-direction [1]

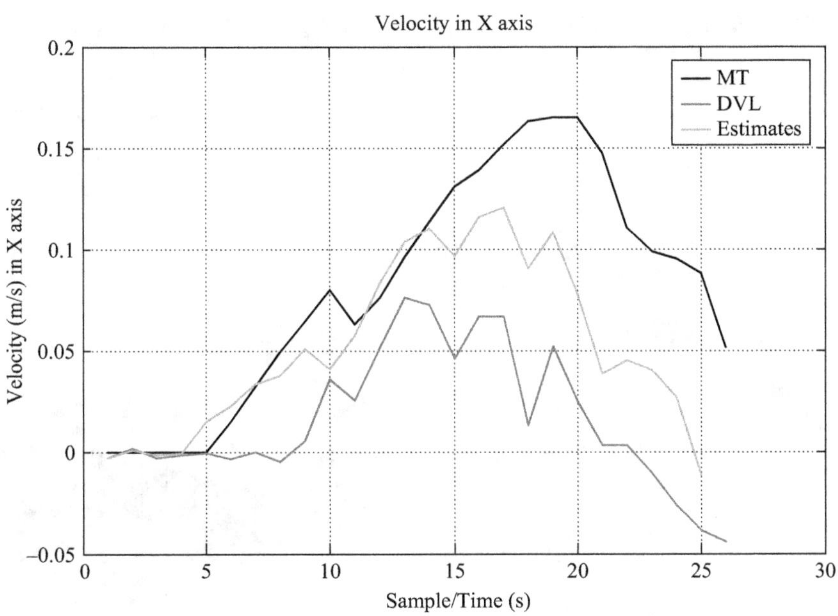

Figure 2.45 Velocity in the X-direction [1]

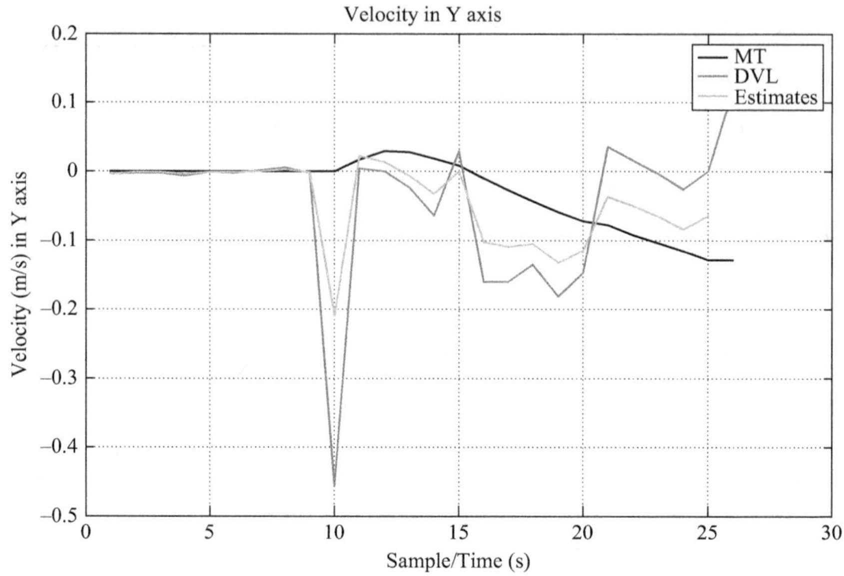

Figure 2.46 Velocity in the Y-direction [1]

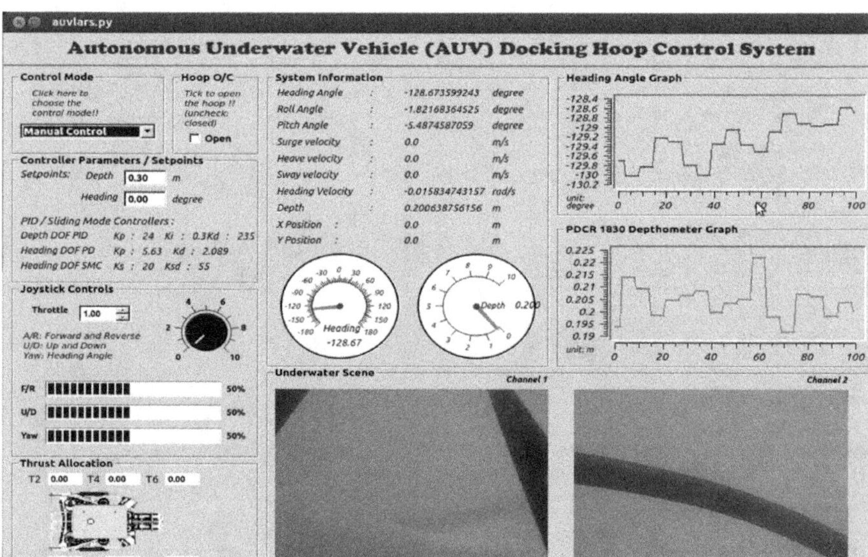

Figure 2.47 GUI control panel layout [1]

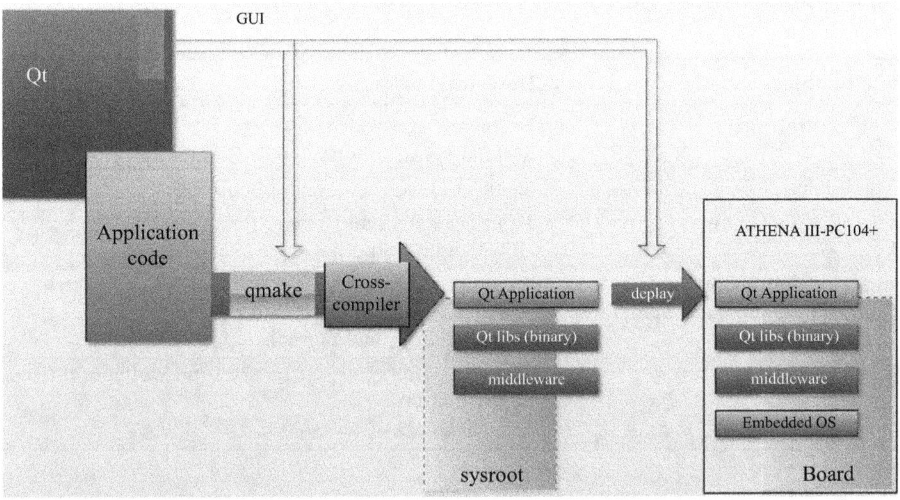

Figure 2.48 Qt user-interface and application flowchart (cited: http://www.qt.io/developers/)

a GUI framework which is available as free and open-source software. PyQt is the Python wrapper for Qt. The two main criteria for a GUI framework are easy to use and cross-platform (Windows and Linux®). PyQt meets the criteria pretty well. To facilitate the software development, an IDE consists of a source code editor, build automation tools and a debugger is needed. Since the Python and Qt were chosen, the IDE written in python and based on the cross-platform Qt GUI toolkit, integrating flexible editor control is highly desired. Thus, Eric4 Python IDE written in Python and based on Qt GUI was selected.

It includes a plugin system which allows easy extension of the IDE functionality with plugins downloadable from the Internet. In summary, the GUI control panel was designed using PyQt4, Python 2.7, and Eric4 Python IDE for the IDE. A brief overall view of the software packages is described. As shown in Figure 2.48, the pre-configured embedded development environment, prebuilt Qt (cited from www.integraatio.fi) optimized software stack for immediate deployment to ATHENA III embedded board, allows users to get up and running, and have a working embedded project prototype. It supports the key requirements for high performance with optimizable native C++ libraries and minimal footprint while providing a flexible, full-framework modular architecture to deliver unparalleled scalability.

There are six function modules (see Table 2.6) for the AUV Docking Hoop GUI. They are control mode, switch control of docking hoop, network connection, sensor readout of remotely operated underwater vehicle, thruster readout, and video image display.

Table 2.6 Function modules for AUV Docking Hoop GUI [1]

Palette object	**Descriptions**
Main control mode	Automatic control mode Manual control mode (using joystick)
Switch control of docking hoop	Open or close the docking hoop
Network connection	TCP: command and sensor readout UDP: video frame data
Sensor readout of remotely operated underwater vehicle	Altitude: instrument panel of yaw, roll, and pitch Angular velocities: p, q, and r Positions: x, y, and z graph
Thruster readout	Thruster parameters: current, torque, voltage, thrusts distribution
Video display	Two channels of video size 320×240

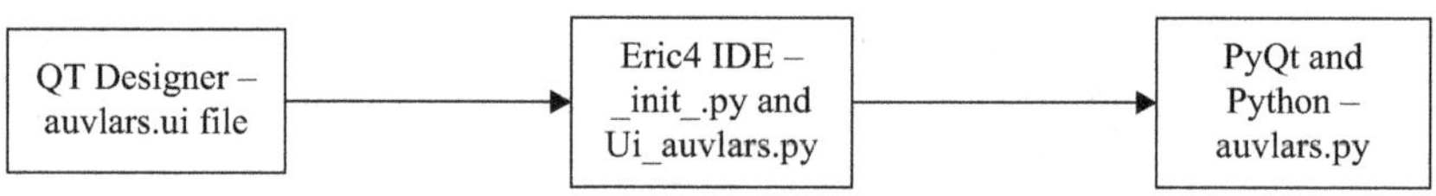

Figure 2.49 Developing process of AUV Docking Hoop control system panel [1]

- Main control mode—to control the AUV Docking Hoop using automatic and manual control;
- Switch control of docking hoop—to control and status of the opening and closing of the docking hoop;
- Network connection—to indicate the network connection status between AUV Docking Hoop and surface ship. We used socket technology to achieve the data transmission using TCP for control command and collect sensor data. UDP is used for rapid image video frame data transferring. UDP uses a simple connectionless transmission model with a minimum of protocol mechanism.
- Sensor readout—to display all the sensor data and plot 2D graphs. It includes position and velocity of AUV Docking Hoop in body-fixed coordinate and position in the earth-fixed coordinate.
- Thruster readout—to display the status of the 6 thrusters. The thrusters' parameters such as current, torque, voltage, and thrust distributions will be displayed.
- Video display—to show the videos captured by 2 USB Logitech Webcams C170 cameras. A USB webcam driver module, Linuxuvc can support Video4Linux2 API.

A brief developing process of control system panel is shown in Figure 2.49. First, the QT designer was used to design and build GUI. Then, we used the Eric4 Python IDE to compile the *ui file into python file named with extension *.py. Finally, the background program was coded using Python language.

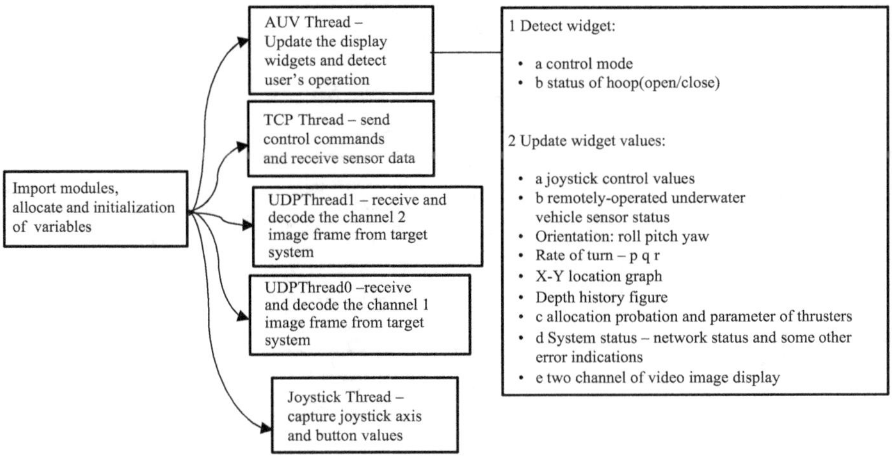

Figure 2.50 Multithreading techniques based on Python language [1]

The AUV Docking Hoop GUI control system control panel was realized using multithreading techniques based on Python Language. As observed in Figure 2.50, the five threads refer to widget detect and update thread, TCP thread, 2 UDP thread, and one joystick thread, respectively. Each thread communicates using global variables. One way is to use message passing between threads via asynchronous queues. This way we can avoid using shared data between threads and only the queues need to be thread-safe. Asynchronous queues were implemented using different synchronization primitives: pipes or sockets, queues protected with a mutex and a condition variable, and nonblocking or lock-free queues.

In the host computer (laptop), the video image UDP thread was responsible for receiving the frame data from the vehicle via UDP socket. The frame data were decoded using FFMPEG decoder. The outputs were displayed in the frame data using the RGB format. Figure 2.51 shows the video image UDP thread process program flowchart.

How to call FFMPEG module (c functions) in Python? First, we need to import ctypes module in Python to support the module. The main thread of the entire GUI control panel serves two functions, namely detect and respond to the user's operation and update all the widgets' values and status. The actual GUI display is shown in Figure 2.52.

The status and values of display widgets are shown in Figures 2.53 and 2.54. The angular velocities in the body-fixed coordinate, Nixie tube and dial using set value were used. Orientations in the earth coordinate (roll, pitch, and yaw) were displayed. Moreover, the depth graph was shown in the GUI. Figure 2.54 shows the values of the thrust, current, and voltage of the six thrusters. The values of these parameters will be logged.

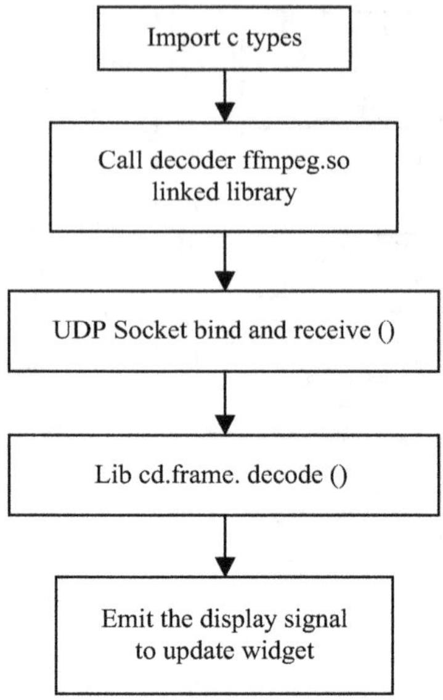

Figure 2.51 Video image UDP thread program flowchart [1]

Figure 2.52 Detect widget for control mode and status of remotely operated underwater vehicle [1]

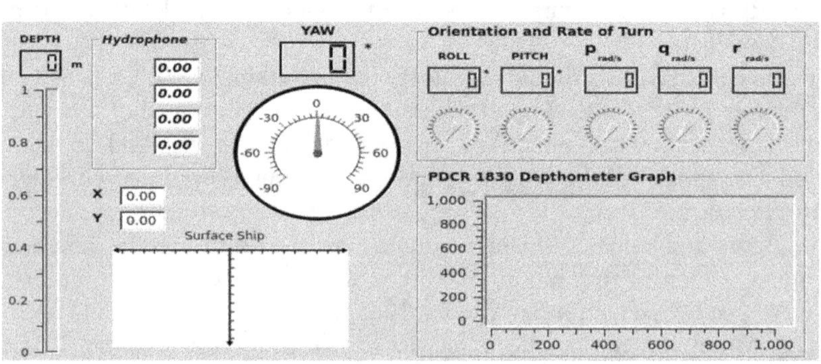

Figure 2.53 Sensor readout display [1]

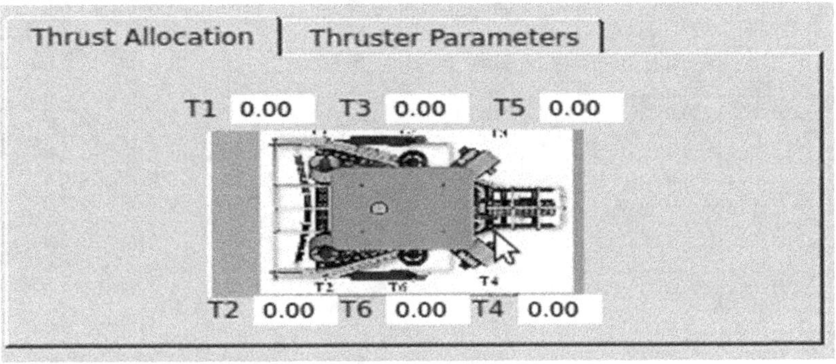

Figure 2.54 Thruster status [1]

References

[1] Lin, WP. Design and Implementation of AUV Docking Hoop Control System, **Master of Philosophy Thesis**, Newcastle University, United Kingdom, February 2016.

[2] Chin, CS, Lin, WP, and Lin, JY. Experimental Validation of Open-Frame ROV model for Virtual Reality Simulation and Control. **Journal of Marine Science and Technology** vol. 23, no. 2, pp. 267–287, June 2018.

[3] Lin, WP, Chin, CS, Looi, LCW, Lim, JJ, and Teh EME. Robust Design of Docking Hoop for Recovery of Autonomous Underwater Vehicle with Experimental Results. **Robotics**. Guest Editors: Prof. Thor I. Fossen and Prof. Ingrid Schjølberg, NTNU, Trondheim, Norway 2015, 4(4), 492–515.

[4] Ribeiro, A, Giannakis, G, and Roumeliotis, S. SOI-KF: Distributed Kalman Filtering with Low-cost Communications Using the Sign of Innovations. **IEEE Transactions on Signal Processing**, vol. 54, no. 12, pp. 4782–4795, December 2006.

[5] Lin, WP and Chin CS. Remote Underwater Dual Cameras Video Image Acquisition System using Linux Based Embedded PC104. **MTS/IEEE OCEANS** 2014, Taipei, Taiwan.

Chapter 3
Modeling and simulation of embedded underwater vehicle system

3.1 Introduction

Numerical modeling and simulation techniques are essential in many engineering applications. This technique has been evolving and has played an essential role in industry and institutions for the past few decades. The applications of numerical modeling and simulation techniques on marine vehicles such as underwater robotic vehicle have been gradually increased in the past few decades. Many of them were used for subsea exploration and installation. The remotely operated vehicles (ROVs) are the major workhorses to carry out several tasks in deeper and risker areas where the use of human divers is impractical. However, there are numbers of challenges in operating the vehicles precisely, such as unpredictable disturbance like current and wave during its operation. The manipulation and maneuverability become the most essential tasks in designing the vehicle. A marine vessel control system has three independent blocks denoted as guidance, navigation, and control (GNC) system. A dynamic model is always needed in designing a good GNC system for an underwater vehicle.

3.2 Overview of remotely operated underwater vehicle

The remotely operated underwater vehicle (see Table 3.1) can maneuver in six degrees of freedoms (DOFs). There are six thrusters to control the six DOFs. Each thruster can generate a maximum thrust of 100 N. The 220 V AC power is supplied from vessel through an umbilical cable or tether to the vehicle.

Modeling of vehicle or plant is often the initial step before performing the numerical simulation. In this section, the hydrodynamic damping and added mass parameters of the remotely operated underwater vehicle are obtained by computational fluid dynamics (CFD) method [1]. The design of the vehicle is discussed followed by modeling and simulation. The remotely operated underwater vehicle (see Figure 3.1) weights 135 kg in air. It has an overall dimension of 1,455 mm (length) × 950 mm (width) × 1,710 mm (height).

Table 3.1 Specification of remotely operated underwater vehicle

Descriptions	Values
Size	1,455 mm (L) × 950 mm (W) × 1,710 mm (H)
Mass	135 kg
Propulsion	Six thrusters
Power	220/24 V

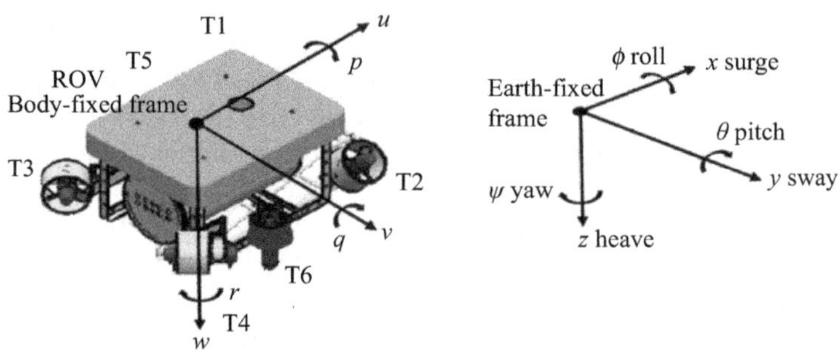

Figure 3.1 Model of the remotely operated underwater vehicle [2]

The remotely operated underwater vehicle was designed for launch and recovery of the smaller vehicle. Six underwater thrusters are used to drive the remotely operated underwater vehicle to six DOF), namely, surge, sway, heave, roll, pitch, and yaw velocity. However, it can be observed that there is a high degree of cross-coupling. A few sensors for velocity and position measurements such as XSENS MTi 10-series, PDCR1830 Depth Sensor, and NavQuest Micro DVL (600 kHz) were used in the remotely operated underwater vehicle.

The derivation of remotely operated underwater vehicle dynamic equation is based on the following assumptions to simplify the remotely operated underwater vehicle model.

- Operating at slow speed (less than 1 m/s);
- rigid body and fully submerged in water (no wave and current disturbance);
- neutrally buoyant by design;
- tether (or umbilical cable) dynamics attached to remotely operated underwater vehicle is not considered.

The notations (see Table 3.2) used for the remotely operated underwater vehicle motions follow the Society of Naval Architects and Marine Engineers (SNAME).

Table 3.2 Notations used for remotely operated underwater vehicle [3]

DOF	Motion descriptions	Forces and moments	Linear and angular velocity	Positions and orientations
1	Motion in the x-direction (surge)	X	u	x
2	Motion in the y-direction (sway)	Y	v	y
3	Motion in the z-direction (heave)	Z	w	z
4	Rotation about the x-axis (roll)	K	p	ϕ
5	Rotation about the y-axis (pitch)	M	q	θ
6	Rotation about the z-axis (yaw)	N	r	ψ

3.3 Dynamics modeling of remotely operated underwater vehicle

The details of the dynamics modeling of remotely operated underwater vehicle used in this section were documented in the following references [1,7]. The Newtonian mechanics are the most common approach to model the rigid body remotely operated underwater vehicle on its body-fixed reference frame [4]. It can be written as follows.

$$\mathbf{M_{RB}}\dot{\mathbf{v}} + \mathbf{C_{RB}}(\mathbf{v}) = \tau_{\mathbf{RB}} \tag{3.1}$$

where $\mathbf{M_{RB}} \in \Re^{6\times 6}$ is the mass matrix, $\mathbf{C_{RB}}(\mathbf{v}) \in \Re^{6\times 6}$ is the Coriolis and centripetal matrix, $\tau_{\mathbf{RB}} \in \Re^{6}$ is a vector of external forces and moments, and $\mathbf{v} = [u \quad v \quad w \quad p \quad q \quad r]^{\mathrm{T}} \in \Re^{6}$ is the linear and angular velocity vector.

According to Fossen [4], the mass inertia matrix in (1) can be written as:

$$\mathbf{M_{RB}} = \begin{bmatrix} m & 0 & 0 & 0 & mz_G & -my_G \\ 0 & m & 0 & -mz_G & 0 & mx_G \\ 0 & 0 & m & my_G & -mx_G & 0 \\ 0 & -mz_G & my_G & I_x & -I_{xy} & -I_{xz} \\ mz_G & 0 & -mx_G & -I_{yx} & I_y & -I_{yz} \\ -my_G & mx_G & 0 & -I_{zx} & -I_{zy} & I_z \end{bmatrix} \tag{3.2}$$

where the component m in this matrix represents the mass of the vehicle (where I_x, I_y, I_z I_{xy}, I_{xz}, I_{yz} represent the mass moments of inertia of remotely operated underwater vehicle). The coordinates of the center of gravity of remotely operated underwater vehicle are x_G, y_G, and z_G.

The Coriolis and the centripetal matrix are used to describe the angular motion of the vehicle as follows.

$$\mathbf{C_{RB}}(\mathbf{v}) = \begin{bmatrix} \mathbf{0}_{3\times3} & \mathbf{C}_{12}(\mathbf{v}) \\ -\mathbf{C}_{12}^{\mathrm{T}}(\mathbf{v}) & \mathbf{C}_{22}(\mathbf{v}) \end{bmatrix} \tag{3.3}$$

and

$$\mathbf{C}_{12}(\mathbf{v}) = \begin{bmatrix} m(y_G q + z_G r) & -m(x_G q - w) & -m(x_G r + v) \\ -m(y_G p + w) & m(z_G r + x_G p) & -m(y_G r - u) \\ -m(z_G p - v) & -m(z_G q + u) & m(x_G p + y_G q) \end{bmatrix} \tag{3.4}$$

$$\mathbf{C}_{22}(\mathbf{v}) = \begin{bmatrix} 0 & -I_{yz}q - I_{xz}p + I_z r & I_{yz}r + I_{xy}p - I_y q \\ I_{yz}q + I_{xz}p - I_z r & 0 & -I_{xz}r - I_{xy}q + I_x p \\ -I_{yz}r - I_{xy}p + I_y q & I_{xz}r + I_{xy}q - I_x p & 0 \end{bmatrix} \tag{3.5}$$

As shown in (3.1), $\tau_{\mathbf{RB}}$ includes three different hydrodynamics forces and moments. The first term consists of drag, added mass, and restoring forces that are named as $\tau_{\mathbf{H}}$ and the propulsion forces generated by the thrusters are named as τ.

$$\tau_{\mathbf{RB}} = \tau_{\mathbf{H}} + \tau \tag{3.6}$$

In this section, the hydrodynamic forces and moments $\tau_{\mathbf{H}}$ are determined. These hydrodynamics forces and moments are represented by the following equation:

$$\tau_{\mathbf{H}} = -\mathbf{M_A}\dot{\mathbf{v}} - \mathbf{C_A}(\mathbf{v})\mathbf{v} - \mathbf{D}(\mathbf{v}) - \mathbf{g}(\boldsymbol{\eta}) \tag{3.7}$$

By substituting (3.7) into (3.6) and back into (3.1), the following equation of motions for the remotely operated underwater vehicle is formed.

$$\mathbf{M}\dot{\mathbf{v}} + \mathbf{C}(\mathbf{v})\mathbf{v} + \mathbf{D}(\mathbf{v}) + \mathbf{g}(\boldsymbol{\eta}) = \tau \tag{3.8}$$

where $v = [u \quad v \quad w \quad p \quad q \quad r]^{\mathrm{T}}$ is the body-fixed velocity vector and $\boldsymbol{\eta} = [x \quad y \quad z \quad \phi \quad \theta \quad \psi]^{\mathrm{T}}$ is the earth-fixed vector comprising the position vector. $\mathbf{M} = \mathbf{M_{RB}} + \mathbf{M_A} \in \Re^{6\times6}$ is the inertia matrix for rigid body and added mass, respectively. The gravitational and buoyancy vector acting on remotely operated underwater vehicle in water is denoted by $\mathbf{g}(\boldsymbol{\eta})$. Since the remotely operated underwater vehicle is neutrally buoyant ($W = B$) and the X–Y coordinates of the CB coincide with the X–Y coordinate of the CG (by placing additional mass on vehicle). Since $(z_G - z_B) = -0.397$ m, the resulting $\mathbf{g}(\boldsymbol{\eta})$ can be expressed as $\mathbf{g}(\boldsymbol{\eta}) = [0\,0\,0 - 0.397\,\mathrm{W}\cos\theta\sin\varphi - 0.397\,\mathrm{W}\sin\theta\;0]^{\mathrm{T}}$. The Coriolis and centripetal matrix for rigid body and added mass can be defined as: $\mathbf{C}(\mathbf{V}) = \mathbf{C_{RB}}(\mathbf{v}) + \mathbf{C_A}(\mathbf{v}) \in \Re^{6\times6}$, $\mathbf{D}(\mathbf{v}) \in \Re^{6\times6}$ is the damping matrix due to the surrounding fluid. The input force and moment vector $\tau = \mathbf{Tu} \in \Re^6$ relate the thrust output vector $\mathbf{u}$ with the thruster configuration matrix $\mathbf{T}$. As observed in the design of remotely operated underwater vehicle, it is a fully actuated system with six thrusters. The thruster configuration matrix**T** based on the layout of the thrusters is defined as follows:

$$\mathbf{T} = \begin{bmatrix} 0 & 0 & 0 & 0 & \cos\alpha & \cos\alpha \\ -\cos\beta & \cos\beta & 0 & 0 & \sin\alpha & -\sin\alpha \\ \cos\beta & \cos\beta & 1 & 1 & 0 & 0 \\ 0.155\cos\beta & -0.155\cos\beta & -0.275 & 0.275 & 0 & 0 \\ 0.394\cos\beta & 0.394\cos\beta & -0.035 & -0.035 & 0.430\cos\beta & 0.430\cos\beta \\ -0.394\sin\beta & 0.394\sin\beta & 0 & 0 & -0.660\sin\beta & 0.660\sin\beta \end{bmatrix} \tag{3.9}$$

where $\alpha = 45^o$ is the inclination angle for T5 and T6 while $\beta = 45°$ is the orientation angle for T1 and T2. Note that T3 and T4 are the two vertical thrusters shown in Figure 3.1. The computation of hydrodynamic-added mass and damping coefficients is discussed in Sections 3.3.1 and 3.3.2.

3.3.1 Hydrodynamic damping model

The hydrodynamic damping coefficients of the remotely operated underwater vehicle are computed in the report [1]. The vehicle has a mass of 75 kg, a volume of 0.05 m^3, and surface area of 4.75 m^2. The moment of inertia of the vehicle is shown in Table 3.3.

As a result, the rigid-body remotely operated underwater vehicle mass inertia in (1) can be written as:

$$\mathbf{M_{RB}} = \begin{bmatrix} 75.00 & 0 & 0 & 0 & 0 & 0 \\ 0 & 75.00 & 0 & 0 & 0 & 0 \\ 0 & 0 & 75.00 & 0 & 0 & 0 \\ 0 & 0 & 0 & 2.51 & 0 & 0 \\ 0 & 0 & 0 & 0 & 3.38 & 0.01 \\ 0 & 0 & 0 & 0 & 0.01 & 1.73 \end{bmatrix} \tag{3.10}$$

Table 3.3 Moment of inertia properties of the remotely operated underwater vehicle [1]

Moment of inertia (kg·m^2)					
I_{xx}	2.51	I_{xy}	0	I_{xz}	0
I_{yx}	0	I_{yy}	3.38	I_{yz}	0.01
I_{zx}	0	I_{zy}	0.01	I_{zz}	1.73

A submerged body experiences lift-and-drag effect while moving through the fluid. This drag component includes frictional and pressure drag. The frictional drag due to the boundary layers depends on the surface area in contact with the fluid. The damping function is a linear function of velocity, a quadratic function of velocity, a sum of both linear and quadratic terms (as used in this section) and with higher-order forms. The steady drag force experienced by the vehicle in its reference state is well known to depend on the square of the velocity and a coefficient that depends on Reynolds number (for a body sufficiently submerged in a fluid). The variation of that force on the body experienced during small perturbations to that motion has repeatedly been found to be better modeled by a linear function of velocity. The linear coefficients are therefore adequate to represent the strength and moments due to inviscid part of the flow for a low-speed remotely operated underwater vehicle.

The remotely operated underwater vehicle hydrodynamic damping matrix **D** can be further simplified. The off-diagonal elements [4] in the hydrodynamic damping matrix **D(v)** are small compared to those diagonal elements on the underwater vehicle. Therefore, **D(v)** becomes a diagonal matrix: $\mathrm{diag}\left[\{X_u, Y_v, Z_w, K_p, M_q, N_r\}\right]$. A turbulence model with the unsteady 3-dimensional flow was built for the Reynolds number flow condition greater than 1.0×10^6. The shear-stress-transport (SST) model in CFD software STAR CCM+ was used. The k-ω SST model is one of the two common models for predicting the flow separation under adverse pressure gradient. It provides a highly accurate prediction of the amount of the flow separation under adverse eddy viscosity. To take advantage of the SST model, the boundary layer should be resolved with at least 10 mesh nodes. This is done by inspecting the $y+$ value on the surface of the vehicle that must be around 1.

The movement in the fluid domain is expected to be turbulent and isothermal. The temperature is fixed at 20 °C, and the water is modeled as an incompressible fluid. It is impractical to set the fluid domain to be infinitely large to analyze damping force acting on the remotely operated underwater vehicle in CFD. As a result, a dimension of approximately 15 times greater than the dimension of the vehicle (see Figure 3.1) to ensure the accuracy [5] of the actual flow domain is used. The properties of the flow domain are required to be defined to ensure the domain has high fidelity. The turbulence properties are difficult to define without turbulent kinetic energy and dissipation data at the inlet flow. The turbulent intensity was used to describe the turbulent properties in the simulation. The turbulent intensity is selected to be 0.1% for the external flow over the vehicle. The static pressure is used at the outlet boundary. It is required to use the log-law method to predict the velocity profile in the turbulent wall-bounded flows. The side, top, and bottom of the fluid domain are modeled using free-slip wall boundary conditions. The boundary condition of the vehicle surface is defined to be nonslip due to fluid viscosity, and the surface speed of the vehicle is near to zero.

The initial variable values are required for STAR CCM+™ solver to commence the steady-state calculation. The appropriate settings of the solver control are essential to facilitate the convergence of the simulation results. The selection of

Table 3.4 Initial condition and solver control settings [1]

Setting	Value
Initial conditions > Turbulence intensity	0.1%
Stopping criteria > Maximum steps	60
Physical timescale control	5 [s]

appropriate time step size helps to obtain good convergence rates in STAR CCM+ solver. The convergence is achieved by using the physical time steps which provide sufficient relaxation of the nonlinearity. A reasonable estimation of this time step is based on one-third of the fluid domain length (*L*) and initial velocity (*U*). Therefore, $t = \frac{L}{3U} = \frac{15}{3*1} = 5$ s. Table 3.4 shows the initial and solver control settings.

Before performing the CFD, the mesh size needs to be defined properly as the shape of the boundary is important in creating pressure gradients which greatly influence the boundary layer. The mesh size is preferred to be sufficiently small to capture the geometry of the remotely operated underwater vehicle. The flow near boundary layer flow can be captured using the layer inflation technique. A surface wrapping technique was used for geometry preparation before the surface meshes to ensure mesh error is kept minimal to improve the meshing quality.

The number of elements in the volume is approximately 2,069,270. The fluid domain has around 546,712 elements. Figure 3.2 shows the 3D of the volumetric mesh of the flow domain around the vehicle. The mesh downstream of the body is finer in the wake region than near the domain boundary.

Different configurations can be computed with a single set of grids. The grid quality will not be affected by changing of remotely operated underwater vehicle's orientation. It allows different principle motions of the vehicle to be simulated with the same flow direction (negative x-axis) in the domain. However, the direction of gravity has to be changed in different principle motion simulation. Table 3.5 shows the mesh statics and settings used in the simulation.

Figure 3.3 illustrates the flow around the vehicle in four DOFs. The formation of the wake at the rear creates a low-pressure region. The high-pressure region at the front resists the motion of the vehicle. As observed, there is certainly flow separation on the remotely operated underwater vehicle. Figure 3.4 illustrates the drag force exerting on the vehicle and its corresponding drag coefficient. It shows that the drag forces converge to a steady-state value at around 40 iterations.

As observed in Figure 3.3, the wakes are formed at the rear of the remotely operated underwater vehicle body when the turbulent flow past the vehicle. The wakes represent the increased in the vortices of the flow separation. The nonlinear damping effects on the vehicle are caused by this flow separation phenomenon in the turbulent model. The hydrodynamic damping forces were then determined by integrating the pressure on the vehicle surface. The three translation motions, namely, surge, sway, and heave are plotted against the velocity in Figure 3.5. As observed from the plots, the vehicle has the most extensive drag in the heave direction due to

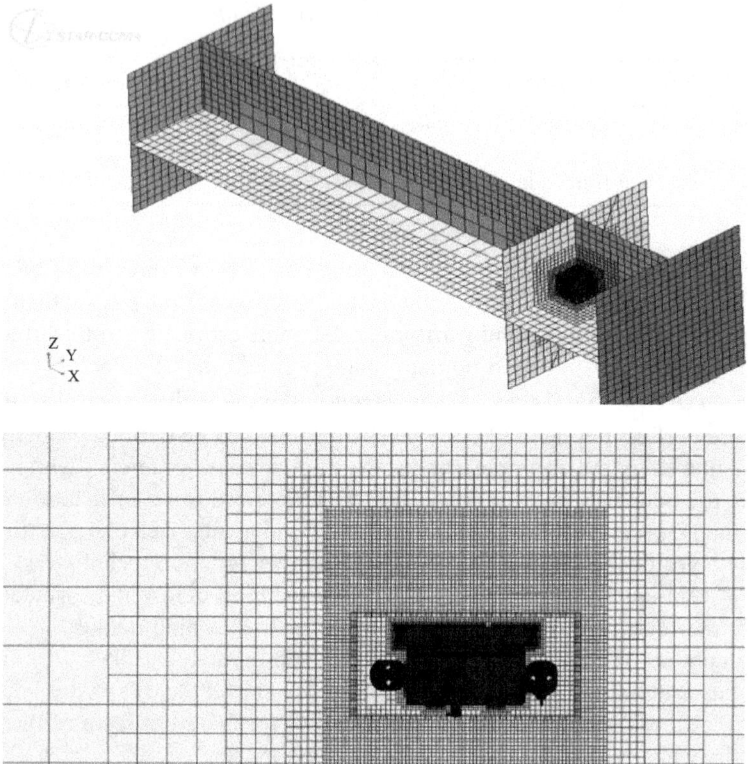

Figure 3.2 Three-dimensional view of volumetric meshing for flow domain (top) and close view of remotely operated underwater vehicle (bottom)

Table 3.5 Mesh settings [1]

Mesh setting	**Value**
No. of mesh elements	Around 2.6 million
Mesh spacing	0.0005 [m^2]
Boundary growth rate (i.e., rate at which the boundary layer thickness grows)	Low
Volumetric mesh type	Polyhedral, trimmer

its most significant frontal area normal to the flow direction of about 0.6 m^2. The drag force in sway is slightly larger than the drag in surge direction due to its small frontal area. The damping coefficient for different directions is obtained using the second-order polynomial fit as shown in Figures 3.5 and 3.6.

Figure 3.3 Vector plot of remotely operated underwater vehicle in STAR CCM+ [1]

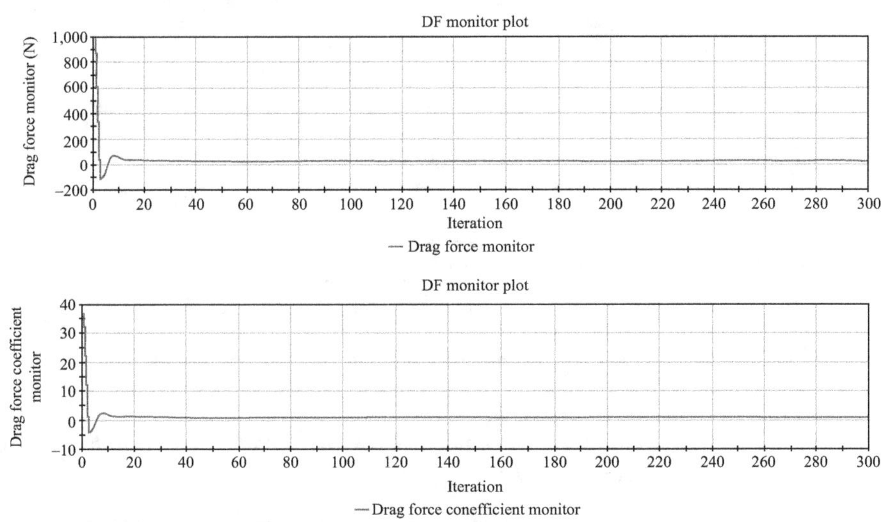

Figure 3.4 Iteration history of drag force and its coefficient [1]

Table 3.6 shows the resulted drag coefficient of vehicle in its four principle motions. The flow in the domain was simulated at different speeds using STAR CCM+. Table 3.7 tabulates the drag moment versus angular velocity in the yaw direction.

Figure 3.6 depicts the drag moment as a function of square angular velocity. The linear damping in the yaw direction is ignored as only quadratic damping is present. The resulted hydrodynamic damping for the remotely operated underwater vehicle is thus written as:

$$\mathbf{D_L} = -\mathrm{diag}\{[3.221 \quad 3.291 \quad 5.682 \quad 0. \quad 0 \quad 0]\} \tag{3.11}$$

$$\mathbf{D_Q} = -\mathrm{diag}\{[105.300 \quad 139.600 \quad 273.800 \quad 0 \quad 0 \quad 6.079]\} \tag{3.12}$$

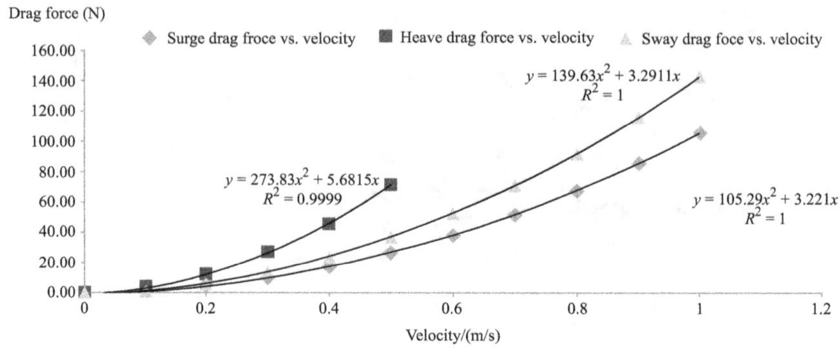

Figure 3.5 Drag force as the function of velocity in surge, sway, and heave [1]

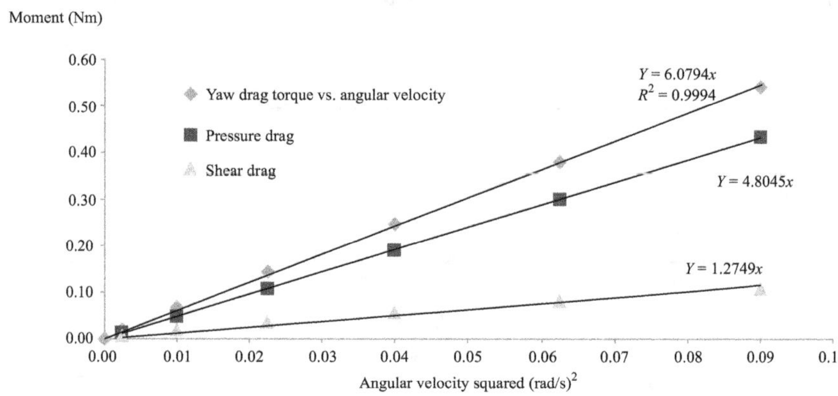

Figure 3.6 Drag torque as the function of velocity in yaw [1]

Table 3.6 Damping coefficients of remotely operated underwater vehicle in four principle directions [1]

Damping coefficients	**Surge**		**Sway**		**Heave**		**Yaw**	
	K_L	K_Q	K_L	K_Q	K_L	K_Q	K_L	K_Q
Values	3.221	105.3	3.291	139.6	5.682	273.8	0	6.079

Table 3.7 Drag moment versus angular velocity in the yaw direction [1]

Angular velocity (rad/s)	**Yaw moment (Nm)**	**Yaw moment coefficient**
0.100	0.067	0.204
0.200	0.248	0.188
0.300	0.542	0.183

In summary, the results show that the lowest damping occurs in surge direction while the heave motion has the largest drag force due to its larger surface area in contact with water. The values of the linear damping coefficients are smaller than the nonlinear damping terms due to its square velocity term.

3.3.2 Hydrodynamic-added mass model

The hydrodynamic added coefficients of the vehicle are now analyzed in the report [1]. The added mass and inertia are independent of the circular wave frequency for a fully submerged vehicle. The added mass coefficients matrix for an remotely operated underwater vehicle can be written as follows:

$$\mathbf{M_A} = \begin{bmatrix} X_{\dot{u}} & X_{\dot{v}} & X_{\dot{w}} & X_{\dot{p}} & X_{\dot{q}} & X_{\dot{r}} \\ Y_{\dot{u}} & Y_{\dot{v}} & Y_{\dot{w}} & Y_{\dot{p}} & Y_{\dot{q}} & Y_{\dot{r}} \\ Z_{\dot{u}} & Z_{\dot{v}} & Z_{\dot{w}} & Z_{\dot{p}} & Z_{\dot{q}} & Z_{\dot{r}} \\ K_{\dot{u}} & K_{\dot{v}} & K_{\dot{w}} & K_{\dot{p}} & K_{\dot{q}} & K_{\dot{r}} \\ M_{\dot{u}} & M_{\dot{v}} & M_{\dot{w}} & M_{\dot{p}} & M_{\dot{q}} & M_{\dot{r}} \\ N_{\dot{u}} & N_{\dot{v}} & N_{\dot{w}} & N_{\dot{p}} & N_{\dot{u}} & N_{\dot{r}} \end{bmatrix} \tag{3.13}$$

where $X_{\dot{u}}$ is the added mass along the x-axis due to an acceleration in the x-direction, $Y_{\dot{v}}$ is the added mass along the y-axis due to an acceleration in the y-direction. On the other hand, the corresponding added Coriolis and centripetal matrix are represented as follows:

$$\mathbf{C_A}(\mathbf{v}) = \begin{bmatrix} 0 & 0 & 0 & 0 & -a_3 & a_2 \\ 0 & 0 & 0 & a_3 & 0 & -a_1 \\ 0 & 0 & 0 & -a_2 & a_1 & 0 \\ 0 & -a_3 & a_2 & 0 & -b_3 & b_2 \\ a_3 & 0 & -a_1 & b_3 & 0 & -b_1 \\ -a_2 & a_1 & 0 & -b_2 & b_1 & 0 \end{bmatrix} \tag{3.14}$$

where the respective elements in the matrix are written as follows:

$$\begin{aligned} a_1 &= X_{\dot{u}}u + X_{\dot{v}}u + X_{\dot{w}}w + X_{\dot{p}}p + X_{\dot{q}}q + X_{\dot{r}}r \\ a_2 &= X_{\dot{v}}u + Y_{\dot{v}}v + Y_{\dot{w}}w + Y_{\dot{p}}p + Y_{\dot{q}}q + Y_{\dot{r}}r \\ a_3 &= X_{\dot{w}}u + Y_{\dot{w}}v + Z_{\dot{w}}w + Z_{\dot{p}}p + Z_{\dot{q}}q + Z_{\dot{r}}r \\ b_1 &= X_{\dot{p}}u + Y_{\dot{p}}v + Z_{\dot{p}}w + K_{\dot{p}}p + K_{\dot{q}}q + K_{\dot{r}}r \\ b_2 &= X_{\dot{q}}u + Y_{\dot{q}}v + Z_{\dot{q}}w + K_{\dot{q}}p + M_{\dot{q}}q + M_{\dot{r}}r \\ b_3 &= X_{\dot{r}}u + Y_{\dot{r}}v + Z_{\dot{r}}q + K_{\dot{r}}p + M_{\dot{r}}q + N_{\dot{r}}r \end{aligned} \tag{3.15}$$

The surface-based computer-aided design (CAD) software MULTISURF™ modeled the geometry of the remotely operated underwater vehicle. The software MULTISURF aims to work with WAMIT™ for exporting the necessary files

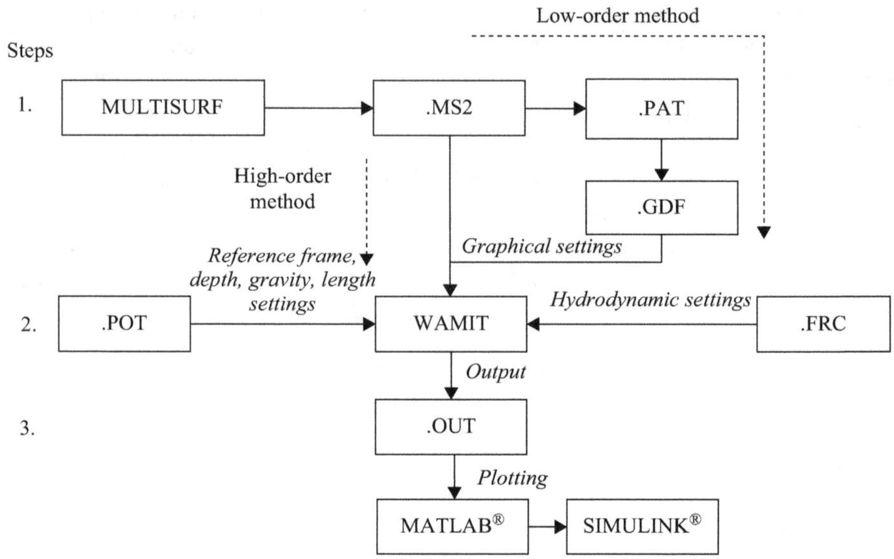

Figure 3.7 Program flowchart for computing hydrodynamic-added mass coefficients [6]

Table 3.8 Description of records used in WAMIT solver

File extension	Description
.GDF	Geometry data file used to define the geometry in panel form
.POT	Potential control file used to define input parameters in POTEN (length, gravity)
.FRC	Force control file used to define input parameters in FORCE (desired hydrodynamic parameters)
.OUT	Output from WAMIT
readAM.m	MATLAB® m-file used to read added mass matrix

for analysis. The vehicle is made of multibodies that can be created using MULTISURF and later use WAMIT to compute the added mass matrix as shown below. The geometry file was imported to WAMIT by MATLAB®. The widely used low-order panel method was adopted in WAMIT. The output file created can be imported into MATLAB® to obtain the hydrodynamic-added mass. Figure 3.7 illustrates the procedure for computing the hydrodynamic-added mass coefficients.

Table 3.8 shows the descriptions of each file used in determining the added mass coefficients. All the necessary parameters required are shown in the following input files. Before using the WAMIT to test the vehicle, a study on the empirical results of a 2 m diameter sphere (see Figure 3.8) was used to verify the program

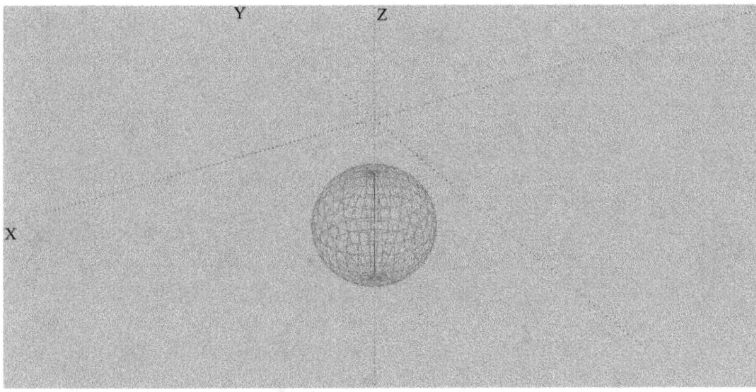

Figure 3.8 Finite surface panels generation of sphere using MULTISURF

Table 3.9 Low-order method for sphere [1]

Panel number	Numerical results			Theoretical results		
	Surge (m^3)	Sway (m^3)	Heave (m^3)	Surge (m^3)	Sway (m^3)	Heave (m^3)
256	2.085	2.085	2.073	2.094	2.094	2.094
512	2.084	2.084	2.087			
1024	2.083	2.083	2.091			
Errors				−0.5%	−0.5%	−0.1%

Table 3.10 Added mass of sphere at different depths [1]

Depth (m)	Added mass (kg)
0	2.5910
1	2.1419
10	2.0838
100	2.0835

setup and parameters. In Table 3.9, the theoretically added mass of a sphere is $A = 2/3\pi\rho r^3$ for the three translational motions: surge, sway, and heave. The added mass of the sphere can be written as $A = 2/3\pi r^3$ (after normalizing with density). Table 3.9 shows that the low-order method results of the sphere are quite close to the theoretical value of 2.094.

Additionally, it is required to define the depth of the submerged body in WAMIT. The impact of different water depth has been identified by using the same sphere. As shown in Table 3.10, the results of the added mass converged at approximately 10 m. As a result, the computation will be performed at 10 m.

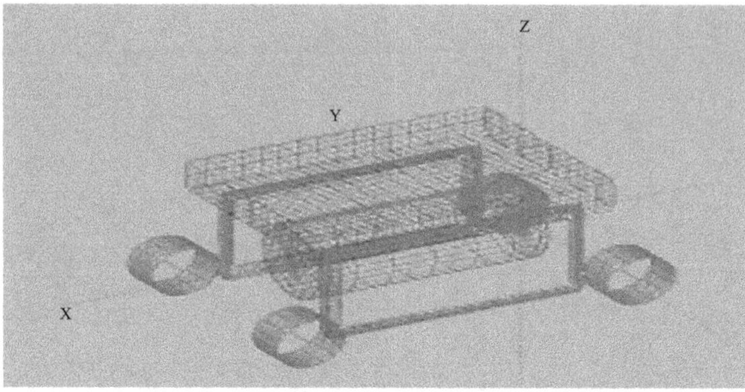

Figure 3.9 Remotely operated underwater vehicle model from WAMIT

The numerical results obtained from WAMIT low-order method have a small difference compared with the theoretical results. Nevertheless, it helps to ensure that the input parameters settings used are appropriate to calculate the added mass coefficient of the remotely operated underwater vehicle. The vehicle modeled using MULTISURF is then imported into the WAMIT to solve the problem using the low-order panel method. As shown in Figure 3.9, the main components of the vehicle are drawn to reduce the complexity of the geometry.

Figures 3.10 and 3.11 show the added mass coefficient components calculated by WAMIT. The convergence tests of the added mass for different panel numbers converged in approximately 3,000 to 4,000 panels. It shows that the minimum required panel numbers should be at least 3,000.

The off-diagonal terms of the added mass are less than the diagonal components as the vehicle has near to three planes of symmetry [4]. Hence, the added mass matrix obtained can be further simplified.

$$\mathbf{M}_\mathrm{A} = -\begin{bmatrix} 20.392 & 0 & 0 & 0 & 0 & 0 \\ 0 & 53.435 & 0 & 0 & 0 & 0 \\ 0 & 0 & 126.144 & 0 & 0 & 0 \\ 0 & 0 & 0 & 2.802 & 0 & 0 \\ 0 & 0 & 0 & 0 & 13.703 & 0 \\ 0 & 0 & 0 & 0 & 0 & 5.263 \end{bmatrix} \quad (3.16)$$

The added mass matrix $\mathbf{M}_\mathrm{A}$ of the remotely operated underwater vehicle must be positive. The following relations are observed from the matrix: $m_{11} < m_{22} < m_{33}$. A smaller added mass in the surge direction, i.e., m_{11} has been observed. It is due to

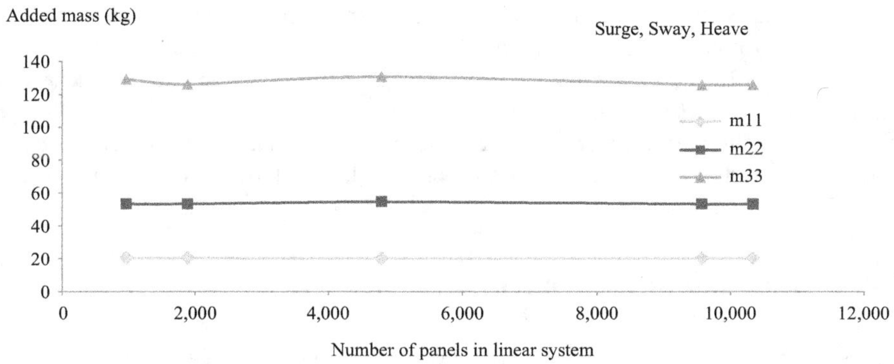

Figure 3.10 Convergence plot for surge, sway, and heave direction [6,7]

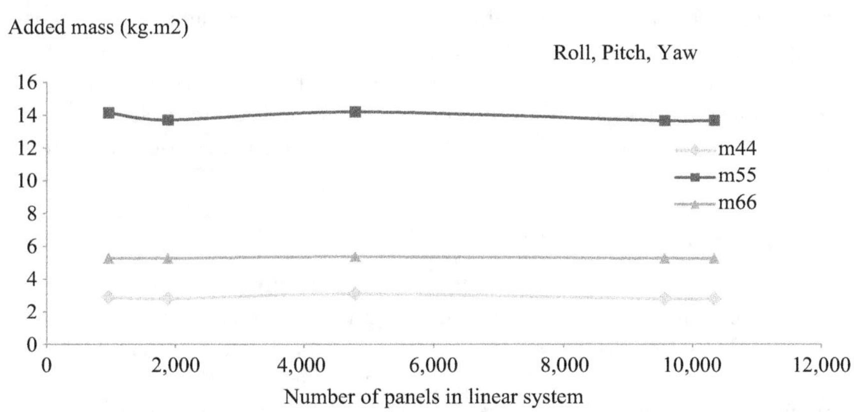

Figure 3.11 Convergence plot for roll, pitch, and yaw direction [6,7]

the vehicle has the smallest projection area in surge direction. The corresponding Coriolis- and centripetal-added mass matrix in (3.14) can be rewritten as:

$$\mathbf{C_A}(\mathbf{v}) = \begin{bmatrix} 0 & 0 & 0 & 0 & -126.144w & 53.435v \\ 0 & 0 & 0 & 126.144w & 0 & -20.392u \\ 0 & 0 & 0 & -53.435v & 20.392u & 0 \\ 0 & -126.144w & 53.435v & 0 & -5.263r & 13.703q \\ 126.144w & 0 & -20.392u & 5.263r & 0 & -2.802p \\ -53.435v & 20.392u & 0 & -13.703q & 2.802p & 0 \end{bmatrix} \tag{3.17}$$

In summary, the added mass coefficients of the remotely operated underwater vehicle are around 20 kg in the surge direction, 53 kg in the sway direction, and followed by 126 kg in the heave direction. The remotely operated underwater vehicle has the larger added mass in the heave, followed by sway and surge directions. In summary, the computations of the hydrodynamic damping and added mass were performed. The proposed numerical simulation can determine the hydrodynamic parameters.

3.4 Validation of experimental results

The details of the results used in this section were documented in the following reference [3]. The results obtained from the simulations were validated with the experimental data in a water tank. The heave and yaw motion was only validated due to the limited reliable sensor installed on the remotely operated underwater vehicle. A depth sensor, DVL and IMU were jointly used to measure the depth, velocity, and acceleration of the vehicle, respectively. The water tank dimensions are 10 m (L) × 4 m (W) × 1.8 m (D). It is equipped with an overhead crane to load and unload the vehicle. The sample rate of 100 Hz (or sample period of 0.01 s) was used to sample the raw data from the sensor. The raw data are plotted against the sample count using a MATLAB® script. Note that one sample count is equivalent to 0.01 s.

3.4.1 Heave model identification

During the depth validation, the prototype vehicle (see Figure 3.12) was maintained at a fixed heading angle before submerging to a depth of 1 m as shown in Figure 3.12. The remotely operated underwater vehicle was commanded to move vertically into the water with little acceleration measured by IMU. The dead-reckoning data for acceleration in the *Z*-direction is not zero as shown in Figure 3.13. The vehicle is not moving at constant heave rate as observed in the heave velocity from the DVL in Figure 3.14. The thrust outputs from T3 and T4 (for heave direction) are approximately 10 N (see Figure 3.15) to allow the vehicle to reach a depth of 1 m as depicted in Figure 3.16.

Figure 3.12 Image of the remotely operated underwater vehicle (top) and motion pictures (bottom) captured during heave test in water [3]

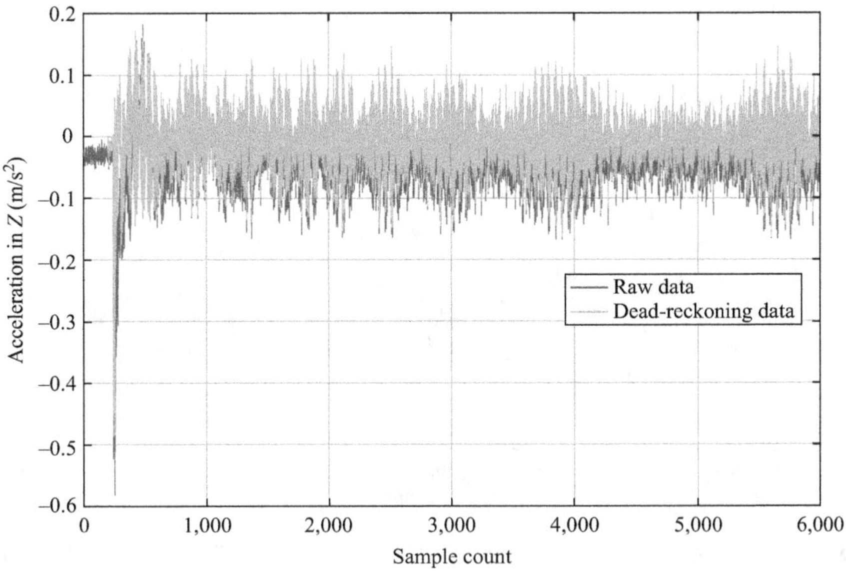

Figure 3.13 Acceleration in the heave direction (1 sample count = 0.01 s) [3]

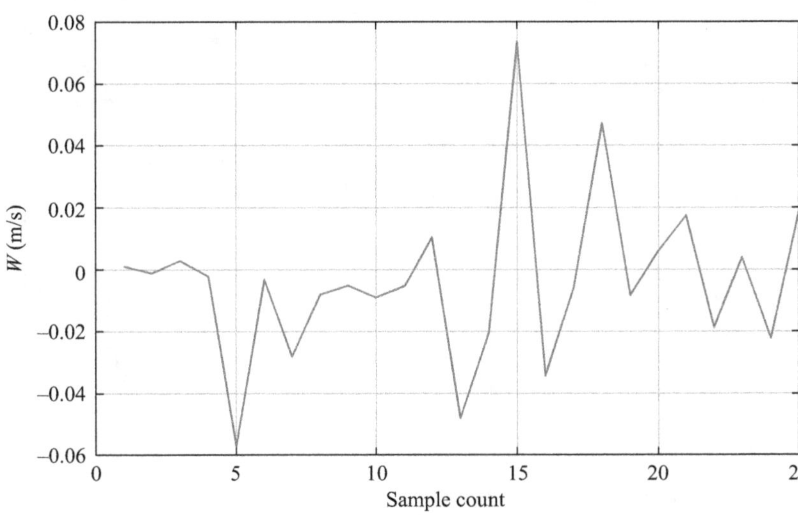

Figure 3.14 Heave velocity response (1 sample count = 0.01 s) [3]

The added mass $Z_{\dot{w}}$ and damping coefficients (linear, Z_w) due to the hydrodynamic force are defined in a body-fixed frame with m as the body mass of the remotely operated underwater vehicle. The heave model can be simplified as:

$$(m + Z_{\dot{w}})\dot{w} + Z_w w = Z \tag{3.18}$$

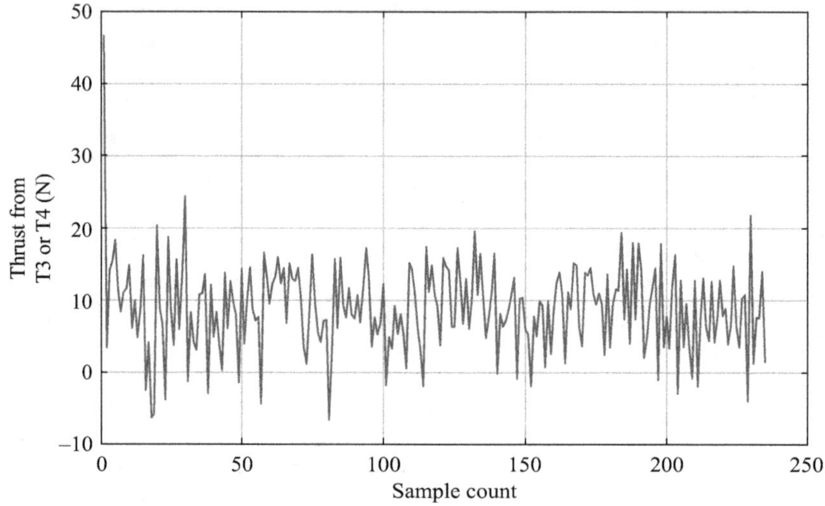

Figure 3.15 Thrust output from thruster T3 or T4 (1 sample count = 0.01 s) [3]

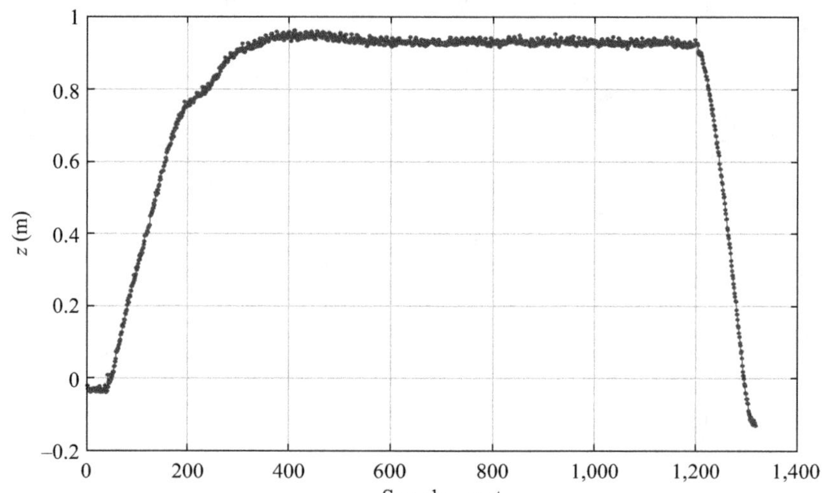

Figure 3.16 Depth response (1 sample count = 0.01 s) [3]

The remotely operated underwater vehicle is designed to be neutrally buoyant, and the centripetal and Coriolis terms are not present in the equation as the remotely operated underwater vehicle was commanded to move in the heave direction only. Rearranging the preceding equation gives:

$$\dot{w} = \frac{Z}{(m + Z_{\dot{w}})} - \frac{Z_w}{(m + Z_{\dot{w}})} w \tag{3.19}$$

In matrix form,

$$\overbrace{\dot{w}}^{\boldsymbol{\varphi}} = \overbrace{\begin{bmatrix} Z & -w \end{bmatrix}}^{\mathbf{H}} \overbrace{\begin{bmatrix} \alpha \\ \beta \end{bmatrix}}^{\hat{\boldsymbol{\theta}}} \tag{3.20}$$

where

$$\alpha = \frac{1}{(m + Z_{\dot{w}})}; \beta = -\frac{Z_w}{(m + Z_{\dot{w}})} \tag{3.21}$$

The least-square method below is used to obtain the estimated values.

$$\underbrace{\begin{bmatrix} \dot{w}_1 \\ \dot{w}_2 \\ \vdots \end{bmatrix}}_{\boldsymbol{\varphi}} = \underbrace{\begin{bmatrix} Z_1 & -w_1 \\ Z_2 & -w_2 \\ \vdots & \vdots \end{bmatrix}}_{\mathbf{H}} \cdot \underbrace{\begin{bmatrix} \alpha \\ \beta \end{bmatrix}}_{\boldsymbol{\theta}} + error \tag{3.22}$$

where subscript i represents the number of samples collected from the experiment. The standard least-square solution is given by the Moore–Penrose pseudoinverse:

$$\hat{\boldsymbol{\theta}}_{\mathbf{LS}} = (\mathbf{H}^{\mathbf{T}}\mathbf{H})^{-1} \cdot \mathbf{H}^{\mathbf{T}} \cdot \boldsymbol{\varphi} \tag{3.23}$$

where the standard deviation, $\mathrm{cov}(\hat{\boldsymbol{\theta}}_{\mathbf{LS}}) = \sigma^2(\mathbf{H}^{\mathbf{T}}\mathbf{H})^{-1}$ and note that $\mathbf{H}$ has to be a full rank matrix. The recursive least square (RLS) is used to compute the parameters $\hat{\boldsymbol{\theta}}_{\mathbf{RLS}}(t)$ in (3.23). The parameters update $\hat{\boldsymbol{\theta}}_{\mathbf{RLS}}(t)$ include a correction term to the previous estimate $\hat{\boldsymbol{\theta}}_{\mathbf{RLS}}(t-1)$ as seen in (3.24). The typical RLS algorithm comprises the following recursive equations computed in sequence. For example, $\hat{\boldsymbol{\theta}}_{\mathbf{RLS}}(t-1)$ in (3.24) requires the computation of (3.25) to (3.27).

$$\hat{\boldsymbol{\theta}}_{\mathbf{RLS}}(t) = \hat{\boldsymbol{\theta}}_{\mathbf{RLS}}(t-1) + \mathbf{K}(t)\varepsilon(t) \tag{3.24}$$

$$\boldsymbol{\varepsilon}(t) = \dot{\mathbf{w}}(t) - \hat{\boldsymbol{\theta}}^{\mathbf{T}}_{\mathbf{RLS}}(t-1)\mathbf{H}(t) \tag{3.25}$$

$$\mathbf{K}(t) = \frac{\lambda^{-1}\mathbf{P}(t-1)\mathbf{H}(t)}{1 + \lambda^{-1}\mathbf{H}^{\mathrm{T}}\mathbf{P}(t-1)\mathbf{H}(t)} \tag{3.26}$$

$$\mathbf{P}(t) = \lambda^{-1}\mathbf{P}(t-1) - \lambda^{-1}\mathbf{K}^{\mathrm{T}}(t)\mathbf{H}(t)\mathbf{P}(t-1) \tag{3.27}$$

The above equations need an initial value for $\hat{\boldsymbol{\theta}}_{\mathbf{RLS}}(t)$ and the error covariance matrix $\mathbf{P}$. The initial value of $\hat{\boldsymbol{\theta}}_{\mathbf{RLS}}(t)$ is set as 0 and $\mathbf{P}$ is $100\mathbf{I}_2$, where $\mathbf{I}_2$ is the identity matrix of dimension two. The forgetting factor that specifies how fast the RLS forget the past sample information is set to 0.001. For example, using $\lambda = 1$, it specifies an infinite memory. By comparing the RLS with a different forgetting factor, the following plot in Figure 3.17 is obtained. The RLS with a lower forgetting

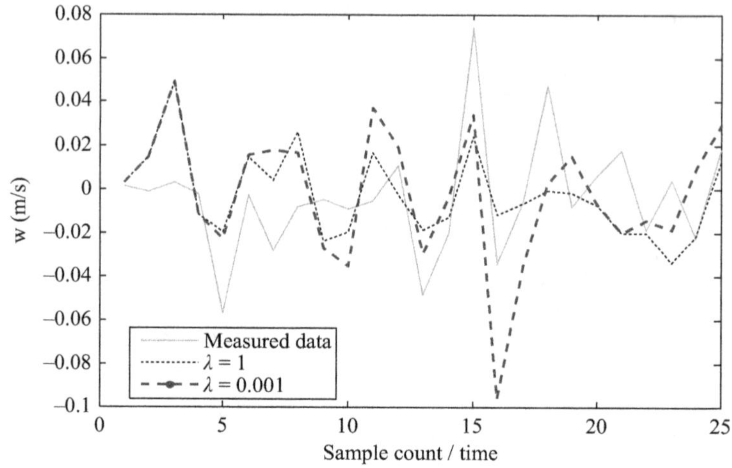

Figure 3.17 Comparison of heave velocity response using RLS [3]

Table 3.11 Comparisons between simulation and experimental results in the heave direction [3]

Descriptions	**Simulation**	**Experiment (Recursive least square)**	**Error**
Added mass (kg)	126.14	136.40	8%
Linear damping coefficient	5.6800	4.0874	28%
Quadratic damping coefficient	273.805	–	–

factor has a closer behavior to the measured data. In general, it can be observed that the differences in the performance are quite large for the damping in the heave direction as seen in Table 3.11. It may be due to the effect of the rotating propeller of the thrusters are not included during the simulation.

3.4.2 Yaw model identification

A simplified description of the yaw model of the remotely operated underwater vehicle is defined as

$$N_{\dot{r}}\dot{r} + N_{|r|r}|r|r + N_r r = N \tag{3.28}$$

$$\dot{\psi} = r \tag{3.29}$$

where $N_{\dot{r}}$ is the added inertia $(= I_r + N_{\dot{r}})$ in the yaw direction, N_r is the linear component of drag force in the yaw direction, and $N_{|r|r}$ is the quadratic component of drag force in the yaw direction. In addition, r is the heading velocity, $\dot{r}$ is the heading acceleration, N is the moment input to the remotely operated underwater vehicle due to the thrusters (T1, T2, T5, and T6), and ψ is the heading angle.

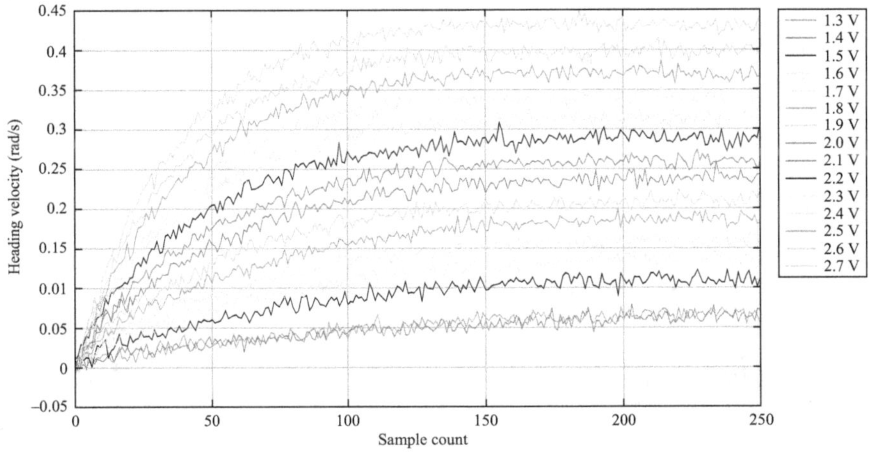

Figure 3.18 Velocity profile at constant control command to thrusters T1, T2, T5, and T6 [3]

The constant heading velocity refers to $\dot{r}$ heading acceleration equal to 0, the yaw equation of motion becomes:

$$N_{|r|r}|r|r + N_r r = N \tag{3.30}$$

The test was performed to estimate the drag coefficients $N_{|r|r}$ and N_r in the heading equation. Fixed velocity commands are sent to the four thrusters T1, T2, T5, and T6, and the steady-state heading velocity responses were recorded using the DVL sensor. The moment in the yaw direction was computed using the thrusts generated by the thrusters multiplied by the moment arm (i.e., distance) from the center of gravity. Figure 3.18 gives the steady-state angular velocity as a result of the input command to the thrusters (1.3 V to 2.7 V). The plot of the moment versus the angular velocity is shown in Figure 3.19, where the linear and quadratic damping terms can be determined from the curve fitting. The value of $N_{|r|r}$ and N_r are estimated at 5.4815 and 4.6114, respectively.

The added mass of the remotely operated underwater vehicle in the yaw direction is then identified as shown. A sinusoidal control command (maximum torque corresponding to the heading velocity) is applied to the thrusters T1, T2, T5, and T6 of the vehicle. A sinusoidal control signal with 10 Hz frequency is applied to the thrusters with a value corresponding to +2.7 V to −2.7 V (for both forward and reverse directions).

As shown in Figures 3.20 and 3.21, the maximum heading velocity is observed to be approximately 0.385 rad/s. With 2.7 V control signal into the thruster, the constant heading velocity is approximately 0.435 rad/s that corresponds to 31.52 Nm to achieve this heading speed of 0.435 rad/s. However, the maximum sinusoidal angular speed is around 0.385 rad/s. As the measured heading velocity is smaller than the speed measured in the water tank test, it shows that the vehicle needs addition moment to overcome the resistance due to its body mass and the added mass. If the

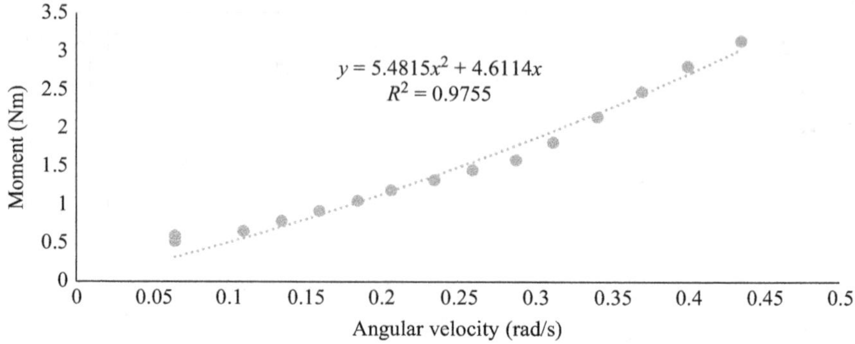

Figure 3.19 Angular velocity rate versus moment in the yaw direction [3]

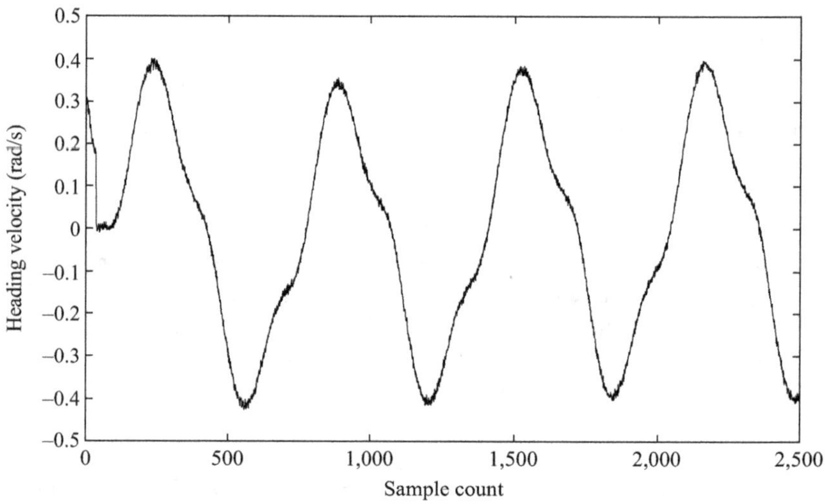

Figure 3.20 Heading velocity response subjected by sine signal command input (2.7 V) for thrusters T1, T2, T5, and T6 [3]

amplitude of the sine input is increased to around 0.435 rad/s, +2.9 V to −2.9 V sine control signal (see Figure 3.22) has a torque approximately 35.17 Nm. It indicates that the vehicle needs an additional of 35.17 − 31.52 = 3.65 Nm to overcome both the body inertia and added mass during the motion.

The next step is to compute the angular acceleration using the +2.9 V to −2.9 V sine control signal. The maximum heading acceleration of approximately 0.85 rad/s^2 was measured (see Figure 3.23) by IMU sensor.

$$N_{\dot{r}}\dot{r} = \Delta N \tag{3.31}$$

where ΔN is the difference at the moment and $N_{\dot{r}}$ is the hydrodynamic-added mass in the yaw direction.

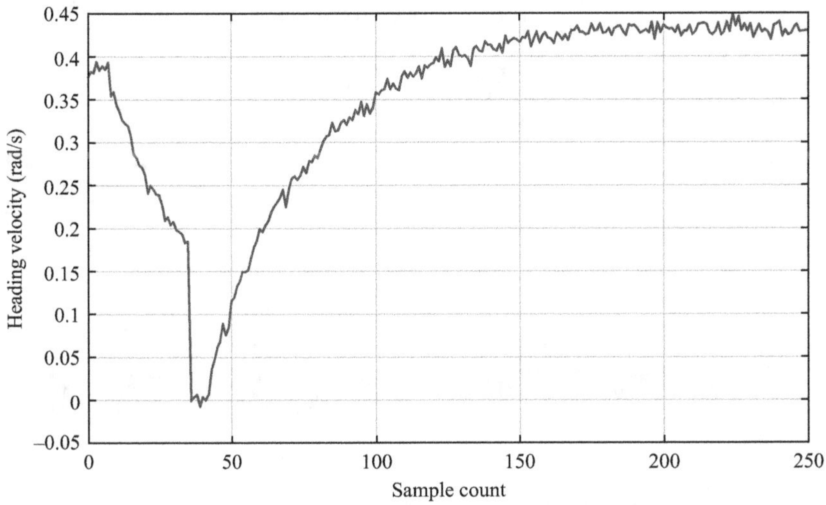

Figure 3.21 Heading velocity due to constant control command (2.7 V) for thrusters T1, T2, T5, and T6 [3]

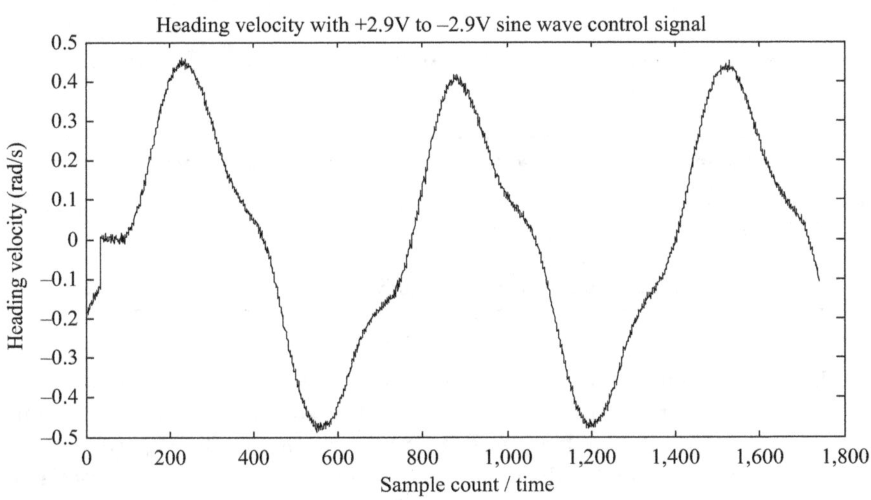

Figure 3.22 Heading velocity of sine input command of 2.9 V for thrusters T1, T2, T5, and T6 [3]

The moment due to the body inertia and added mass is computed as follows:

$$N_{\dot{r}}\dot{r} = \Delta N = 3.65 \tag{3.32}$$

Substituting the peak acceleration equals 0.85 rad/s^2, $N_{\dot{r}}$ obtained from (3.31) becomes 4.3 kg·m^2. As a result, the heading model can be written as:

$$4.3\dot{r} + 5.4815|r|r + 4.6114r = N \tag{3.33}$$

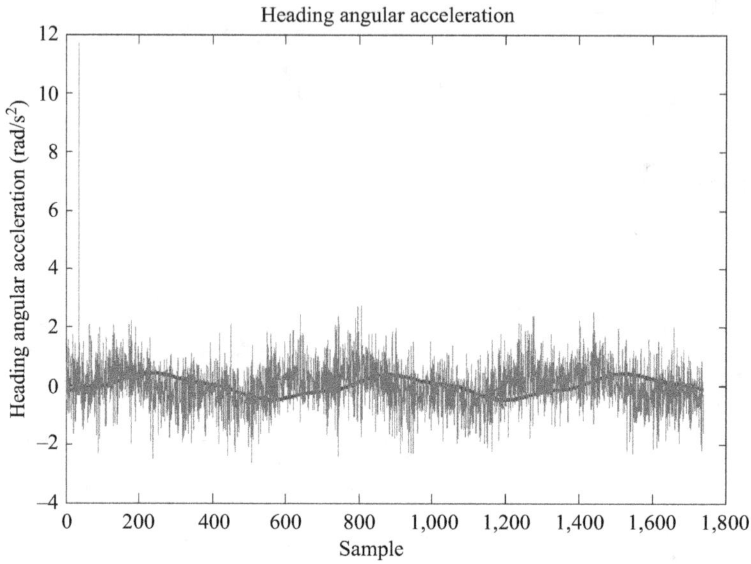

Figure 3.23 Angular acceleration of sine input signal of 2.9 V [3]

Table 3.12 Comparisons between simulation and experimental results in the yaw direction [3]

Hydrodynamic coefficients (in the yaw direction)	Simulation	Experiment	Error
Added mass (kg)	5.263	4.300	18%
Linear damping (N·s/rad)	–	4.611	–
Quadratic damping (Nm·s^2/rad^2)	6.079	5.481	9.8%

The results tabulated in Table 3.12 show that the quadratic damping has a closer value as compared to the experiments. The added mass does not seem to compare well with the simulation result. It may be due to the experimental test setup in a water tank that influences the reading. The effects of the interaction effect between rotating propellers and propeller-to- remotely operated underwater vehicle's hull have not been included in the simulation.

3.5 Simulation of remotely operated underwater vehicle model

The identified remotely operated underwater vehicle parameters from the previous sections are used to model the remotely operated underwater vehicle virtually in MATLAB®/Simulink® environment. The detailed results can be seen in the following reference [3]. The differential equation of the vehicle is solved by ordinary differential equation solver such as Dormand–Prince solver. The six inputs

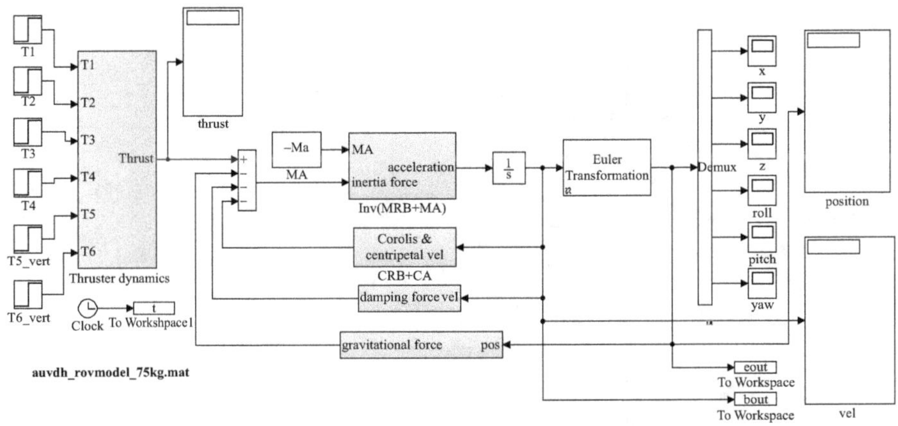

Figure 3.24 Nonlinear remotely operated underwater vehicle model [3]

(T1 to T6) on the left-hand side are the thruster inputs. For example, the vehicle is commanded to move in the *Z*-direction using only T3 and T4 thrusters. A remotely operated underwater vehicle simulator developed in MATLAB®/Simulink® environment is shown in Figure 3.24.

The thrust allocation on the vehicle is set by the thrust configuration matrix. The tether force is treated as external disturbance acting against the thrust inputs in the *X*, *Y*, and *Z* directions. Using integration function, the velocities can be calculated and send to the next iteration until the simulation time has reached. The external environmental forces due to the wave and sea current are acting on the remotely operated underwater vehicle. The thruster model is shown in Figure 3.25. The thrusters' dynamics (Tecnadyne Model 520 and 540) are obtained from the 2D lookup tables. To determine the thrust inputs to the six thrusters, an inverse operation on T is performed. Note that the multiports switches are used to trigger the hoop to open or close.

A virtual underwater world environment was developed to enhance the user interface in Figure 3.26. A virtual reality world for the remotely operated underwater vehicle uses the output signals from the vehicle model to move the vehicle virtually. It allows the movement and position data to be animated and displayed during the dynamic positioning. First, the CAD model is exported into Virtual Reality Modeling Language file format for the V-Realm editor to edit the model in Figure 3.26. The V-Realm Builder has an extensive object library where the user can import the 3D background sceneries and objects to create a virtual world. The backgrounds such as the sea and offshore structures are then imported from the library. As shown in Figure 3.26, the VRML editor has hierarchical tree style showing all elements in the virtual world. These structural elements are known as nodes. The tree allows changing and modifying of the virtual world and allows changing of some viewpoints to improve the view of the virtual reality world. Necessary lighting and the proper coordinate system could also add to make the background and remotely operated underwater vehicle more realistic.

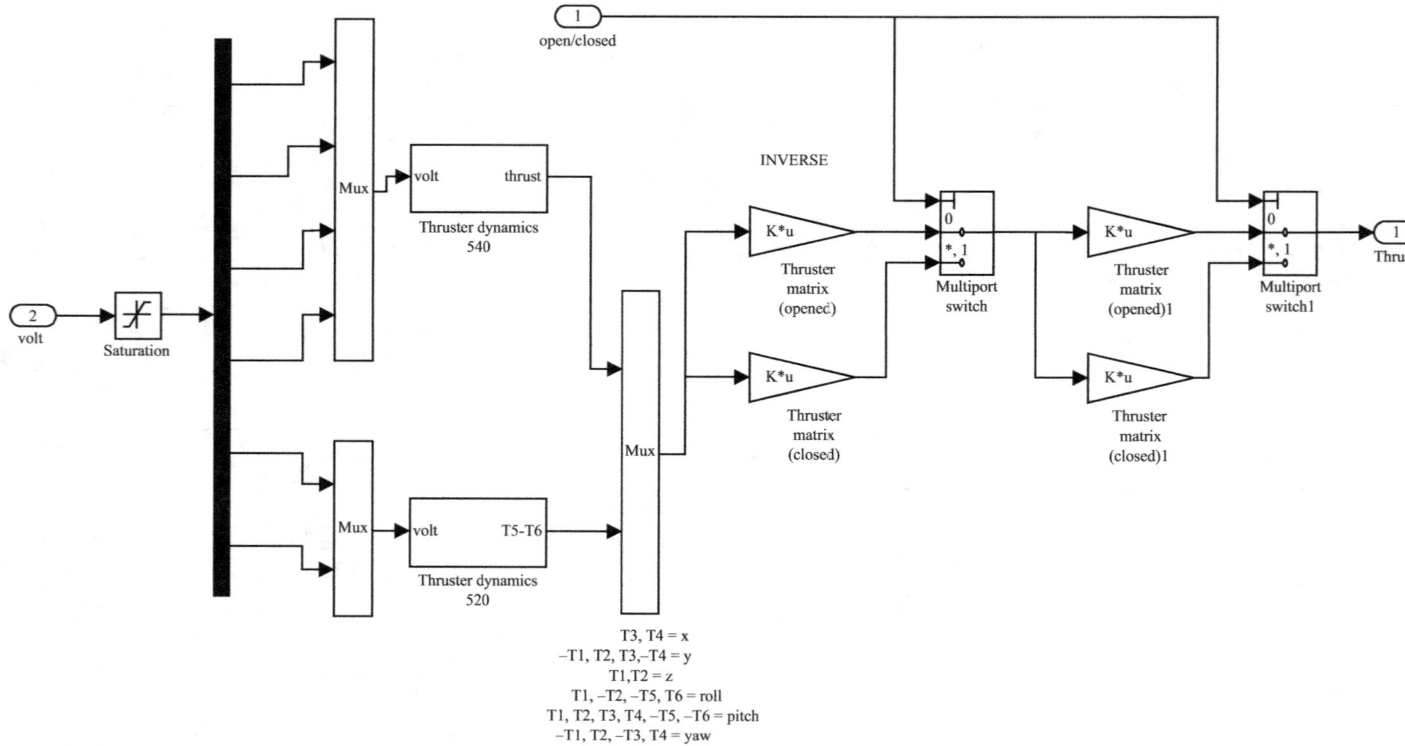

Figure 3.25 Thrust inputs model in Simulink® [3]

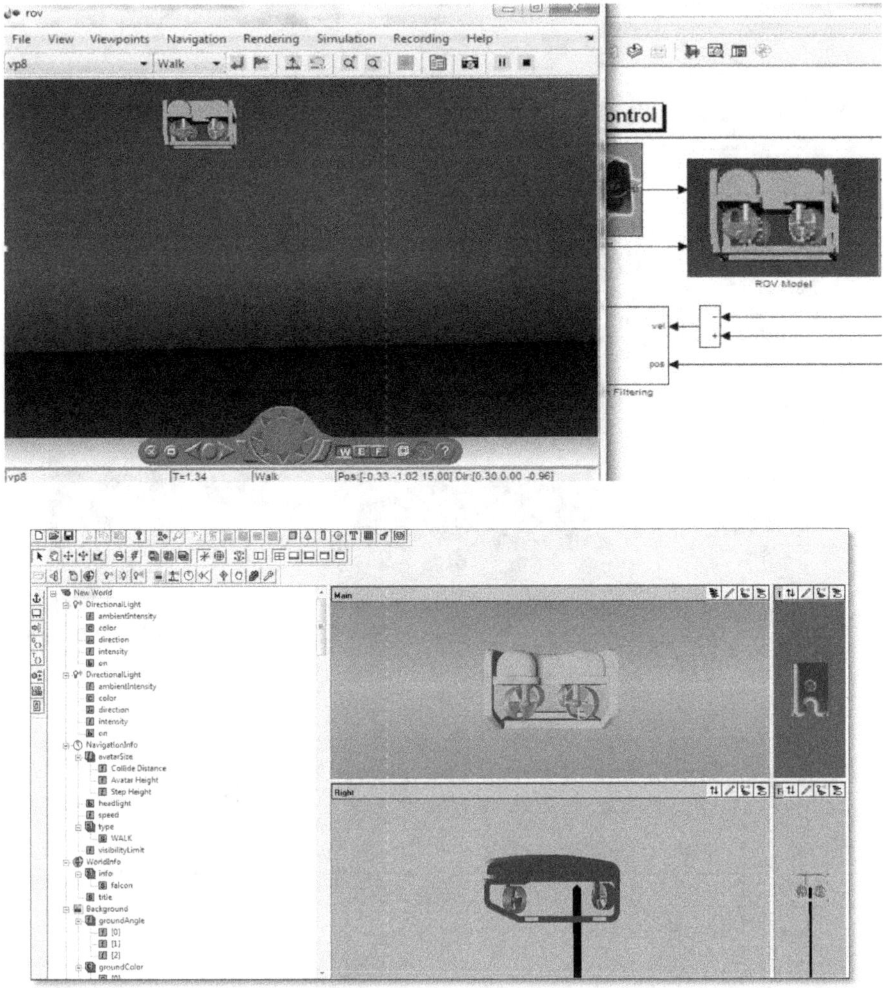

Figure 3.26 V-Realm builder screenshot (top) and virtual reality world of the remotely operated underwater vehicle (bottom) [3]

Second, the Simulink® model of the remotely operated underwater vehicle is connected to the virtual world through VR- Sink block. By connecting the model to the virtual world, the output data from the Simulink® model can be used to control and animate the virtual world as shown in Figure 3.26. The translational vehicle's motion outputs and the rotational motion outputs use the Euler's transformation to animate the position of the remotely operated underwater vehicle in the virtual reality world. Third, the vehicle can be controlled by a joystick using the user interface design the Qt GUI in Ubuntu 14 operating system as shown in Figure 3.27.

The simulated time response of the heave and yaw model of the vehicle is compared with experimental results in the water tank as shown in Figure 3.26

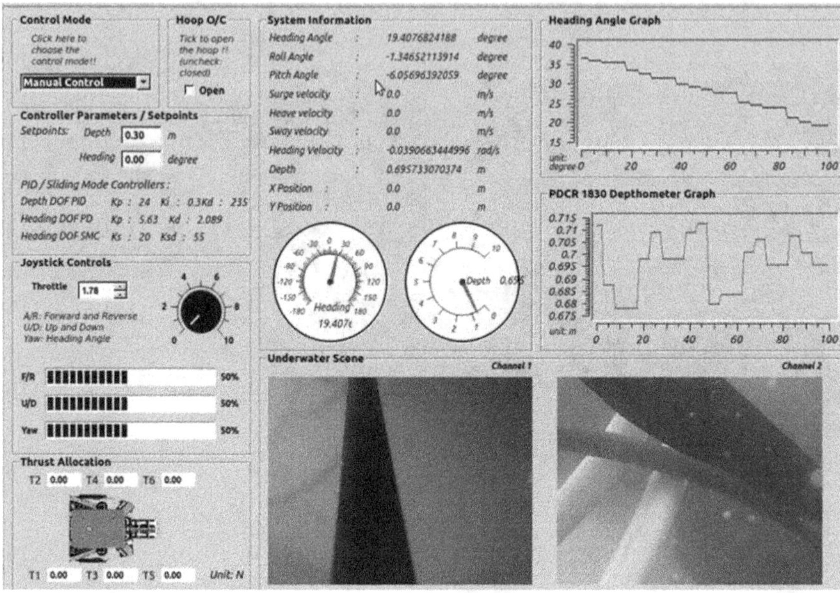

Figure 3.27 User interface for controlling remotely operated underwater vehicle during the test [3]

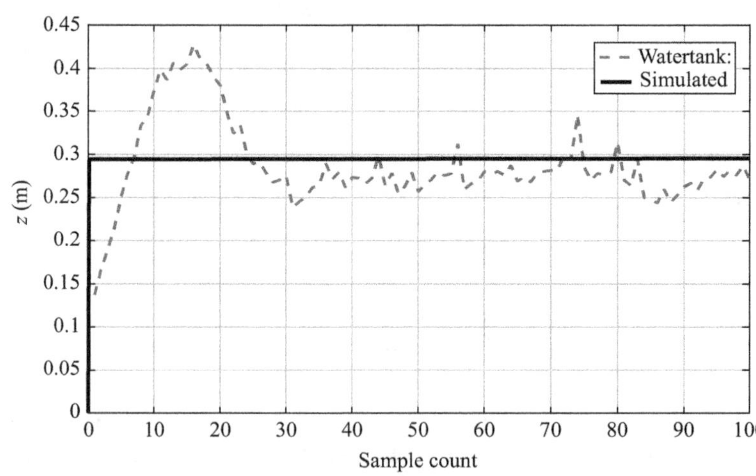

Figure 3.28 Heave response between the simulated and experiment results of remotely operated underwater vehicle [3]

and 3.27. Due to the constraints, only the heave and yaw models are verified. During the tests, the remotely operated underwater vehicle was commanded to move to a target depth of 0.3 m and 140° yaw angle. As shown in Figure 3.28, the remotely operated underwater vehicle can settle to a steady-state value of 0.3 m via

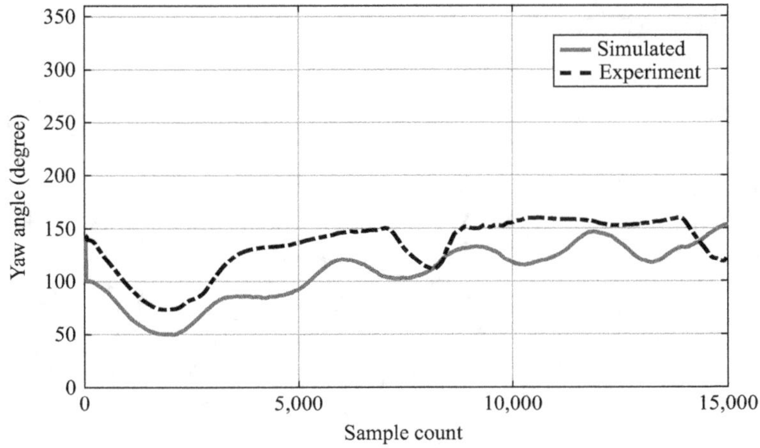

Figure 3.29 Yaw response between the simulated and experiment results of the remotely operated underwater vehicle [3]

the propulsion force generated by mainly T3 and T4 thruster. The vehicle takes around 30 samples to settle to its steady-state value. The difference in the response is due to the thrusters dynamics that are not included in the simulation. As shown in Figure 3.29, the vehicle can regulate itself at a constant yaw rate with the steady-state yaw angle maintained at 140° for almost 15,000 samples (or 150 s). In summary, the experimental validation demonstrates that the numerical model in heave and yaw models behaves reasonably close to the actual vehicle responses. Although the numerical error due to the CFD is 28% and 18% for the heave and yaw directions, respectively, it gives a fairly sufficient model for the initial control system design.

3.6 Simulating external disturbance for remotely operated underwater vehicle model

The environmental disturbances [4] are disturbances caused to the system due to the action of sea current and wave effects. For a fully submerged object like a remotely operated underwater vehicle, the disturbance caused by the wind can be considered as negligible. The wave disturbance is only significant when the vehicle is near to the surface (< 10 m). On the other hand, the current disturbance is considered. The current velocity is included as relative velocity.

$$\mathbf{v_r} = \mathbf{v} - \mathbf{v_c} \tag{3.34}$$

where $\mathbf{v_c} = \begin{bmatrix} u_c & v_c & w_c & 0 & 0 & 0 \end{bmatrix}^{\mathrm{T}}$ is the current velocity vector in body-fixed frame. As the earth-fixed current velocity vector is given by $\mathbf{v_c^E} = \begin{bmatrix} u_c^E & v_c^E & w_c^E & 0 & 0 & 0 \end{bmatrix}^{\mathrm{T}}$,

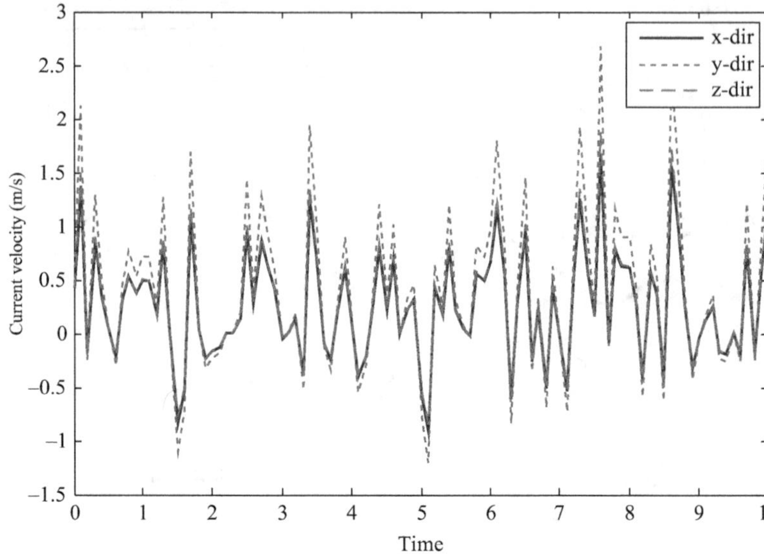

Figure 3.30 Sea current response in earth-fixed frame (at 0.5 m/s)

the body-fixed velocity is determined by the Euler's transformation on the following earth-fixed velocity.

$$u_c^E = u_c \cos \alpha_c \cos \beta_c \tag{3.35}$$

$$v_c^E = v_{\mathbf{c}} \sin \beta_c \tag{3.36}$$

$$w_c^E = w_{\mathbf{c}} \sin \alpha_c \cos \beta_c \tag{3.37}$$

where α_c is the angle of attack and βc is the sideslip angle.

Both the angle of attack and sideslip angle are set to 45°. The current speed is simulated at 0.5 m/s. However, these settings can be adjusted to reflect the actual current experienced by the remotely operated underwater vehicle. The body-fixed velocity in the surge, sway, and heave is seen in Figure 3.30. The current speed under band-limited white noise consists of both sine and cosine wave at 45° acting on the vehicle body-fixed frame. The current speed was set to 0.5 m/s with the power spectral density of 0.1 and sampling time of 0.1, which can be varied.

Linear wave model [4] approximations are used owing to their simplicity and applicability. The high-frequency motion of a dynamically positioned vehicle in the surge, sway, and yaw is expressed as:

$$h(s) = \frac{K_w s}{s^2 + 2\varsigma w_o s + w_o^2} \tag{3.38}$$

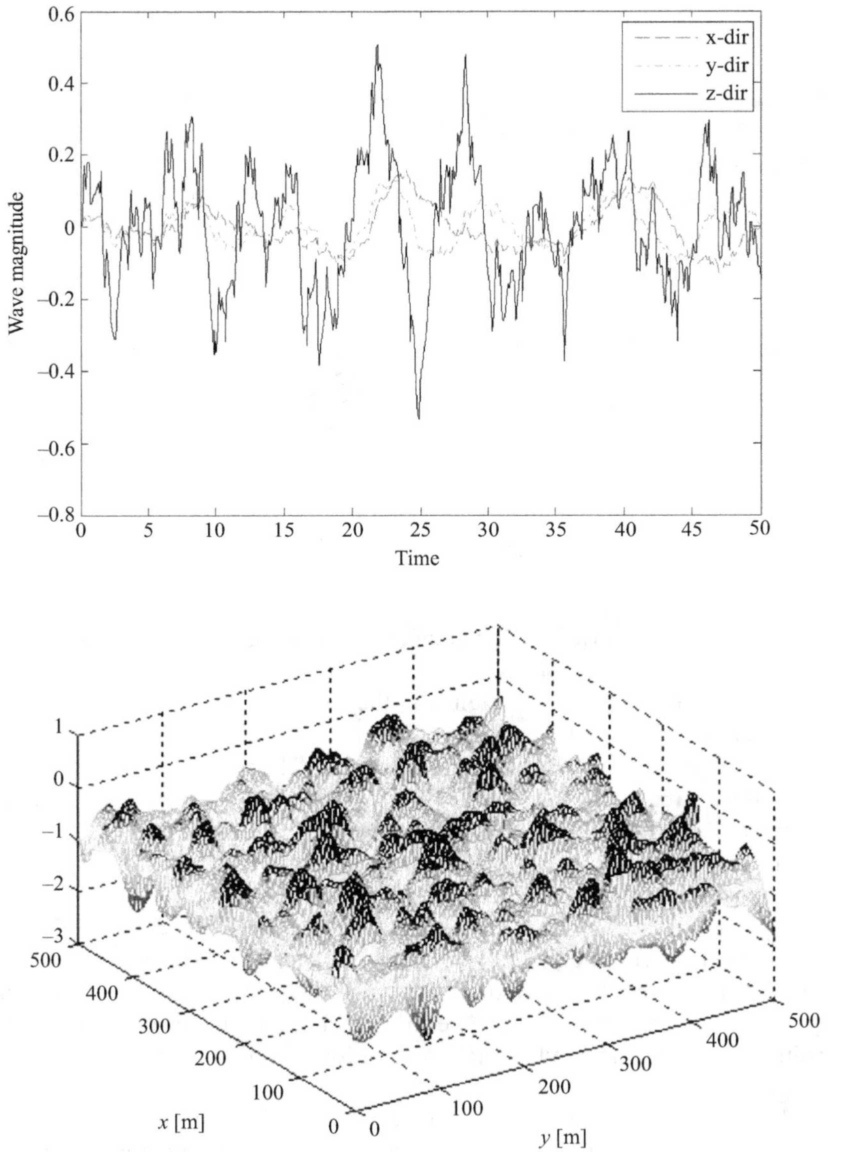

Figure 3.31 Wave magnitude (top) and 3D wave (bottom) in earth-fixed frame

where it is convenient to define the wave spectrum constant (= sea state 2) according to:

$$K_w = 2\varsigma w_o \sigma_w \tag{3.39}$$

Here σ_w is a fixed value describing the wave intensity, ς is a damping coefficient, and w_o is the dominating wave frequency. As shown in Figure 3.31, the wave

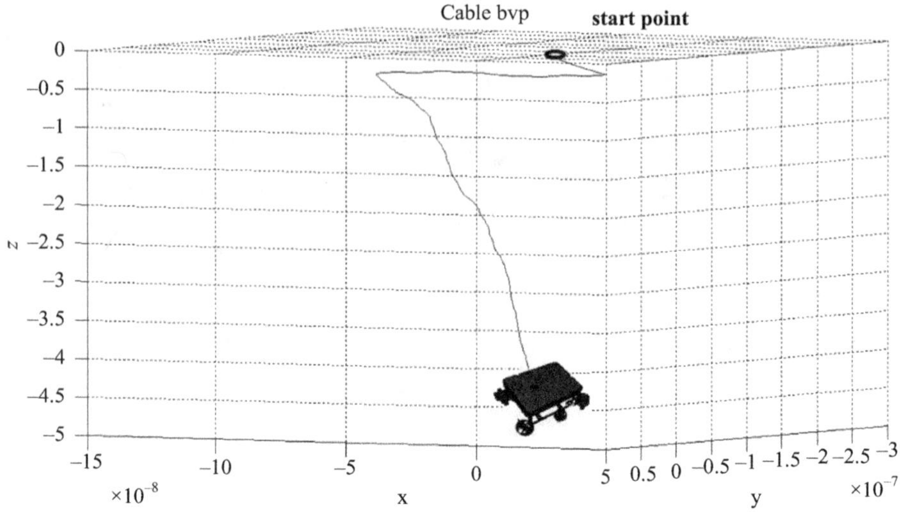

Figure 3.32 Three-dimensional catenary boundary value problem solution [7]

is propagated in three different directions with randomized wave frequency, wave intensity, and damping coefficients.

A tether of length 50 m is fastened in the start coordinate (0,0,0). The tether weight is approximately 1.1 N/m, the diameter is 0.0177 m, and the Young's modulus is 0.025 GPa for polyurethane elastomer. The tether forces are treated as an external disturbance to the vehicle. We solve the 3D catenary boundary value problem (BVP) using the MATLAB® routine [7,8]. An estimate of the end force given the endpoint distance or the remotely operated underwater vehicle final earth-fixed position can be computed. The BVP result can be seen in Figure 3.32. The end forces in the X, Y, and Z directions are approximately 0, 0, and 5.06 N, respectively. This routine will be used to calculate the force exerted on the vehicle simulation model while maneuvering in the water. However, it is entirely computational consuming to compute the tether forces for every change in the endpoint coordinates. Instead, a backpropagation neural network (NN) is used to model the tether forces.

NN models the tether dynamics force at the end point of the vehicle. The training function used is Levenberg–Marquardt (trainlm) and the performance mean square error (MSE) is used. A total of 10 neurons and one hidden layer are used. The inputs are the X, Y, and Z of the end point of the tether and the outputs are the respective forces on the body-fixed coordinate of the remotely operated underwater vehicle at the CG. As shown in Figure 3.33, the NN model (difference in the target and the output) has mostly an absolute error of 3.594, and they are also closely correlated as seen in Figure 3.33. Thus, the NN model gives a close result to the actual computed values from 3D Catenary BVP.

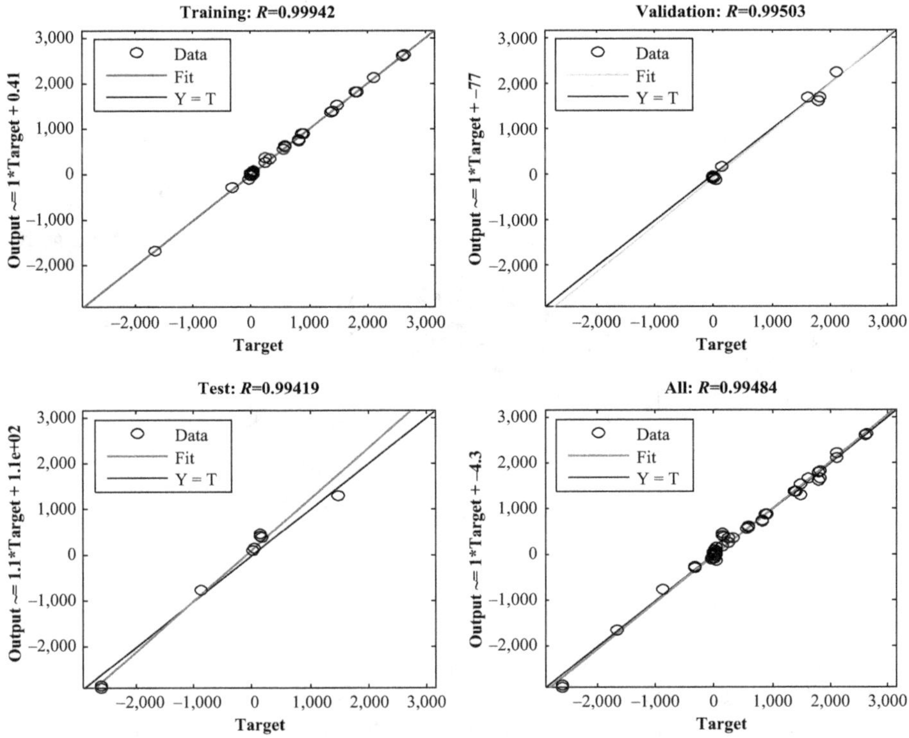

Figure 3.33 Results of NN model for the tether model [7]

3.7 Launch and recovery process model

The contents in this section were obtained from the reference [9]. The launch and recovery process consists of deploying the vehicle into the water. The mission begins once the remotely operated underwater vehicle reaches the mission depth and steers away from the aft of the launch vessel. After the mission of recovering the smaller vehicle is completed, the remotely operated underwater vehicle ascends to the surface that is within the range of the hand-operated rope winch. The vehicle is recovered using the winch, and that concludes the mission.

A robust control system such as sliding-mode controller (SMC) will be used to control the remotely operated underwater vehicle in the recovery of the vehicle stage. The control strategy is to steer the vehicle away from the aft before submerging and align the remotely operated underwater vehicle to the oncoming vehicle. However, it can be difficult to determine the time as the current speed and direction changes with time. Likely scenario of path misalignment can be seen in Figure 3.34. As observed, the heading or/and y-offset in either direction can happen during the path alignment. Definitely, another scenario when AUV passes the

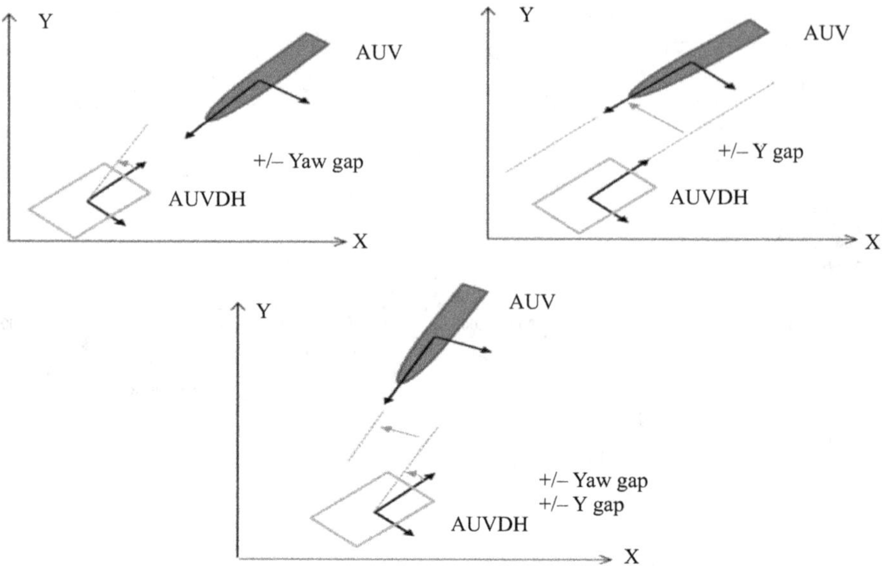

Figure 3.34 Detection and recovery scenarios [9]

vehicle. This is unlikely to happen as the control strategy is tasked to detect the presence of the AUV first and try aligning itself to the incoming vehicle. If it fails, the second attempt will be activated.

To accomplish the control strategy, the following control architecture is used. As observed, the control system uses the nonlinear remotely operated underwater vehicle model determined in the earlier part of the works. The first part of the control is to detect the presence of the AUV using the detection mentioned above method. Path planning (with sufficient time) and speed correction (insufficient time) are required to provide robustness due to the uncertainty in the time allocated for hoop triggering that depends very much on the current speed.

3.8 Control systems design

The contents used in this section were obtained from the following reference [2]. The controller controls both the position and velocity of the vehicle such that the path is maintained (i.e., aligned with the correct heading and path of the incoming). The control signal then triggers the respective thrusters to control the vehicle to its desired position and speed.

A robust SMC incorporating evolutionary fuzzy-based scheme is intended to tune the SMC parameters despite the model uncertainties and the external disturbance. The external disturbances are usually bounded and attenuated by the controller. Recall the following definitions where $\mathbf{v} = [u \quad v \quad w \quad p \quad q \quad r]^{\mathrm{T}}$ is the velocity of the fluid in body-fixed coordinate and $\mathbf{v_r} = \mathbf{v} - \mathbf{v_c}$ is the relative velocity between the fluid and ROV. As fluid is assumed to be irrotational, the fluid velocities are given as follows.

$$\mathbf{v_c} = \begin{bmatrix} u_c & v_c & w_c & | 0_{1\times 3} \end{bmatrix}^{\mathrm{T}} \tag{3.40}$$

The ROV dynamics in [4] includes the current disturbance and model uncertainty in the following form:

$$\mathbf{M}\dot{\mathbf{v}} + \mathbf{C}(\mathbf{v})\mathbf{v} + \mathbf{D}(\mathbf{v_r}) + \mathbf{H}(\mathbf{v}, \mathbf{v_c}, \dot{\mathbf{v}}_\mathbf{c}) = \tau \tag{3.41}$$

where the term $\mathbf{H}(\mathbf{v}, \mathbf{v_c}, \dot{\mathbf{v}}_\mathbf{c})$ contains the hydrodynamic forces between the submerged ROV and the motion of the fluid. The term $\mathbf{D}(\mathbf{v_r})$ is the hydrodynamic damping force characterized as the sum of the linear and quadratic function of the relative velocity between the ROV and the fluid. Here, a known nominal model (denoted by the subscript "o") of the vehicle dynamics has the following form:

$$\mathbf{M_o}\dot{\mathbf{v}} + \mathbf{C_o}(\mathbf{v})\mathbf{v} + \mathbf{D_o}(\mathbf{v_r}) = \tau_\mathbf{o} \tag{3.42}$$

Equation (3.41) is rewritten as:

$$\mathbf{M_o}\dot{\mathbf{v}} + \mathbf{C_o}(\mathbf{v})\mathbf{v} + \mathbf{D_o}(\mathbf{v_r}) + \Delta(\mathbf{v}, \mathbf{v_c}, \dot{\mathbf{v}}_\mathbf{c}) = \tau \tag{3.43}$$

where $\mathbf{M_o}$ is a diagonal matrix,

$$\Delta(\mathbf{v}, \mathbf{v_c}, \dot{\mathbf{v}}_\mathbf{c}) = \Delta\mathbf{M}\dot{\mathbf{v}} + \Delta\mathbf{C}(\mathbf{v})\mathbf{v} + \Delta\mathbf{D}(\mathbf{v_r}) + \mathbf{H}(\mathbf{v}, \mathbf{v_c}, \dot{\mathbf{v}}_\mathbf{c}) \tag{3.44}$$

and

$$\Delta\mathbf{M} = \mathbf{M} - \mathbf{M_o}; \Delta\mathbf{C} = \mathbf{C}(\mathbf{v}) - \mathbf{C_o}(\mathbf{v}); \Delta\mathbf{D} = \mathbf{D}(\mathbf{v_r}) - \mathbf{D_o}(\mathbf{v_r}) \tag{3.45}$$

Note that the parameters used in (3.42) can be obtained from nominal model except for $\Delta(\mathbf{v}, \mathbf{v_c}, \dot{\mathbf{v}}_\mathbf{c})$ that defines the lumped uncertainty due to the model uncertainties and external underwater current is bounded by the following properties.

$$\exists m_1, m_2 > 0 : m_1\|\mathbf{v}\|^2 \leq \mathbf{v}^\mathbf{T}\mathbf{M_o}\mathbf{v} \leq m_2\|\mathbf{v}\|^2, \forall \mathbf{v} \in \Re^6 \tag{3.46}$$

$$\exists \mu_1 > 0 : 0 \leq \mathbf{v}^\mathbf{T}\Delta\mathbf{M}\mathbf{v} \leq \mu_1\|\mathbf{v}\|^2 \tag{3.47}$$

$$\exists \mu_2 > 0 : \|\Delta\mathbf{C}(\mathbf{v})\|\mathbf{v} \leq \mu_2\|\mathbf{v}\| \tag{3.48}$$

$$\exists d_i > 0, i = 1, \ldots 3 : \|\mathbf{D}(\mathbf{v_r})\| \leq \sum_{i=1}^{3} d_i\|\mathbf{v}\|^i + \sum_{i=1}^{3} d_i\|\mathbf{v}_c\|^i \tag{3.49}$$

$$\exists \beta_i > 0, i = 1, \ldots 3 : \|\mathbf{H}\| \leq \beta_i\|\mathbf{v}\|\|\mathbf{v_c}\| + \beta_i\|\mathbf{v_c}\|^2 + \beta_i\|\dot{\mathbf{v}}_\mathbf{c}\|^2 \tag{3.50}$$

The disturbances are bounded signals, $\mathbf{u_c} \in L^{\infty}_{[0,\infty)}$, $\dot{\mathbf{u}}_\mathbf{c} \in L^{\infty}_{[0,\infty)}$ for $\mathbf{v_c}$ and $\dot{\mathbf{v}}_\mathbf{c}$, respectively. The estimated bounds for the model uncertainties are obtained from ROV physical characteristics and experimental tests that indicate around more than 20% error or uncertainties in the nominal model in Section 3.3. Once the upper bound for $\mathbf{u_c}$ and $\dot{\mathbf{u}}_\mathbf{c}$ is fixed, the effect of the disturbance becomes a bounded perturbation on the nominal ROV model.

3.8.1 Sliding-mode control

The mathematical proof for the stability and convergences of the SMC for the evolutionary fuzzy-based scheme is shown. In SMC, a sliding surface in terms of error is defined. Any state that reaches the sliding surface implies convergence to the desired earth-fixed coordinate defined as $\boldsymbol{\eta}_\mathbf{d} = [x_d \quad y_d \quad z_d \quad \phi_d \quad \theta_d \quad \psi_d]^\mathrm{T}$.

Let the earth-fixed coordinate error denoted by $\tilde{\boldsymbol{\eta}} = \boldsymbol{\eta} - \boldsymbol{\eta}_\mathbf{d}$ with the sliding surfaces expressed as:

$$\mathbf{s} = \dot{\tilde{\boldsymbol{\eta}}} + \lambda\tilde{\boldsymbol{\eta}} \tag{3.51}$$

where $\boldsymbol{\eta} = [x \quad y \quad z \quad \phi \quad \theta \quad \psi]^\mathrm{T}$ is the actual earth-fixed coordinate and $\lambda > 0$ is the bandwidth. A common Lyapunov function is written as:

$$V(t) = \frac{1}{2}\mathbf{s}^\mathbf{T}\mathbf{M_o}\mathbf{s} \tag{3.52}$$

The time derivative yields:

$$\dot{V}(t) = \mathbf{s}^\mathbf{T}\mathbf{M_o}\dot{\mathbf{s}} \tag{3.53}$$

With $\mathbf{M_o}\dot{\mathbf{s}} = \mathbf{M_o}(\ddot{\boldsymbol{\eta}} + \lambda\dot{\boldsymbol{\eta}})$, $\ddot{\boldsymbol{\eta}}_\mathbf{d} = 0, \dot{\boldsymbol{\eta}}_\mathbf{d} = 0$ and $\ddot{\boldsymbol{\eta}} = \mathbf{J}(\boldsymbol{\eta})\dot{\mathbf{v}} + \dot{\mathbf{J}}(\boldsymbol{\eta})\mathbf{v}$, $\dot{V}(t)$ can be written as:

$$\begin{aligned}\dot{V}(t) = {} & \mathbf{s}^\mathbf{T}\mathbf{M_o}\mathbf{J}\mathbf{M_o^{-1}}[\tau - \mathbf{C_o}(\mathbf{v})\mathbf{v} - \mathbf{D_o}(\mathbf{v_r}) - \Delta(\mathbf{v}, \boldsymbol{\eta}, \mathbf{v_c}, \dot{\mathbf{v}}_\mathbf{c})] \\ & + \mathbf{s}^\mathbf{T}\mathbf{M_o}(\dot{\mathbf{J}}\mathbf{v} + \lambda\dot{\boldsymbol{\eta}})\end{aligned} \tag{3.54}$$

Remark 1: *The conditions* $\ddot{\boldsymbol{\eta}}_\mathbf{d} = 0$ *and* $\dot{\boldsymbol{\eta}}_\mathbf{d} = 0$ *imply that ROV is regulating its velocity around zero with almost zero acceleration for most of the time. In this case, the ROV is designed to maneuver slowly when it approaches its desired position.*

From the properties in (3.46) to (3.50), there exist positive constants, ς_1, ς_2, and ς_3 such that $\Delta(\mathbf{v}, \mathbf{v_c}, \dot{\mathbf{v}}_\mathbf{c})$ in (3.54) is bounded

$$\Delta(\mathbf{v}, \mathbf{v_c}, \dot{\mathbf{v}}_\mathbf{c}) \le \mu_1\|\dot{\mathbf{v}}\| + \varsigma_1\|\mathbf{v}\| + \varsigma_2\|\mathbf{v}\|^2 + \varsigma_3 = \delta(\mathbf{v}, t) \tag{3.55}$$

The right-hand side of the inequality (3.55) is a function (referred to $\delta(\mathbf{v}, t)$). It can be computed using the feedback law and the bounds established for the model uncertainties. The proposed control input $\tau = \mathbf{Tu}$, where the control input $\mathbf{u}$ can be written as:

$$\mathbf{u} = -\mathbf{T}^{+}\left\{\mathbf{M_o}\mathbf{J}^{-1}\left[K_{sp}\mathrm{sgn}(\mathbf{s}) + K_{si}\mathbf{s} + K_{sd}\mathbf{v} + \dot{\mathbf{J}}\mathbf{v} + \lambda\dot{\boldsymbol{\eta}}\right] - \mathbf{C_o}(\mathbf{v})\mathbf{v} - \mathbf{D_o}(\mathbf{v_r}) - \delta(\mathbf{v}, t)\right\} \tag{3.56}$$

where K_{sp}, K_{si}, K_{sd} are the adjustable SMC's gains and $\mathbf{T}^{+} = (\mathbf{T}^\mathbf{T}\mathbf{T})^{-1}\mathbf{T}^\mathbf{T}$ is the Moore–Penrose pseudoinverse. Substituting the control law back into $\dot{V}$ yields:

$$\dot{V}(t) \le \mathbf{s}^\mathbf{T}\mathbf{M_o}\left[-K_{sp}\ \mathrm{sgn}(\mathbf{s}) - K_{si}\mathbf{s} - K_{sd}\mathbf{v}\right] \le 0 \tag{3.57}$$

This is because the control gains K_{sp}, K_{si} and K_{sd} are chosen to be large enough such that $\dot{V}(t) \leq 0$. Notice that $\dot{V} \leq 0$ implies $V(t) \leq V(0)$, and therefore that **s** is bounded. This, in turn, implies that $\ddot{V}$ is bounded and $\dot{V}$ must be uniformly continuous. The application of Barbalat's lemma [10] then shows that $\mathbf{s} \to 0$ gives $\dot{\tilde{\boldsymbol{\eta}}} + \lambda\tilde{\boldsymbol{\eta}} = 0$, and thus $\tilde{\boldsymbol{\eta}}(t) = e^{-\lambda t}\tilde{\boldsymbol{\eta}}(0)$ implies the position error, $\tilde{\boldsymbol{\eta}}(t)$ converges to zero exponentially as $t \to \infty$. Hence, the system is asymptotically stable.

Remark 2: *If the bound is chosen to be too large, the sign function in (3.57) will cause chattering in the control signals. However, if it is selected to be too small, the stability condition will not be satisfied and lead to the unstable control system.*

To ensure the stability under the uncertainties and decrease the requirements of the parameters in the control system, a genetic algorithm (GA) with fuzzy inference mechanism is used in the following section. Within the boundary layer, the fuzzy logic is used to reduce the chattering phenomenon by smoothing the switch signal.

3.8.2 Proposed fuzzy-based genetic algorithm for SMC

The GA is embedded in SMC to produce the optimal control effort. The coefficients of the crossover and mutation step are selected using fuzzy inference mechanics, to give the right evolutionary direction and the suitable evolutionary step. The proposed control architecture for tuning SMC is shown in Figure 3.35. As shown in Figure 3.35, there are three parts including SMC mentioned previously. In the GA, each chromosome is a point in population which consists of a finite number of chromosomes. The chromosome is chosen by an evolutionary mechanism to obtain the fitness value. New population will form, and some members of the new population will produce new solutions. After some generations, the fittest

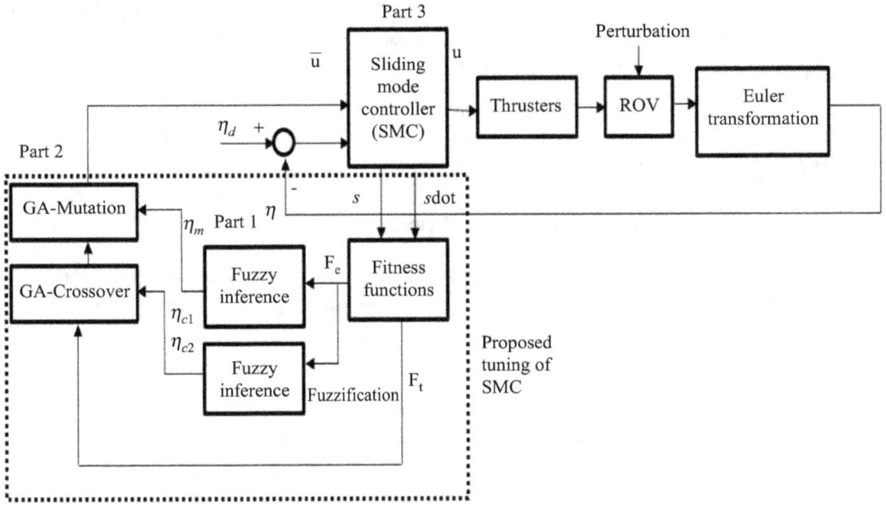

Figure 3.35 Proposed control architecture for tuning SMC [2]

control signal can be achieved. To decide on the correct evolutionary direction and stability in the system dynamic, the fuzzy-based evolutionary is used. The fuzzy logic is also used to compensate for the chattering phenomenon inherent in the SMC with less discontinuous switching across.

The ultimate evolutionary target of GA can be regarded as obtaining $\mathbf{s}(t) = 0$ and $\dot{\mathbf{s}}(t) = 0$ (i.e., the desired position is achieved). A fitness function used for tuning the coefficients in crossover and mutation for fuzzy interference is defined as

$$\mathbf{F_e}(\mathbf{s}) = e^{-\sigma(\mathbf{s}^2+\dot{\mathbf{s}}^2)} \in [0, 1] \tag{3.58}$$

where $\mathbf{s}$ is a sliding surface defined in (3.30) and σ is the rate of decay. The basic genetic operation is shown below.

1. *Elites selection in GA*: The control voltage is treated as the chromosome in GA. The reproduction is utilized to decide the chromosomes into the mating pool for subsequent genetic operations. Equation (3.58) is used to decide the chromosomes that will be selected from the mating pool for further genetic operations. SMC's gains (K_{sp}, K_{si} , K_{sd}) for (3.57) are randomly selected from the positive operational region that ranges from $K_{\min} = 0$ to $K_{\max} = 50$. The sliding surface $\mathbf{s}$ and its derivative $\dot{\mathbf{s}}$ are used to compute the fitness value in (3.58). If the fitness of the new control input is greater than all previous values, it will be the new elite. The highest fitness value is regarded as the elite. By utilizing the elite as a parent, the generated offspring is the new control signal in the next control interval. If the fitness value of the new control signal is greater than the previous one, it will become the next elite. Therefore, it adopts the survival of the fittest.
2. *Fuzzy-based crossover and mutation in GA*: Crossover operation is the primary method to create perturbations in GA. It can generate offspring by varying the signature of the parent. The direction-based crossover operation is used in this section. The concept is that the greater or further away from the sliding surface and its derivative, the greater the variation (or the control input) between the offspring and the parent. Conversely, the control input will be less or remain constant if the system dynamic is within the sliding surface. Hence, the sliding surface $\mathbf{s}$ and its derivative $\dot{\mathbf{s}}$ are integrated into the crossover operation. The offspring generated from the parent can be expressed as:

$$\mathbf{u}_{\mathbf{GA,new}} = \mathbf{u}_{\mathbf{GA,old}} + |\eta_{c_1}|\mathbf{s} + |\eta_{c_2}|\dot{\mathbf{s}} \tag{3.59}$$

where $\mathbf{u}_{\mathbf{GA,old}}$ is the selected chromosome of K_{sp}, K_{si}, K_{sd} from the previous generation and $\mathbf{u}_{\mathbf{GA,new}}$ is the newly created offspring using η_{c_1} and η_{c_2} for $\mathbf{s}(t)$ and $\dot{\mathbf{s}}(t)$, respectively. The tuning coefficients η_{c_1} and η_{c_2} are vital for crossover operation. For example, the tuning coefficients are chosen to be too large (small); the tuning step from generation to generation will be large (small) to enable the system dynamic to divergence quickly (become unstable). In this section, the fuzzy crossover inference mechanism is used to select the right tuning coefficients.

The mutation in GA is then used to prevent the issue of the local minimal. A new chromosome can be included in the population. As compared to the current random mutation operation, there is no assurance that the offspring perform better than their parents that may affect the system dynamics performance in ROV. Hence, the direction-based mutation operation is used. The idea is that the mutation will be adjusted according to the gap between the sliding surface ($\mathbf{s} = 0$) and the system dynamics. If the control effort is unable to achieve close to the sliding surface, the mutation will force the ROV's dynamic to maintain nearer to the sliding surface. The offspring after mutation operation can be expressed as:

$$\bar{\mathbf{u}} = \mathbf{u}_{\mathbf{GA,new}} + |\eta_m| \tag{3.60}$$

where η_m is the adjustment of mutation operation from the fuzzy inference mechanism and $\bar{\mathbf{u}}$ is the offspring after mutation operation.

The occurrence of mutation operation in GA is performed using the mutation probability that may affect the system stability. Hence, a specified positive fitness value (set by the user) named FIT is utilized to check whether the mutation has occurred or not. If the fitness value is less than FIT, the mutation occurs. On the other hand, if the FIT is more than the specified value, the mutation will not happen. Also, the robustness of the GA is ensured by introducing the incremental uncertainties on the ROV model and external disturbance during the GA operation. In summary, the proposed fuzzy-based genetic algorithm on SMC (FGA_SMC) can be written in the following processes.

Step 1: Establish fitness function in (3.58).
Step 2: Choose from the initial population (K_{sp}, K_{si} , K_{sd}) for (3.56).
Step 3: Increment the uncertainty (up to 20% of the nominal ROV model and 1 m/s of underwater current disturbance).
Step 4: Evaluate fitness value in (3.58) and select the elite $\mathbf{u}_{\mathbf{GA,old}}$.
Step 5: Start fuzzy crossover operation to produce $\mathbf{u}_{\mathbf{GA,new}}$ in (3.59).
Step 6: If the fitness value is less than FIT, proceed to the next process or go to Step 8.
Step 7: Start fuzzy mutation operation to produce $\bar{\mathbf{u}}$ in (3.60).
Step 8: Compute control signal $\mathbf{u}$ in (3.56). It will be treated as the parent of the next generation as it is the most appropriate chromosome for the current situation.

The fuzzy inference mechanism is used in the evolutionary procedure. It also decides how to tune the above coefficients $(\eta_{c_1}, \eta_{c_2}, \eta_m)$ and the adjustment of mutation operation. The input variables are B (Big), MI (Middle), and S (Small). The output variables are H (High), ME (Medium), and L (Low). For fuzzy crossover, X represents $\mathbf{F_e}(\mathbf{s})$, and Y represents the coefficients (η_{c_1}, η_{c_2}) as seen in Figure 3.35. For fuzzy mutation, X represents the sliding surface and Y represents the coefficients η_m. According to the process stated in fuzzy-based genetic algorithm (FGA), the evolutionary procedure comprising IF-THEN rules is as follows.

1. Rule 1: IF X is B; THEN Y is L;
2. Rule 2: IF X is MI, THEN Y is ME;
3. Rule 3: IF X is S, THEN Y is H.

The computational effective triangular membership functions for each input and Singleton membership functions for the output (using product inference and average center defuzzifier) are used. The membership functions of the input and output fuzzy sets have x_1, x_2, and x_3 as the centers of the membership functions S, MI and B; y_1, y_2, and y_3 are the centers of the membership functions L, ME, and H, respectively. The fuzzy inference mechanism output is then derived as follows:

$$\bar{y} = \frac{w_1 y_3 + w_2 y_2 + w_3 y_1}{w_1 + w_2 + w_3} = w_1 y_3 + w_2 y_2 + w_3 y_1 \tag{3.61}$$

where $0 \le w_1 \le 1, 0 \le w_2 \le 1, 0 \le w_3 \le 1$ are the weight of Rules 1 , 2, and 3, respectively. The relation $w_1 + w_2 + w_3 = 1$ is valid according to the triangular membership function. Besides, the input and output fuzzy sets for the three parameters $(\eta_{c_1}, \eta_{c_2}, n_m)$ are different. They can be determined by similar fuzzy rule and inference mechanism. A type 2-fuzzy logic controller (FLC) can be used to better cope with modeling uncertainties in the ROV.

In the first simulation, the basic SMC is used to control the remotely operated underwater vehicle to the desired position. It has the following parameters: $\lambda = 111; \Phi = 4; k_s = 20; k_{sv} = 1 \times 10^{-4}; k_{sd} = 55$. For avoiding chattering, a boundary layer around the sliding surface with small value Φ is adopted. The vehicle was commanded to move in the vertical plane followed by horizontal plane, that is, $z_d = 5$ m, $\psi_d = -45°$. The simulation results of the SMC, i.e., the position and velocity time responses curves, are shown in Figures 3.36 and 3.37. The 2D animation of vehicle was done offline after the simulation was done. It basically consists of many frames of vehicle motions which are combined to form an animation of the vehicle from the start to the final AUV recovery as seen in the right-hand side of the figures. As observed in Figure 3.37, the position response gives good tracking without high discontinuous switching as seen in the velocity plots. The performance of boundary layer acted well.

The vehicle was tasked to move sideways, change its heading, and then submerge to a certain depth. To accomplish this, the inputs used are as follows: $y_d = 5$ m, $z_d = 5$ m, $\psi_\mathrm{d} = 45°$. As seen in Figure 3.38, the position response yields good tracking in all directions except for one overshoot in the heave motion which the vehicle tried to adjust to the desired yaw angle. It can reach the desired positions with small velocity response (see Figure 3.39) at steady state. As observed, the roll and pitch DOF is quite stable with no control actions.

To improve on the visual effect with the wave, the wave propagation on the remotely operated underwater vehicle is included as shown in Figure 3.40. The wave was superimposed on the vehicle during the recovery process. This provides some information on the wave impact on the simulation.

In the second simulation, the vehicle was controlled to move diagonally in the horizontal plane from its zero initial position to ($x_d = y_d = 0.5$ m) using the FGA-SMC. The controller gains (K_{sp}, K_{si} , K_{sd}) used for the control are tuned using the

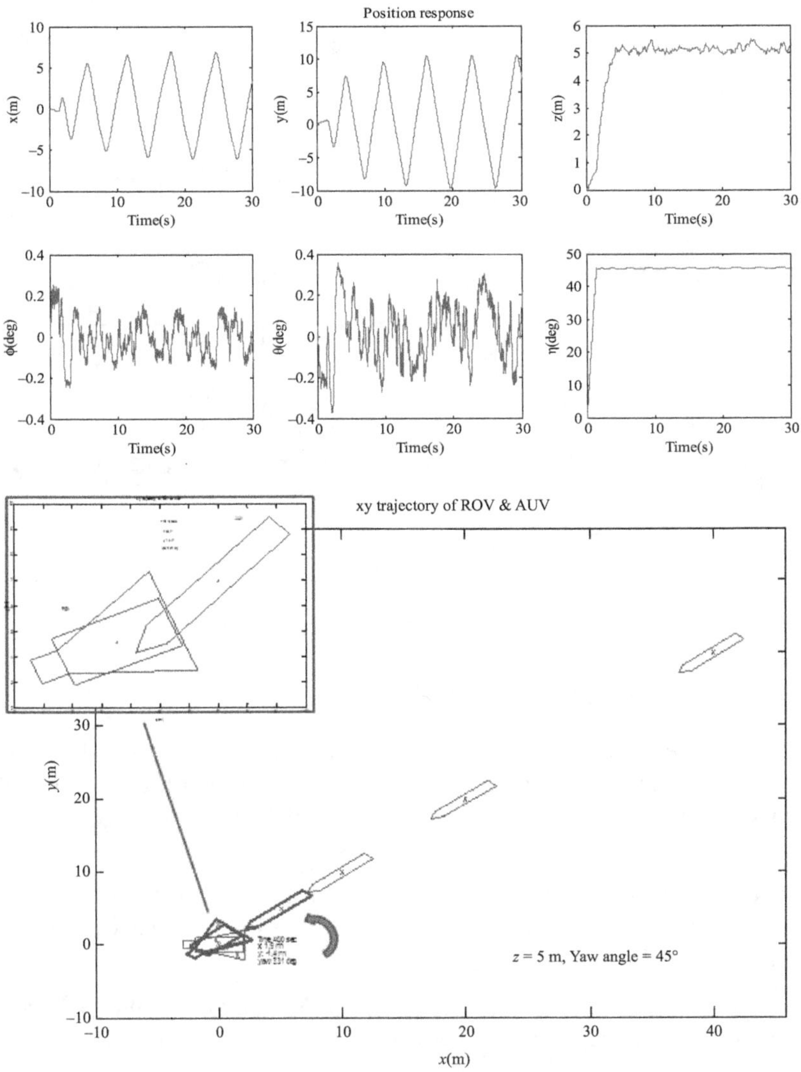

Figure 3.36 Position response for z = 5 m, yaw angle = 45°

proposed tuning approach. The outputs of the FLC are η_{c_1}, η_{c_2}, and η_m. The nominal model can be obtained from the previous sections. Note that the mass of the vehicle used is 135 kg with the docking hoop as seen in Figure 3.41. The updated hydrodynamic damping and added mass coefficients are shown in Tables 3.13 and 3.14. It is evident that there exists an error of at least 20% in the hydrodynamic damping and added mass coefficients. Hence, it creates some uncertainties in the modeling and measurement process that require a robust controller to cope with the uncertainties.

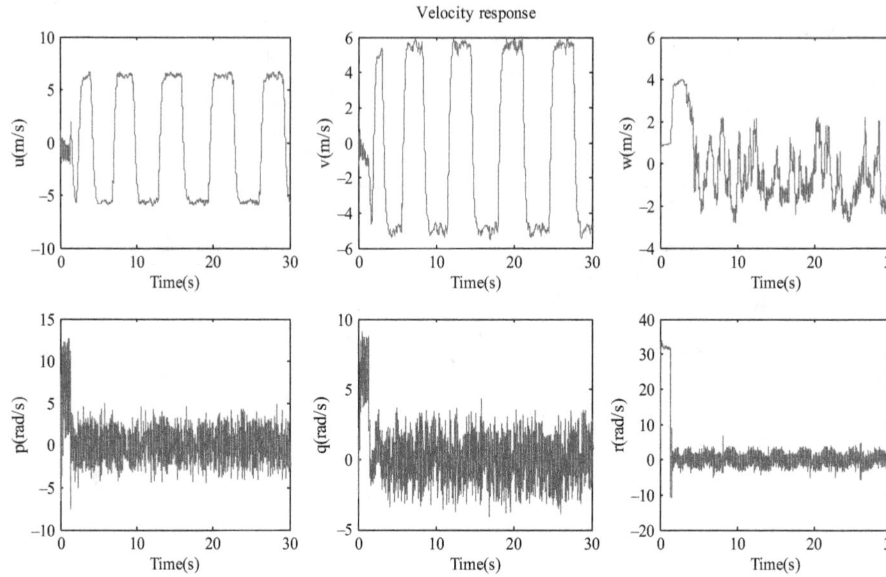

Figure 3.37 Velocity response for z = 5 m, yaw angle = 45°

For comparison, there are two simulation cases.

Case 1: Simulate the proposed FGA_SMC is first carried out at the different perturbation level. The control system is tested up to 20% (or 1.2 times the nominal ROV model) under the current underwater speed of 1 m/s.

Case 2: Compare the proposed method with SMC, fuzzy logic and proportional-integral-derivative (PID) at 20% perturbation on the ROV model with the underwater current speed to 1 m/s.

The following two cases are carried out in MATLAB®/Simulink® package, and their control parameters are discussed in the following sections.

3.8.3 Proportional-integral-derivative

The resulted PID controller gains were automatically tuned PID Tuner (in MATLAB®/Simulink®) to meet the steady-state value with small overshoot, short settling time, short rise time, higher gain and phase margin for closed-loop stability. The PID gains are as follows:

$$K_p = 14, K_i = 5, K_d = 15 \tag{3.62}$$

3.8.3.1 Conventional SMC

The conventional SMC parameters used are as follows: $K_{sp} = 26$, $K_{si} = 42$, $K_{sd} = 37$, $\lambda = 0.01$.

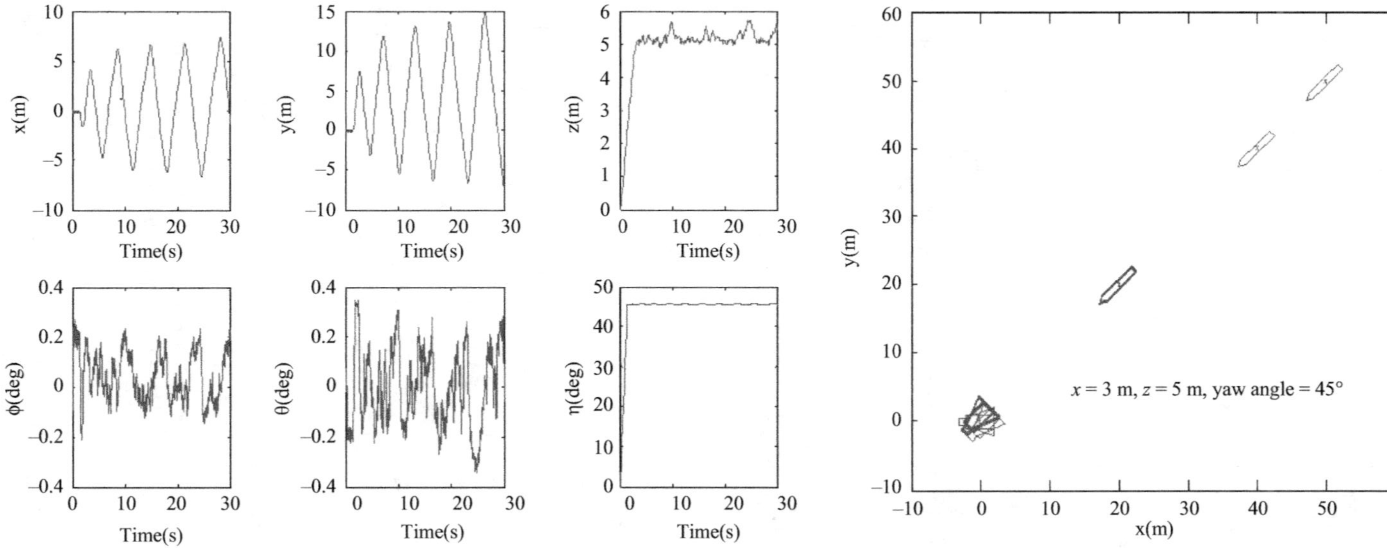

Figure 3.38 Position response at y = 3 m, z = 5 m, yaw angle = 45°

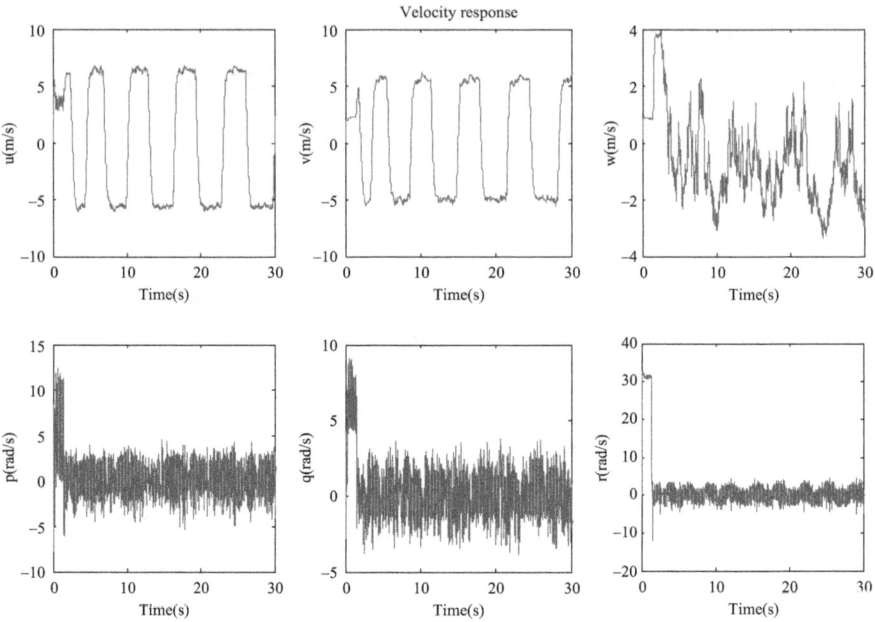

Figure 3.39 Velocity response at y = 3 m, z = 5 m, yaw angle = 45°

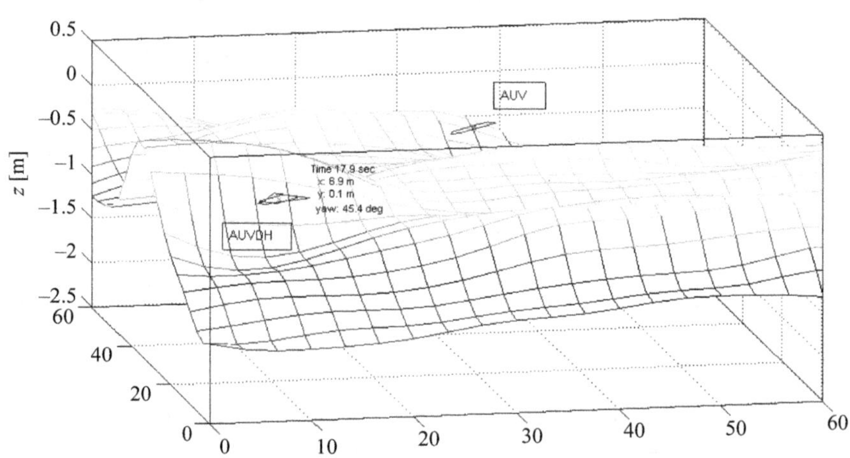

Figure 3.40 Visualization of wave effect on the remotely operated underwater vehicle (named: AUVDH) during the recovery of AUV

Fuzzy logic controller

The parameters of the FLC without the proposed fuzzy-based evolutionary process are $\eta_{c_1} = \eta_{c_2} = 1$ with random mutation.

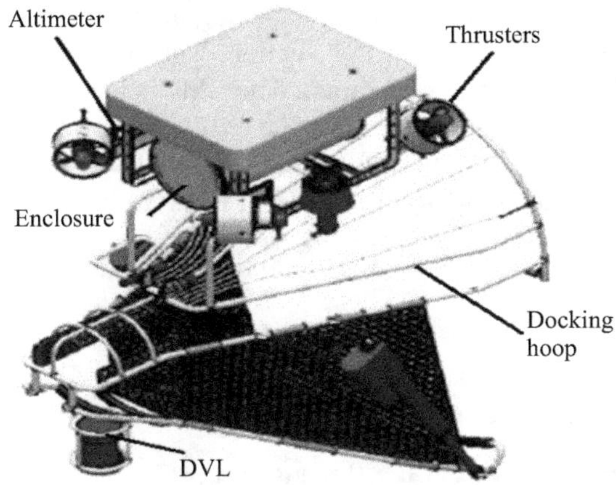

Figure 3.41 Final remotely operated underwater vehicle

Table 3.13 Hydrodynamic damping and added mass coefficients for the linear direction [2]

Descriptions	**Surge**		**Sway**		**Heave**	
Damping coefficients	X_u 3.22	$X_{u\|u\|}$ 105	Y_v 3.29	$Y_{v\|v\|}$ 139	Z_w 5.68	$Z_{w\|w\|}$ 273
Added mass coefficients	$X_{\dot{u}}$ 20.4		$Y_{\dot{v}}$ 53.4		$Z_{\dot{w}}$ 126	

Table 3.14 Hydrodynamic damping and added mass coefficients for the angular direction [2]

Descriptions	**Roll**		**Pitch**		**Yaw**	
Damping coefficients	K_p 0.68	$K_{p\|p\|}$ 15.5	M_q 5.32	$M_{q\|q\|}$ 33.2	N_r 12.3	$N_{r\|r\|}$ 26.7
Added mass coefficients	$K_{\dot{p}}$ 2.80		$M_{\dot{q}}$ 13.7		$N_{\dot{r}}$ 5.26	

Proposed FGA for SMC (FGA_SMC)

The maximum number of generations is 100, and the population size is 50. As the aim is to cause the population to convergence, crossover (for convergence operation) should occur more frequently than the mutation (for divergence operation). However, choosing the right probability can be entirely subjective in different applications. Hence, a balanced approach to setting both the crossover and mutation probability to 20% was adopted in this section. The fitness chromosome will be retained and duplicated in each generation. The SMC's gain distribution across the entire generation consists of 5,000 points (i.e., 100 generations × 50 populations). The lowest error norm (or highest fitness, $\mathbf{F}_t(t)$) is at 96th generation. The final fittest control parameters and variables used in FGA-SMC are as follows:

$$K_{sp} = 21, K_{si} = 5.9, K_{sd} = 15, \sigma = 0.1, \lambda = 0.01, K_{\min} = 0, K_{\max} = 50, \text{FIT} = 0.3 \tag{3.63}$$

The output and input parameters used in the fuzzy-based evolutionary process are selected as follows:

$$y_{1,1} = y_{1,2} = 0.3, y_{2,1} = y_{2,2} = 0.6, y_{3,1} = y_{3,2} = 1 \tag{3.64}$$

$$y_{1,m} = -1, y_{2,m} = 0, y_{3,m} = 1 \tag{3.65}$$

$$x_{1,1} = x_{1,2} = 0, x_{2,1} = x_{2,2} = 0.5, x_{3,1} = x_{3,2} = 1 \tag{3.66}$$

$$x_{1,m} = -0.5, x_{2,m} = 0, x_{3,m} = 0.5 \tag{3.67}$$

In Case 1, the simulation results of FGA-SMC can be seen in Figure 3.42. The position and velocity time responses at different perturbed condition are plotted. The position responses indicate quite robust behavior to changing perturbations except for heave, roll, pitch, and yaw angle at higher perturbation. As the limit of the uncertainty increases to more than 8% and underwater current disturbance greater than 1 m/s, FGA_SMC continues to show quite robust performance with little change in the output response. The velocity responses also reflect a good regulation about zero velocities. The proposed control is robust only in low perturbations while in higher perturbation, only two of the six DoF display behavior of robustness to uncertainties.

In Case 2, the controller parameters are used for the computer simulation to compare with other controllers (see Figure 3.43) at 20% and 1 m/s. The proposed FGA-SMC controller exhibits a smaller error in the surge and sway position (where the set point is: $x_d = y_d = 0.5$ m) than PID, SMC, and FLC. The PID has a larger steady error as compared to all controllers. On the other hand, FGA-SMC exhibits less chattering phenomena as compared to SMC due to the fuzzy inference mechanism used in the evolutionary procedure. The FLC can compensate for the chattering phenomenon inherent in the conventional SMC. It resulted in less oscillatory behavior with a higher error in yaw position. On the other hand, SMC shows the inherent oscillatory behavior in all DoFs. As observed, the pitch and roll angle is small and quite self-stabilizing as compared to PID.

In summary, SMC, PID, and FLC exhibit less robustness to the perturbation as seen in the overshoot, steady-state error and oscillatory responses as compared to the proposed FGA-SMC approach. The 2-norm position error also shows FGA-SMC has the lowest error as compared to the other controllers (FGA_SMC: 200; SMC: 600; PID: 1500; Fuzzy: 500). It implies that FGA-SMC proposed on ROV can handle the predetermined uncertainty to achieve the robustness of the position control.

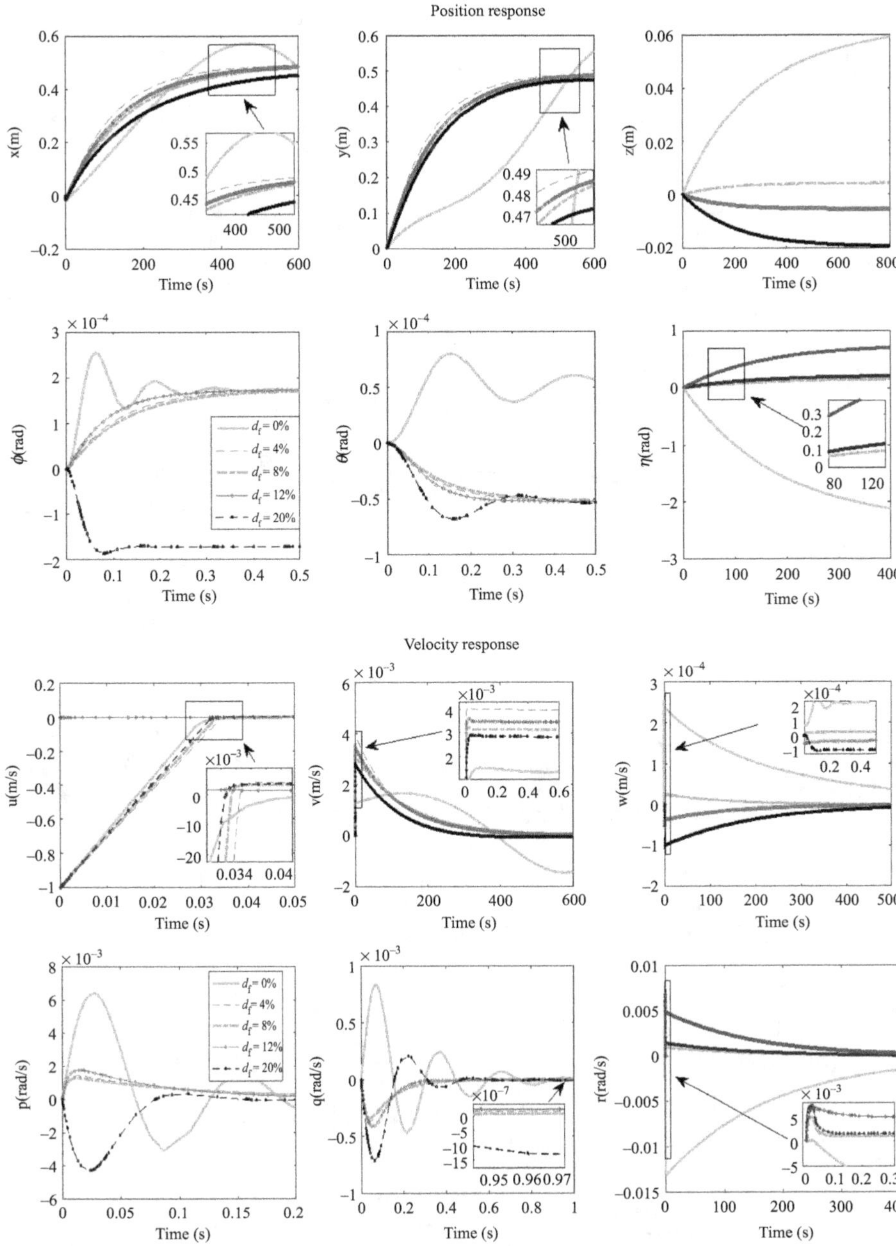

Figure 3.42 Case 1: Time-domain plots of position (top) and velocity (bottom) at various perturbed conditions ($x_d = y_d = 0.5$ m) using FGA-SMC [2]

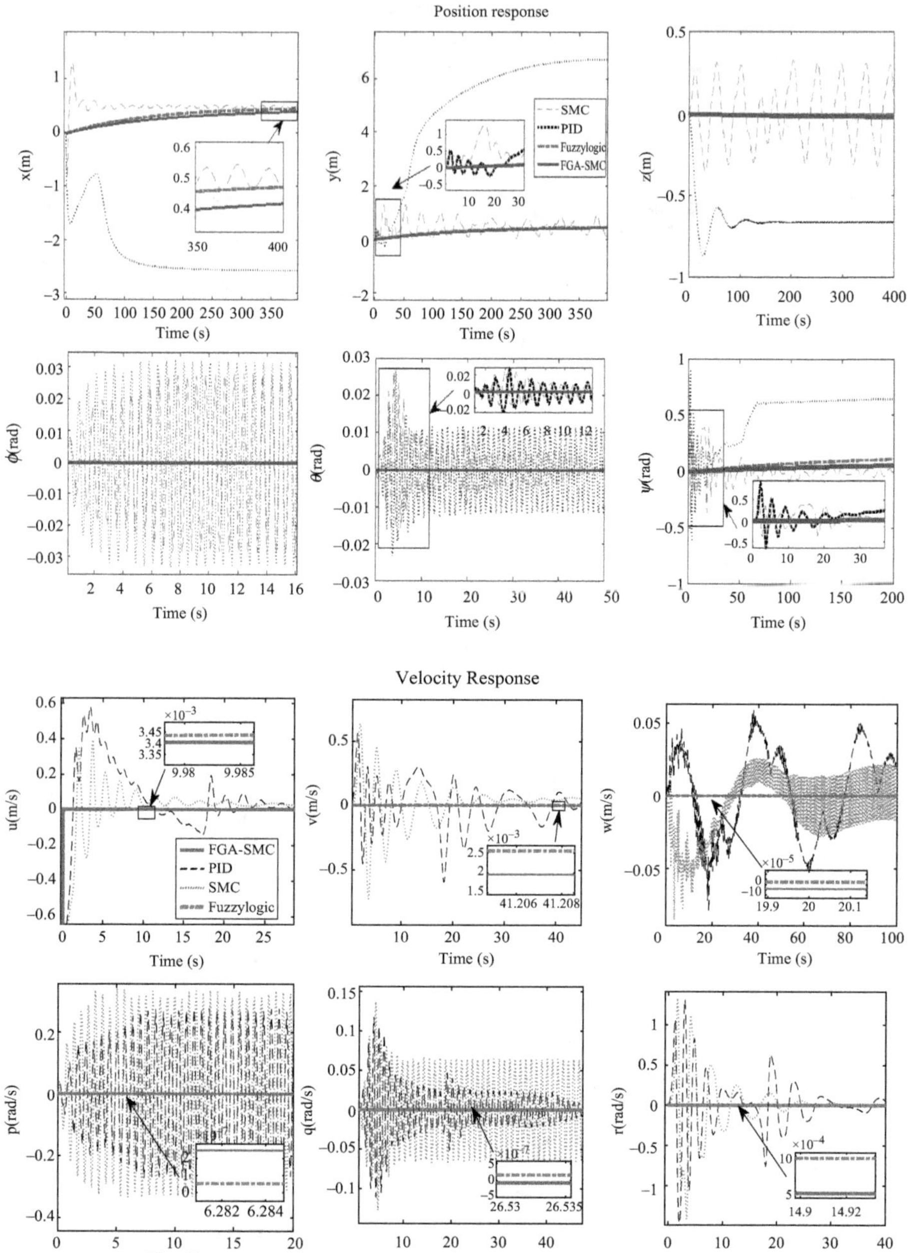

Figure 3.43 Case 2: Controller comparison of position (top) and velocity (bottom) performance at 20% model perturbation and 1 m/s underwater current [2]

3.9 Remotely operated underwater vehicle sea trial

The experimental test of remotely operated underwater vehicle (see Figure 3.44) was conducted near Singapore Yacht Club. The joystick was first used to move the vehicle from a launch position. The following two tests, namely, heave and yaw tests were performed. The command signal and controller parameters were initialized using the graphical-user-interface designed by Python-Qt open-source codes. All data were sampled at 100 Hz. Due to time constraint, FGA-SMC and PID controller were programmed into the target PC104 in ROV. During the experimental tests, the control parameters of FGA-SMC was further tuned to give a better response in real-time. It is attributed to the uncertain external factors such as different significant wave height (approximately 1 m) and sea current (around 2 knots) from Meteorological Service Singapore (http://www.weather.gov.sg/home/) that affect the results and performance.

The actual ROV was trying to regulate its depth of 0.6 m as shown in Figure 3.44 (right). In the experiment, the original PID controller was unable to attenuate the disturbances. PID controller parameters was retuned to $K_p = 31$, $K_i = 27$, $K_d = 34$. It was then used to compare with FGA-SMC. The controller gains used was tuned online with $K_{sp} = 26$, $K_{si} = 11$, $K_{sd} = 20$ and other parameters remain.

As seen in Figure 3.45 (top), PID controller is unable to attenuate the disturbances due to the external wave and current on the ROV. PID controller tends to follow the wave perturbation. However, the FGA-SMC exhibits less oscillation with a small steady-state heave error of approximately 0.7%, and the chattering response is much alleviated. FGA-SMC produces much smaller fluctuation in the yaw angle (except during the transient stage) as compared to PID controller during heave test. Similar behavior can be seen in the roll and pitch angle where less overshoot was observed in FGA-SMC than PID controller in Figure 3.45 (top). The results showed that FGA-SMC could robustly position the ROV at the desired depth despite the uncertainties and the external disturbances.

Figure 3.44 ROV at launch position (left), yaw (middle) and heave (right) tests at position (2) using FGA-SMC [2]

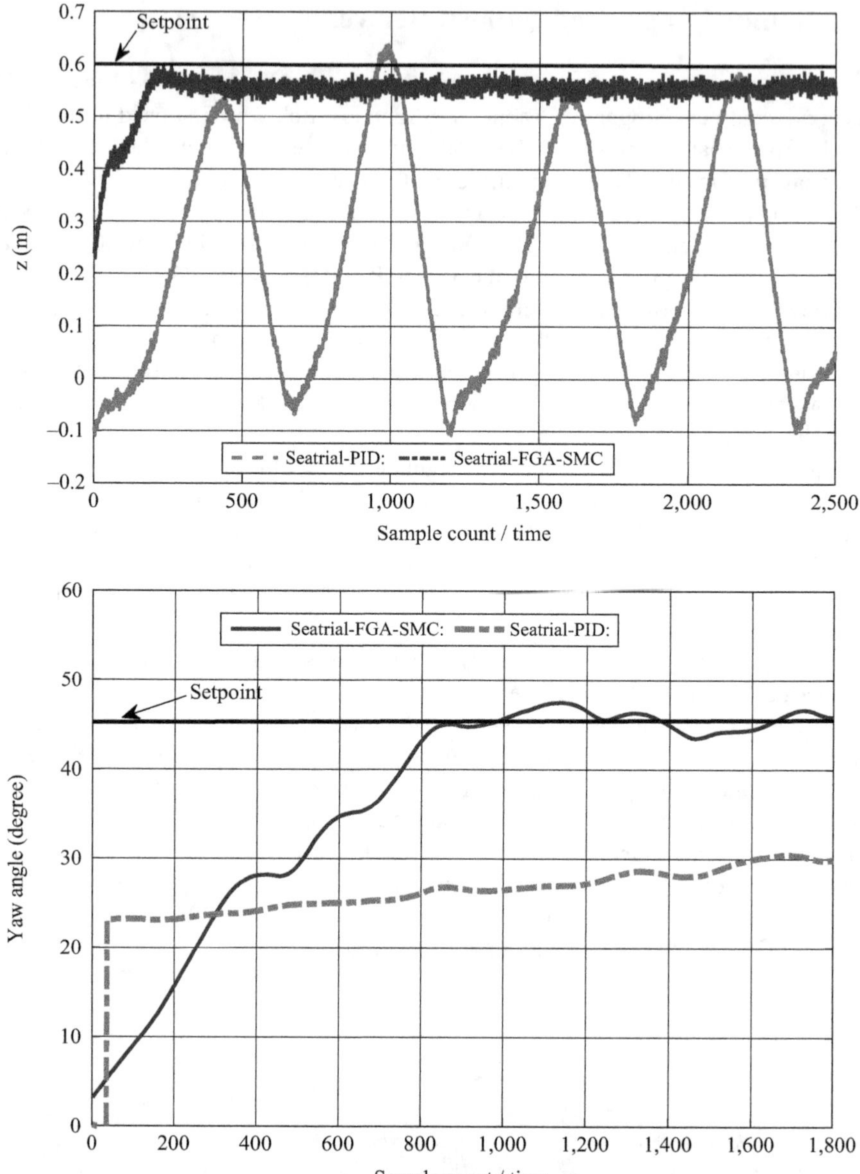

Figure 3.45 Heave (top) and yaw (bottom) response between PID and FGA-SMC [2]

After the depth control, the results of the yaw control (see Figure 3.45 (bottom)) were compared at 45°. The yaw angle of the FGA-SMC settles to the steady-state value with a slow response (a similar behavior as seen during the simulation). Despite increasing the proportional gain, PID controller still exhibits an

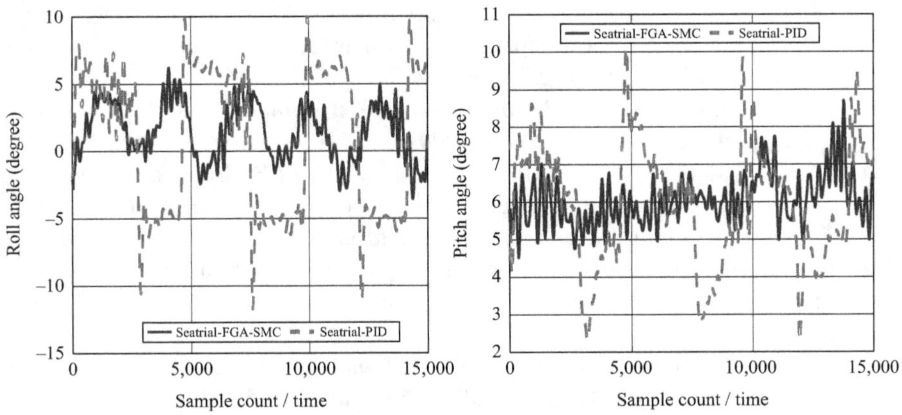

Figure 3.46 Roll (left) and pitch (right) angle response between PID and FGA-SMC [2]

error of 15° in the yaw angle. Although PID has a faster rise and settling time, FGA-SMC fairs better in terms of the steady-state yaw angle despite its sluggish response. The sea trial reveals that ROV could maintain the desired yaw angle and the depth despite the model uncertainties and the external disturbances. As seen in Figure 3.46, the roll and pitch angle of FGA-SMC are smaller as compared to PID controller.

In summary, the proposed tuning of FGA-SMC is first applied on ROV to achieve a more efficient, and systematic architecture for SMC's tuning and better time-domain responses for deployment. With SMC integrated into GA on a real ROV platform with acceptable position responses with less chattering despite the sluggish response. As observed in the experimental results, the steady-state error in these results is attributed to the factor due to the changing uncertainties in an uncertain environment. Note that additional sensor such as surface acoustic wave sensor will be useful to provide measurable input to the controller. As shown in the captured response, the performance of the FGA-SMC designed controller is still better compared to that of the manually tuned system, in terms of steady-state error, less chattering behavior, and robustness.

References

[1] Lin, JY. Computation of Hydrodynamic Damping and Added Mass for a Complex-Shaped Remotely-Operated Vehicle (ROV) for Sufficient Control Purpose. **Undergraduate Report**, Newcastle University, United Kingdom, May 2014.

[2] Chin, CS and Lin, WP. Robust Genetic Algorithm and Fuzzy Inference Mechanism Embedded in Sliding-Mode Controller for Uncertain Underwater Robot, **IEEE/ASME Transactions on Mechatronics** 23(2), pp. 655–666, 2018.

[3] Chin, CS, Lin, WP, and Lin JY. Experimental Validation of Open-Frame ROV model for Virtual Reality Simulation and Control. **Journal of Marine Science and Technology** 23(2), pp. 267–287, 2018.

[4] Fossen, TI. **Guidance and Control of Ocean Vehicles**, John Wiley & Sons Ltd., Chichester and New York, 1994.

[5] Chin, CS and Lau, MSW. Modeling and Testing of Hydrodynamic Damping Model for a Complex-Shaped Remotely-Operated Vehicle for Control, **Journal of Marine Science and Application**, 11(2), pp. 150–163, 2012.

[6] de Barros, EA, Pascoal, A, and de Sa, E. Investigation of a Method for Predicting AUV Derivatives, **Ocean Engineering**, 35(16), pp. 1627–1636, 2008.

[7] Lin, WP, Chin, CS, Looi, LCW, Lim, JJ, and Teh, EME. Robust Design of Docking Hoop for Recovery of Autonomous Underwater Vehicle with Experimental Results. **Robotics**. Guest Editors: Prof. Thor I. Fossen and Prof. Ingrid Schjølberg, NTNU, Trondheim, Norway 2015, 4(4), 492–515.

[8] Sørensen, AJ. Marine Control Systems: Propulsion and Motion Control of Ships and Ocean Structures. Lecture Notes, The Norwegian Institute of Technology, Department of Marine Technology: Trondheim, Norway, 2012; pp. 222–224.

[9] Lin, WP, Chin, CS, and Mesbahi, E. Remote Robust Control and Simulation of Robot for Search and Rescue Mission in Water. **MTS/IEEE OCEANS** 2014, Taipei, Taiwan.

[10] Slotine, J-JE and Li, W. **Applied Nonlinear Control**, 1st ed. Englewood Cliffs, NJ, USA: Prentice-Hall, 1991.

Chapter 4

xPC-Target embedded system design

4.1 Introduction

The most efficient way to develop an embedded system is to connect the control systems to the real plant. Very often, trial-and-error method is used to verify the functionality of the plant. However, it is time-consuming and leads to modifications or costly replacement of the hardware. To circumvent the problem, an emulator that consists of an embedded system (system under test) that provides a control signal to plant is needed. The sensor reading is read by the system (feedback) and output actuator control signals to the plant. Many iterations of the control system will perform using the graphical programming software till the specifications are met. The method mentioned above is known as a hardware-in-the-loop simulation (HILS). HILS is a technique that is used in the development and test of complex real-time embedded systems using graphical programming software. The apparent benefit to the graphical programming tool is the elimination of writing code such as C/C++, as seen in Chapter 2. It is not to say that the students do not "program"—in fact, they do direct the embedded computer to follow their control algorithm. However, the details of real-time embedded systems are left up to the code generation tool. It allows the student to create software that can be executed in a real-time environment without requiring the student to learn the intricacies of writing specific multirate, real-time, embedded code. With that, the HILS provides an efficient platform by adding the complexity of the plant under control to the test platform. As shown in Figure 4.1, it includes the following steps.

- hardware interfacing;
- rapid prototyping simulation;
- final implementation in the HILS.

A typical microprocessor-based mechatronics application is shown in Figure 4.1. It consists of the following subsystems.

- Target system such as an underwater robotic vehicle constitutes the device being controlled. The device itself may involve some individual or interrelated mechanisms with their subsystems.
- Host PC is our desktop computer that is installed with the simulation software tools such as MATLAB®/Simulink® and xPC-Target block sets. Besides, Window Operating System such as Window 98/XP/Vista/7 has to be installed

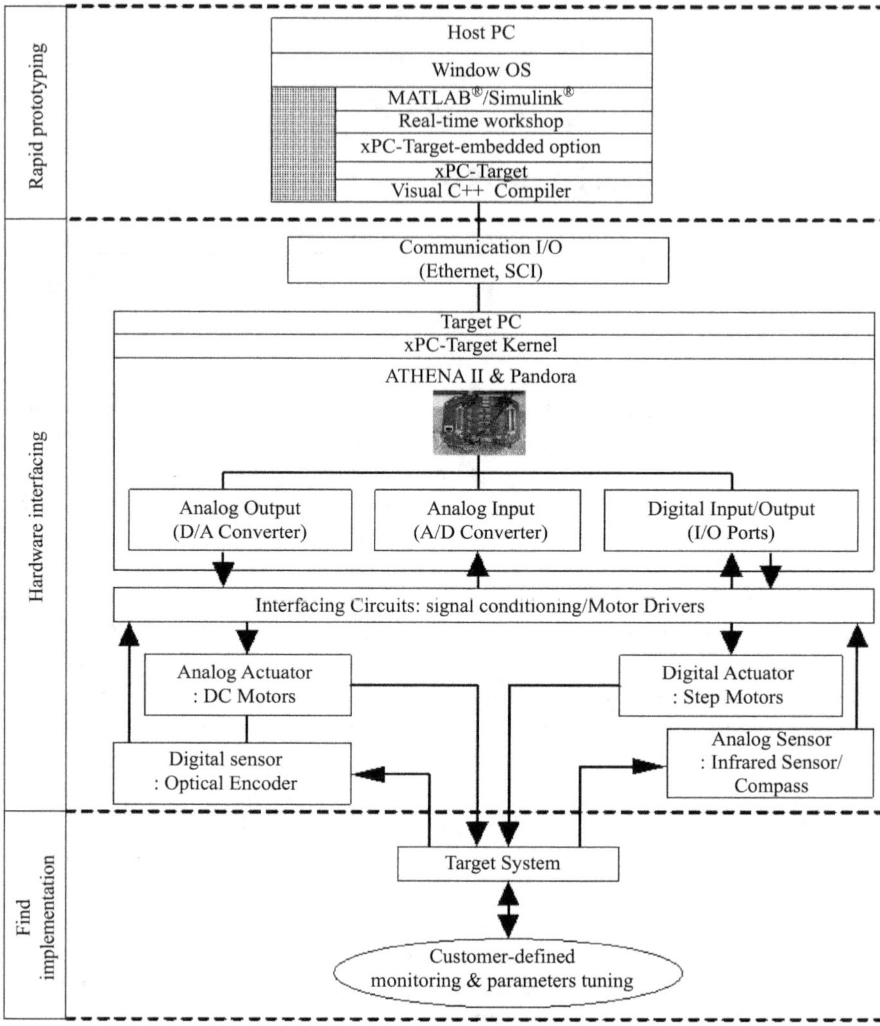

Figure 4.1 HILS solution for rapid prototyping to final production

before the use of this software. The Host PC communicates with the Target PC by an Ethernet or RS232 connection.

- Target PC is an embedded microprocessor such as PC104 that has DOS installed to allow the real-time kernel to run and execute set of commands that are downloaded through the communication port such as Ethernet or RS232 connection.
- A sensor such as infrared sensors, magnetic compass sensors, and encoders measures the controlled variables. Sensors of all types provide the computer with information it needs to perform monitoring and control tasks.

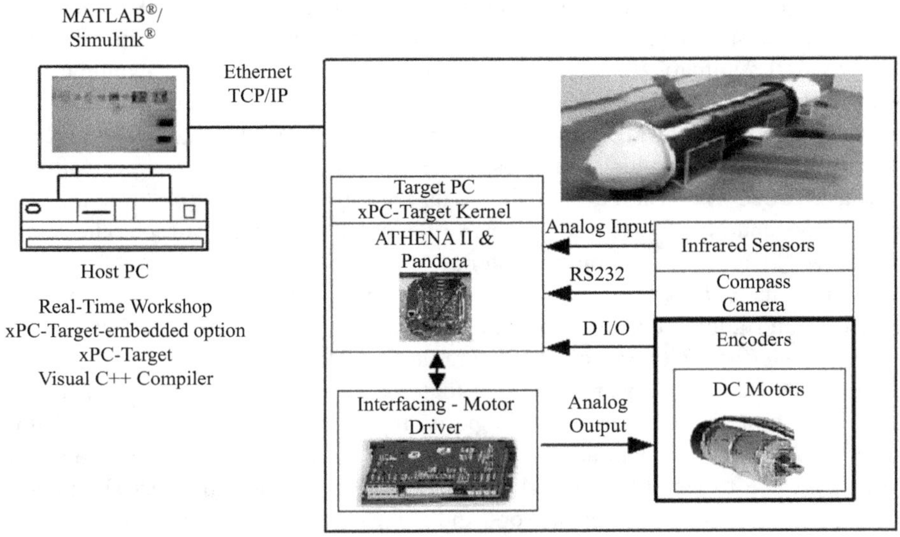

Figure 4.2 Example of HILS solution on underwater robotic vehicle

- Actuators such as DC motors, stepper motors, and linear actuators convert digital signals into physical as required to correct the errors in meeting the required performance.
- Interfacing circuit involves all necessary signals conditioning circuit such as motor drivers to interface actuators, sensors, and devices to the microprocessor.

An example of the HILS environment (see Figure 4.2) can be developed to support and validate the hardware and software development of an underwater robotic vehicle. The HILS incorporates a high-fidelity dynamic model that includes the sensor and actuator models, from the identified parameters from experiments. A user-friendly graphical interface such as a 3-D visualization of the vehicle's motion completes the simulation environment that is an indispensable tool for rapid certification of both the hardware and the control software while performing simulated tests with minimal cost and effort.

In summary, the overview of hardware interfacing to simulation testing will be introduced. It is followed by the hardware interfacing using the PC104, operational amplifier, and PWM H-Bridge motor driver. xPC-Target in the MATLAB® software is explained followed by the applications example on the HILS example on an underwater robotic vehicle.

4.2 Overview of hardware interfacings for simulations testing

A graphical-simulation and control system design tool like Simulink® is easily programmed for execution on the real hardware: real-time controllers and communication with sensors and motors. It is not easy to get lost in coding software and miss out on essential concepts in controller design and real-time system operation. Recently, the Mathworks, Inc. has developed an environment to design, simulate, analyze, and automatically generate real-time embedded code. For example, The Mathworks' MATLAB® has featured Real-TimeWorkshop for some years. This software takes a graphical control block in Simulink® and compiles it for execution under various environments such as protected mode DOS™. A new software package, called xPC Target, is available as a toolbox for Simulink® versions 6.0 and greater and allows users to move from simulation to hardware on The Mathworks' works real-time operating system.

Additionally, the xPC-Target package includes numerous libraries of supported hardware and allows users to compile graphical Simulink® models that include blocks to interface with specific hardware devices. The xPC-Target software fully supports the PC104 hardware, meaning that analog to digital inputs—A/D inputs, digital I/O, timers, serial ports, parallel ports, and watchdog timers—are included in the Simulink® environment as model blocks—eliminating the need for students to write drivers for these functions. Additional hardware which includes DC motors with optical encoders, infrared sensors, and compass can be supported. They can be connected to the digital I/O, analog inputs, and RS232 on the PC104. The DC motors can be interfaced to the analog output of the PC104 via an H-bridge amplifier.

The graphical Simulink® models are downloaded to the Target hardware running the xPC-Target operating system. xPC Target's development blocks for Simulink® and its real-time operating system allow rapid development of controllers in the progression from the previous controller design and simulation, to the final implementation. The xPC-Target compiled the block diagrams into an executable program that is downloaded from the Internet to the Target embedded computer. This code is capable of executing subsystems at multiple sampling rates, has integrated ordinary differential equation solvers and manages processor usage, memory allocation, and system input and output. Furthermore, data collected when running an experiment is in MATLAB® format for convenient post-processing and offline analysis. Once a model is compiled and downloaded to the Target, model parameters such as proportional, integral, and derivative gains of the controllers or even the commanded inputs can be changed.

4.3 Hardware interfacings

In this section, the hardware and its interfacing options are introduced. There are many types of PC104 that can be used for embedded control applications. In common, they consist of various inputs and outputs communications such as digital, analog, RS232, USB, Ethernet, and VGA. It eliminates the need for students to write drivers and design tedious terminal blocks for these functions to communicate with the hardware. In this course, the Athena II will be used. Besides, the PWM motor driver from ADVANCED MOTION CONTROLS, the infrared sensor from SHARP, and magnetic compass from HONEYWELL are shown.

Athena II PC104 computer is a stack of small 3.5" square cards connected by an embedded ISA communication bus (consisting of 104 pins) is used in this chapter. Note that Athena II processor can be used with xPC-Target. On the other hand, Athena III is not tested with xPC-Target. As seen in Figures 4.3 and 4.4, there are two cards in this stack: Athena II CPU (or Prometheus CPU) card and DAQ card with the panel I/O board. The Athena II card has the processor and all of the connectors to various peripherals on it (keyboard, Ethernet, VGA) and system I/O that includes a 10/100BaseT Ethernet port, two RS-232 ports, two RS-232/485 ports, four USB 1.1 ports, Ultra DMA IDE controller with support for two drives, and PS/2 keyboard and mouse support.

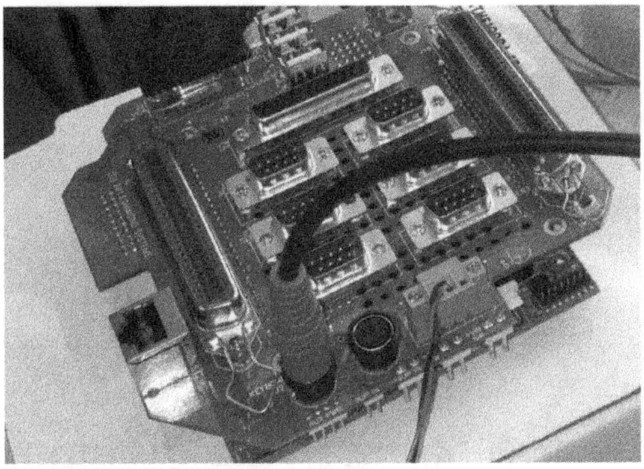

Figure 4.3 Athena II– Prometheus CPU card

Figure 4.4 Athena II– Prometheus panel I/O card

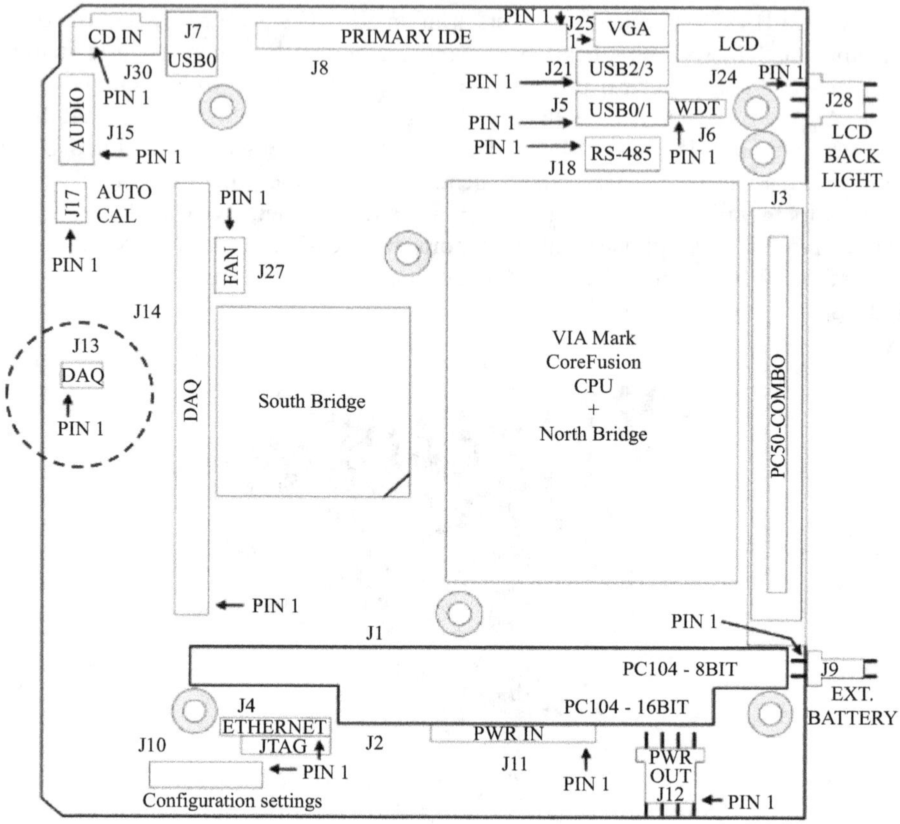

Figure 4.5 Athena CPU card with the location of jumper setting (Cite: http:// www.diamondsystems.com/)

Athena II includes an autocalibrating analog and digital I/O circuit. It has sixteen 16-bit analog inputs with 100 kHz sample rate, backed by a 2048-sample FIFO with a programmable threshold. Programmable input that ranges from a wide range of ± 10 V down to 0–1.25 V are provided. The analog circuitry also includes 4 D/A channels with 12-bit resolution and jumper-selected output ranges. Multi-range autocalibration on both A/D and D/A ensures maximum accuracy over time and temperature and enables reliable, maintenance-free performance. On the digital side, Athena II provides 24 digital I/O lines with programmable direction, as well as two counter/timers for A/D sample rate control, pulse counting, frequency generation, or other applications. Its advanced control logic has the features and flexibility to fit almost any application, such as internal/external A/D clock source, scan, and single-sample modes. To configure the A/D and D/A on the Athena II, the location of various jumpers indicated by the letter "J" needs to be configured. The users should orientate the CPU board as shown in Figure 4.5 to locate the J13 for A/D and D/A configurations.

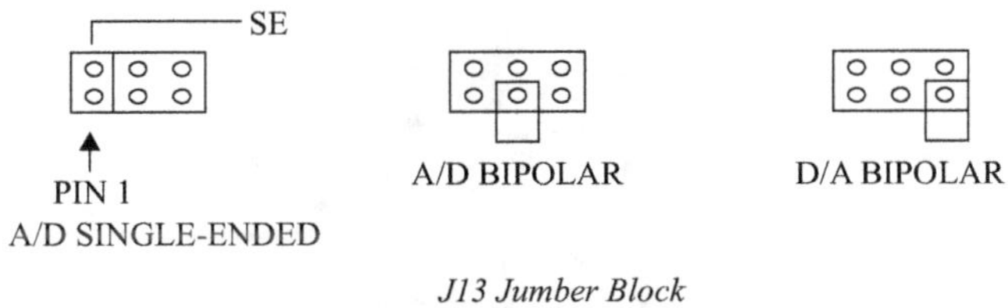

Figure 4.6 J13 Jumper block setting

Jumper block J13 (as shown in Figure 4.6) is used to configure the A/D and D/A circuits. Athena II can accept both single-ended and differential inputs. A single-ended input uses two wires: input and ground. The measured input voltage is the difference between these two wires. A differential input uses three wires: input (+), input (−), and ground. The measured input voltage is the difference between the (+) and (−) inputs. Differential inputs are frequently used either when the grounds of the input device and the measurement device (Athena II) are at different voltages, or when a low-level signal is being measured that has its ground wire. A differential input also has higher noise immunity than a single-ended input because most noise affects both (+) and (−) input wires equally to cancel out in the measurement. The disadvantage of differential inputs is that only half as many are available because two input pins are required to produce a single differential input. Athena II can be configured for either 16 single-ended inputs or eight differential inputs, as shown below. The default setting is single-ended mode.

- A/D—Analog input configuration settings: The analog inputs can be configured for either unipolar (positive input voltages only) or bipolar (both positive and negative input voltages). For bipolar inputs, omit the jumper. The default configuration is bipolar mode (jumper out).
- D/A—Analog output configuration settings: The four analog outputs can also be configured for unipolar (positive voltages only) or bipolar (both positive and negative output voltages). In unipolar mode, the outputs range between 0 and 10 V. In bipolar mode, the outputs range between –10 and +10 V. For the bipolar mode, the outputs should be configured for a mid-scale reset because 0 V is halfway between −10 V and +10 V, for the ±10 V range.

To increase the assembly process, reduce the number of internal cables, and improve the reliability, the CPU and I/O stacks are inserted into the enclosure and bolted into place with direct connections to both the top and bottom of the enclosure for extra rigidity. The rear/bottom panel provides convenient holes for mounting the stack as shown in Figure 4.7. The enclosure is known as Pandora. The Pandora enclosure is available in multiple lengths up to ten inches to support stacks of arbitrary size and is compatible with the Diamond Systems' SBCs: Athena II, Helios, Athena, Prometheus, and Elektra. The Pandora enclosure system can also

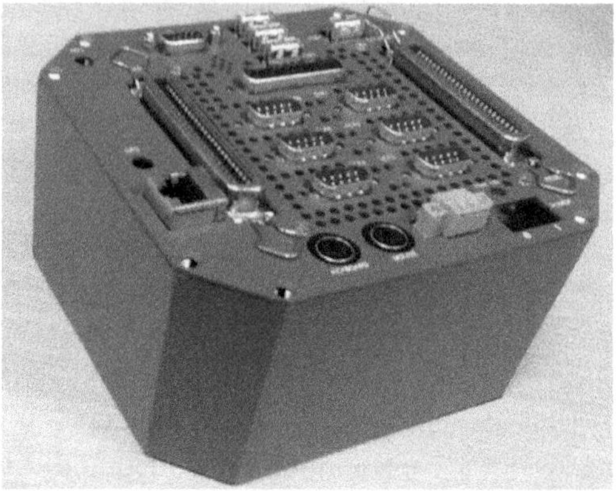

Figure 4.7 Pandora enclosure that houses CPU and panel I/O card

be used with other PC104 SBCs. This generic version provides a flexible enclosure for arbitrary PC104 stacks. It consists of the following parts:

- ATHM500-256A (Athena II SBC, 500 MHz, 256 MB RAM, DAQ);
- PB-EAP-300-K (3″ Pandora Enclosure with Panel I/O board, Panel I/O front panel and Endcap);
- CF-ADA-CF adaptor board 40P,44P, and FDD type power input connector without cables);
- PS-5V-04 (45 W AC/DC Universal Power Adaptor (5 V DC/9 A).

The details of the configuration and assembly of Athena II and Pandora can be found in https://www.diamondsystems.com/files/binaries/Pandora%20Enclosure%20Assembly%20Instructions%20v2.0.pdf

The signals handled by microprocessors need to be conditioned for efficient processing. The signals from sensors may be nonlinear, noise-filled, or weak in magnitude. Such signals should be modified and magnified to suit the specification rating of the microprocessors. These signal-conditioning functions are most commonly done using operational amplifiers of various types. The microprocessors should be protected from excessive voltage and reverse the polarity of the signal using proper electronic protection circuits. The signal conditioning is accomplished by the interface systems inserted between the sensors/actuators and the microprocessors. Most of the interfacing circuit uses the operating amplifier to condition the signals.

An H-bridge (see the schematic in Figures 4.8 and 4.9) is an electronic circuit which enables a voltage to be applied across a load in either direction. These circuits are often used in applications to allow DC motors to run forwards and backward. H-bridges are available as integrated circuits-SN754410 or L298 or can be built from discrete components using IRF3205 Power MOSFET, 1N5818 Schottky Diode, and LM7805CV linear 5 V voltage regulator and resistors.

The H-Bridge arrangement is generally used to reverse the polarity of the motor, but can also be used to "brake" the motor, where the motor comes to a sudden stop, as the motor's terminals are shorted, or to let the motor "free run" to a stop, as the motor is effectively disconnected from the circuit. Table 4.1 summarizes the operation.

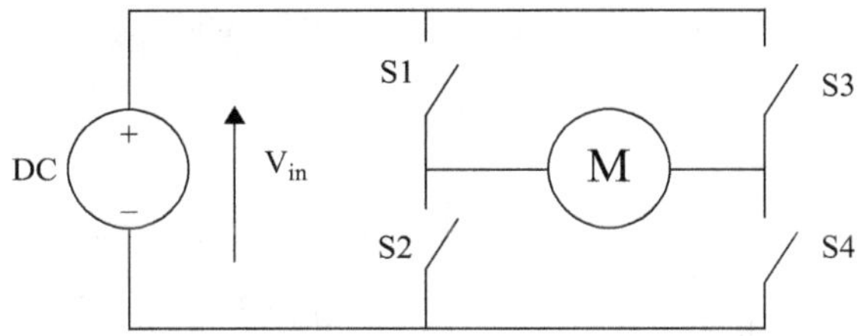

Figure 4.8 Schematic diagram of H-bridge circuit

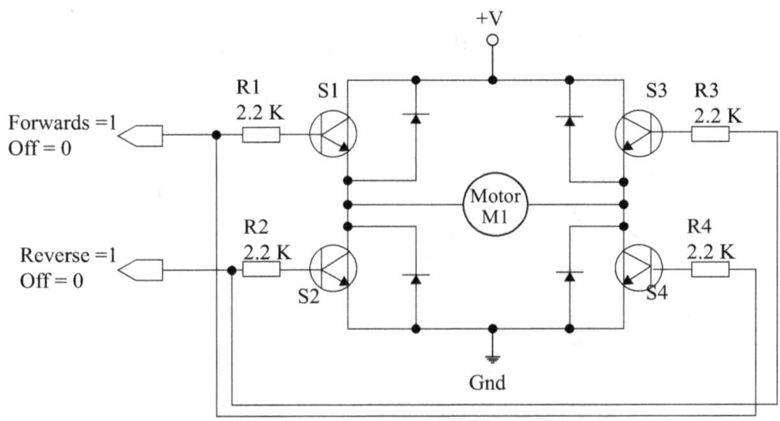

Figure 4.9 H-bridge circuit

Table 4.1 H-bridge operation procedure

S1	S2	S3	S4	Result
1	0	0	1	Motor moves right
0	1	1	0	Motor moves left
0	0	0	0	Motor free runs
0	1	0	1	Motor brakes
1	0	1	0	Motor brakes

Besides directional change, the speed of the motor can be varied using the digital method. There are few ways to implement the digital speed control. The pulse-width modulation method is used to control the speed of the motor. In a nutshell, PWM is a way of digitally encoding analog signal levels. Through the use of high-resolution counters, the duty cycle of a square wave is modulated to encode a specific analog signal level. The PWM signal is still digital because, at any given instant of time, the full DC supply is either entirely on or off. The voltage or current source is supplied to the analog load using a repeating series of on and off pulses. The on-time is the time during which the DC supply is applied to the load, and the off-time is the periods during which that supply is switched off. Given sufficient bandwidth, any analog value can be encoded with PWM. Figure 4.10 shows three

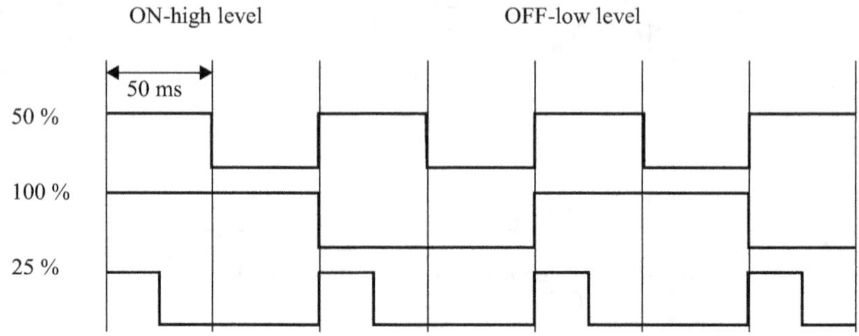

Figure 4.10 PWM output signals

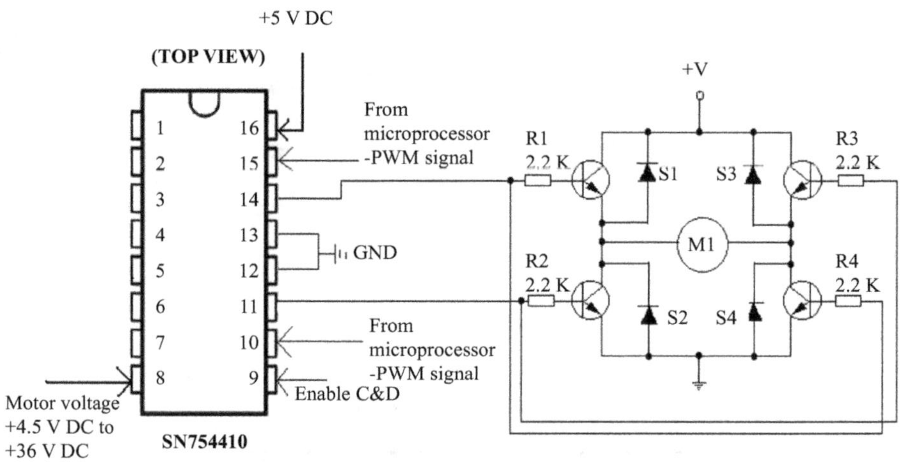

Figure 4.11 DIPIC-SN754410 connection to H-bridge

different PWM signals. These three PWM outputs encode three different analog signal values, at 50%, 100%, and 25% of the full strength.

One of the ways is to implement the PWM on the DC motor (see Figure 4.11) is using a dual inline package (DIP) integrated circuit—SN754410 together with the H-bridge circuit as shown earlier. Sometimes, L298 which is also a dual H-bridge driver for DC brushed motors (and even stepper motors) can be used.

However, it is time-consuming to design the H-bridge circuit for the DC brush-type motor. The 12A8 PWM servo drive is designed to drive brush-type DC motors at a high switching frequency without the hassle of soldering and designing the circuitry. The drive is adequately protected against overvoltage, undervoltage, overcurrent, overheating, and short-circuits across the motor, ground, and power leads. Furthermore, the drive can interface with digital controllers or be used standalone and requires only a single unregulated DC power supply. The DC motor is connected to the 12A8 PWM driver as shown in Figures 4.12 and 4.13. For the

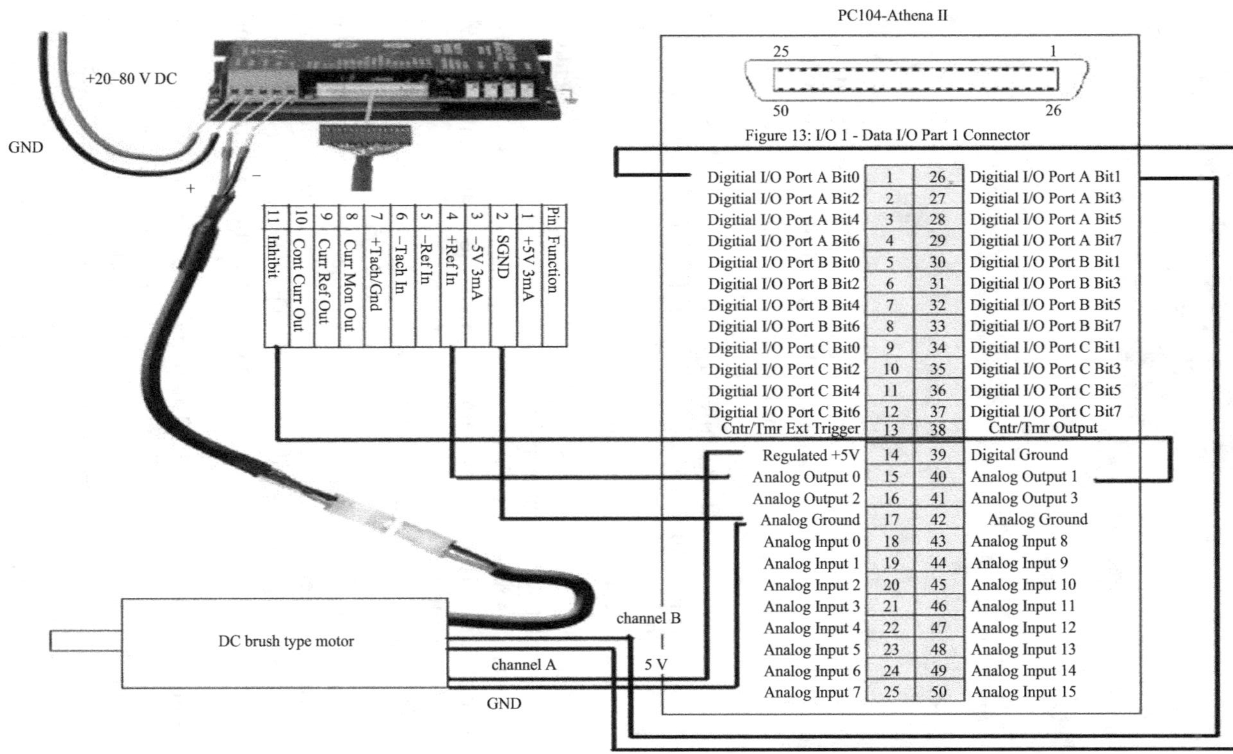

Figure 4.12 12A8 PWM driver connection to DC motor (brush type)

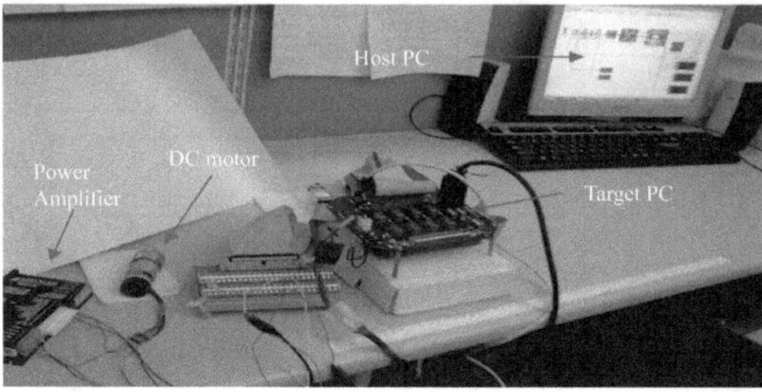

Figure 4.13 Image of HIL setup for DC motor control

12A8 motor driver, it is essential to offset the gain such that at 0 V, the motor is not rotating. The loop gain is set to around 3 to 4.5 clockwise (CW) turns to ensure the loop gain is not too high. In this case, a positive voltage rotates the shaft counterclockwise (CCW) while negative voltage rotates the shaft in CW direction. It is a nonlinear effect called hysteresis (forward is not equal to reserve) and a dead band around 1 V. As observed, as voltage increases, the speed becomes less accurate.

The PINs used (see Table 4.2) on the P1 connector are as follows:

- **Single-ended command signals** (+REF IN, Pin 4). The single-ended command signal is dictated by the capabilities of the controller, for this case is the PC104-Athena II. Analog ± 10 V command signals are used for the velocity or speed control mode. The 0 V gives no rotation to the motor shaft, $+10$ V gives CCW rotation while the -10 V gives CW rotation.
- **Ground** (SIGNAL GND, Pin 2). Almost all signals between the drive and the controller are referenced to signal ground. Without this reference, the drive and the controller would not be able to transmit signals to each other. To ensure that the signals between the drive and the controller are referenced to the same potential, the signal grounds on the controller and the drive must be connected. To avoid ground loops, do not add a connection if there is already continuity between the two grounds.
- **Inhibit line** (INHIBIT IN, Pin 11). The inhibit line is used to turn off power to the motor while the drive is still powered on. Sometimes this is necessary if power to the motor needs to be removed quickly or if the user needs to manually move the load in a freewheeling condition. If your controller has an inhibit function, then we highly recommend that you use it. Note that the inhibit input is configured to disable the drive when pulled low (active low).
- **5 V supply** (+5 V, 3 mA OUT, Pin 1). The +5 V supply can be used for an encoder that requires a +5 V supply to the hall sensor. The ground for the encoder can be connected to the PC104-Analog Ground.

Table 4.2 Pins location on 12A8 PWM driver (cite: http://www.researchconcepts.com/Files/20a14.pdf)

Connector	**Pin**	**Name**	**Description/Notes**	**I/O**
	1	+5 V OUT		O
	2	SIGNAL GND	Internal DC-to-DC converter, outputs regulated voltages of ±5 V @ 3 mA for customer use. Short-circuit protected.	GND
	3	−5 V OUT		O
	4	+REF IN	Differential analog input, maximum ±15 V, 50K input resistance.	I
	5	−REF IN		
	6	−TACH IN	Maximum ±60 VDC, 60K input resistance.	I
	7	+TACH (GND)		
	8	CURRENT MONITOR OUT	This signal is proportional to the actual current in the motor leads. Scaling is 2 A/V for 12A8 and 4 A/V for 25A8, 20 A14, and 20 A20.	O
P1	9	CURRENT REFERENCE OUT	Command signal to the internal current loop. The maximum peak current rating of the amplifier always equals 7.25 V at this pin. See current limit adjustment information below.	O
	10	CONTINUOUS CURRENT LIMIT	Can be used to reduce the factory-preset maximum continuous current limit.	I
	11	INHIBIT	This TTL level input signal turns off all four power devices of the "H" bridge drive when pulled to ground. This inhibit will cause a FAULT condition and a red LED. For inverted inhibit inputs, see section "G".	I
	12	+INHIBIT	Disables the amplifier for the "+" direction only. This inhibit will not cause a FAULT condition or a red LED.	I
	13	−INHIBIT	Disables the amplifier for the "-" direction only. This inhibit will not cause a FAULT condition or a red LED.	I
	14	FAULT OUT (red LED)	TTL-compatible output. It becomes high during output short-circuit, overvoltage, overheating, inhibit, and during "power-on reset". Fault condition is indicated by a red LED.	O
	15	NC	Not connected	
	16			
P2	1	−MOTOR	Motor minus connection	O
	2	+MOTOR	Motor plus connection	O
	3	POWER GROUND	Power ground	GND
	4	POWER GROUND	Power ground	GND
	5	HIGH VOLTAGE	DC voltage input	I

The PINs used on the P2 connector are as follows:

- **−MOTOR** (Pin 1). Connect to the M+ (red cable) of the DC motor;
- **+MOTOR** (Pin 2). Connect to the M- (black cable) of the DC motor;
- **POWER GROUND** (Pin 3). Connect to the negative end of the power supply;
- **HIGH VOLTAGE** (Pin 5). Connect to the positive end of the power supply.

Velocity mode controls the motor such that the motor velocity is proportional to the input command, ±10 V. Switches used for the voltage mode are as follows:

- SW1 On;
- SW2 Off;
- SW3 Off;
- SW4 Off.

Potentiometers setting for the voltage mode are as follows:

- Pot 1. Current limit—no change;
- Pot 2. Loop gain—no change;
- Pot 3. Reference gain—4 turns CW*;
- Pot 4. Offset set for no torque (i.e., motor's shaft not rotating) when the input voltage at +REF = 0.

To maintain consistency in the number of turns, the initial starting point is defined as follows:

- Turn the pot CCW until the pot begins to click on every turn;
- Continue to slowly turn the pot CCW until the next click is heard, then stop;
- Now turn the pot in the CW direction as per the number of turns indicated.

4.4 Hardware-in-the-loop testing using xPC-Target

The xPC-Target is tightly integrated into the MATLAB®/Simulink® environment and allows Simulink® models to be compiled into executable code and run on a proprietary real-time operating system. It is a solution for rapid prototyping, hardware-in-the-loop testing of control systems, and deploying real-time systems using standard PC hardware such as Athena II. Using standard PC hardware and off-the-shelf I/O boards, xPC-Target turns a standard PC into a real-time rapid prototyping Target. "Real-time" means that code is executed at very regular intervals, very useful for control.

Very often, PC104 is also known as Target PC. It separates from a Host PC (where the xPC-Target software is installed) for running real-time applications. The xPC-Target software environment includes the real-time kernel, real-time applications, and I/O drivers. As seen in Figure 4.14, the hardware environment consists of a Host PC, Target PC (or I/O boards in the Target PC), and a serial or Ethernet connection between the Host and Target PC. The real-time kernel and PC hardware enable xPC-Target to provide a cost-effective, high-performance, real-time target.

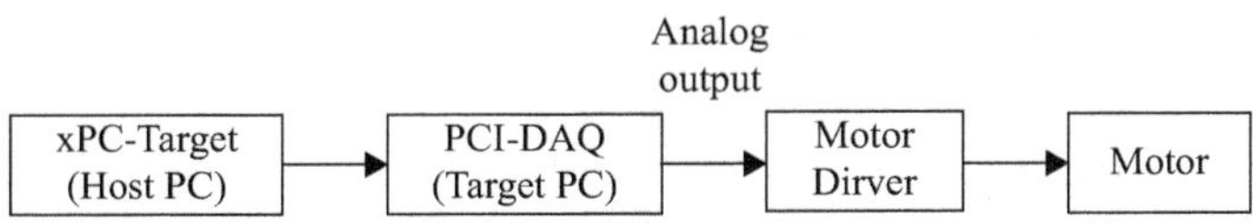

Figure 4.14 xPC-Target (as Host PC) and Desktop PC-NI PCI card (as Target PC)

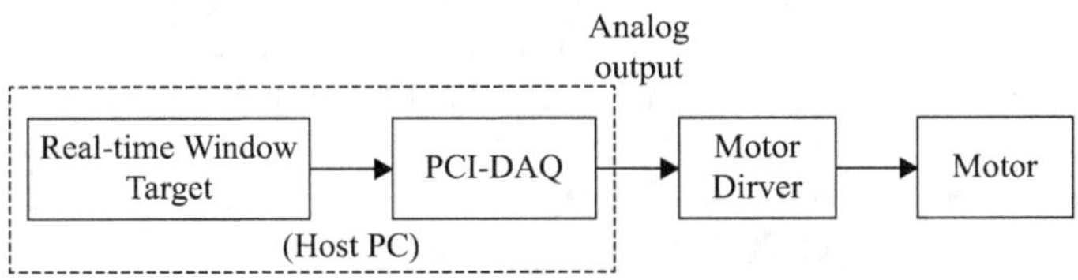

Figure 4.15 Real-time Window Target (as Host PC) and NI PCI card (in Host PC). Note that the user can reduce to just one PC

Alternatively, the user can use a desktop PC with an I/O boards like PCI DAQ card (e.g., NI-PCI 6023E, 6221) inserted into the empty PCI slot as a Target PC in Figure 4.15. Instead of using an extra desktop PC as the Target device, one can use the real-time Window Target. The PCI card can be slotted into the Host PC and thus make the overall control systems more compact to use.

In general, the steps in using the xPC-Target from preparing the xPC-Target using the Simulink® block diagrams and lastly to the downloading and running in the real-time mode are as follows:

- **Create xPC-Target real-time kernel** that requires formatting of the hard disk in the xPC-Target to the DOS partitioned format and ensures that it is bootable using the DOS operating system. It also involves the configurations of the Basic Input/Output System (BIOS) settings. As shown in Figure 4.23, the CompactFlash disk is used in place of a hard disk. It requires the card to be configured as a bootable disk. We will format the card so that it will automatically boot into xPC real-time kernel when the board is powered.
- **Create xPC-Target Simulink® block diagrams** for the HILS systems. In all configurations, a Host PC is installed with MATLAB®. The models will be created through the Simulink® blocks and the xPC-Target block sets. It can add I/O blocks to the model, and then use the Host PC with Real-Time Workshop and a C/C++ compiler to create executable code in the Target PC. As seen in the later part, programs are created through simple drag-and-drop procedures and require minimal "coding" except a few standard setup procedures that is required to perform.
- **Download and run in real time**. The executable code is then downloaded from the Host PC to the Target PC running the xPC-Target real-time kernel. After downloading the executable code, Target application can be run and tested in real time. It can then be triggered to run in standalone or externally using the Simulink® External mode on the Host PC.

4.4.1 Create xPC-Target real-time kernel using desktop PC as target PC

The following is the steps required if the Desktop PC is used as the Target PC. The first step is to format the Target PC. Before formatting the hard drive, ensure that the data are backed up. The setup for the Target PC using the DAQ PCI card is shown in Figure 4.16.

In the computer system's BIOS, select booting up using the CD-ROM Drive. This may be done by restarting the computer and pressing the "F2," "F8," or "delete" key. Then go to the BIOS menu and from there; go to the "advance BIOS" feature and set the CD-ROM boot as the first boot and the hard drive as the second boot. Save the configuration by pressing the "F10" key and pressing "Y" to exit. The Target PC has to be DOS partitioned for at least 5 to 10 GB. To format the hard disk in the Target PC, use the following options.

1. Before starting the system, make sure to insert the Window 98/2000/XP Operating System's installer CD/DVD-ROM drive. Instructions on how to boot from the CD are usually included on the Windows installation program.
2. The hard drive may also be formatted through the Computer Management Program. To do this, click the "start" key, enter "settings," and click on the "control panel" icon shown on the screen. The "Computer Program Management" may be found on the control panel menu. Under the "Computer Program Management," click on "Disk Management" and choose the hard drive that needs reformatting. Right-click on the chosen hard drive and select "Format".
3. The last way to format a hard drive is through a floppy diskette. Make sure to get a bootable floppy diskette. This diskette must be capable of booting in MS-DOS mode. The PC will be in the DOS mode after the bootup process is completed. Type the syntax "format c:" (note that the letter in the command may change depending on the letter of drive the hard drive is located).

Figure 4.16 Step up of the target PC (in the desktop PC)

Next, ensure that the Host PC is installed with MATLAB® consisting of the xPC-Target, xPC-Target embedded option, and Real-Time Workshop software (see Figures 4.17 to 4.20). The current version of xPC-Target is 4.0. Connect the crossover cable from Host PC to Target PC.

1. Double-click on **MATLAB®** icon on your **desktop**.
2. In the MATLAB® command window, Type> **xpcexplr**. Then xPC-Target Explorer window opens.
3. In the **xPC Target Explorer**, expand a **TargetPC1** node. A configuration node appears. Under this are nodes for **Communication, Settings**, and **Appearance**. The parameters for the Target PC node are grouped in these categories.
4. Select> **Communication**.
5. From the **Host Target communication** list, select>**TCP/IP**.
6. You must enter the network properties with the correct values according to your environment.
   - The Target PC Target PC IP address: **192.168.0.222**.
   - LAN subnet mask address—This is the subnet mask address of your LAN. The subnet masks address: **2555.255.255.0**.
   - Select> TCP/IP Target driver: **I8245X** (an Intel 8254X family PCI Fast Gigabit Ethernet).

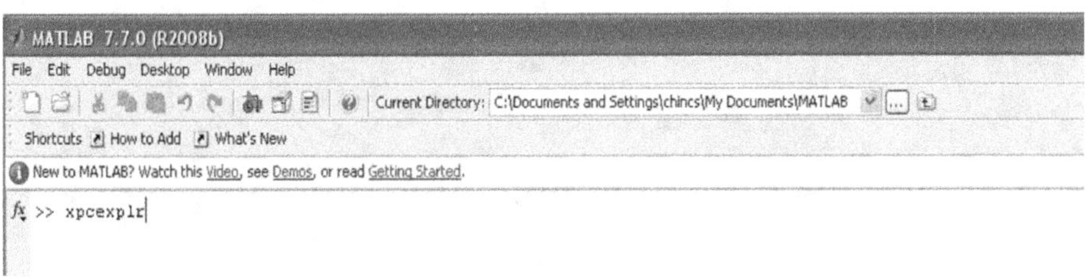

Figure 4.17 Screenshot 1

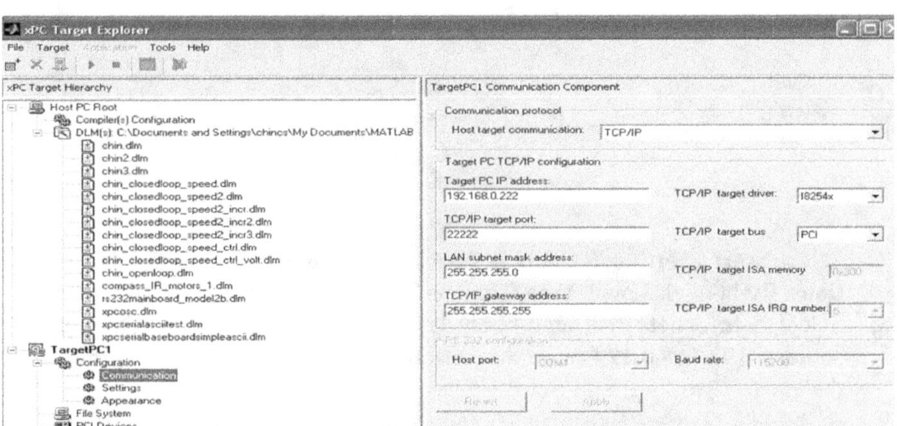

Figure 4.18 Screenshot 2

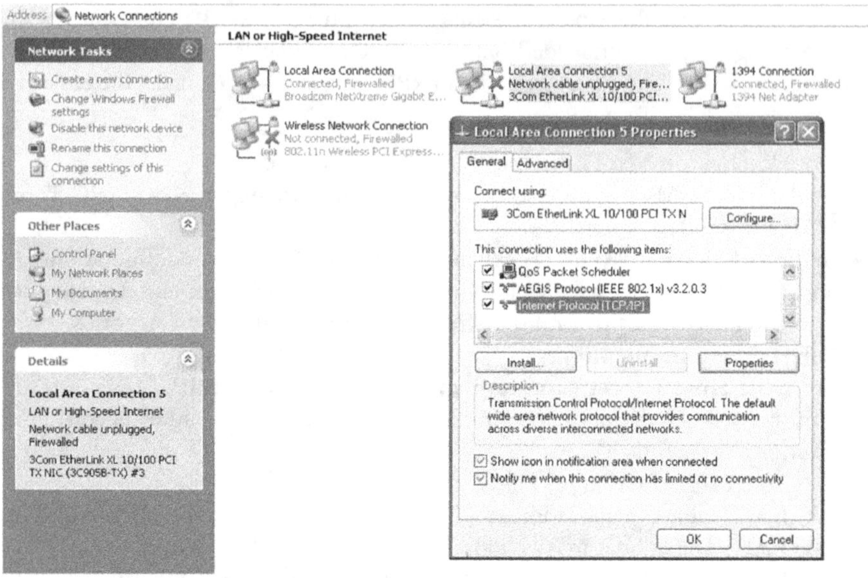

Figure 4.19 Screenshot 3

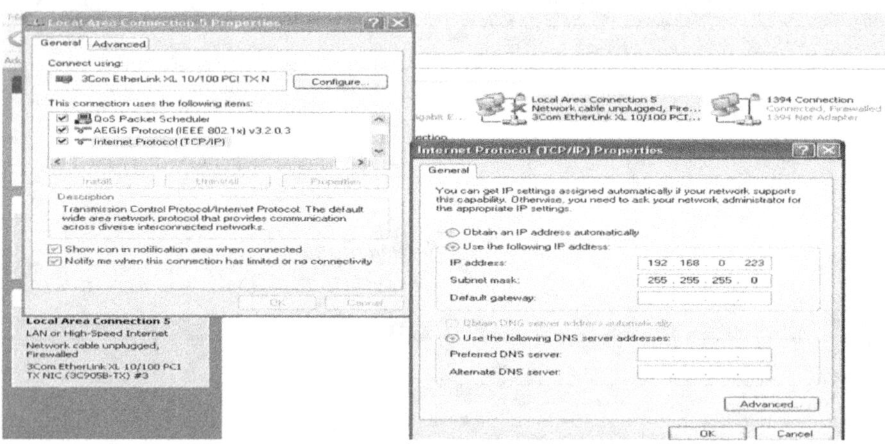

Figure 4.20 Screenshot 4

7. Select **Start>Settings>Control Panel**, then double-click **Network Connections.** Right-click **Local Area Connection**, then Select> **Properties**.
8. Select >**Internet Protocol (TCP/IP)**, then click **Properties**. Set the IP address: **192.168.0.223** (note that differences in the last value as compared to the Host PC setting) and Subnet mask:**255.255.255.0**.
9. In the **xPC Target Explorer** Hierarchy pane, select the C**onfiguration** node for **TargetPC1**.

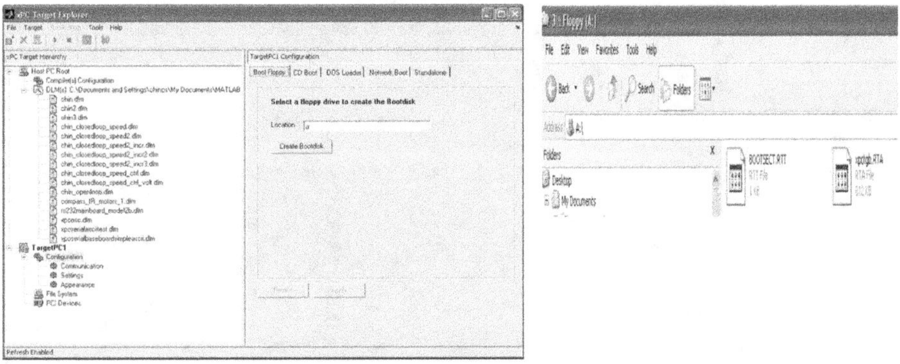

Figure 4.21 Screenshot 5

10. A **TargetPC1 configuration** pane appears in the right-most panes (see Figures 4.20 and 4.21). This pane contains a series of tabs. Select> **Boot Floppy**.
11. Insert the floppy diskette into A:\ and click **Create Bootdisk**.
12. Select> **DOS loader** and save into location in the Host PC folder. For example, enter **C:\cschin\xPCfile**. Click > **Create DOS Loader**.
13. Copy these contents (except the xPCtgo.rtb) into the A:\ .
14. Try to boot the Target PC using the floppy diskette drive (removed the CD if any).
15. Alternatively, to boot using the hard disk, copy the contents in A:\ into Target PC. Include the xPCtgo.rtb into the Target PC as well.
16. Remove the floppy diskette and reboot the Target PC. You will see the following real-time kernel screen. You can only see this screen if the VGA monitor is connected to the VGA port on the Target PC.

Finally, if the users have to remove or change the xPC-Target real-time kernel setup, create a MS-DOS bootable diskette so that the Target PC can boot up to allow the content in the Target PC drive to be changed.

To create the **MS-DOS bootable diskette**, follow the steps below:

- When formatting a floppy diskette, users have the option of creating a MS-DOS startup diskette, follow the below steps to do this;
- Place diskette in the computer;
- Open My Computer, right-click the A: drive and click Format;
- In the Format window, check Create an MS-DOS startup diskette;
- Click Start.

4.4.2 Create xPC-Target real-time kernel using Athena II-PC104 as target PC

The following are the steps (see Figures 4.22 to 4.25) required if Athena II PC104 (and the Compact Flash Card I) is used as the Target PC. If the Compact Flash disk is to be used as a book disk into xPC (or any other OS), the CF card are not set up for this usually. Here, we will format the disk to the proper file format and enable

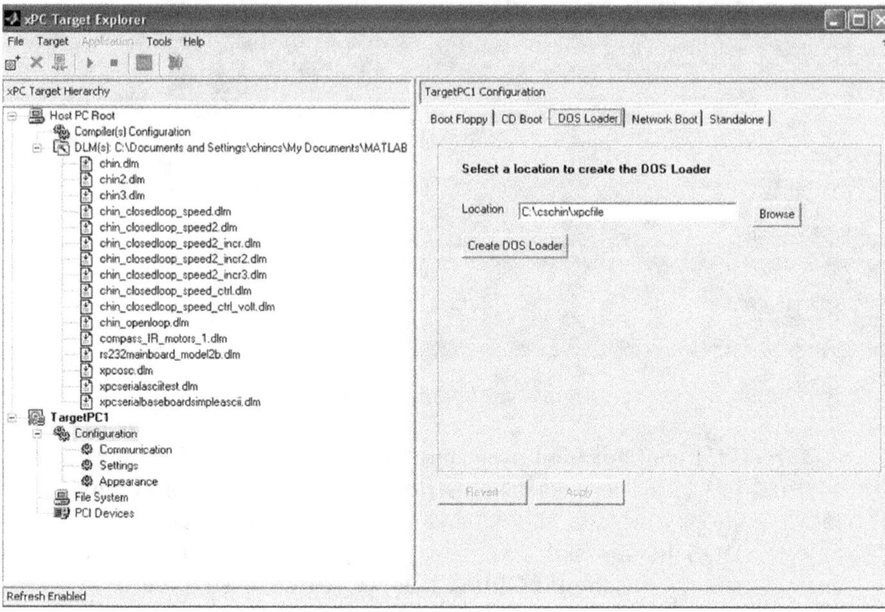

Figure 4.22 Screenshot 6

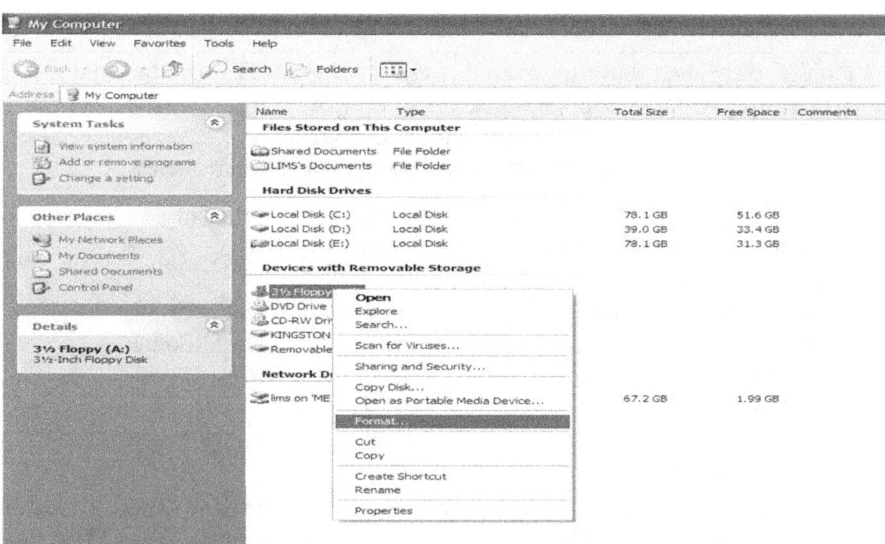

Figure 4.23 Screenshot 7

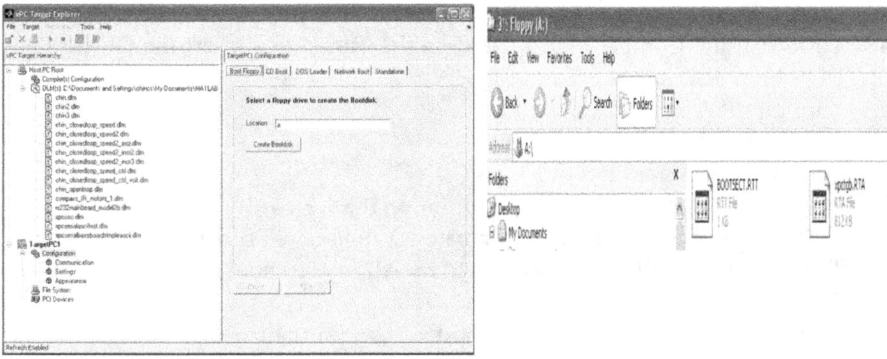

Figure 4.24 Screenshot 8

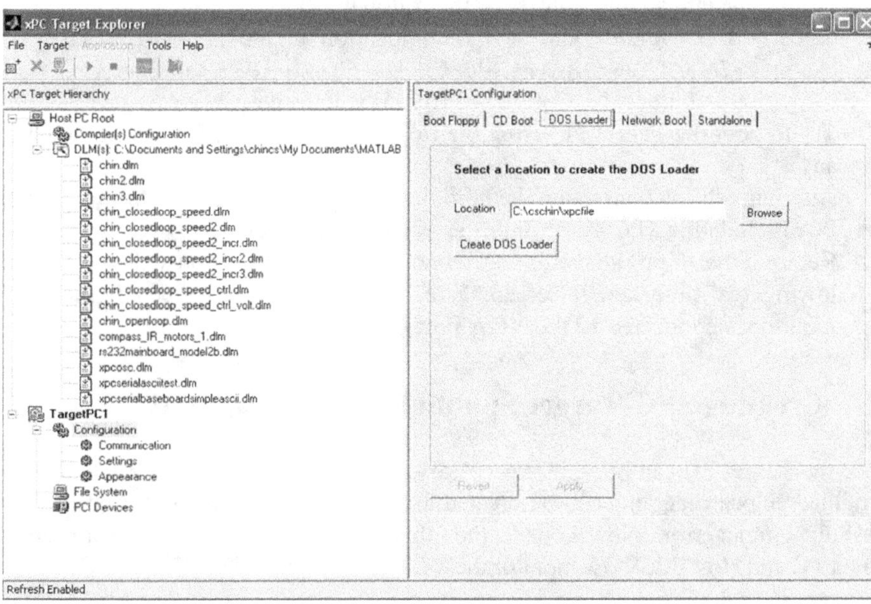

Figure 4.25 Screenshot 9

the Master Boot Record (MBR) allowing us to boot off it. The CF adapter is connected directly to the IDE bus on the Athena II board. It will be drive C:\ (if no other hard drive is present).

1. Create a Bootable Floppy in Windows. Insert a floppy diskette into the Windows PC. Format the floppy by right-clicking on the disk drive in **My Computer** and selecting **Format**. Do not check "quick erase."
2. Run the file boot600.exe. (download from http://www.instructables.com/file/F7CQIP8FLG5OREF/.)

3. Click OK when prompted. Once finished, the floppy is formatted.
4. To format the Compact Flash Card on the Target PC, insert the CF card and DOS boot disk into the Target PC.
5. Turn on the Target PC.
6. Let the Target PC boot into DOS.
7. Enter the FDISK utility: A:\ fdisk.
8. Use the menu to delete every partition on the CF card.
9. After all, partitions are deleted, create a single new Primary DOS partition.
10. The Target PC will have to be restarted. Allow it to boot off the floppy again.
11. Reset the MBR by typing: A:\ fdisk /mbr.
12. Now format the CF as a DOS boot disk: A:\ format c: /s.
13. Remove the floppy disk and reboot to make sure the drive boots correctly.
14. Turn off the Target PC.
15. Insert the floppy disk into A:\. Click on the Boot Floppy. After a while, the floppy diskette should consist of the following files.
16. Select> **DOS loader** and save into location in the Host PC folder. For example, enter **C:\cschin\xPCfile**. Click > **Create DOS Loader**.
17. Copy these contents (except the xPCtgo.rtb) into the A:\.
18. Try to boot the Target PC using the floppy diskette drive (removed the CD if any).
19. Alternatively, to boot using the hard disk, copy the contents in A:\ into Target PC. Include the xPCtgo.rtb into the Target PC as well.
20. Remove the floppy diskette and reboot the Target PC. You will see the following real-time kernel screen. You can only see this screen if the VGA monitor is connected to the VGA port on the Target PC.

4.5 Creating xPC-Target Simulink® block diagrams

Once the Target PC can boot, it is necessary to configure the Host PC such that the Simulink® block diagrams can be created and downloaded into the Target PC successfully without errors. Now, connect the Ethernet cable (crossover type) between the Target PC and Host PC, VGA monitor the VGA port, keyboard to PS/2 on PC104 (see Figures 4.26 to 4.42).

1. Double-click on **MATLAB®** icon on your **desktop**.
2. If xPC-Target Explorer is not already open, in the **MATLAB® Command** Window, Type **xpcexplr**. The xPC-Target Explorer Window appears.
   - xPC-Target Explorer always has a default Target PC node in its Configuration. The default Target PC node is always bold faced. In a multi Target environment, this visual aid helps you easily identify the default Target PC.
3. In the xPC-Target Explorer Window, select> **Compiler(s) Configuration** node. In the right pane, the compiler parameters appear.
4. At the **Select, C Compiler** drop-down list, select the compiler you have installed on the Host PC. The path to the compiler is **C:\Program Files\Microsoft Visual Studio 9.0**.

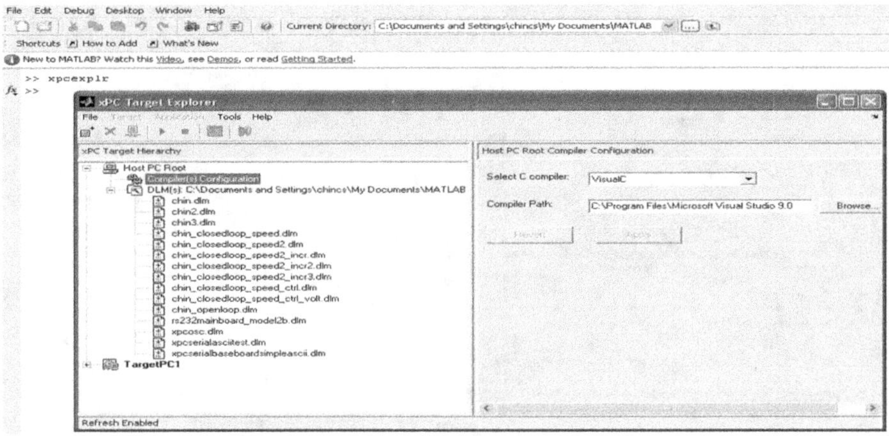

Figure 4.26 Screenshot 10

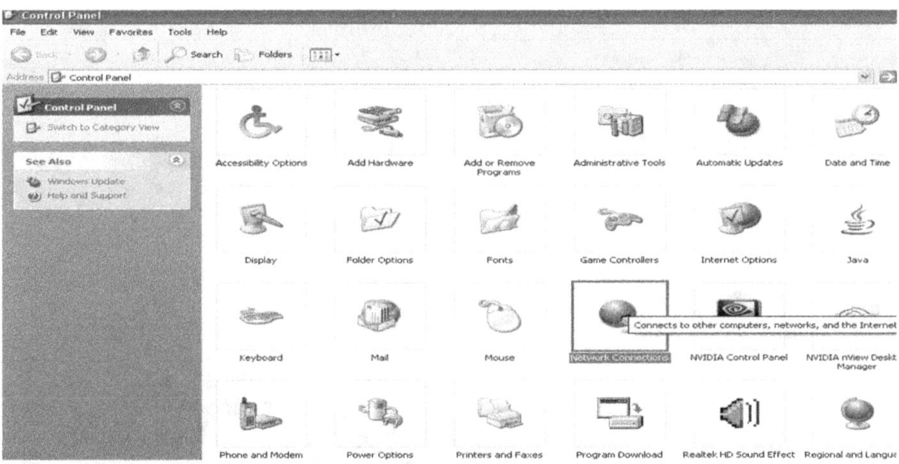

Figure 4.27 Screenshot 11

5. Select **Start>Settings>Control Panel**, then double-click **Network Connections.** Right-click **Local Area Connection**, then Select> **Properties**.
6. Select >**Internet Protocol (TCP/IP)**, then click **Properties**. Set the IP address: **192.168.0.223** (note that differences in the last value as compared to the Host PC setting) and Subnet mask:**255.255.255.0**.
7. In the **xPC Target Explorer**, expand a **TargetPC1** node. A configuration node appears. Under this are nodes for **Communication, Settings**, and **Appearance**. The parameters for the Target PC node are grouped in these categories.

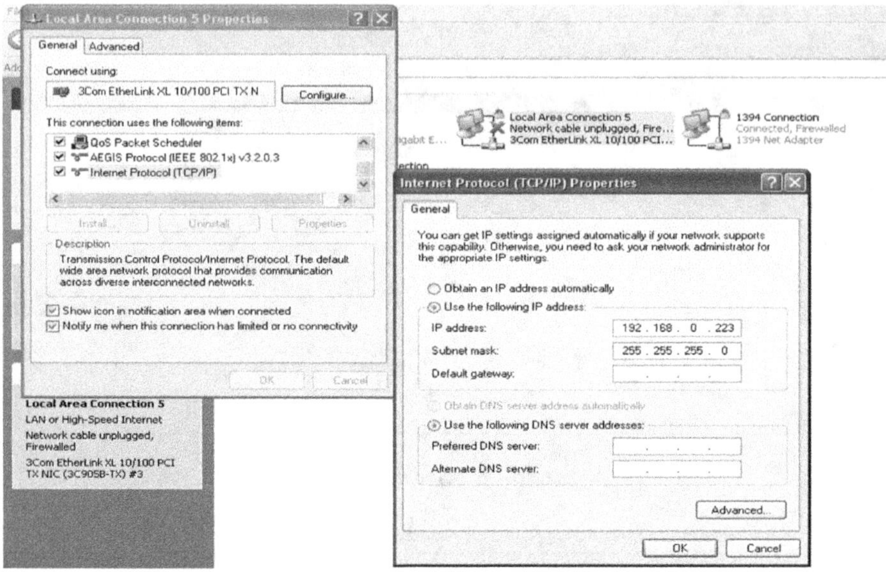

Figure 4.28 Screenshot 12

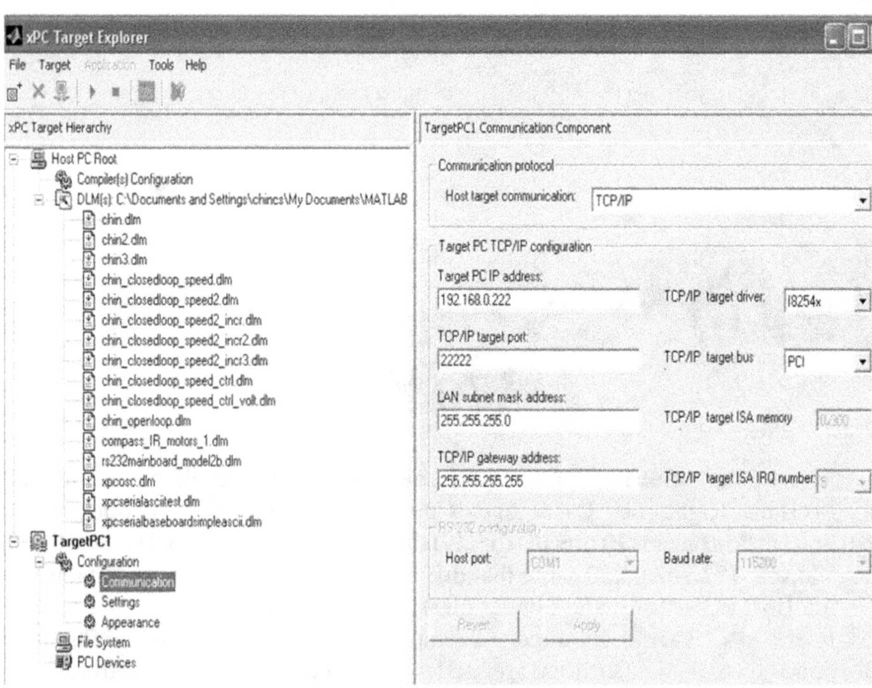

Figure 4.29 Screenshot 13

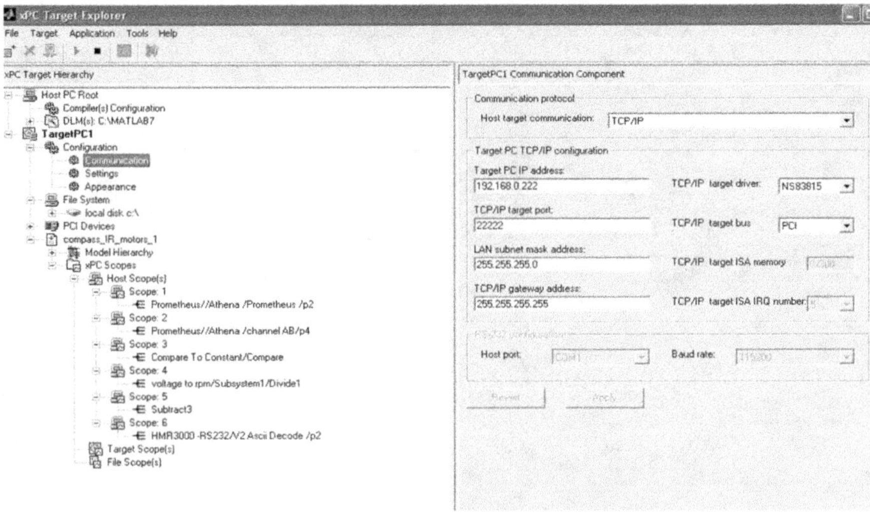

Figure 4.30 Screenshot 14

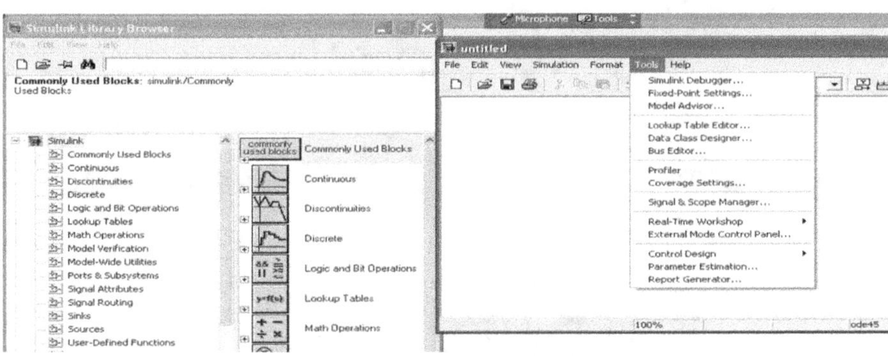

Figure 4.31 Screenshot 15

8. Select> **Communication**.
9. From the **Host Target communication** list, select>**TCP/IP**.
10. You must enter the network properties with the correct values according to your environment.
    - The Target PC Target PC IP address: **192.168.0.222**.
    - LAN subnet mask address—This is the subnet mask address of your LAN. The subnet masks address: **2555.255.255.0**.
    - Select> TCP/IP Target driver: **NS83815.**
11. In MATLAB® window, click on >**Simulink®** icon. Select>**File**>**New**>**Model**. Save the file.

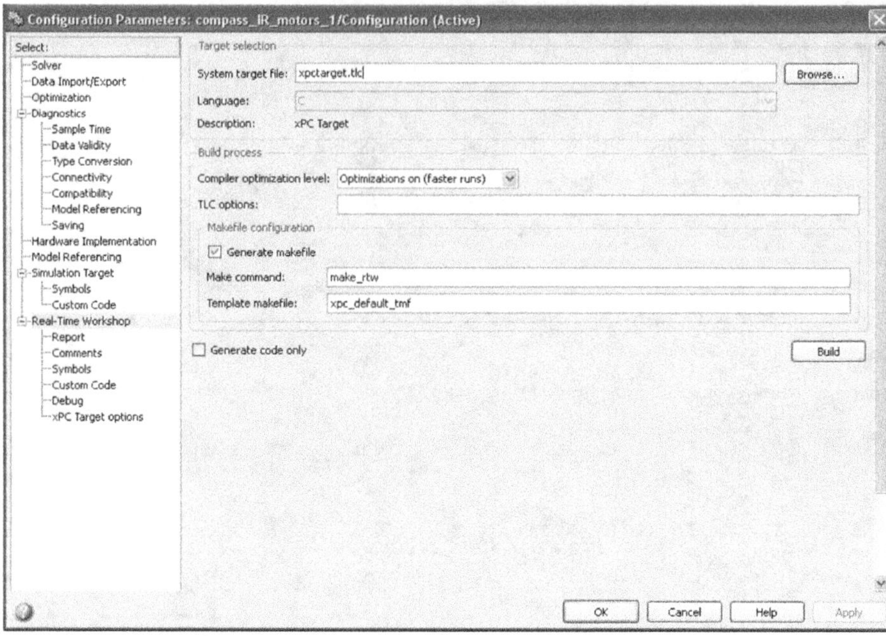

Figure 4.32 Screenshot 16

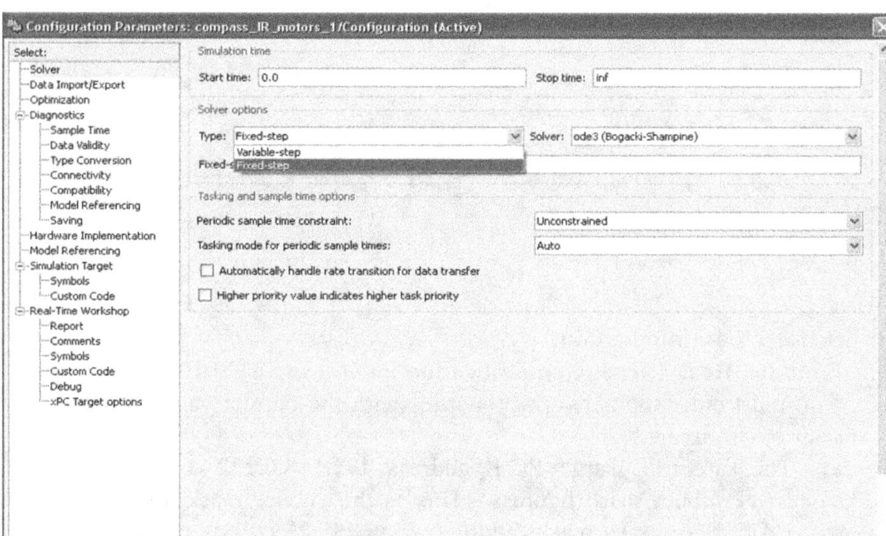

Figure 4.33 Screenshot 17

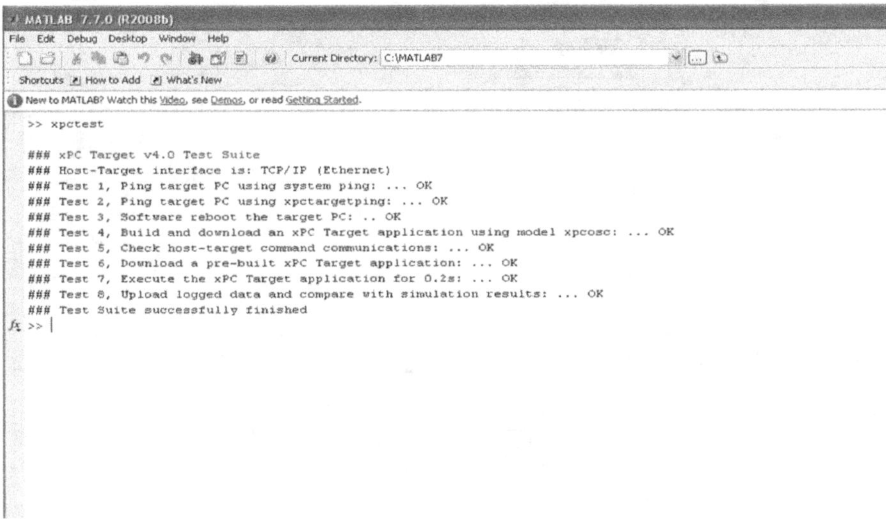

Figure 4.34 Screenshot 18

12. Select **Tools>Real-Time Workshop>Options**. Browse and select system Target file: **xPCTarget.tlc** and select Stop time> **inf.** Then click OK.
13. Under the same pane, select **Solver: Fixed-step**.
14. In the **MATLAB® Command Window**, type> **xpctest**. MATLAB® runs the test scripts for the default Target PC and displays messages indicating the success or failure of a test. If all of the tests succeed, you are ready to build and download a Target application to the Target PC. If any of the tests fails, see the appropriate test section in the "**xPC Target 4 Getting Started Guide.**"
15. Create a Simulink® Model including the encoder model as shown below. The Host/File/Target Scope can be found in the **Simulink® Library>xPC Target>Misc.> Scope (xPC).** Select > **Scope type** to be Host, Target or File. If Target Scope is used, select > **Scope mode> Graphical rolling.**
16. Now the Simulink® is ready to be downloaded and built as executable files. Select>**Tools>Real-Time Workshop>Build Model.**
17. Select> **Tools>Real-Time Workshop>xPC-Target Explorer**. In the pane, right-click a **TargetPC1** and select>**Connect**. You will see the execution time running.
18. Select > **Application** > **Start Application**.
19. In the **xPC Target Explorer**, expand a **TargetPC1** node. **xPC Scopes** node appears. Select > **Host Scope(s)** and right-click **view Scopes**.
20. To change the view mode from graphical to numerical, right-click on the graph. Select>**View Mode>Numerical.**

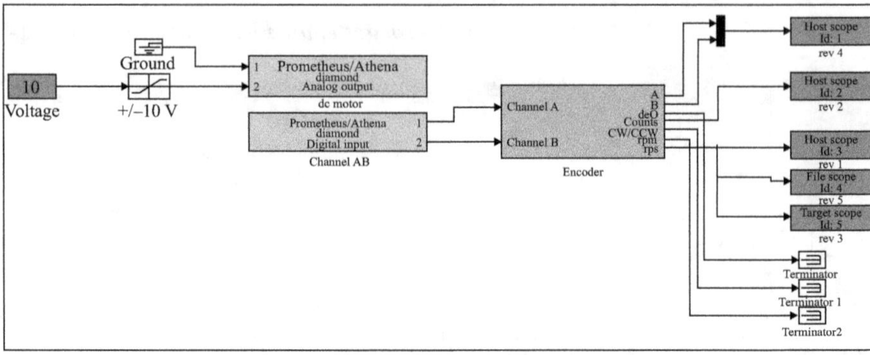

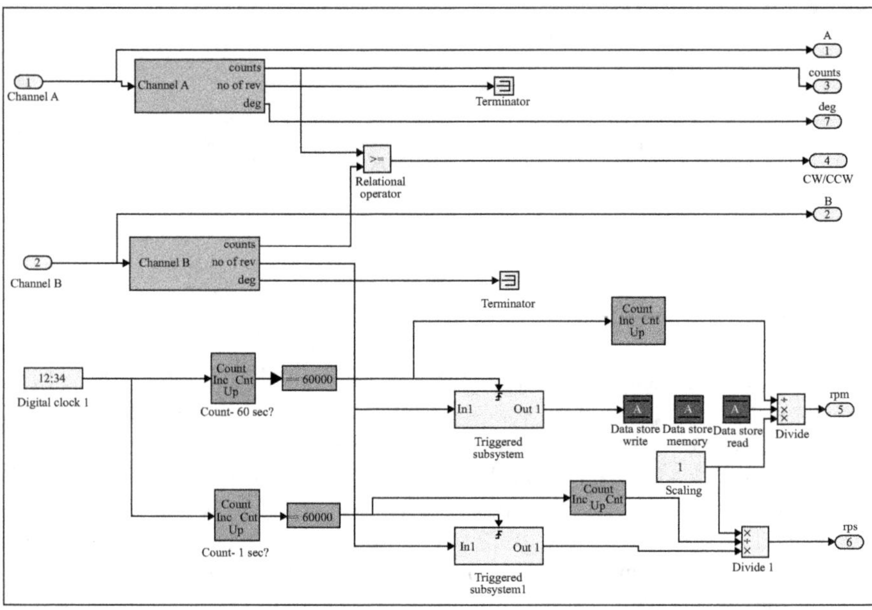

Figure 4.35 Screenshot 19

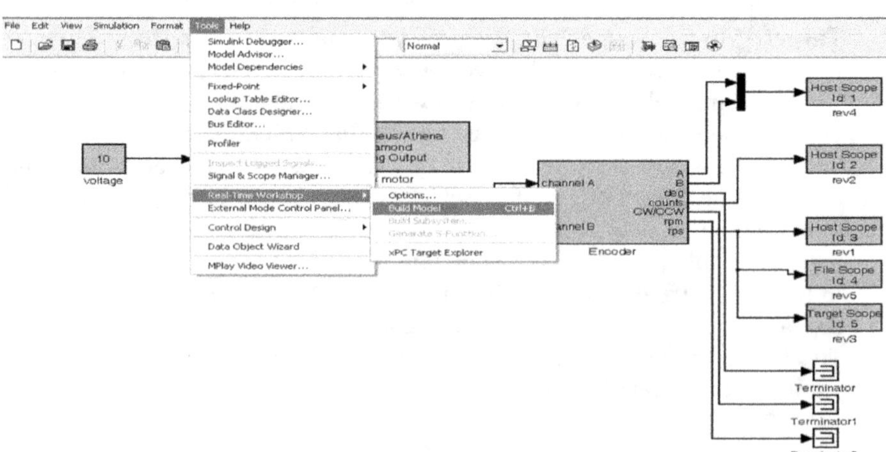

Figure 4.36 Screenshot 20

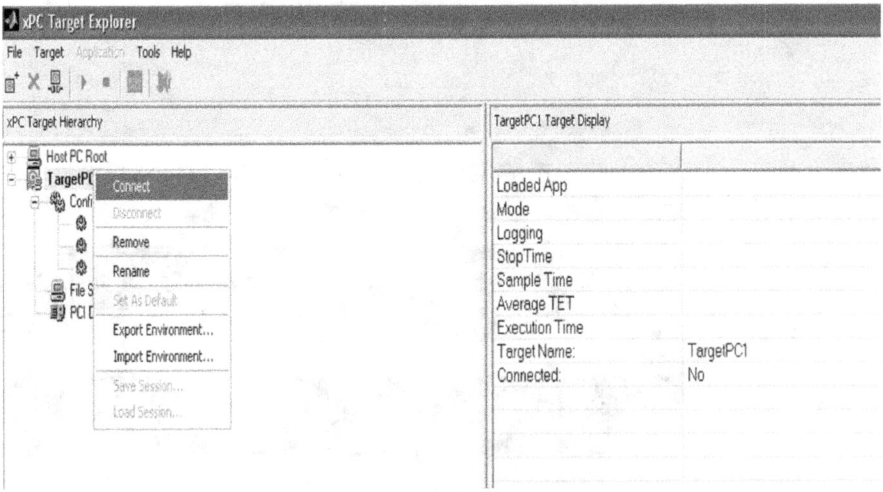

Figure 4.37 Screenshot 21

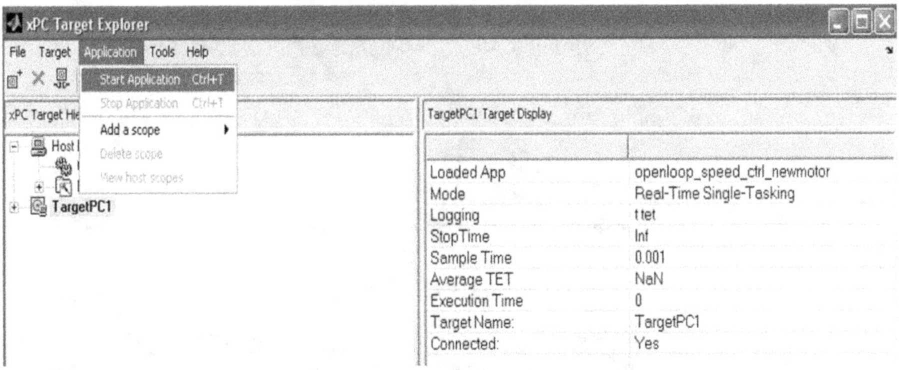

Figure 4.38 Screenshot 22

21. To export the data to MATLAB® Workspace, select > **Application>Stop Application.** In the **xPC Target Explorer**, expand a **TargetPC1** node. **File Scope(s)** node appears. Select > **File Scope(s)>Scope 4** and right-click **Export to workspace**. Accept the default variable name and click **OK** to continue.

    In the MATLAB® Command Window, enter the following commands to plot the graph.

    >> **plot(new_data.data(:,2),new_data.data(:,1),'-') %% [time, data]**
22. To view the Target Scope(s) instead of the Host Scope(s), in the MATLAB® Command Window, enter the following commands>> **xPCTargetspy** to have a snapshot of the Target screen. Alternatively, you can view the Target Scope real-time using the connected VGA monitor screen.

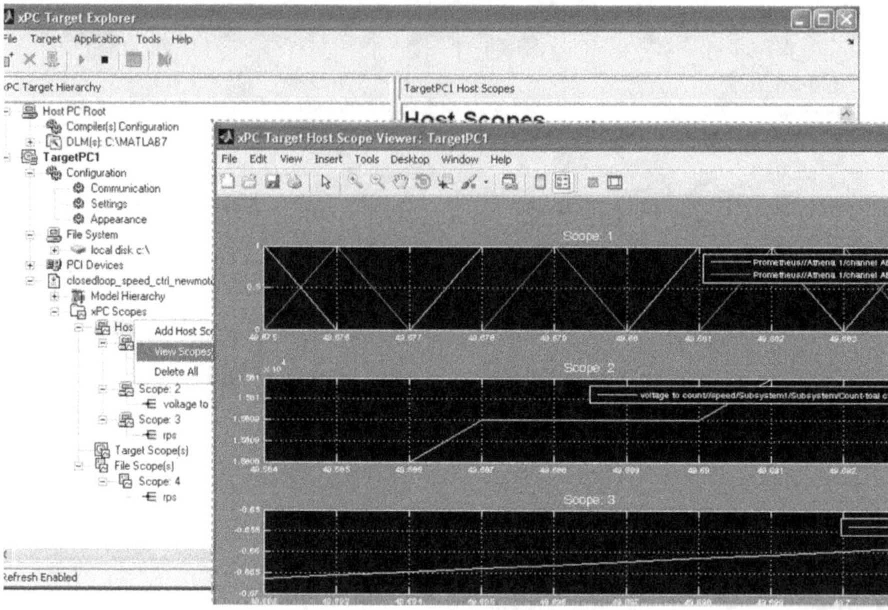

Figure 4.39 Screenshot 23

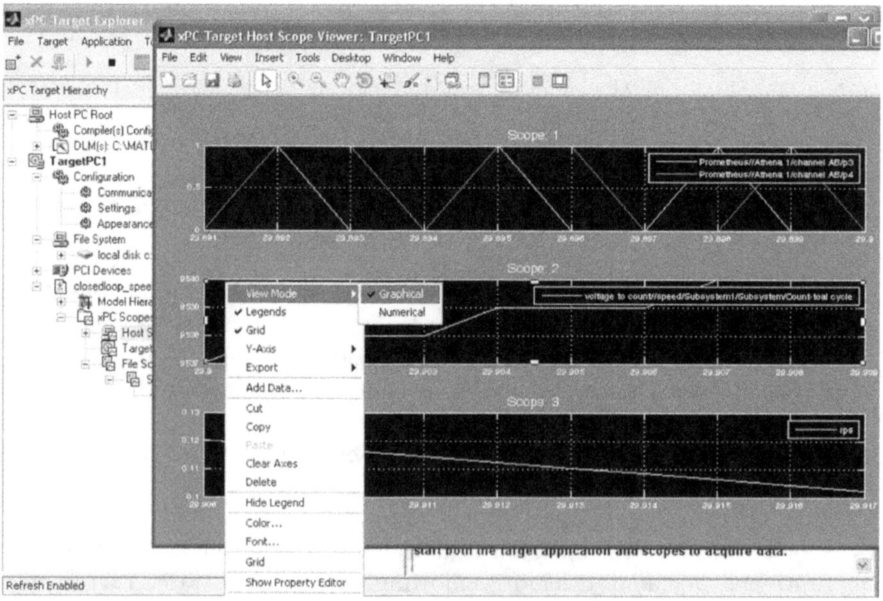

Figure 4.40 Screenshot 24

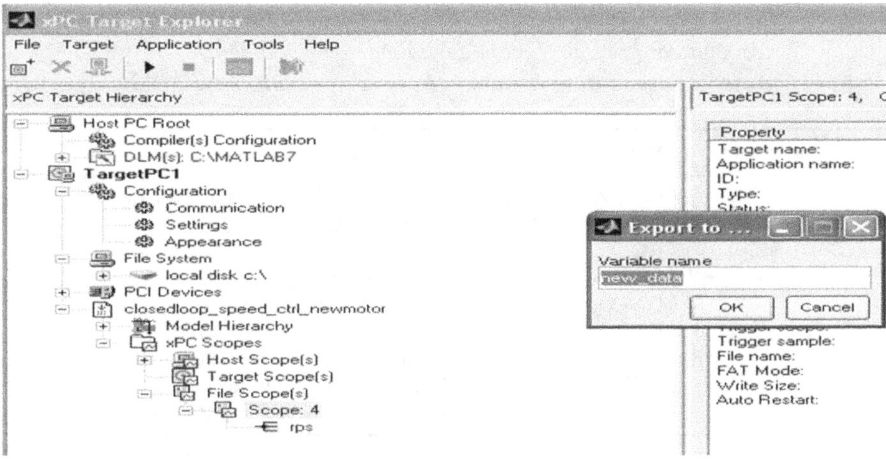

Figure 4.41 Screenshot 25

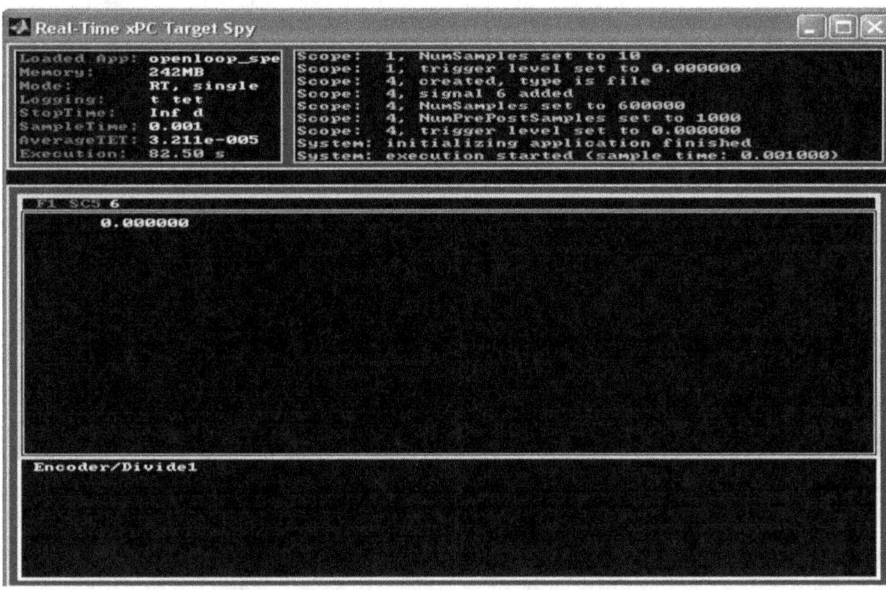

Figure 4.42 Screenshot 26

4.6 Using RS232, analog, and digital I/O in xPC-Target

Table 4.3 shows the input/output pins for HMR3000-RS232 magnetic compass used. The cable is connected to the COM1 (or any other available COM) of the desktop and the other end connected to the compass as shown. The pins information

Table 4.3 Pin assignment of HMR3000Pin assignment of HMR3000 (cite: https://aerocontent.honeywell.com/aero/common/documents/myaerospacecatalog-documents/Missiles-Munitions/HMR3000.pdf)

Pin number	Pin name	Description
1	OP/CAL	Operate/Calibrate (open = operate)
2	TD	Transmit data, RS-485 (B+)
3	RD	Receive data, RS-485 (A-)
4	RDY/SLP	Ready/Sleep input (open = ready)
5	GND	Power and signal ground
6	RN/STP	Run/Stop input (open = run)
7	CT/RST	Continue/Reset (open = continue)
8	+5 V	Regulated power input (+5 V)
9	V+	Unregulated power input (+6 to +15 V DC)

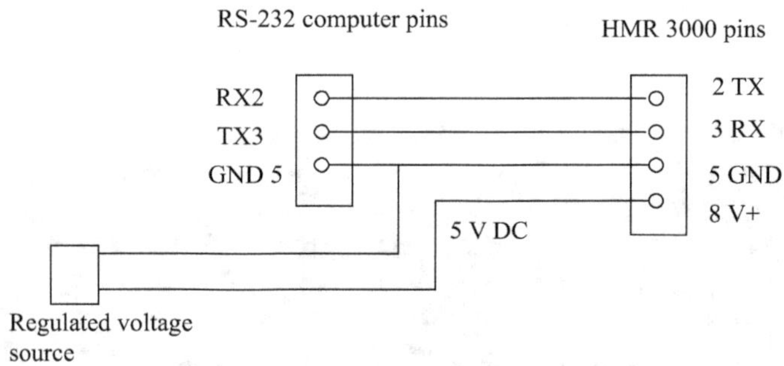

Figure 4.43 RS232 connection to PC

of the compass can be seen in Table 4.3. Since regulated power input of 5 V was used, Pin 8 was connected only. Note that only Pins 2, 3, 5, and 8 were used in the RS232 connection as shown in Figure 4.43.

The compass is oriented in such a way that the heading angle can be registered. In this case, the heading angle is orientated at 20° from the reference (CW direction). The HyperTerminal was used to obtain the raw reading in the heading, roll, and pitch angles. The following settings (as seen in Figure 4.44) are used: Communication Port = COM 1, Bauds rate = 19,200 bits/s, Data bit = 8, Polarity = None, Stop bits = 1, and Flow Control = none. Click "OK" and begin with the data collection as shown in Figure 4.45.

As compared to the software—PCDemo Interface 4.01 given by the vendor, the same readings can be registered as shown.

To avoid port conflict, close the PCDemo Interface 4.01 and use the MATLAB® to verify. Type in the following commands:

```
s1 = serial('COM1', 'BaudRate', 19200)
fopen(s1)
instrfind
fprintf(s1, '$PTNTHPR');idn=fscanf(s1,'%s')
```

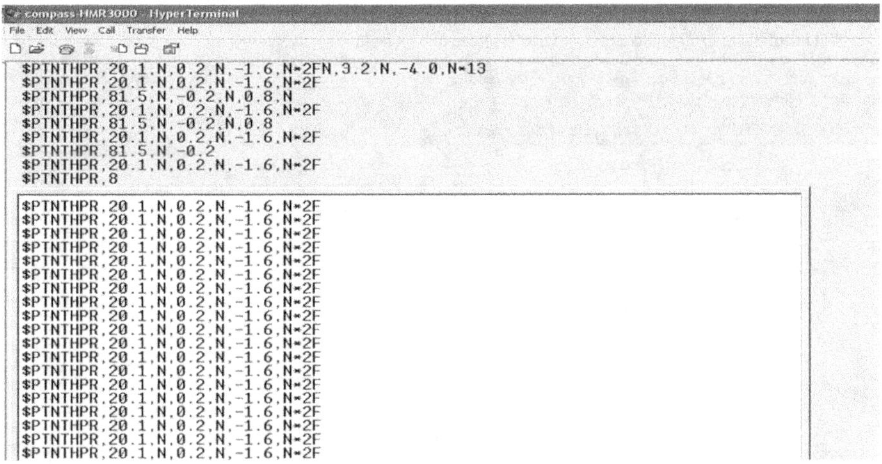

Figure 4.44 Hyper terminal results

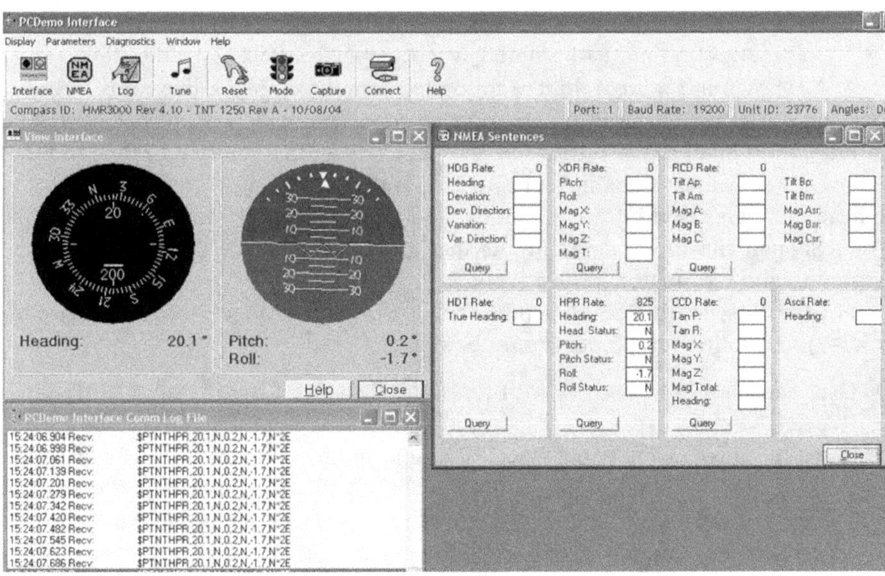

Figure 4.45 PC demo interface program for compass

The result on the MATLAB® command prompt reads: $PTNTHPR,20.1,N,0.2, N,-1.7,N*2E. As observed, it is similar to the results seen earlier. The fprintf command can be reissued to see the latest reading of the compass. If there is an error in the opening of the serial port, exit from the program and try again as the command-fclose does not work. One may ask how to interface with xPC-Target instead of using MATLAB®. The answer is to contact the serial port to the Serial Port 2 on the PC104 as shown. Block diagram for the data acquisition using the serial port is shown in Figure 4.46.

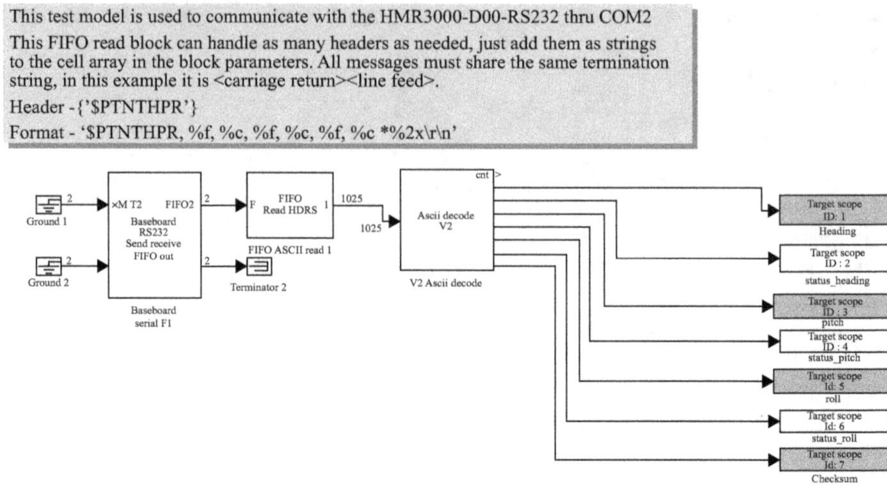

Figure 4.46 xPC-Target block diagram for RS232 data acquisition

There are few points to be noted when using the RS232 in the xPC-Target ver. 4. As shown in Figure 4.46, the following Simulink® block diagrams from the xPC-Target toolbox are used. Users can pull out the block diagrams as shown and configure individual to the settings as shown. Connect the 7 output to the scopes—to show the relevant values in scope 1, 3, 5, and 7 as shown. If one output (that is the heading angle) is required, the block diagram can be simplified to has only one Target Scope. The inputs and outputs that are not used have to be connected to the ground and terminator, respectively.

For the Baseboard RS232 Send/Receive FIFO OUT block diagram (see Figure 4.47), the following points should be noted.

- User can choose to use "w/o "FIFO OUT" that is "Baseboard RS232 Send/Receive" only.
- Since the three connections—RX, TX, and ground—are used, make sure that the auto RTS/CTS checkbox is not enabled. If the control signals are enabled but are not connected, the receiver may ignore all the incoming data.
- Select the **Parameter group** dropdown list. The **Board Setup** and **FIFO Setup**, it should be configured as shown in Figure 4.48.
- Data coming from the sensor are in ASCII format and delimited by a line feed carriage return, hence it has to be set to [10].
- Most serial devices are relatively slow. That means it takes about 100 ms (or even more) before they will answer a command. If your application is running at a sample time of 0.25 ms, it will take 400 (1/0.25 × 100) executions of application until the device replies to the command sent during the first execution (assuming the device needs 100 ms to reply). If the command is at such a high frequency, this may cause an overload of the input buffer of the device or even leads to a crash of the communication controller inside the

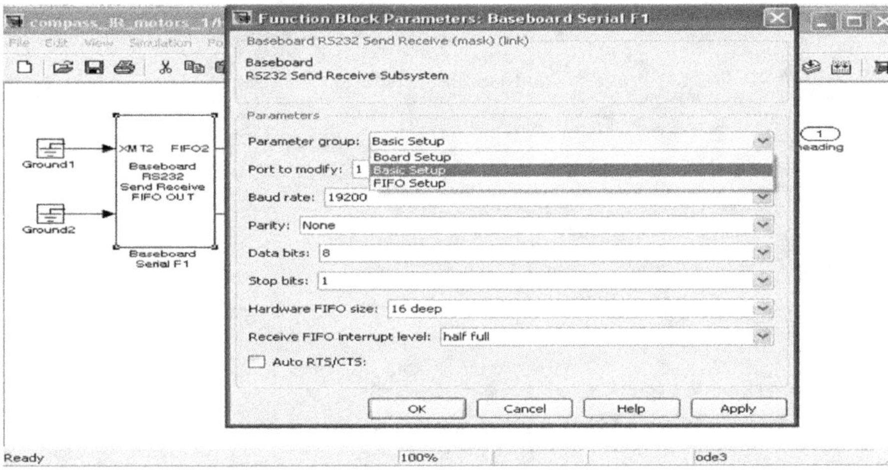

Figure 4.47 Basic setup for Baseboard RS232 Send/Receive

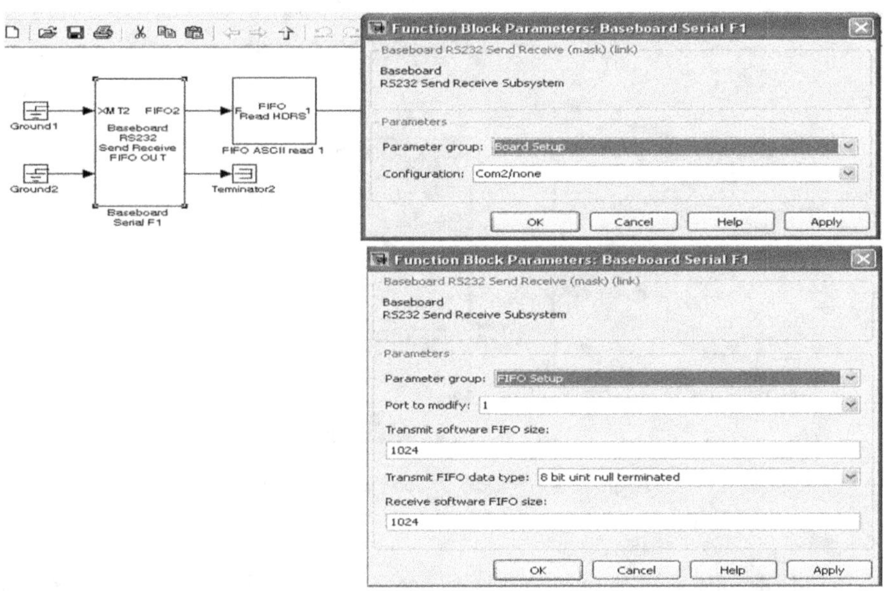

Figure 4.48 Board and FIFO setup for Baseboard RS232 Send/Receive

device. Hence it is not sufficient to send the right values, one should consider the timing too. In this case, the sample time is set as 0.001.

- the format in the ASCII decode v2 block being the single %f while it should be **'$PTNTHPR,%f,%c,%f,%c,%f,%c*%2x\r\n'** (see Figures 4.49 and 4.50).
- Note that the original ASCII decode does not want the format surrounded by single quotes, but the V2 version requires the single quotes.

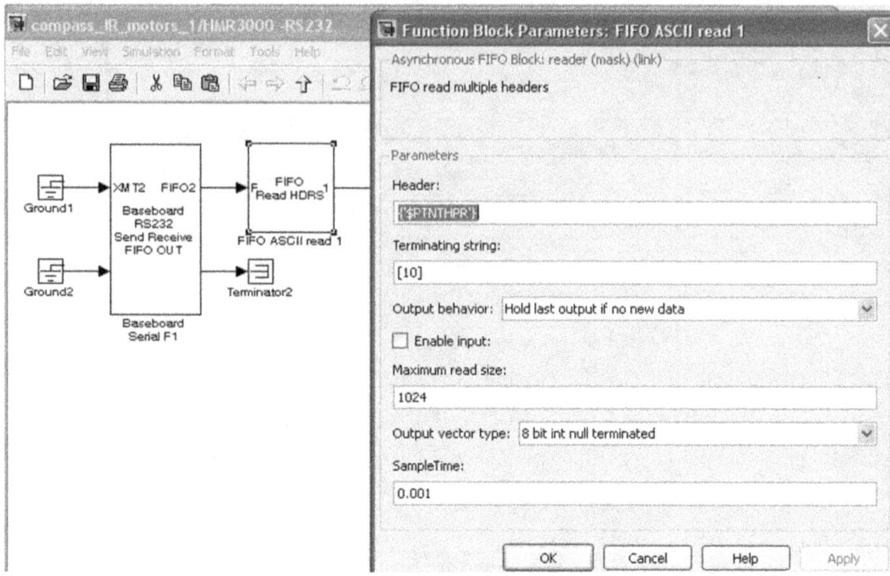

Figure 4.49 FIFO read hdrs setup

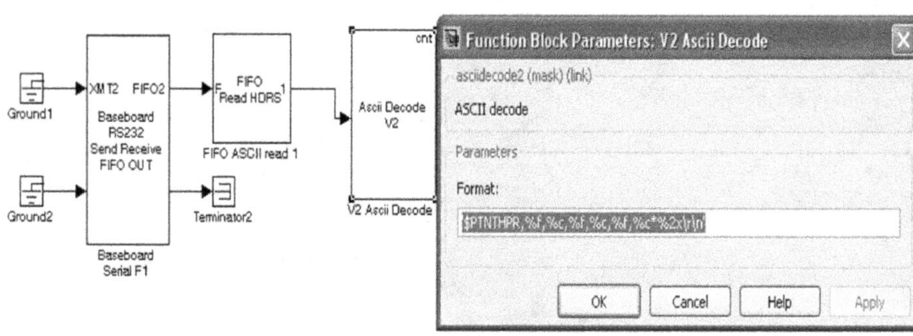

Figure 4.50 ASCII decode V2 setup

- ASCII decode V2 has to be used and not ASCII decode.
- If you are trying to examine the string coming out of the FIFO, read hdrs block by using a Demux block and pick off 10 element sections and send each to a Target scope set to numeric mode. There is an error.

As shown in Figure 4.51, the compass reads 20.1° in the heading direction. It is similar to the previous readings registered using the Hyper Terminal and MATLAB® Command. However, users will find that the xPC-Target is user-friendly as graphical means can be used and easily configured as an input to a control system design (as seen in the later part of the section).

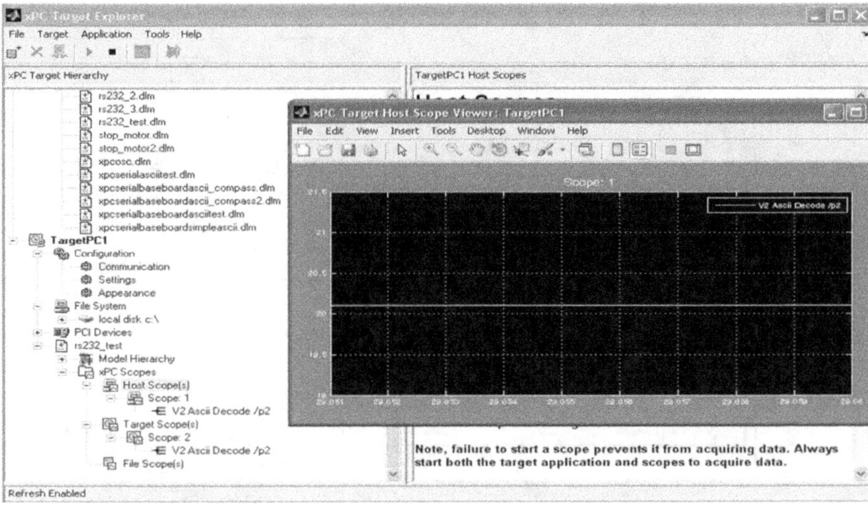

Figure 4.51 xPC-Target host scope viewer

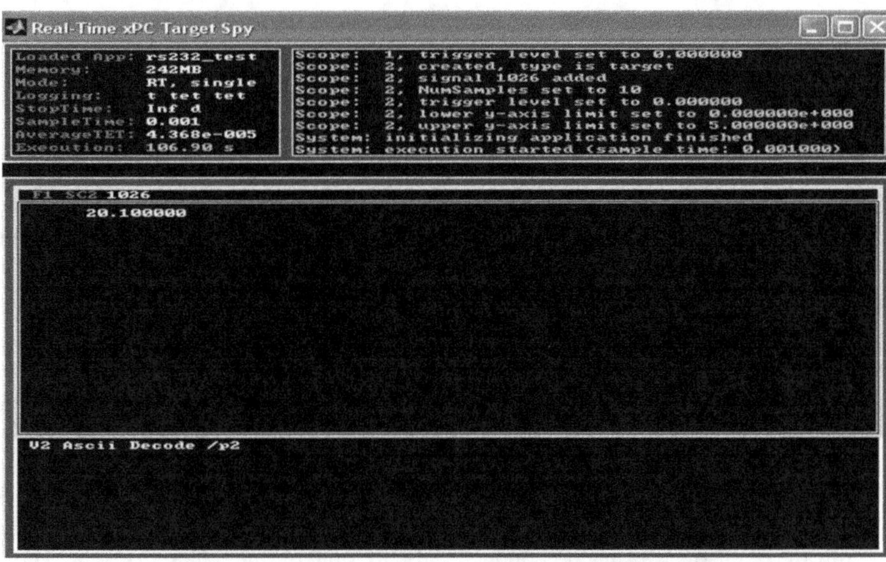

Figure 4.52 Real-time xPC-Target spy screen

To see the Target scope or the scope at the Target PC side, enter: xPCTargetspy at the MATLAB® command prompt. The snapshot of the screen is shown in Figure 4.52 and it can only display the previous reading and it has to be updated continuously to obtain the latest results. The connection diagram can be seen in Figure 4.53.

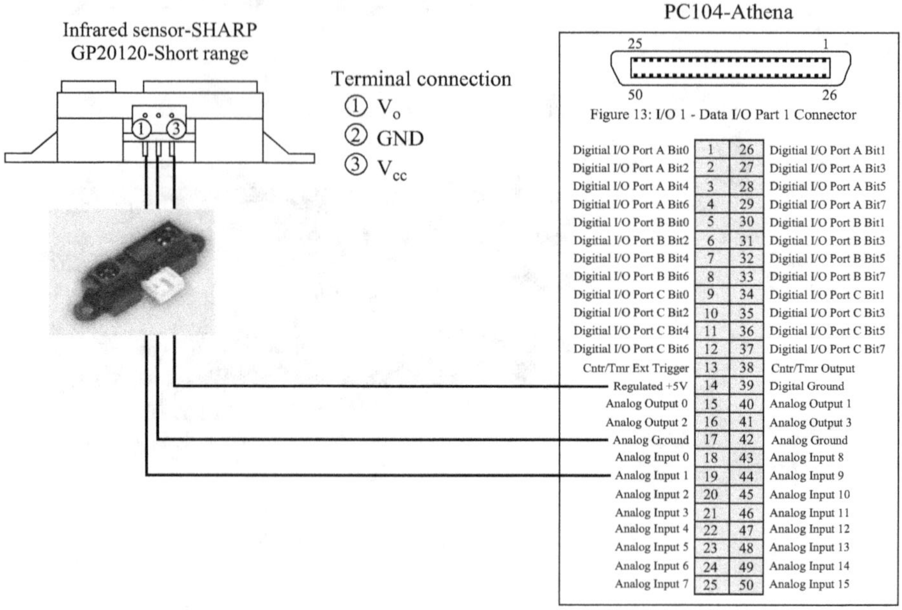

Figure 4.53 SHARP GP2D120 connection to Athena II PC104

4.7 Infrared sensor model

Signal of the infrared sensor was connected to the analog input (VOUT2) of PC104. The single-ended input and the voltage range from 0 to 5 V were used. For obstacle avoidance and to navigate through the arena, infrared sensors provide the control system with distance between the vehicle and the obstacle. The specific infrared sensor type that was purchased was SHARP GP2D120- short range (see https://media.digikey.com/pdf/Data%20Sheets/Sharp%20PDFs/GP2D120.pdf). These detectors offer a small package, a voltage of 5 V DC to operate, minimal current consumption (50 mA), low cost, and output option as an analog input. The operational range is 30–40 cm.

To use the infrared sensor in xPC-Target, the following configurations in Figure 4.54 should be performed. The Analog Input block can be used. Since only one infrared sensor was used, connect only to one of the channels. The Base address is set as 0×280 according to Athena II board specification. The sample time is set as 0.001 s for this case.

One advantage of infrared sensors is that they are immune to ambient light interference. However, the accuracy is affected by the target area's color and texture. Infrared sensors estimate distance in analog form. For example, the distance between the obstacles can be measured using a measuring tape and the corresponding voltage was recorded. Measured voltage versus the distance was plotted

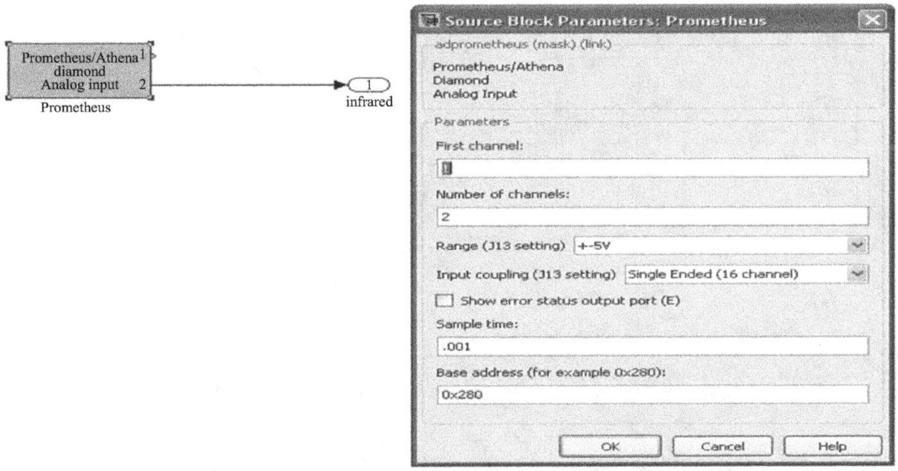

Figure 4.54 Prometheus analog input configuration

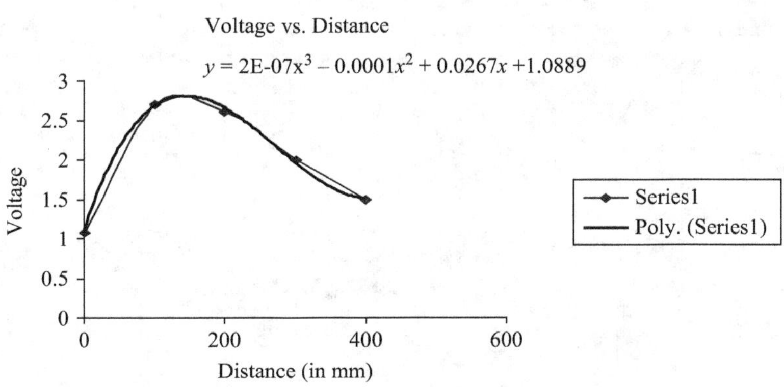

Figure 4.55 Distance vs. voltage sensed

as shown in Figure 4.55. As the distance increases, the voltage decreases. The equation for the infrared sensor can be modeled as a third-order polynomial.

4.8 Incremental encoder model

Optical incremental encoders are a mean for capturing speed and traveled distance on a motor. Counting the pulses (from encoder's channel A and B) indicate how many revolutions of the motor have turned. Rotation velocity can be determined from the number of counts within a given period. By detecting the phase between channel A and B of the encoder, the direction of rotation can be determined. Figure 4.56 shows the incremental encoder attached to the backend of the DC

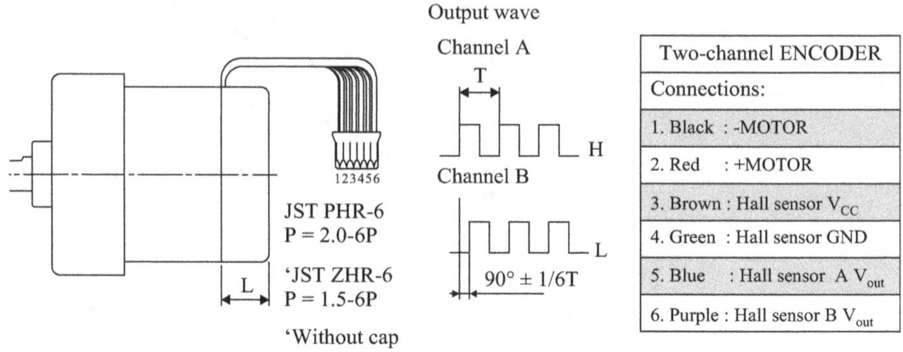

Figure 4.56 Optical incremental encoder connections

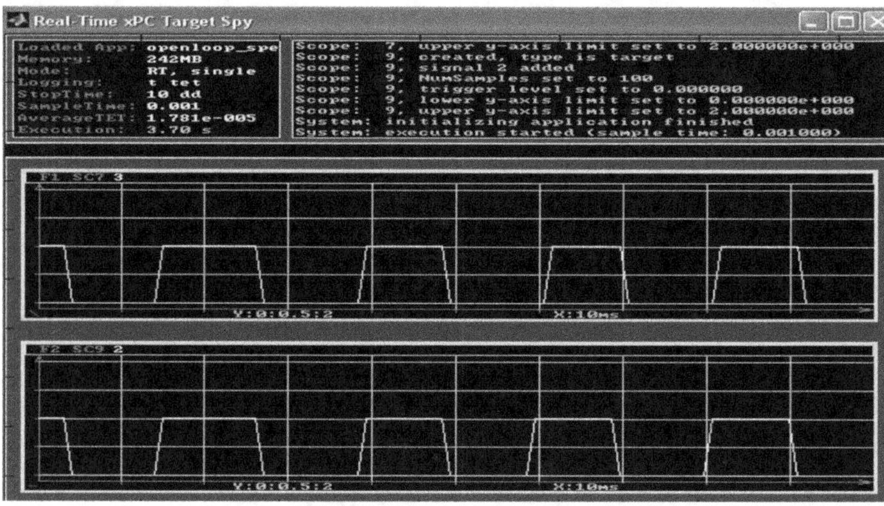

Figure 4.57 Channel A and B outputs from xPC-Target

motor (that is PD3046-12-189-BFEC-Transmotec). The connections for the motor have to be connected correctly. There should be positive and negative connections to the motor.

Since A and B channels are 90° out of phase (see Figure 4.57 captured by the xPC-Target), with one cycle of a line change, we can obtain up to 4 counts by counting the transition in each channel. For a quadrate incremental encoder, the number of counts per revolution (CPR) is four times the number of lines per revolution.

The Transmotec DC motor, PDS4377-12-103-ME (https://transmotec.com/product/pds4377-12-103-me/), is used (as shown in Figures 4.58 to 4.60). The encoder used has 24 CPR for the two channels together. It includes only the leading edge transition. Hence for a single channel, the CPR is 12. If trailing edge is

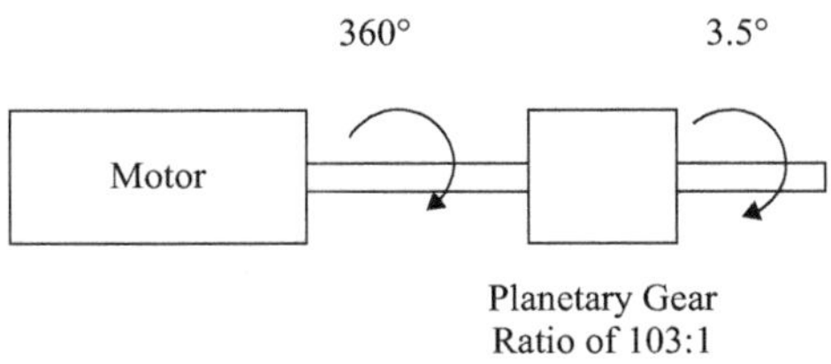

Figure 4.58 Planetary gear reduction of Transmotec DC motor

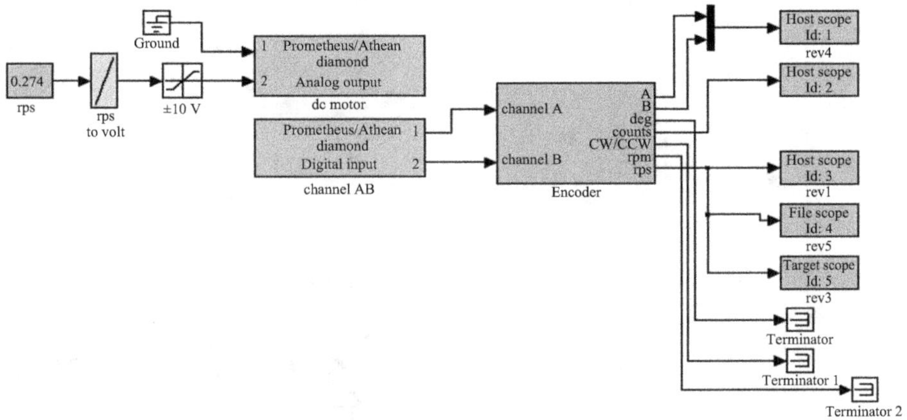

Figure 4.59 Overall simulation diagram for motor with encoder

included, it becomes 24. However, in general, for line encoders, if the CPR is 48, it has $48\times4 = 192$ counts that include the leading and trailing edges transition. However, due to the planetary gear reduction of 103, the speed in rpm is decreased by 103 times or $360/103 = 3.50°$ of the final output rotation per count.

After having the number of CPR, the following Simulink® block diagrams are required. First, create a Simulink® Model including the encoder model as shown below.

As shown in Figure 4.61, **Count Up** block diagrams are needed to count the number of pulses registered in the channel A and B. Since the sample time is 0.001 s, to achieve 1 s, it requires 1,000 counts. Similarly, for 60 s, it requires 60,000 counts. The setting of the Count Up for 1 s and 60 s is shown in Figure 4.62.

Select>**Count event**>**Nonzero sample**. The nonzero sample will trigger a count or reset operation at each sample time when the Inc/Dec or Rst input is not zero.

Next, build and run the model. Observe the CPR. For the motor that was used, the CPR is 1,300. Update the value into the constant block diagram.

Note that the speed to voltage conversion block diagram is required. It is used to convert the corresponding speed (in rps) to voltage applied to the DC motor. A **lookup Table** is used to model the relationship. Click > **Edit** >**Lookup Table Editor** (see Figure 4.63).

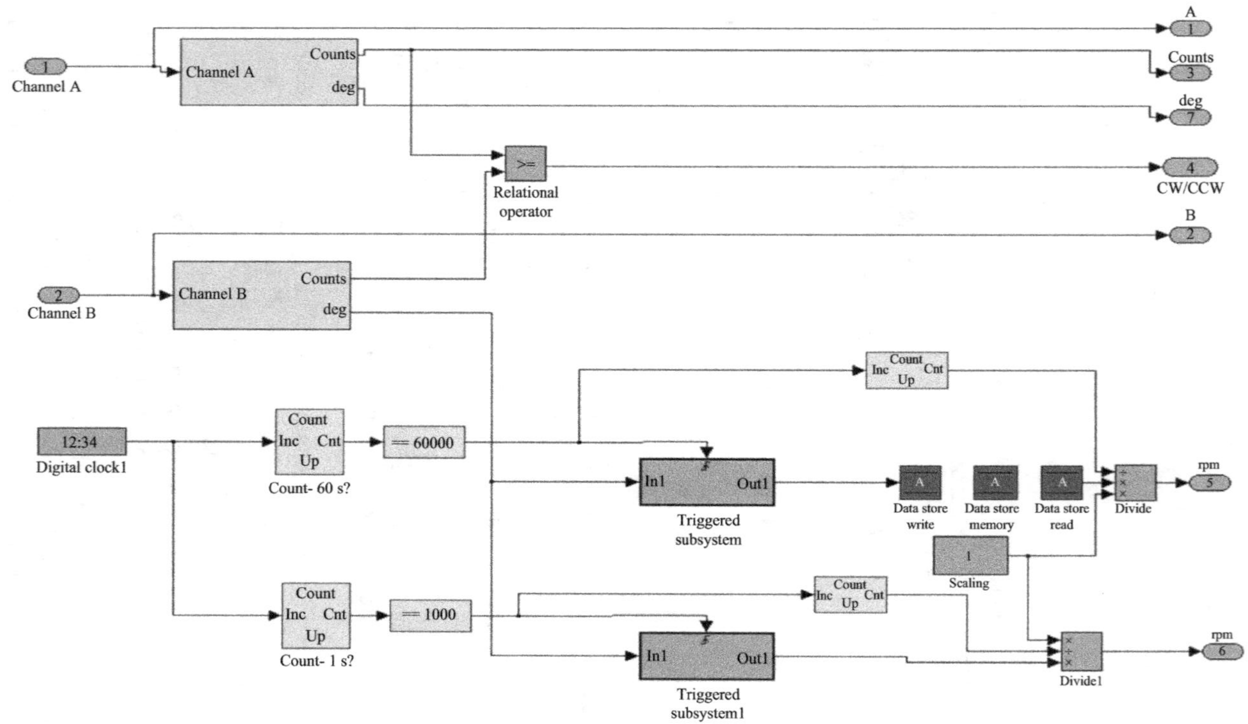

Figure 4.60 Encoder block diagrams

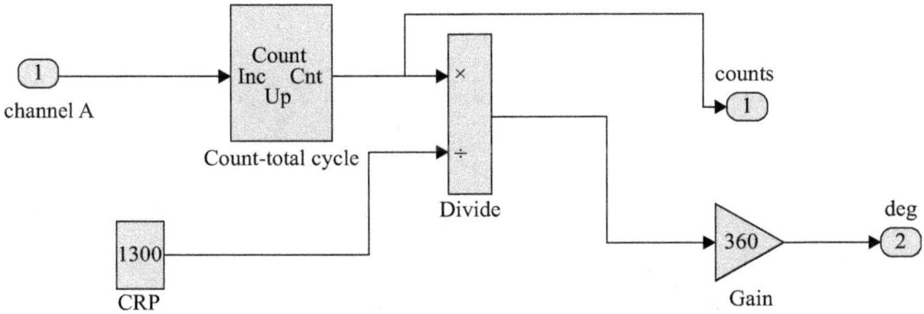

Figure 4.61 Channel A of encoder block diagrams

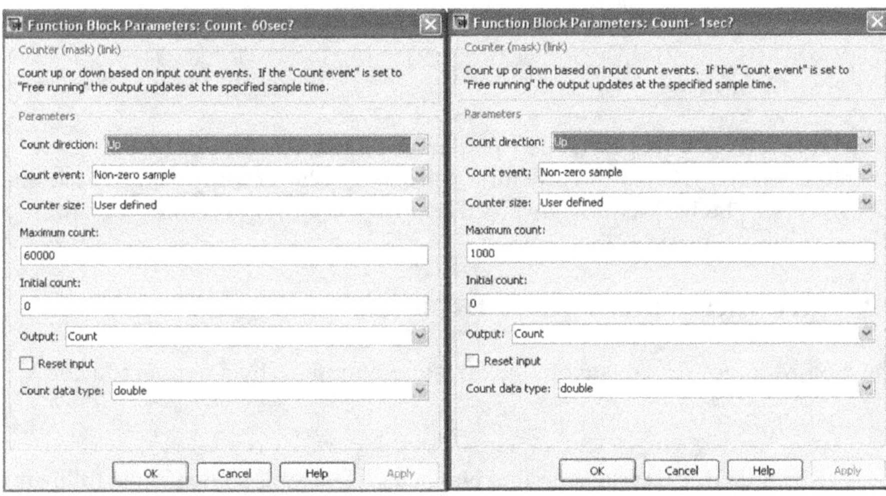

Figure 4.62 Configurations for Count Up for 60 s and 1 s

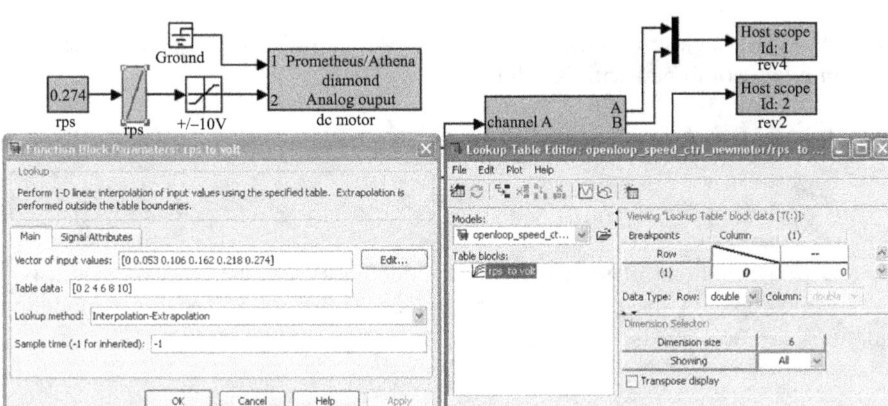

Figure 4.63 Lookup Table Editor

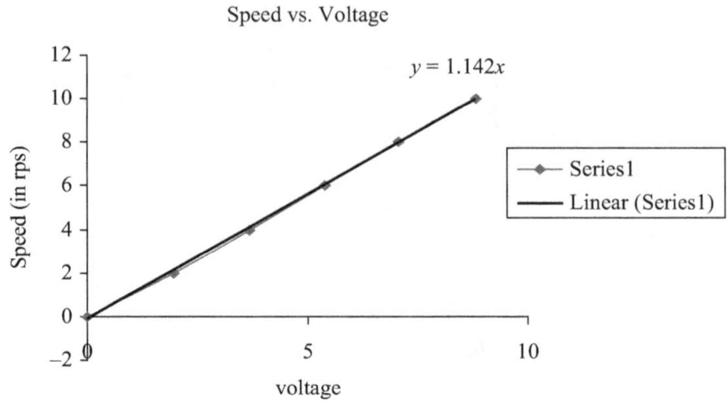

Figure 4.64 Speed vs. voltage curve

The speed vs. voltage curve is plotted as shown (Figure 4.64). The curve is generated when a DC motor is rotating with an angular velocity $\dot{\theta}(t)$, a back-emf voltage $e_b(t)$. The back-emf is governed by $e_b(t) = K_e\dot{\theta}(t)$, where $K_e = 1.142$.

4.9 Identification of a servo DC motor

The motor torque, T, is related to the armature current, i, by a constant factor K_t.

$$T = K_t i \tag{4.1}$$

The back-emf, e, is related to the rotational velocity, as shown in the following equations:

$$e = K_e\dot{\theta} \tag{4.2}$$

where K_e is the back-emf constant. We can write the following equations based on Newton's law combined with Kirchhoff's law:

$$J\ddot{\theta} + K_e\dot{\theta} = K_t i \tag{4.3}$$

$$L\frac{di}{dt} + Ri = V - K_e\dot{\theta} \tag{4.4}$$

where J is the moment of inertia, L is the inductance, R is the resistance, and V is the voltage input to the motor. Using Laplace Transforms, the above modeling equations can be expressed in terms of s.

$$\begin{aligned} s(Js + K_e)\theta(s) &= K_t I(s) \\ (Ls + R)I(s) &= V(s) - K_e s\theta(s) \end{aligned} \tag{4.5}$$

By eliminating $I(s)$, we can get the following open-loop transfer function, where the rotational angle is the output, and the voltage is the input.

$$\frac{\theta(s)}{V(s)} = \frac{1}{s}\left[\frac{K_t}{(Js+K_e)(Ls+R)+K_tK_e}\right] \tag{4.6}$$

alternatively, rotational speed as the output:

$$\frac{\dot{\theta}(s)}{V(s)} = \frac{K_t}{(Js+K_e)(Ls+R)+K_tK_e} \tag{4.7}$$

In the state-space form, the equations above can be expressed by choosing the rotational speed and electric current as the state variables and the voltage as an input. The output is chosen to be the rotational angle.

$$\begin{bmatrix}\ddot{\theta}\\ \dot{I}\end{bmatrix} = \begin{bmatrix}-\frac{K_t}{J} & \frac{K_t}{J}\\ -\frac{K_e}{L} & -\frac{R}{L}\end{bmatrix}\begin{bmatrix}\dot{\theta}\\ I\end{bmatrix} + \begin{bmatrix}0\\ \frac{1}{L}\end{bmatrix}V \tag{4.8}$$

Moreover, the rotational velocity to be measured can be computed using:

$$\dot{\theta} = [1 \quad 0]\begin{bmatrix}\dot{\theta}\\ I\end{bmatrix} \Rightarrow \theta = \int_0^t [1 \quad 0]\begin{bmatrix}\dot{\theta}\\ I\end{bmatrix}dt \tag{4.9}$$

Since L is usually small (or electrical time constant is shorter as compared to the system response time), it can be neglected, and the open-loop transfer function for the speed control becomes:

$$\frac{\dot{\theta}(s)}{V(s)} = \frac{K_m}{\tau_m s+1} \tag{4.10}$$

where $K_m = \frac{K_t}{K_tK_e+K_eR}$, $\tau_m = \frac{JR}{K_tK_e+K_eR}$. The open-loop transfer function was compared with actual speed of the motor. As shown in Figure 4.65, the transfer function was modeled in the Simulink®. For example, an input of 0.274 rps was used.

As shown in Figure 4.65, there is a time delay in the response. To model the motor dynamics accurately, the first-order plus dead-time model is used.

$$\frac{\dot{\theta}(s)}{V(s)} = \frac{K_m}{\tau_m s+1}e^{-\theta s} \tag{4.11}$$

where the time delay, $\theta = 1$ s (as seen in the response time) and the remaining parameters can be tabulated as follows.

The estimated and calculated process parameters for the motor (after speed reduction, 189:1) are as follows (see Figures 4.66 to 4.70). With small J, as the resistance R is decreased, the motor time constant approaches zero, and the motor acts as an ideal integrator.

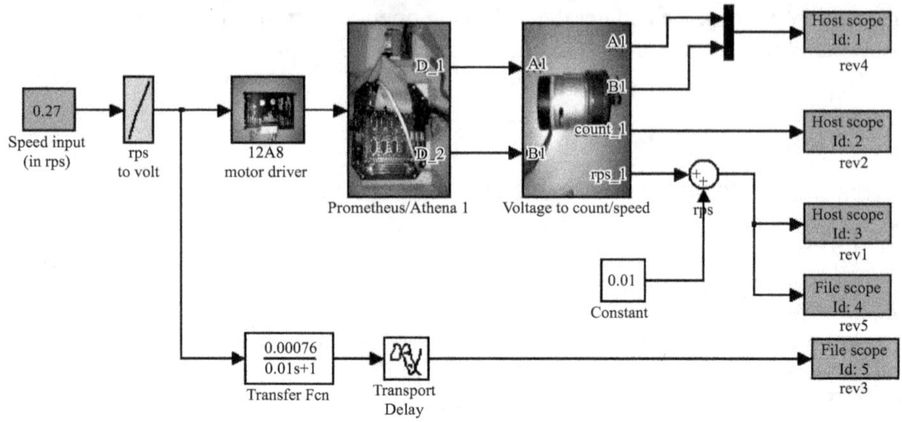

Figure 4.65 Open-loop speed regulation block diagram

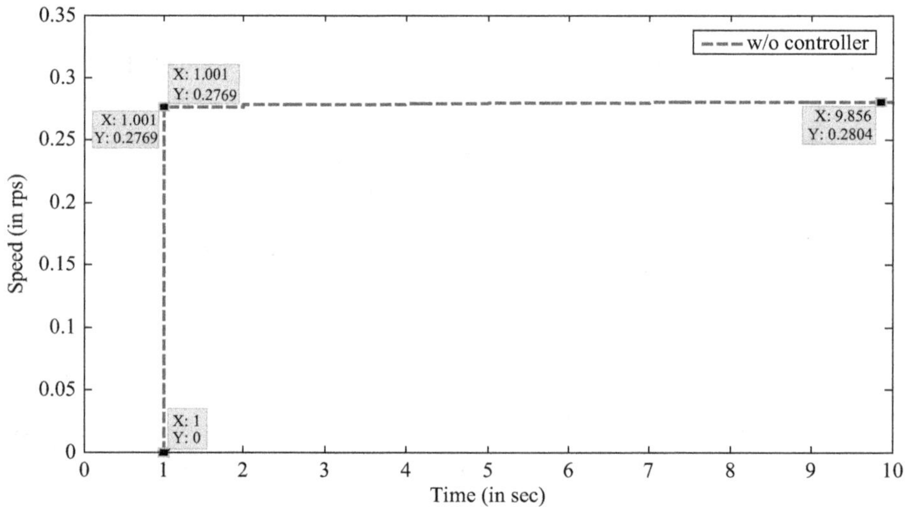

Figure 4.66 Step response of DC motor

- $K_m = \frac{K_t}{K_t K_e + K_e R} = 0.00076$
- $\tau_m = \frac{JR}{K_t K_e + K_e R} = 0.01$ s
- $J_m = \frac{mr^2}{2} = 2.9245 \times 10^{-7}\ \text{kg} \cdot \text{m}^2$
- $R = \frac{12}{150 \times 10^{-3}} = 80\ \Omega$ (at no load condition)
- $K_e = 1.142$ V/S
- $K_t = 0.053$ Nm/A (by substituting J, R, K_e into K_m)

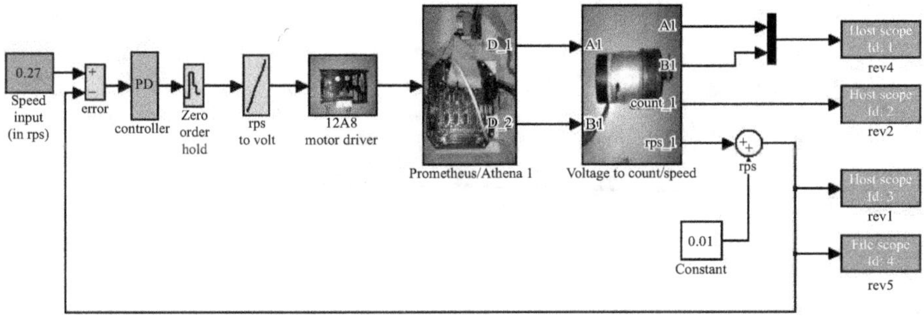

Figure 4.67 PID control of DC motor

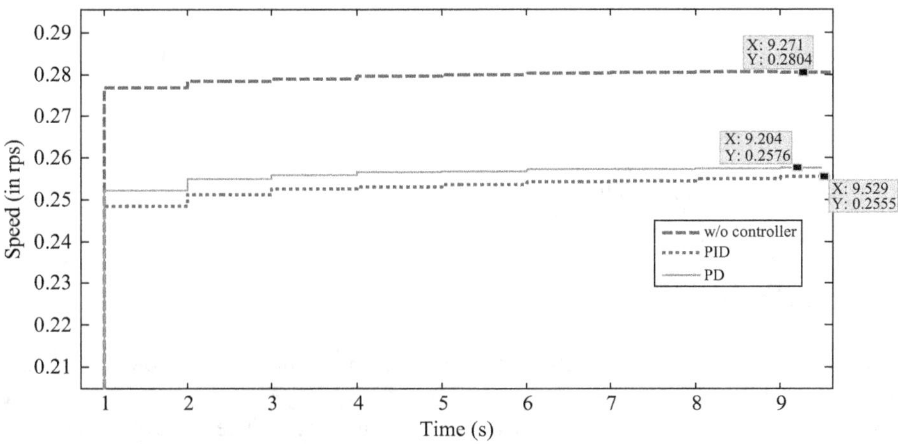

Figure 4.68 Time response of PID control of DC motor

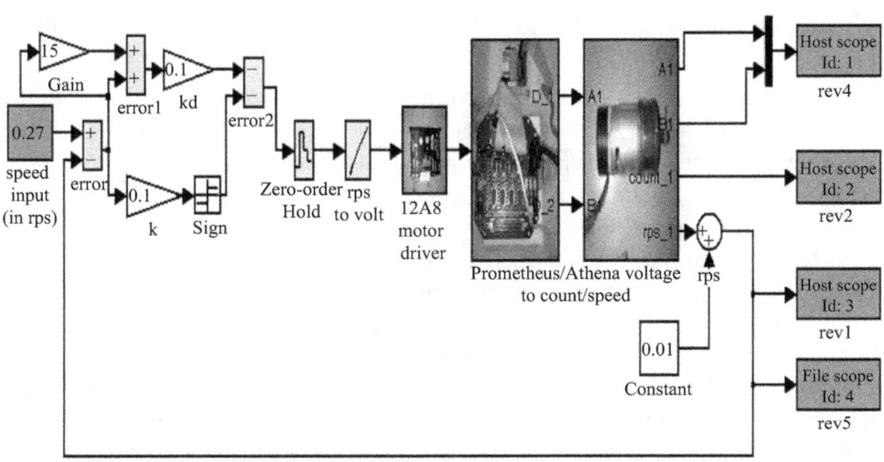

Figure 4.69 Sliding mode control of DC motor

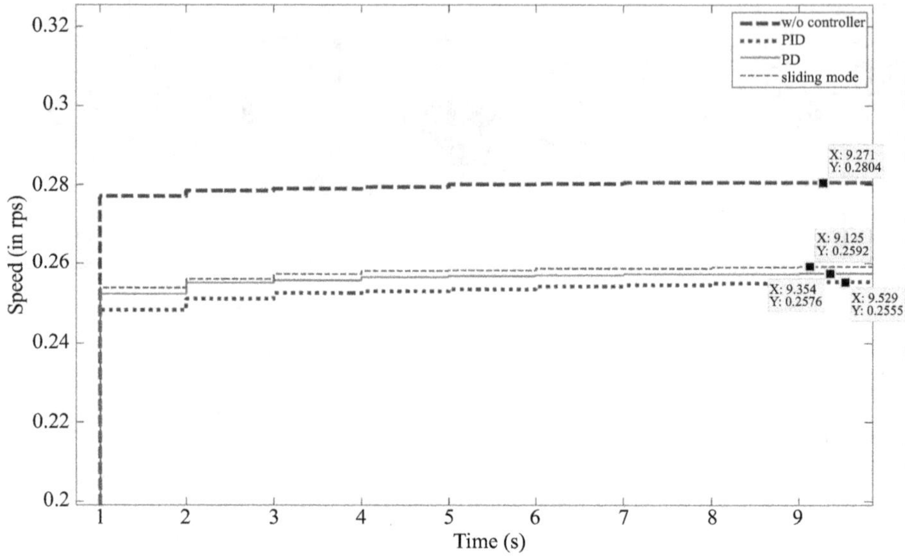

Figure 4.70 Time response of SMC of DC motor

4.10 PID speed control of servo DC motor

A closed-loop system with a PID controller is represented by the block diagram shown. The plant is the motor's speed to be controlled. The encoder monitors the motor output and provides feedback to the control system. The controller takes the input and feedback signals and produces a control action to the motor.

The transfer function of a PID controller can be written as follows:

$$G_c(s) = \left(K_p + \frac{K_p}{\tau_i s} + K_p \tau_d s\right) e(s) \tag{4.12}$$

The proportional gain (K_p) determines the amplitude (magnitude) of the control action. However, if the proportional gain is increased too far, the closed-loop system will become unstable. Integral control action (τ_i) eliminates steady-state error or offset. The smaller the integral constant, the greater the integral control action. Derivative control action (τ_d) adds effective damping to the closed-loop system, thereby reducing the settling time and allowing a higher proportional gain. The larger the derivative constant, the higher will be the derivative control action.

In 1942, Ziegler and Nichols published procedures for setting the parameters of PID controllers based on system response characteristics. Their recommendations are based on the "quarter decay" closed-loop response where the step response exhibits a maximum overshoot of 25% of the step. In the first method, the control parameters are set based on the open-loop step response of the plant alone, which may be obtained analytically, numerically, or experimentally. If the step response of the plant is known, the following characteristics lag time (L) and response slope (R), Ziegler and Nichols recommend the following controller parameter values:

- $K_p = 1.2/RL\,; \tau_i = 2.0L;\ \tau_d = 0.5L$ for PID controllers;
- $K_p = 0.9/RL;\ \tau_i = 3.3L;\ \tau_d = 0$ for PI controllers;
- $K_p = 1.0/RL;\ \tau_i = \infty L; \tau_d = 0$ for P controllers.

The strength of this method is that the control parameter values may be estimated from the analytical plant model. From the open-loop speed response curve, the response slope $R = 0.28/0.001 = 280$ and the response delay $L = 1$ s can be observed. Hence, the parameters used in the PID speed control are computed as follows:

$$K_p = \frac{1.2}{280 \times 1} = 0.004; \tau_i = 2.0; \tau_d = 0.5$$

$$K_i = \frac{K_p}{\tau_i} = \frac{0.004}{2} = 0.002; K_d = K_p \tau_d = 0.004 \times 0.5 = 0.002$$

Similarly, the parameters used in the PD speed control can be computed as follows:

$$K_p = \frac{1.2}{280 \times 1} = 0.004; \tau_i = 0; \tau_d = 0.5$$

$$K_i = 0; K_d = K_p \tau_d = 0.004 \times 0.5 = 0.002$$

4.11 Sliding-model speed control of servo DC motor

For notational simplicity, let us write the motor model according to:

$$\tau_m \ddot{\theta} + \dot{\theta} = K_m V \tag{4.13}$$

where $\tau_m > 0, K_m > 0$. We assumed that both $\dot{x}(=\dot{\theta})$ and $x(=\theta)$ are measured. Define a scalar measure of tracking:

$$s = \dot{\tilde{x}} + \lambda \tilde{x} \tag{4.14}$$

where $\tilde{x} = x - x_d$ is the tracking angle error and $\lambda > 0$ is the control bandwidth. For $s = 0$, this expression describes a sliding surface with exponential dynamics:

$$\tilde{x}(t) = \exp[-\lambda(t - t_0)]\tilde{x}(t_0) \tag{4.15}$$

which ensures that the tracking error $\tilde{x}(t)$ converges to zero infinite time when $s = 0$ (sliding mode). In fact, the error trajectory will reach the time-varying sliding surface in finite time for any initial condition $\tilde{x}(t_0)$ and then slide along the surface toward $\tilde{x}(t) = 0$ exponentially. Hence, the control objective is reduced to finding a nonlinear control law which ensures that:

$$\lim_{t \to \infty} s(t) = 0 \tag{4.16}$$

In the design of the sliding control law, it is convenient to define a virtual reference x_r satisfying:

$$\dot{x}_r = \dot{x}_d - \lambda\tilde{x} \Rightarrow s = \dot{x} - \dot{x}_r \tag{4.17}$$

Hence, the following expression for $\tau_m \dot{s}$ is obtained.

$$\begin{aligned} \tau_m \dot{s} &= \tau_m(\ddot{x} - \ddot{x}_r) \\ &= (K_m V - \dot{x}) - \tau_m \ddot{x}_r \\ &= -|\dot{x}|s + (K_m V - \tau_m \ddot{x}_r - |\dot{x}|\dot{x}_r) \end{aligned} \tag{4.18}$$

Consider the scalar Lyapunov-like function candidate:

$$V_L(s,t) = \frac{1}{2}\tau_m s^2, \tau_m > 0 \tag{4.19}$$

Differentiating V_L with respect to time (assuming $\dot{\tau}_m = 0$) yields:

$$\dot{V}_L = \tau_m s\dot{s} = -|\dot{x}|s^2 + s(K_m V - \tau_m \ddot{x}_r - |\dot{x}|\dot{x}_r) \tag{4.20}$$

Taking the control law to be:

$$V = 1/K_m(\tau_m \ddot{x}_r + |\dot{x}|\dot{x}_r - K_d s - K\,\mathrm{sgn}(s)) \tag{4.21}$$

where

$$\mathrm{sgn}(s) = \begin{cases} 1 & \text{if } s > 0 \\ 0 & \text{if } s = 0 \\ -1 & \text{otherwise} \end{cases} \tag{4.22}$$

yields,

$$\dot{V}_L = -|\dot{x}|s^2 + s(-K_d s - K\mathrm{sgn}(s)) \tag{4.23}$$

Conditions on the switching gain K are found by requiring that $\dot{V}_L \leq 0$. This implies that:

$$\dot{V}_L \leq -(|\dot{x}| + K_d)s^2 - K\mathrm{sgn}(s) \leq 0 \tag{4.24}$$

Note that $\dot{V}_L \leq 0$ implies that $V_L(t) \leq V_L(0)$, and therefore that s is bounded. This implies that $\ddot{V}_L$ is bounded. Hence, $\dot{V}_L$ must be uniformly continuous. Finally, application of Barbalat's lemma then shows that $s \to 0$ and thus $\tilde{x} \to 0$ as $t \to \infty$.

4.12 Linear quadratic regulator

The concept of linear quadratic regulator (LQR), or commonly known as optimal state-feedback design, uses a set of weighting matrices to penalize the states and control inputs of the system. This requirement is set in the cost function to be minimized as shown below. The weighting matrices Q and R, which are applied to the states and inputs respectively, are tuned iteratively until the desired performance is achieved. Inevitably, this becomes an ad-hoc process.

LQR problem aims to minimize the weighted sum of the energy of the states and control inputs in the form of the cost function:

$$J = \int_0^T \left[x^T(t)Qx(t) + u^T(t)Ru(t)\right]dt \tag{4.25}$$

with respect to the input vector u(t). For the optimization, Q must be a symmetric positive semi-definite matrix ($Q^T = Q \geq 0$) and R must be a symmetric positive definite matrix ($R^T = R > 0$). Using for instance, the Pontryagin's minimum principle, the solution is of the form of a time-varying control law:

$$u(t) = -F(t)x(t) \tag{4.26}$$

where

$$F(t) = R^{-1}B^T P(t) \tag{4.27}$$

and $P(t)$ is a solution of the Riccati Differential Equation (RDE)

$$A^T P(t) + P(t)A + Q - P(t)BR^{-1}B^T P(t) = -\frac{d}{dt}P(t) \tag{4.28}$$

To implement the LQR controller in (4.26), all states $x(t)$ are required to be measurable. If the time horizon is indefinite ($T \to \infty$) and an optimal solution exists, then $P(t)$ tends to a constant matrix P. The control law is then;

$$u(t) = -Fx(t) \tag{4.29}$$

where

$$F = R^{-1}B^T P \tag{4.30}$$

The matrix P is a solution of the Algebraic Riccati Equation (ARE);

$$A^T P + PA + Q - PBR^{-1}B^T P = 0 \tag{4.31}$$

Note the time dependent parameter becomes independent and the right-hand side of (4.31) becomes zero. For the positive-definite matrix P, the corresponding closed-loop system with LQR controller in (4.29) is asymptotically stable.

LQR-designed controllers ensure good stability margin and sensitivity properties. Using this approach, a minimum phase margin of 60° and infinite gain margin can be achieved. By the Kalman identity, the return difference matrix, $I + L(s)$ is always greater or equal to one as shown below:

$$\begin{aligned} \left[I + F(sI - A)^{-1}B + D\right] &\geq 1 \\ [I + L(s)] &\geq 1 \end{aligned} \tag{4.32}$$

where F is the optimal state feedback gain matrix. This implies good disturbance rejection and tracking properties and hence robustness is guaranteed for the LQR approach. To achieve the results presented above, the following design steps were taken. Solve the Algebraic Riccati Equation for P using (4.31).

1. Determine the optimal state feedback gain, F (LQR.m).
2. The weighting matrix Q is typically selected as C^TC since this choice reflects the weighting of the states in the output since $y^Ty = x^TC^TCx$. For the weighting matrix R, this is selected to be a diagonal matrix of equal weights on the control inputs. It was found that R equal to the identity matrix gave a good compromise result on the control input over the 50% and 0% load conditions. $Q = k_cC^TC$, $R = I$, where k_c is a scalar used to decrease the states interaction. It has a value here of 0.0001.
3. The simulation block diagram is formulated using the SIMLINK facility in Figure 4.71 and the response can be seen in Figure 4.72. Note that an initial scaling block is used to reduce the size of the output. This is applied to all subsequent controller designs.

Now, using the inhibit function (assemble a switching control) that is activated when the motor shaft has reached 360° (i.e. setpoint). The block diagram can be seen in Figure 4.73. As shown in Figure 4.74, the position of the motor's shaft reached 360° faster than the conventional approach without the inhibit function.

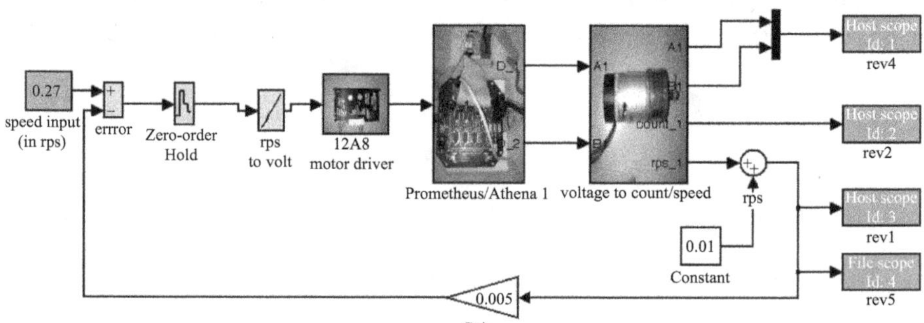

Figure 4.71 Block diagram of LQR control of DC motor

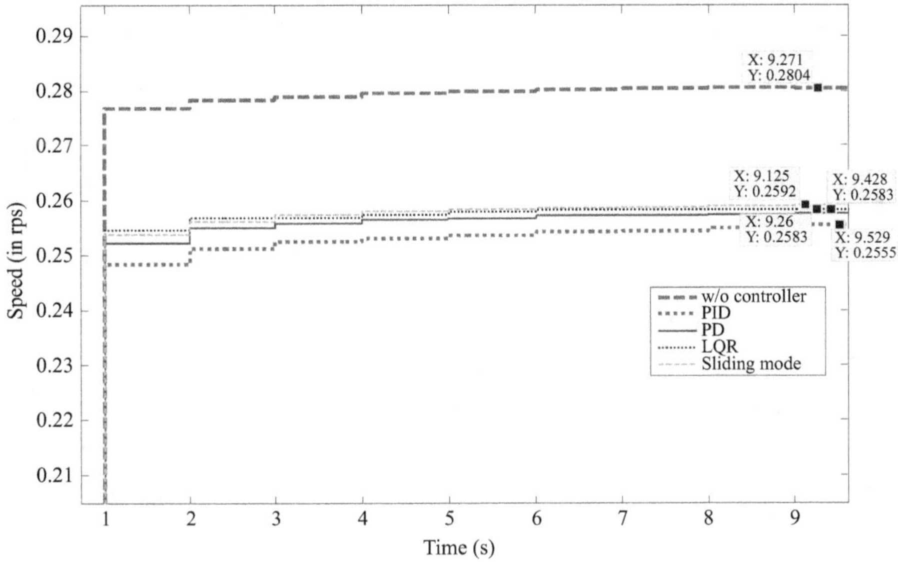

Figure 4.72 Time response of LQR control of DC motor

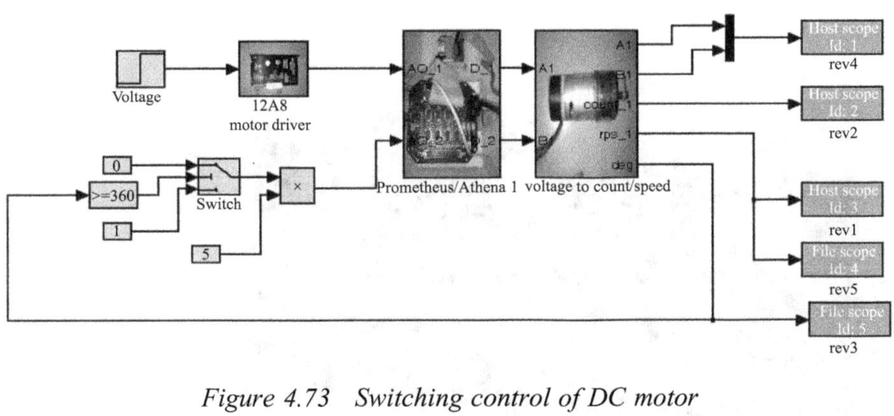

Figure 4.73 Switching control of DC motor

Now, we can check the robustness of the controllers by injecting a 1.5 kHz sinusoidal disturbance (see Figure 4.75) into the system as shown in Figure 4.76. Without a controller, the steady-state error increases. As seen in the SMC in Figure 4.77, the steady-state error is the least with faster response time. The LQR and PID have almost the same response.

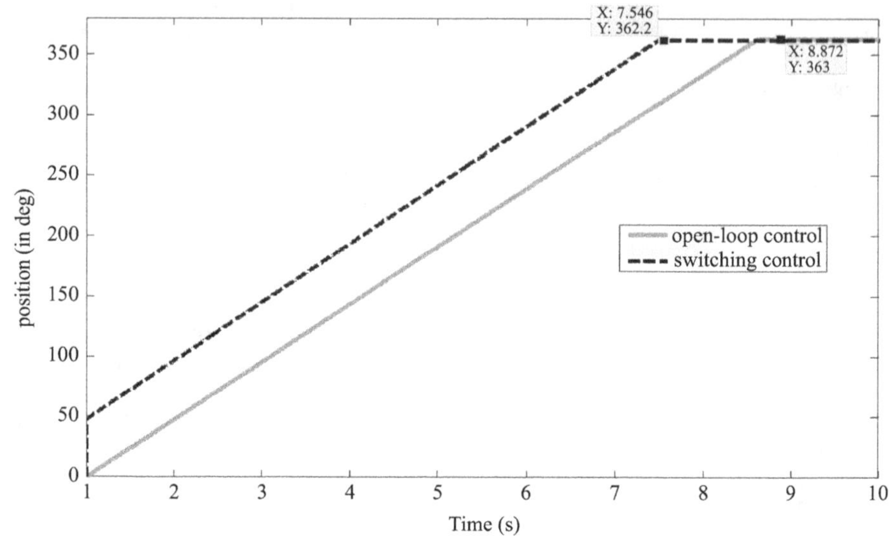

Figure 4.74 Response to switching control of DC motor

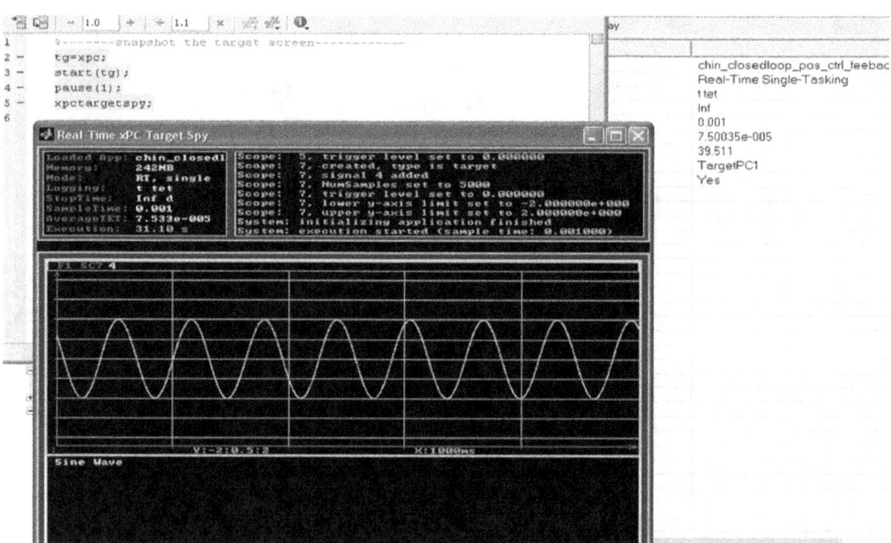

Figure 4.75 1.5 kHz sinusoidal disturbance into the system

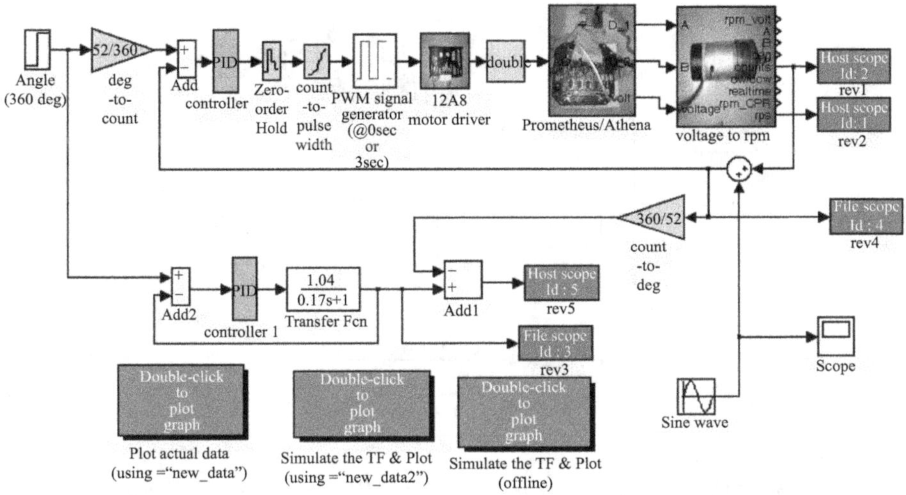

Figure 4.76 Block diagram of switching control of DC motor

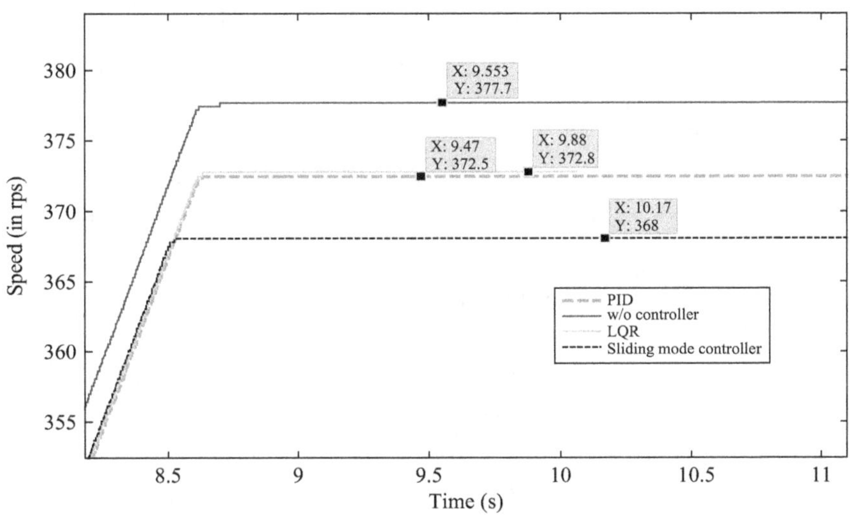

Figure 4.77 Response time of switching control of DC motor under disturbance

4.13 Digital speed control of servo DC motor

Instead of using a continuous PID controller, a discrete-time PID controller (see Figure 4.78) can be used to replace the PID block in Figure 4.76. The same procedure can be applied to perform the simulation. The discrete time response can be seen in Figure 4.79. Here, the desired speed response is 0.3 rps. The SMC followed by PID and LQR remains quite robust with good transient and steady-state response

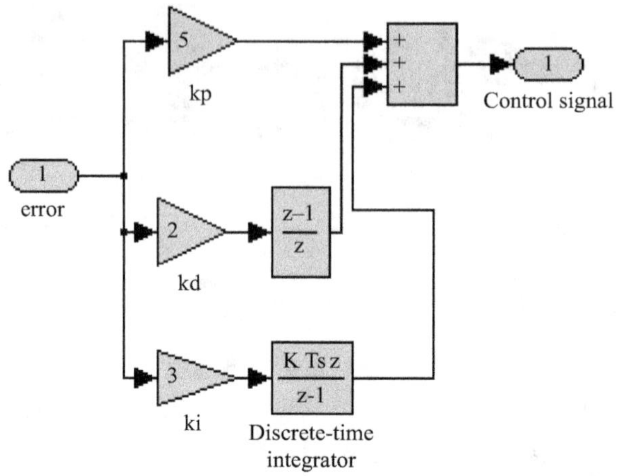

Figure 4.78 Block diagram for digital PID control of DC motor

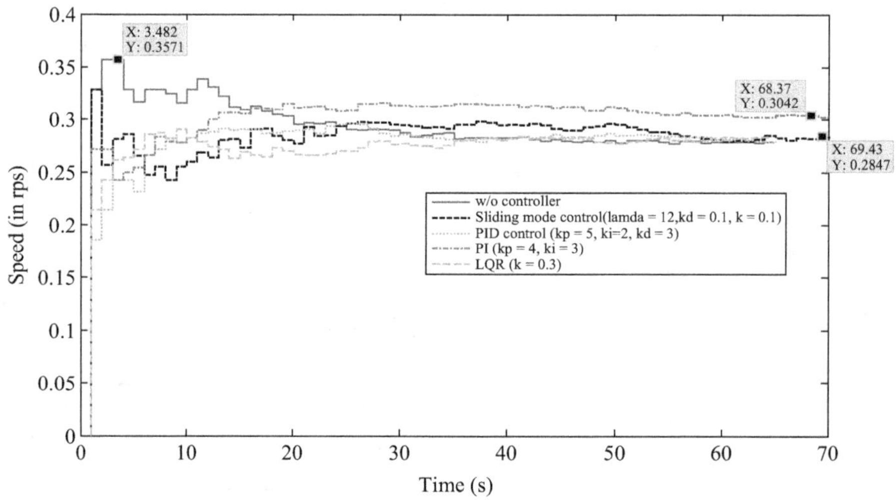

Figure 4.79 Response time of digital PID control of DC motor under disturbance

despite the 1.5 Hz sinusoidal disturbance. The response without controller did show poor response with significant overshoot. Hence, the SMC will be used in the actual implementation of the embedded control system in water.

The rapid control systems prototyping using hardware-in-the-loop (see Figure 4.80) have been used to develop controllers. As mentioned, the real-time windows target, real-time embedded target, and xPC-Target were used to obtain the block diagrams and results. It provides the tools for automatic generation of real-time code from graphical models instead of program coding, without having to write and learn the intricacies of writing detailed embedded code. These can be

Figure 4.80 Rapid control systems prototyping using hardware-in-the-loop [1]

further seen in the full range of applications from automotive system design, process control, and robotic systems.

4.14 Case study: marine robotic vehicle with uncertainties using xPC-Target system

The prototyping of the embedded system could vary from applications to applications. Noticeably, it has not covered much on the underwater robotic vehicles (underwater robotic vehicle) such as an autonomous underwater vehicle development using xPC-Target, in particular, the rapid control system prototyping involving dynamic modeling, control systems design, and hardware-in-the-loop testing. The details of the works done in this section can be seen in the following references [1,2].

Besides, most underwater robotic systems development [3,4], just to name a few, used are QNX, Unix-OS/90, and LinuxMZ runtime environment in their control systems design. These methods may not be easy for engineers and/or students to use as it consists of extensively C/C++ programming codes and inputs–outputs circuitry for interfacing prior to the dynamics modeling, and also the testing of the hydrodynamic parameters for the control systems design. In recent years, there has been much interest in the development of underwater robotic vehicle and the practical application of underwater robotic vehicles that involves major challenges of precise track-keeping and fast obstacle avoidance [5]. Recently, underwater robotic vehicle development capacity is enhanced by the use of the computational modeling in the design, manipulation, and control of underwater robotic vehicles. xPC-Target computer models provide a means of modeling and determining the performance of a proposed vehicle during the design phase, and to devise and optimize a controller for enhancing the vehicle's performance characteristics iteratively using hardware-in-the-loop concept.

To facilitate the dynamic modeling and subsequent motion control of the underwater robotic vehicle with the installed thrusters, the works are broken down into three subtasks: the determination of the hydrodynamic coefficients of the vehicle, the simulation of the control systems design, for a known set of control inputs, to obtain the ensuing motion of the vehicle, and the testing of the underwater robotic vehicle in the field. The uncertainty involved in the determination of the hydrodynamic coefficients usually induces the errors in the model or final result. However, with robust control scheme [6] that can compensate for these modeling errors and uncertainties, the accuracy of the hydrodynamic coefficients would not pose an enormous problem in most underwater robotic vehicle control systems design. The work aims to use the xPC-Target and the proposed embedded hardware on the underwater robotic vehicle from dynamics modeling, control systems design to final testing in water.

4.14.1 System design and architecture

An autonomous underwater vehicle (underwater robotic vehicle) was constructed. The underwater robotic vehicle is intended to serve as a test-bed for systems research, and as a general research and development tool for years to come. It is neutrally buoyant in water, axi-symmetric, and has a cylindrical-like body as shown in Figure 4.81. The vehicle has an overall length (L) of 0.6 m, a diameter (D_a) of 0.25 m, and a dry weight (m) of approximately 14 kg. The center of buoyancy is located at (0, 0, zB). The thruster is made up of a brush-type DC motor and a three-bladed propeller (with a diameter equal to 0.18 m). The underwater robotic vehicle has a pair of side fins to stabilize the underwater robotic vehicle in water. A pair of

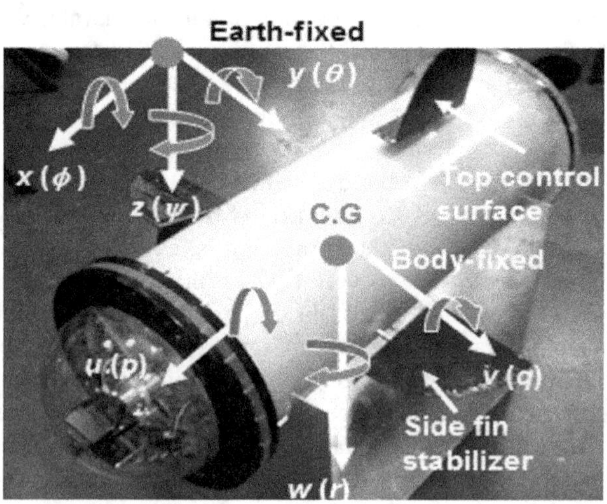

Figure 4.81 A test-bed underwater robotic vehicle with coordinate systems [1]

top and bottom control surfaces near its end part generates yawing moments for horizontal movement. The necessary power is supplied by packs of lithium-ion batteries, which can be sustained to operate at the speed of maximum 1 m/s. The vehicle is equipped with an altimeter sensor for measuring distance and a magnetic compass for measuring heading. The encoders at the motors are used to measure the angular position and the angular velocity of the motor shaft.

For underwater robotic vehicle modeling and control verification using the hardware-in-the-loop concept, the xPC-Target software and hardware as shown in Figure 4.82 are used. The schematic diagram of the system setup is composed of:

- an industrial PC104- Athena II CPU (or Prometheus CPU) card;
- a Pandora enclosure with DAQ card;
- two brush-type Transmotec DC motor with encoder;

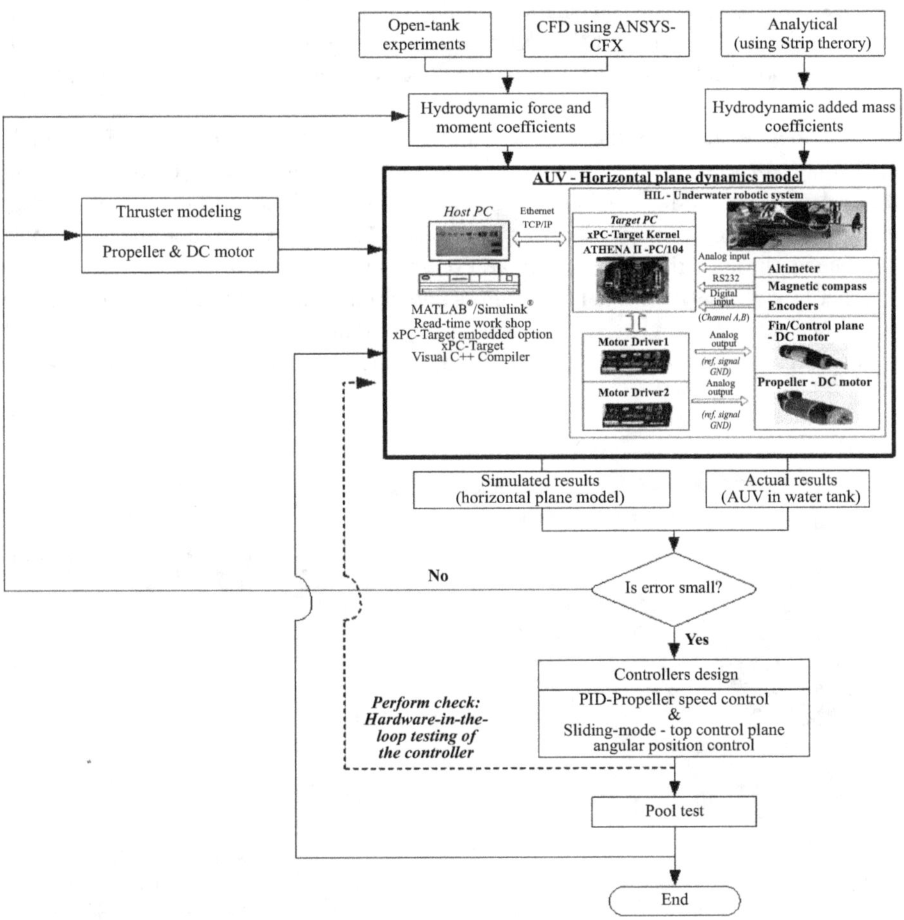

Figure 4.82 Rapid control system prototyping process flow for test-bed underwater robotic vehicle [1]

- a three-bladed propeller;
- four Omega strain gages with Kyowa amplifier;
- two advanced power amplifiers;
- a Honeywell magnetic compass; and
- a Tritech PA200 digital precision altimeter.

The motion of the vehicle is controlled by an onboard industrial PC104 computer in target PC, which runs with the help of the xPC-Target software and Target kernel. Ethernet link is used for communication with the host PC that has MATLAB® and Simulink® packages installed. The PC104 computer, Athena II is connected to power amplifier, DC motor, altimeter, and magnetic compass. When the executable code is running in xPC-Target kernel, the commanded voltage signals are generated from the I/O board analog output channels and go into the power amplifiers to drive the respective DC motors. As shown in Figure 4.82, the encoders are connected to digital channel of the I/O board while the altimeter is connected to the analog input of the I/O board. The magnetic compass is connected to the RS232 port and was calibrated against the software provided by the vendor as shown in Figure 4.83.

To operate and test the test-bed underwater robotic vehicle, the Simulink® block diagram that contains the proposed control architecture is compiled (using Microsoft Visual Studio V9.0 compiler). It is then downloaded to the target PC that runs the highly optimized xPC-Target kernel. The host-target communication is established through the Ethernet link. For real-time monitoring and analysis, the Ethernet link allows the model parameters and command inputs to be changed while the underwater robotic vehicle is running. For standalone control or during the actual test in water, the Ethernet cable could be disconnected from the target PC.

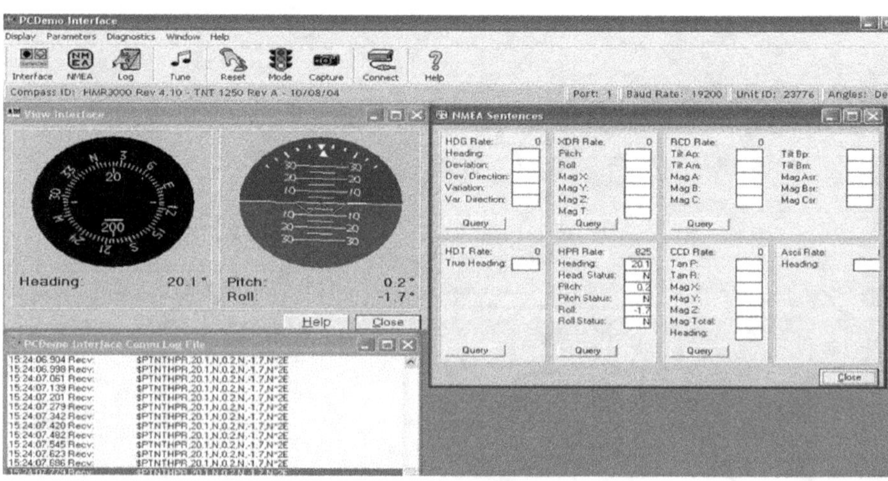

Figure 4.83 PC demo interface program for magnetic compass [1]

Figure 4.82 shows the overall procedure of the rapid control system prototyping of the test-bed underwater robotic vehicle. The modeling involved process of obtaining the hydrodynamic forces and moment coefficients from the measured three DOF or the horizontal plane motion of the underwater robotic vehicle using an open-tank test and a CFD-ANSYS CFX method. The hydrodynamic-added mass coefficients are determined analytically using the strip theory. From the measured position, the velocities in the body-fixed coordinate are induced and the hydrodynamic forces and moment are estimated. The thruster dynamics consist of a propeller and a DC motor is modeled. The control input variables consist of the top control surface angle and the control input voltage of the thruster. At this point, a comparison is made between the simulated data and actual results obtained; the process repeats until the output data matches the simulated results within a reasonable bound. After obtaining the open-loop underwater robotic vehicle dynamics model, the SMC is designed to control the angular position of the top control surface while the PID controller is used to control the propeller angular speed. As seen in Figure 4.82, to verify the overall control systems design and the hydrodynamic coefficients obtained from the experiment and the CFD simulation, the underwater robotic vehicle was conducted in a swimming pool.

4.14.2 Underwater robotic vehicle dynamic model

Considering the coordinate systems shown in Figure 4.81, the body-fixed frame coincides with the principal axes of the vehicle and is free to translate and rotate with respect to the earth-fixed frame. The origin of the body-fixed frame is the center of mass of the underwater robotic vehicle. The positive x-axis is oriented in the forward direction of the vehicle along the longitudinal centerline, and the positive z-axis is oriented downwards.

The dynamic model of a nonlinear underwater robotic vehicle [7] is used for control systems design. In general, there are two methods to derive the dynamic model: the Lagrange equations using the variation method and the Newtonian's equation using Newton's Second Law of Motion. The latter method gives an equation using the angular velocities about the body-fixed axes, unlike the generalized variables used in the Lagrange equation, the angular velocities cannot be integrated to obtain angular displacements about these axes and are therefore unsatisfactory to describe adequately the orientation of a rigid body in three-dimensional (3D) space.

However, the body-fixed angular velocities can be solved and transformed to give the orientation of the vehicle in 3D space using the Euler angles. The Euler's transformation (or commonly known as kinematics equation) provides the relationship between the dynamics derived from the earth-fixed and the body-fixed coordinates. There are practical advantages to derive the dynamics in the body-fixed coordinate for the following reasons. First, most sensors and actuators mounted on the underwater robotic vehicle measure the body parameters such as the underwater robotic vehicle's velocities and provided propelling forces. Second, the dynamic equation is inherently parameterized and hence less complicated since this parameterized form is not Euler's angles dependent.

Hence, the rigid body underwater robotic vehicle dynamic equation is commonly expressed in the body-fixed frame since the control forces and measurement devices are quickly and intuitively related to this body frame of reference. Using the Newtonian approach, the motion of a rigid body with respect to the body-fixed reference frame at the origin can be obtained. A detailed discussion about the underwater robotic vehicle modeling was presented in [7] and the hydrodynamic forces and moments modeling could be found in [8,9]. The assumptions made in deriving the underwater robotic vehicle's dynamic equation are (a) the vehicle is a rigid body and is fully submerged once in water; (b) the vehicle is slow moving for operation in underwater; and (c) the earth-fixed frame of reference is inertial.

In (4.1), the external force and moment vector includes the hydrodynamic forces and moments due to damping, inertial of surrounding fluid, and the restoring force and moment. These forces and moments tend to oppose the motion of the underwater robotic vehicle. The equations of motion can be described by a set of six differential equations in the body-fixed frame based on the application of Newton's second law in the earth-fixed frame [7]. The similar expression to model the vehicle dynamics can be seen in (3.8) in Chapter 3.

The transformation using the Euler angles provides an important transformation between the dynamics expressed in the body-fixed frame in (3.8) to the earth-fixed frame. As the accelerations of a point on the surface of the Earth can be neglected, for slow-moving marine vehicles, the earth-fixed frame can be considered to be an inertial frame. The kinematics equations [7] can be written as:

$$\dot{\boldsymbol{\eta}} = \mathbf{J}(\boldsymbol{\eta})\boldsymbol{\nu} \tag{4.33}$$

where the Euler transformation matrix, $\mathbf{J}(\boldsymbol{\eta})$ is derived by successive rotation of the Euler's angles about Z, Y, and X axes to give:

$$\mathbf{J}(\boldsymbol{\eta}) = \begin{bmatrix} \mathbf{J}_1(\boldsymbol{\eta}_2) & 0 \\ 0 & \mathbf{J}_2(\boldsymbol{\eta}_2) \end{bmatrix} \tag{4.34}$$

where

$$\mathbf{J}_1(\boldsymbol{\eta}_2) = \begin{bmatrix} c(\psi)c(\theta) & -s(\psi)c(\phi) + c(\psi)s(\theta)s(\phi) & s(\psi)s(\phi) + c(\psi)c(\phi)s(\theta) \\ s(\psi)c(\theta) & c(\psi)c(\phi) + s(\phi)s(\theta)s(\psi) & -c(\psi)s(\phi) + s(\theta)s(\psi)c(\theta) \\ -s(\theta) & c(\theta)s(\phi) & c(\theta)c(\phi) \end{bmatrix} \tag{4.35}$$

$$\mathbf{J}_2(\boldsymbol{\eta}_2) = \begin{bmatrix} 1 & s(\phi)t(\theta) & c(\phi)t(\theta) \\ 0 & s(\phi) & -s(\phi) \\ 0 & \dfrac{s(\phi)}{c(\theta)} & \dfrac{c(\phi)}{c(\theta)} \end{bmatrix} \tag{4.36}$$

and s = sin(.), c = cos(.), and t = tan(.).

Some methods have been proposed to determine the hydrodynamic coefficients. They can be broadly classified into test-based and predictive methods. Test-based methods involve experimental determination of the hydrodynamic parameters directly based on wind-tunnel or towing-tank model tests, as well as the testing of full-sized vehicles [10–12]. An overriding disadvantage of the above methods is the need of an accurate physical model of the vehicle or the vehicle itself, as well as laboratory or in-field testing facilities. These are often not available, either for reasons of cost or simply because the vehicle has not yet been constructed. Predictive methods offer an attractive alternative to test-based methods when the vehicle is still in the design stages or when cost prohibits a full-scale testing program.

The most basic of the predictive methods are purely analytical methods, but these are prone to yielding unrealistic results for complex bluff bodies. A detailed discussion and evaluation of three of the existing methods [13] was proposed. Empirical and semi-empirical methods are the most widely used of the predictive methods and have been shown to yield reasonable results [14,15]. Hence, the above predictive methods are likely to yield reasonable results on a streamlined vehicle since their dynamics are more easily predicted. Few researchers [16,17] estimated the hydrodynamic coefficients of a streamlined underwater robotic vehicle via two nonlinear observers: a sliding mode observer and an extended Kalman filter, in which a model-based estimation algorithm was used. This method can obtain 15 linear damping coefficients [18].

Since the underwater robotic vehicle is not a complex block structure, suggesting a less complete approach may be required to quantify the hydrodynamic coefficients. An alternative semipredictive approach that uses the tank and the CFD with semi-empirical method to validate the CFD process are used. This method is becoming increasingly tractable due to the development of CFD packages and the advancement of computer technology such as xPC-Target to facilitate the determination of the parameters obtained from the tank test. However, regardless of the method used, it is always subjected to experimental and numerical errors. Thus, due to these error or uncertainties, the model derived by either method has to depend on some form of robust control scheme such as the SMC (as seen in later section) to control the underwater robotic vehicle. The following sections illustrate the proposed methodology of rapid control system prototyping on the underwater robotic vehicle.

4.14.3 Steady-state thruster's dynamics

The thruster consists of a DC motor connected to a three-bladed propeller (with a diameter equal to 0.18 m). To determine the thrust developed by the thruster at different voltage input, the dynamics of the thruster need to be studied. As shown below, the dynamics of the DC motor is obtained followed by the propeller's dynamics. Through experiment, it was observed that thruster's dynamics at low speed could be modeled as a simple linear function of the propeller's thrust and the input voltage while the propeller angular speed can be obtained via the steady-state DC motor model.

It is known that the motor's shaft speed (assuming small electrical time constant as compared to the mechanical time constant) can be obtained using the following equation. Here, we briefly discuss the model and how it is incorporated into the simulation.

$$J_m\dot{\Omega} + \frac{K_m K_e}{R_m}\Omega = \frac{K_m}{R_m}\bar{u} - Q \tag{4.37}$$

$$\bar{u} = R_m I_t + L_m \frac{dI_t}{dt} + K_e\Omega \tag{4.38}$$

Moreover, the propeller's thrust and torque model [19,20] at steady state (without considering the axial fluid velocity at propeller) can be obtained using the following equation:

$$u = K_{Td}\Omega^2 \tag{4.39}$$

$$Q = K_{Qd}\Omega^2 \tag{4.40}$$

where $\bar{u}$ is the voltage input to the thruster in V, R_m is the armature resistance in Ohms, I_t is current to the armature in Ampere, L_m is the inductance in Henry, K_m is the motor torque constant $N \cdot m/A$, K_e is the motor back-emf in $V \cdot s/rad$, J_m is the rotor moment of inertia in $Nm \cdot s^2/rad$, Ω is the rotational speed of the propeller in rad/s, Q is the propeller's torque in $N \cdot m$, u is the thrust from the propeller in N, K_{Td} is the thrust constant in $N \cdot s/rad$, and finally, K_{Qd} is the torque constant in $N \cdot m \cdot s/rad$.

From the experiment in the water tank, an approximated linear relationship between the thrust, u and the commanded voltage input, $\bar{u}$ for the thruster was obtained. In Figure 4.84, the underwater robotic vehicle was running at a slow

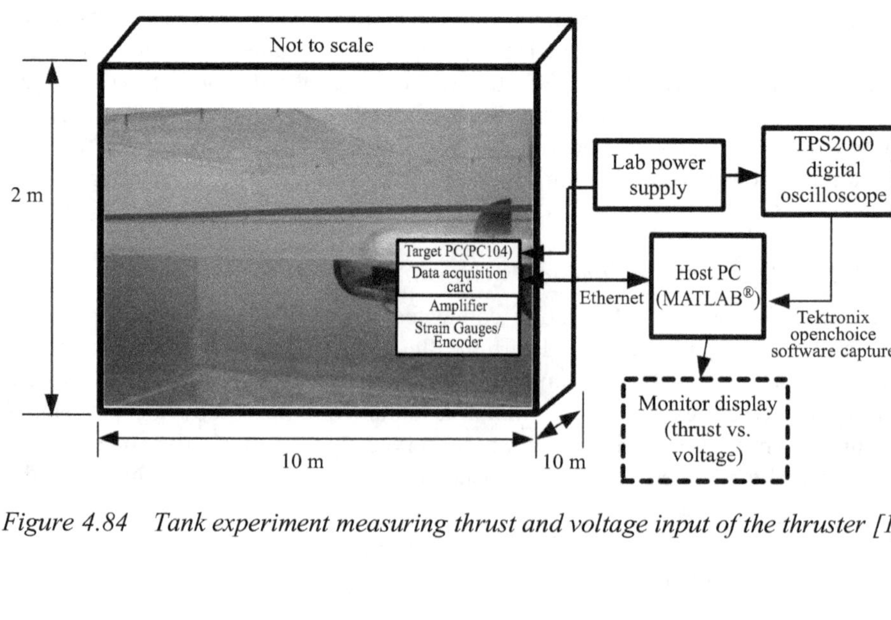

Figure 4.84 Tank experiment measuring thrust and voltage input of the thruster [1]

forward speed due to the space limitation of the tank (10 m × 10 m × 2 m). The underwater robotic vehicle platform was concealed using a water-tight aluminum casing. The torque was measured using a series of strain gages attached to the motor shaft, and the calibration was executed in the air with known masses. The test results showed linear behavior according to mass variations. The readings from the torque sensor were obtained via the interface connected to the computer. The applied voltage to the thruster was measured using a digital oscilloscope. The experiments were repeated for a few times. After the experiments, the thrusts were stored and computed offline by dividing the torque values from the strain gages by the pitch radius of the propeller. The experimental plot for the thrusts versus the applied voltage is shown in Figure 4.85. The plot gives a reasonable correlation between the two variables $\bar{u}$ and u as R^2 (that is the correlation factor) falls in the typical range of 0.7–1.

With the experimental results in Figure 4.85, the thrust and the voltage input relationship can be written as

$$u = f_T \bar{u} \tag{4.41}$$

where $f_T = 4.03$ N/V is from Figure 4.85. The input force and moment vector become $\tau = 4.03\mathbf{T}\,\bar{\mathbf{u}}$. As shown below, the thrust configuration matrix, **T,** provides the forces and moment distribution based on the top control surface and the thruster orientation on the underwater robotic vehicle platform. It is obtained by resolving the forces and moment about the underwater robotic vehicle's center of gravity.

$$\tau_x = f_T \bar{\text{u}} \tag{4.42}$$

$$\tau_y = f_T \bar{\text{u}} \sin \varphi \tag{4.43}$$

$$\tau_\psi = L/2 f_T \bar{\text{u}} \sin \varphi \tag{4.44}$$

where φ is the angle of the top control surface, f_T is 4.03 N/V, and L is the length of the underwater robotic vehicle.

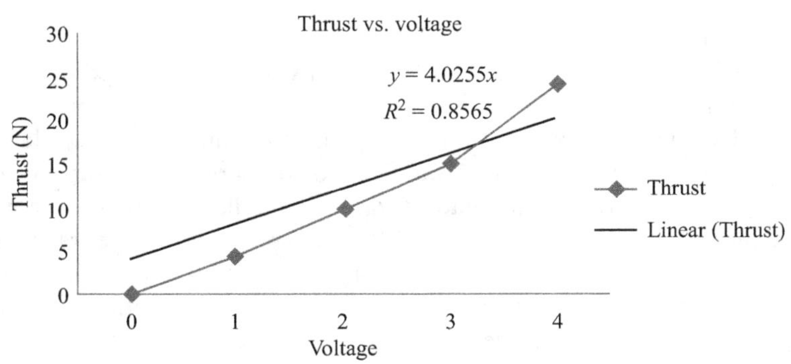

Figure 4.85 Thrust vs. voltage input to a thruster (forward motion) [1]

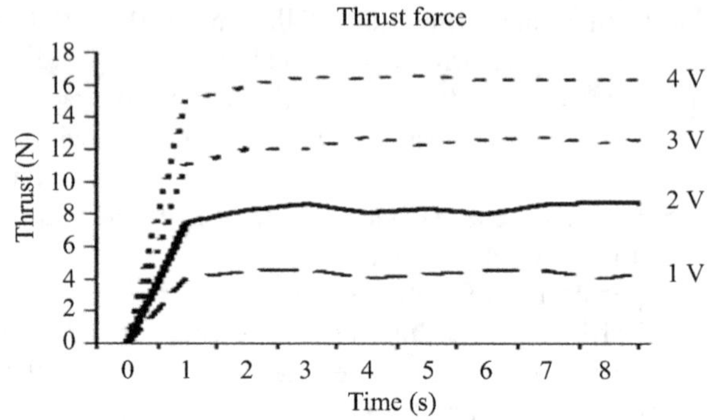

Figure 4.86 Thrusts over time at different input voltages in the water-tank test [1]

By manipulating the equations in (4.42) to (4.44) into matrix form, it becomes:

$$\begin{bmatrix} \tau_x \\ \tau_y \\ \tau_\psi \end{bmatrix} = f_T \begin{bmatrix} 1 & 0 & 0 \\ 0 & \sin\varphi & 0 \\ 0 & 0 & L/2.\sin\varphi \end{bmatrix} \bar{\mathrm{u}} \tag{4.45}$$

where **T** is:

$$\mathbf{T} = \begin{bmatrix} 1 & 0 & 0 \\ 0 & \sin\varphi & 0 \\ 0 & 0 & L/2.\sin\varphi \end{bmatrix} \tag{4.46}$$

As shown in Figure 4.86, time responses of the thrusts were captured during the experiment. The plot shows the transient and steady-state behavior of the thrust in water. Due to the tank size, the underwater robotic vehicle was left running for only 8 s before it reached the other side of the tank. The results have shown that the thrusts are able to settle to the steady-state values and the settling time is around 2 s for different voltage inputs.

The setup in Figure 4.87 is used to obtain the relationship between the angular speed of the propeller shaft and the input voltage. The speed was captured via the encoder mounted on the motor and the voltage was supplied by the laboratory DC power supply. With the speed and input voltage relationship obtained, xPC-Target in Figure 4.87 was used to compare the actual measurement results with the xPC-Target simulated data. The angular speeds were computed using the motor's encoder that has 24 CPR (for each leading edge transition) for the two channels together. With the planetary gear reduction of 103, the angular speed was reduced. As observed in the xPC-Target, it is useful to control the propeller's angular speed via the input voltage to the thruster. For control purpose, the angular propeller speed versus the input voltage is modeled as a first-order system with process delay. Since the electrical inductance is usually small,

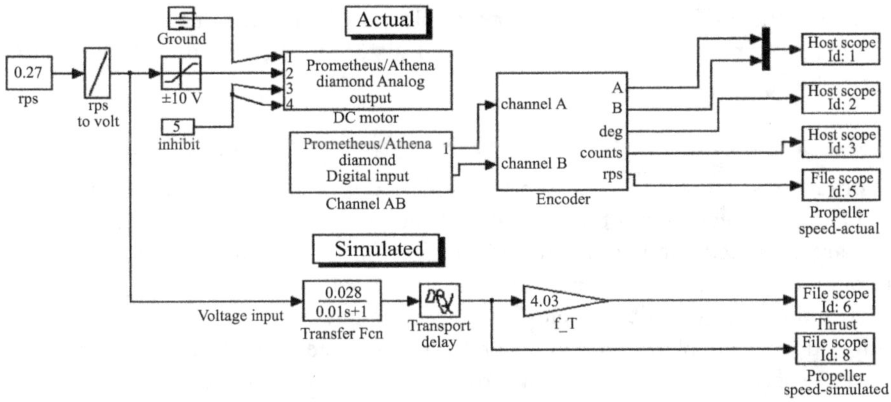

Figure 4.87 xPC-Target block diagram to determine propeller's speed vs. input voltage [1]

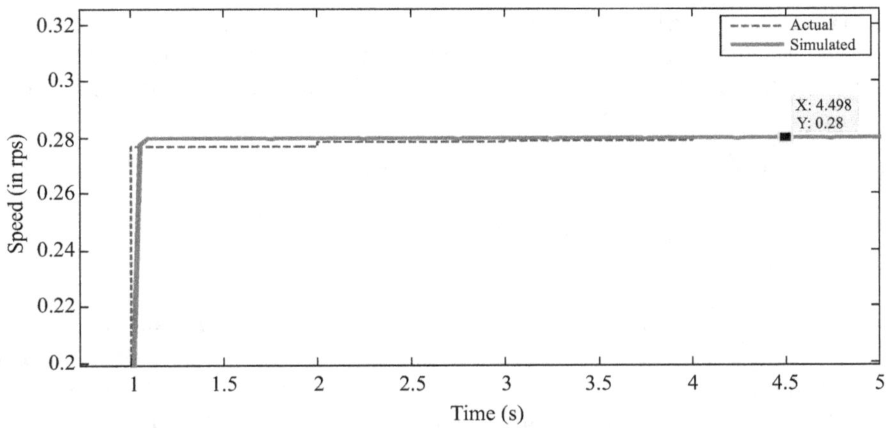

Figure 4.88 Simulated and actual propeller's speed over time in the water-tank test [1]

it is neglected. Thus, the open-loop transfer function for the speed $\Omega(s)$ over the input voltage $\bar{u}(s)$ becomes:

$$\frac{\Omega(s)}{\bar{u}(s)} = \frac{K_m}{\tau_m s + 1} e^{-s} \tag{4.47}$$

where the process time delay is observed to be 1 s (as seen in the time response plots in Figure 4.88), the motor gain K_m is around 0.028 and the motor time constant τ_m is about 0.01 s. As shown in Figure 4.87, the xPC-Target is used to compare the actual and simulated speed output of the thruster. By comparing the simulated result with the actual test (at a certain voltage input), there is a good match in the steady state except for the slight deviation during the transient stage.

However, the difference is quite negligible and hence the model is useful in obtaining the propeller angular speed (and also the corresponding thrust output using (4.45)).

4.14.4 Underwater robotic vehicle—horizontal subsystem model

As observed in the underwater robotic vehicle dynamics in (3.8), it is highly coupled and nonlinear in motions. In most of the underwater robotic vehicles design, they are designed for horizontal and vertical [4,7] plane motion. It reduces the coupling in motions as the underwater robotic vehicle moves in one plane at a time. Most underwater robotic vehicles adopt a decoupled maneuver scheme: moving with a horizontal plane motion whereby the vehicle is first steered at constant forward speed along a straight line to the desired target point. Thus considering only a few DOF at a time decouples the motions. This reduces the number of controllable DOF in subsequent control systems design.

In the preliminary stage, the underwater robotic vehicle is designed to operate in the horizontal plane. The roll, pitch, and heave velocities are bounded as the underwater robotic vehicle is moving along the X-axis with a constant heave (that is $w = \bar{w} = 0$) and at a constant surge velocity that defines the horizontal plane dynamic. The horizontal plane dynamics used in the underwater robotic vehicle can be written as:

$$(m - X_{\dot{u}})\dot{u} + X_{|u|u}|u|u + (Y_{\dot{v}} - m)vr = \tau_x \tag{4.48}$$

$$(m - Y_{\dot{v}})\dot{v} + Y_{|v|v}|v|v + (m - X_{\dot{u}})ur = \tau_y \tag{4.49}$$

$$(I_z - N_{\dot{r}})\dot{r} + N_{|r|r}|r|r + (-Y_{\dot{v}}u + X_{\dot{u}}u)v = \tau_\psi \tag{4.50}$$

where τ_x, τ_y, and τ_ψ are the input forces and moment. $X_{\dot{u}}, Y_{\dot{v}}$, and $N_{\dot{r}}$ are the hydrodynamic-added mass coefficients. $X_{|u|u}, Y_{|v|v}$, and $N_{|r|r}$ are the nonlinear hydrodynamic damping coefficients. The underwater robotic vehicle mass, m is around 14 kg and the inertia about the Z-axis was tabulated using the Pro/ENGINEER software as shown in Figure 4.89. The moment of inertia, I_r, is about 0.093 kg $\cdot$ m^2.

In the horizontal plane motion, the kinematics equation becomes:

$$\dot{x} = u \cos \psi - v \sin \psi \tag{4.51}$$

$$\dot{y} = u \sin \psi + v \cos \psi \tag{4.52}$$

$$\dot{\psi} = r \tag{4.53}$$

As shown in the dynamic equations, the hydrodynamic-added mass and damping coefficients are the unknown and need to be determined. Besides the thruster open-tank test, two groups of experiments were performed: (a) steady motion experiments with the underwater robotic vehicle to acquire the curves of the forces and moments versus the velocity variables and to verify the validity of CFD simulation using similar approaches; and (b) constant thrust tests to verify the

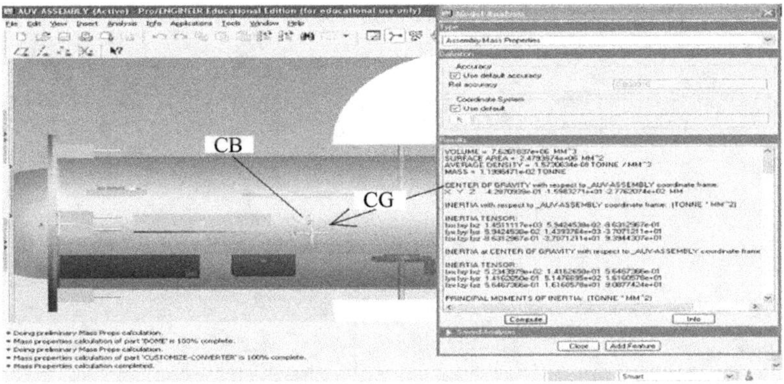

Figure 4.89 Pro/ENGINEER software to compute the moment of inertia and center of gravity [1]

values of hydrodynamic coefficients acquired by the CFD method and controller design in the water.

The flow passes over simple geometric shapes such as the underwater robotic vehicle have less complicated flow fields compared to the flow that passes over complex shapes object such as a remotely operated vehicle (remotely operated underwater vehicle). In an underwater robotic vehicle, the resultant flow field is mainly a function of the Reynolds number of the vehicle. The Reynolds number of a rigid body is the ratio of the inertial effects to the viscous effects within the fluid (at 20 $^{\circ}$C). The Reynolds number of the underwater robotic vehicle was 6×10^5, which was obtained as follows:

$$\mathrm{Re} = \frac{uL}{\nu} = \frac{1.0 \times 0.6}{1 \times 10^{-6}} = 6 \times 10^5 \tag{4.54}$$

When a body moves through a fluid, an interaction between the body and the fluid creates some induced hydrodynamic loads on the underwater robotic vehicle. The induced hydrodynamic loads are quantified by the added mass or inertial coefficients and the damping coefficients. The added mass coefficients which arise are due to potential flow phenomena for acceleration in translation and rotation, and the other are the damping coefficients due to viscous phenomena for translation and rotation of the body.

In most streamlined underwater robotic vehicle design, the added mass coefficients $X_{\dot{u}}, Y_{\dot{v}}, N_{\dot{r}}$ in (4.48)–(4.50) are computed analytically using the principle of Strip theory [7] as shown in (4.55). It involves dividing the submerged part of the vehicle into some strips. The two-dimensional hydrodynamic coefficients for the added mass can be computed for each strip and summarized over the length of the body to yield 3D coefficients. The underwater robotic vehicle is considered as a slender body. The added mass coefficients for each direction can be computed as follows.

$$\frac{\partial X}{\partial \dot{u}} = -X_{\dot{u}} = -\int_{-L/2}^{L/2} A_{11}^{(2D)}(y,z)dx = 0.1 \tag{4.55}$$

$$\frac{\partial Y}{\partial \dot{v}} = -Y_{\dot{v}} = \int_{-L/2}^{L/2} A_{22}^{(2D)}(y,z)dx = -\pi\rho(0.5D_a)^2 L \tag{4.56}$$

$$\frac{\partial N}{\partial \dot{r}} = -N_{\dot{r}} = \int_{-D_a/2}^{D_a/2} y^2 A_{11}^{(2D)}(x,z)dy - \int_{-L/2}^{L/2} x^2 A_{22}^{(2D)}(y,z)dx$$

$$= 1/24\left[\pi\rho(0.5D_a)^2 L^3 + 0.1{D_a}^3\right] \tag{4.57}$$

where $L = 0.6$ m and $D_a = 0.25$ m are the length and diameter of the underwater robotic vehicle, respectively. $\rho = 1{,}000$ kg/m^3 is the density of water at 20 °C, and by substituting the values into the equations, the added mass coefficients for each direction are obtained as $X_{\dot{u}} = 0.1, Y_{\dot{v}} = 29.4$, and $N_{\dot{r}} = 0.1$.

As shown in Figure 4.90, a set of experiments were carried out in a 10 m × 10 m × 2 m open tank (tank is not shown for clarity) to investigate the motion characteristics of the underwater robotic vehicle in the surge, sway, and yaw in the positive direction on the horizontal plane. Each experiment was repeated for many times at each velocity and the average damping force readings (the longitudinal

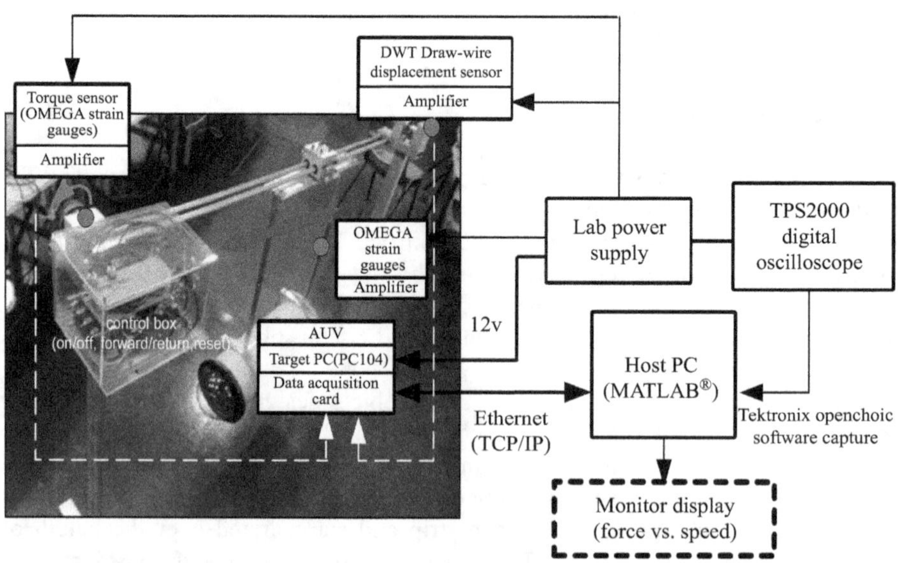

Figure 4.90 Rig for drag testing in the X-direction (water tank not shown for clarity) [1]

Table 4.4 Comparison between experiment and ANSYS-CFX nonlinear damping coefficients for underwater robotic vehicle [1]

Velocity (m/s)	**Experiment (*L*/*D* = 2.5)**			**ANSYS-CFX (*L*/*D* = 2.5)**		
	$X_{\|u\|u}$ **(Ns²/m²)**	$Y_{\|v\|v}$ **(Ns²/m²)**	$N_{\|r\|r}$ **(Ns²/rad²)**	$X_{\|u\|u}$ **(Ns²/m²)**	$Y_{\|v\|v}$ **(Ns²/m²)**	$N_{\|r\|r}$ **(Ns²/rad²)**
0	0	0	0	0	0	0
0.1	0.62	0.81	0.45	0.52	0.51	0.35
0.2	0.71	0.95	0.98	0.61	0.75	0.78
0.3	0.85	1.19	1.21	0.75	1.04	1.01
0.4	0.83	1.21	1.23	0.78	1.11	1.13
0.5	0.86	1.21	1.29	0.76	1.15	1.18
0.6	0.88	1.24	1.35	0.84	1.19	1.23
0.7	0.91	1.25	1.28	0.87	1.21	1.25
0.8	0.91	1.32	1.29	0.94	1.26	1.26
0.95	0.92	1.34	1.32	0.93	1.31	1.3
Grand average	**0.75**	**1.05**	**1.04**	**0.70**	**0.95**	**0.95**
Nondimensional results [21] on Kempf model (using ANSYS-FLUENT, *L*/*D* = 6)				Not available	**0.0767**	**0.0185**
Present non-dimensional results on Kempf model (using ANSYS-CFX, *L*/*D* = 6)				Not available	**0.0811 (5.7% error)**	**0.0192 (3.8% error)**

force $X_{|u|u}$, the transverse force $Y_{|v|v}$, and moment $N_{|r|r}$ are the surge force, sway force, and yaw moment, respectively) were tabulated in Table 4.4. With the vehicle surge or sway set to a constant velocity, the thrust can be considered to be equal to the hydrodynamic force, and the moments provided by thrusters can be considered to be equal to the hydrodynamic moments when it yaws at a constant angular velocity. Therefore, the curves (or the data in Table 4.4) of the thrust versus the motion variables are assumed to be same as the curves of hydrodynamic loads versus these motion variables. As a result, the curves of the thrust versus the motion variables u, v, and r could be obtained in Figure 4.91.

The test equipment consists of a DWT draw-wire displacement transducer (string pot) that was used to measure the displacement of the vehicle in water with respect to time. The terminal velocities of the vehicle were determined. The underwater robotic vehicle was orientated such that the displacement in each direction can be measured during the test. For example, Figure 4.90 shows the testing of the damping force in the X-direction. To measure the damping force in

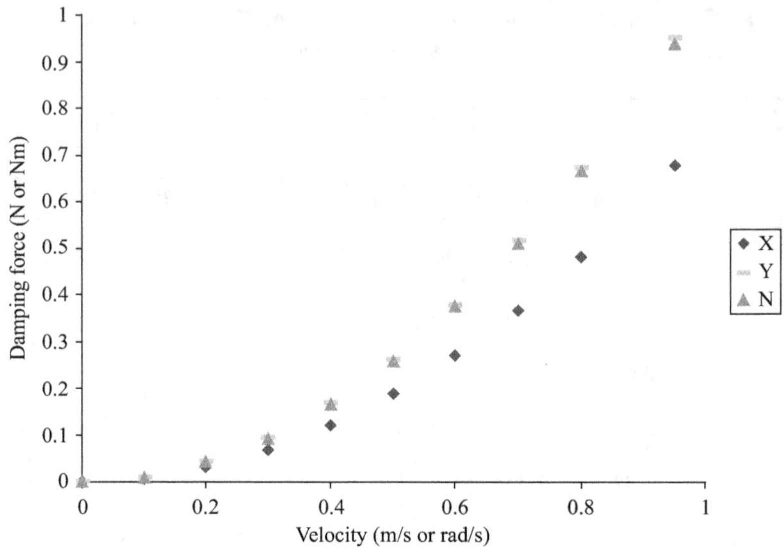

Figure 4.91 Nonlinear damping force in X, Y, and moment about Z from the experiment [1]

the *Y*-direction, the underwater robotic vehicle has to rotate 90° (clockwise) so that the lateral side faces the longitudinal direction. The torque sensor (or pre-wired strain gage) mounted on the motor shaft measured the torque generated and later converted to the translational force in the underwater robotic vehicle's *X*- and *Y*-direction. The pre-wired strain gages on the beam surfaces captured the moment generated during the yaw motion (moment about the *Z*-axis). The net forces and moments to move the underwater robotic vehicle were then determined.

To save the computational cost, the equations of motion can be simplified based on the following assumptions: (a) the vehicle moves only along the direction of thrust, (b) the hydrodynamic moment is negligible, and (c) the attitude angles or the pith angle are always zero. The vehicle is moving linearly in each direction and the maximum speeds at each direction (that correspond to zero acceleration) were measured. The damping coefficients in the *X*- and *Y*-directions were determined as follows.

For drag force in the *X*-direction,

$$\tau_x - X_{|u|u}|u|u = 0 \Rightarrow X_{|u|u} = \tau_x/u_{\max}^2 \tag{4.58}$$

For drag force in the *Y*-direction,

$$\tau_y - Y_{|v|v}|v|v = 0 \Rightarrow Y_{|v|v} = \tau_y/v_{\max}^2 \tag{4.59}$$

For drag force about the *Z*-axis, the underwater robotic vehicle was allowed to rotate freely so that the top control surface could change the underwater robotic vehicle's heading direction. Assuming maximum displacement (or zero velocity) in the *X*- and *Y*-directions, the thrust becomes:

$$\tau_\psi - (N_{|r|r}|r|r) = 0 \Rightarrow N_{|r|r} = \tau_\psi / r_{\max}^2 \tag{4.60}$$

The estimated damping coefficients of the underwater robotic vehicle with a velocity that varies from 0 to 0.95 m/s with an increment of 0.1 m/s are tabulated in Table 4.4 while the curves of the thrust versus the motion variables *u*, *v*, and *r* are shown in Figure 4.91. To compare the experimental results obtained from the water-tank test with CFD simulation, the flow solver ANSYS-CFX is used.

In most underwater robotic vehicle operations, considering its scale and speed, the calculated Reynolds number in equation (4.61) indicates turbulence in the flow field. In turbulence flow, the fluid motion is characterized by a highly random, unsteady 3D flow. The turbulent length and time scales are much smaller than the smallest finite volume meshes that are used in the most numerical analysis. To generate such a small mesh is beyond the computing power currently available in our laboratory. The problem is solved by employing turbulence model. The ANSYS CFX was used because of the availability of the SST model. Currently, the k-ω-based SST model is the most prominent two-equation models to predict flow transition from laminar to turbulent. It was designed to give highly accurate predictions of the onset and the amount of flow separation under adverse eddy viscosity.

Successful computations of turbulent flows require some consideration during the mesh generation. Due to the strong interaction of the mean flow and turbulence, the numerical results for turbulent flows tend to be more susceptible to grid dependency than those for laminar flows. To resolve the regions where the mean flow changes rapidly and there are shear layers with a large mean rate of strain, one must generate the sufficiently fine mesh. The SST turbulence model with automatic wall function treatment will be used because of its highly accurate predictions of flow separation. The size of the mesh [21] around the underwater robotic vehicle body was 3.66×10^{-5} *L*, which was obtained as follows:

$$y^+ = 0.172\left(\frac{y}{L}\right)\mathrm{Re}^{0.9} \Rightarrow y = \frac{1}{0.172 \times (6 \times 10^5)^{0.9}} L = 3.66 \times 10^{-5} L \tag{4.61}$$

where *L* is the length of the underwater robotic vehicle body.

The main difficulty is to determine the fluid domain dimension as shown in Figure 4.92 (left). To study damping force acting on the underwater robotic vehicle in an unbound fluid domain, an infinite large fluid domain is needed. However, this is not practical both in CFD and experiment. The fluid domain dimensions of around 20 times the length, width, and height of the underwater robotic vehicle as shown in Figure 4.92 are chosen. The uncertainty analysis had been carried out by increasing the domain size to see any variations in results; however, the deviations in the results are negligible as it reaches the 11 m × 3 m × 5.25 m. The region of disturbed flow (usually turbulent) downstream of the underwater robotic vehicle body moving through a fluid, caused by the flow of the fluid around the body, is formed. As shown in the velocity distribution in Figure 4.92 (right), the region

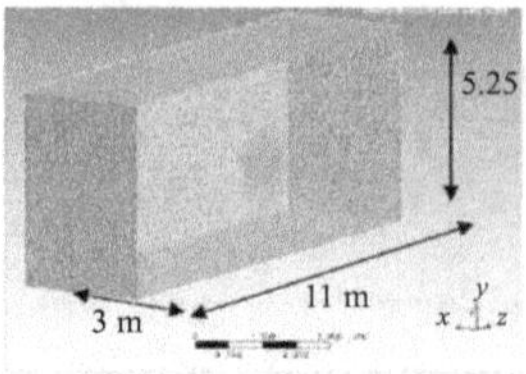

Setting	Value
Basic settings > Fluids list	Water
Domain models > Pressure > Reference pressure	1 [atm]
Heat transfer > Option	Isothermal
Heat transfer > Fluid temperature	20 [C]
Turbulence > Option	Shear stress transport

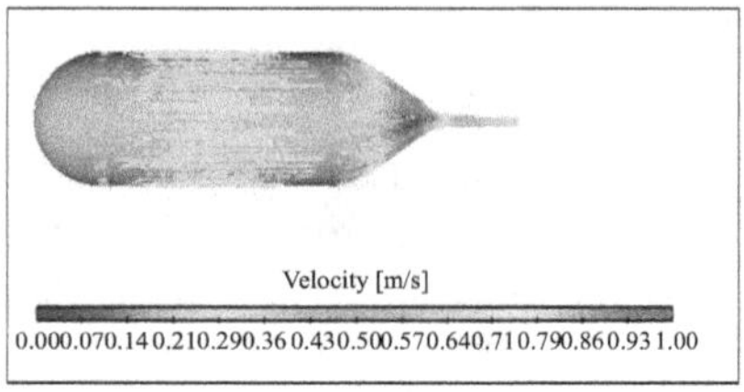

Figure 4.92 Meshing for flow domain around underwater robotic vehicle (left) and streamlines plot of the underwater robotic vehicle (right) [1]

creates high-velocity (or low-pressure) region at the rear end and thus low velocity (or high pressure) in the front resists the vehicle motion. This exhibits similar phenomena as compared to the streamline plots for the sphere.

In ANSYS-CFX, the flow in the domain surrounding the vehicle was simulated for various speeds ranging from 0 to 0.95 m/s (see Table 4.4). The hydrodynamic loads or the damping forces were obtained by integrating the pressure on its surface, which was obtained via the simulation. As observed in Table 4.4, the forces show good agreement when the vehicle surges and sways; however, the calculated moments are smaller than the experimental results. There are two main reasons for the differences in the moments. The first is that the thruster is not considered in the CFD model, which may lead to an error in the simulated hydrodynamic center of rotation. The other is that some small changes were made to adjust the center of gravity and buoyancy in the experiments due to the fixture that held the underwater robotic vehicle during the test, which implies that they may be different from their design points. Both of these factors have some effect on the dynamic of the vehicle, whereas their effect on the rotational dynamics is small as compared to the translational dynamics. In addition, it must be noted that this vehicle was operated along rail of the rig with some pitching effect during the experiment, and therefore the pitching may increase the experimental values as compared to the CFD results. This can be seen in the slight increase over the CFD results.

To gain confidence in CFD computed results, comparison with available data in literature based on computation and semi-empirical formulae for geometries of comparable shapes have to be performed. Many of the available empirical relations provide only the linear damping coefficients, which are presumed to hold good for small value of angle attack or in this case the pitch angle (defined as the angle between velocity and longitudinal axis of the body) for the remotely operated underwater vehicle but not for the underwater robotic vehicle which moves at a larger pitch angle during diving (assumed zero pitching due to the current underwater robotic vehicle design limitation). However, the present computations show that the nature of these forces and moments are essentially nonlinear and therefore for adequate representation for these forces and moments, a linear model is insufficient even over a small-velocity range.

The geometry in the literature [22] has been used. The model is axisymmetric with diameter, a hemispherical nose, and a sinusoidal stern with tail radius 0.1 of diameter. However, the underwater robotic vehicle has no stern radius and the results can be different. Moreover the available data on hydrodynamic coefficients for submerged vehicles are extremely scarce in literature [21,23]. Hence, a direct comparison of the computed derivatives with available results could be difficult. As such, since even for same geometry, the coefficient values obtained from different sources show large scatter [24], it can only be said that the estimated coefficient values presently are at best accurate within some bound. The main objective of the present work is to show that the CFD results are within the acceptable range and more importantly to show that CFD results essentially predict the forces or moments as nonlinear variation of the velocity using the same model. ANSYS-CFX is used as FLUENT (commonly used in the literature) was not available, but the model adopted for the analysis is similar. In this case, the Kempf model in [21,22] with $L/D = 6$ is used. As shown in Table 4.1, the results seem to be higher than the one stated in the literature. This may be due to the boundary, meshing, outflow and inflow settings in the software that are not mentioned. However, the results exhibit around 5% error as compared to the results from the literature.

Hence, ANSYS-CFX is able to predict similar results and therefore the results are at best accurate within some reasonable bound. Besides, the key nonlinear damping coefficients were obtained; it is also possible to obtain linear damping coefficients using appropriate initial conditions. However, since the nonlinear damping is more difficult to obtain and important to the underwater robotic vehicle, it is commonly designed to pitch in order to reach a greater depth (although the test has less pitching).

In summary, all of these findings offer a reference point for determining which coefficients need to be investigated when considering the proposed underwater robotic vehicle that has a different shape as compared to the streamlined semi-empirical or empirical model at different conditions. The values of the parameters obtained for the underwater robotic vehicle are shown in Table 4.5.

Table 4.5 Measured values for test-bed underwater robotic vehicle [1]

Descriptions	**Values**
D_a	0.25 m
L	0.60 m
m	14 kg
I_r	0.093 kg·m^2
$Y_{\dot{v}}$, $Y_{\|v\|v}$	29.4, 1.05 $\mathrm{N \cdot m^{-2} s^2}$
$\mathrm{N}_{\dot{r}}$, $\mathrm{N}_{\|r\|r}$	0.1, 1.04 $\mathrm{N \cdot m^{-2} s^2}$
$X_{\dot{u}}$, $X_{\|u\|u}$	0.1, 0.75 $\mathrm{N \cdot m^{-2} s^2}$

With the hydrodynamic parameters obtained in a horizontal plane, the dynamics (denoted by subscript "h") can be written in the compact form:

$$\mathbf{M_h}\dot{\boldsymbol{\nu}}_\mathbf{h} + \mathbf{C_h}(\boldsymbol{\nu}_\mathbf{h})\boldsymbol{\nu}_\mathbf{h} + \mathbf{D_h}|\boldsymbol{\nu}_\mathbf{h}|\boldsymbol{\nu}_\mathbf{h} = \tau \tag{4.62}$$

where $\boldsymbol{\eta}_\mathbf{h} = [x \quad y \quad \psi]^T$, $\boldsymbol{\nu}_\mathbf{h} = [u \quad v \quad r]^T$, $\tau = f_T \mathbf{T}\,\bar{\mathbf{u}} = [\tau_x \quad 0 \quad \tau_\psi]^T$, and φ is the angular position of the top control surface.

$$\mathbf{M_h} = \begin{bmatrix} m - X_{\dot{u}} & 0 & 0 \\ 0 & m - Y_{\dot{v}} & 0 \\ 0 & 0 & I_r - N_{\dot{r}} \end{bmatrix}; \; \mathbf{C_h}(\boldsymbol{\nu}_\mathbf{h}) = \begin{bmatrix} 0 & (Y_{\dot{v}} - m)r & 0 \\ 0 & 0 & (m - X_{\dot{u}})u \\ 0 & -(Y_{\dot{v}} - X_{\dot{u}})u & 0 \end{bmatrix};$$

$$\mathbf{D_h} = \begin{bmatrix} X_u & 0 & 0 \\ 0 & Y_v & 0 \\ 0 & 0 & N_r \end{bmatrix}; \; \mathbf{T} = \begin{bmatrix} 1 & 0 & 0 \\ 0 & \sin\varphi & 0 \\ 0 & 0 & 0.3.\sin\varphi \end{bmatrix}; \; f_\mathrm{T} = 4.03 \tag{4.63}$$

The kinematics equations can be expressed as:

$$\dot{\boldsymbol{\eta}}_\mathbf{h} = \mathbf{J_h}(\boldsymbol{\eta}_\mathbf{h})\boldsymbol{\nu}_\mathbf{h} \tag{4.64}$$

where

$$\mathbf{J_h}(\boldsymbol{\eta}_\mathbf{h}) = \begin{bmatrix} \cos\psi & -\sin\psi & 0 \\ \sin\psi & \cos\psi & 0 \\ 0 & 0 & 1 \end{bmatrix} \tag{4.65}$$

Equation (4.62) can be rewritten in state-space form:

$$\dot{\mathbf{x}}_\mathbf{h} = \bar{\mathbf{f}}_\mathbf{h}(\mathbf{x_h}, t) + \bar{\mathbf{g}}_\mathbf{h}(\bar{\mathbf{u}}, t) \tag{4.66}$$

where $\mathbf{x_h} = [\boldsymbol{\eta}_\mathbf{h}\boldsymbol{\nu}_\mathbf{h}]^T$

$$\bar{\mathbf{f}}_\mathbf{h}(\mathbf{x_h}, t) = \begin{bmatrix} \mathbf{J_h}(\boldsymbol{\eta}_\mathbf{h})\boldsymbol{\nu}_\mathbf{h} \\ -\mathbf{M}_\mathbf{h}^{-1}[\mathbf{C_h}(\boldsymbol{\nu}_\mathbf{h}) + \mathbf{D_h}]\boldsymbol{\nu}_\mathbf{h} \end{bmatrix}; \; \bar{\mathbf{g}}_\mathbf{h}(\bar{\mathbf{u}}, t) = \begin{bmatrix} \mathbf{0}_{3\times 1} \\ [\bar{\mathbf{g}}_{\mathbf{h}1} \quad \bar{\mathbf{g}}_{\mathbf{h}2} \quad \bar{\mathbf{g}}_{\mathbf{h}3}]\mathbf{T_h}\mathbf{F_T}\bar{\mathbf{u}} \end{bmatrix};$$

$$\bar{\mathbf{g}}_{\mathbf{h}_1} = \begin{bmatrix} \frac{1}{m - X_{\dot{u}}} & 0 & 0 \end{bmatrix}^T; \; \bar{\mathbf{g}}_{\mathbf{h}_2} = \begin{bmatrix} 0 & \frac{1}{m - Y_{\dot{v}}} & 0 \end{bmatrix}^T; \; \bar{\mathbf{g}}_{\mathbf{h}_3} = \begin{bmatrix} 0 & 0 & \frac{1}{I_r - N_{\dot{r}}} \end{bmatrix}^T$$

The constant thrust tests were used to verify the values of hydrodynamic coefficients acquired by the CFD method. The respective thrust inputs 1.2 N (0.3 m/s), 2 N (0.5 m/s), and 3.8 N (0.95 m/s) are used in these simulations. As shown in Figure 4.93, xPC-Target was used to simulate the horizontal dynamics of the underwater robotic vehicle. The dynamics of the underwater robotic vehicle for the respective x, y, and yaw directions can be seen in the subsystem block diagrams. It generates the simulated time responses to compare with the experiment

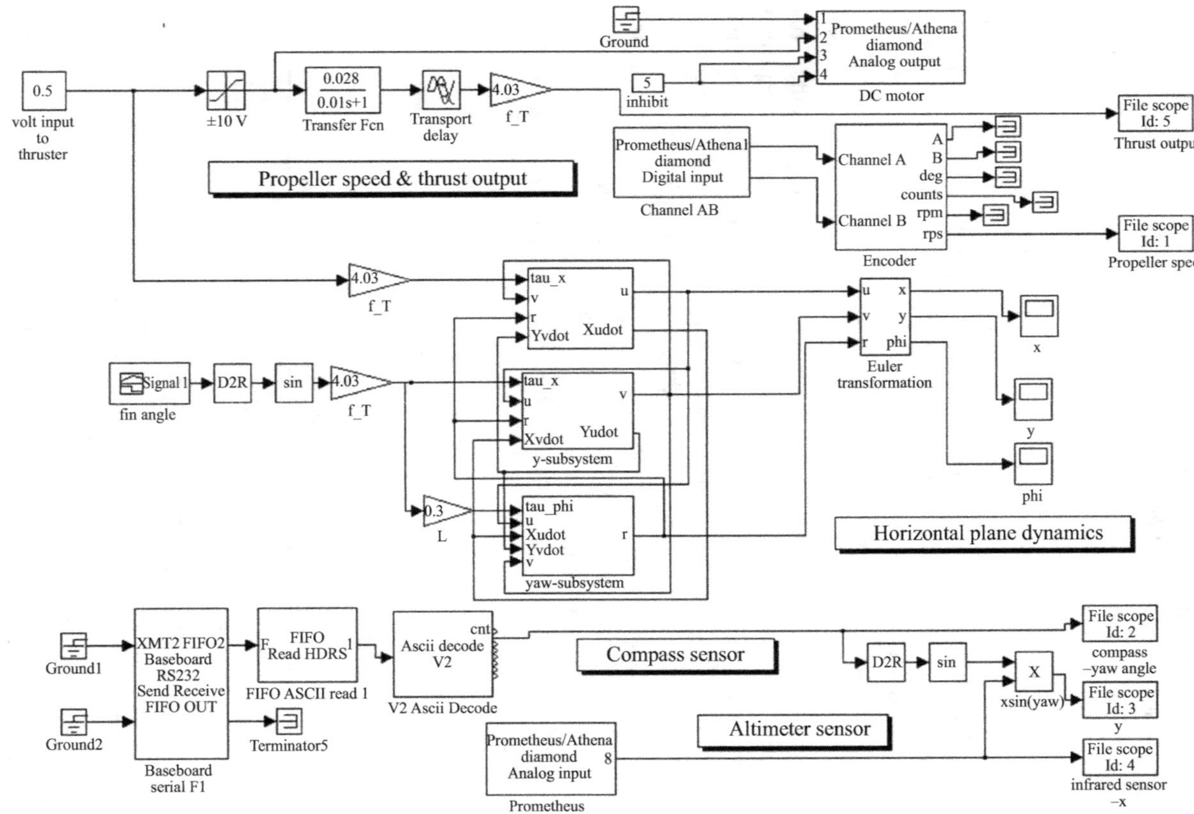

Figure 4.93 xPC-Target in horizontal plane used in the water-tank test [1]

results obtained from the respective sensors. The thruster is used for the longitudinal motion while the heading of the underwater robotic vehicle is controlled by the angular position of the top control surface, φ. The sway position or the deviation of the underwater robotic vehicle from its longitudinal axis is indirectly computed via the heading angle as shown in the use of sine function in the block diagram. The compass that reads the yaw or the heading angle gives the output in ASCII format. Through the Baseband RS232 Send Receive, FIFO ASCII Reader, and ASCII Decode subsystems, the data was decoded into acceptable format for control purpose.

In conjunction with the physical vehicle, the computer numerical method using xPC-Target in Figure 4.93 is used to compare with the hydrodynamic coefficients obtained from the CFD and the experiments. A fixed-step solver was used in the simulation, and each simulation was run for 70 s to ensure the obstacle avoidance was performed adequately. Each simulation was repeated a number of times. Due to space constraints, we will present the results for the underwater robotic vehicle moving at 0.95 m/s. As shown in Figure 4.94, there are some differences between the dynamic predicted by the equations of motion and the experiment results. This is to be expected since the horizontal equations of motion have been simplified based on the assumptions made in the equations. As observed in the plots, the time responses in x, y positions, and yaw angle are quite oscillatory due to the top control surface that tried to overcome the opposing drag force during the experiment. Nevertheless, the simulated hydrodynamic coefficients (obtained through the experiments and numerical methods) do provide a good representation of the overall dynamics of the vehicle in water.

However, the hydrodynamic coefficients can be further improved by repeating the entire process.

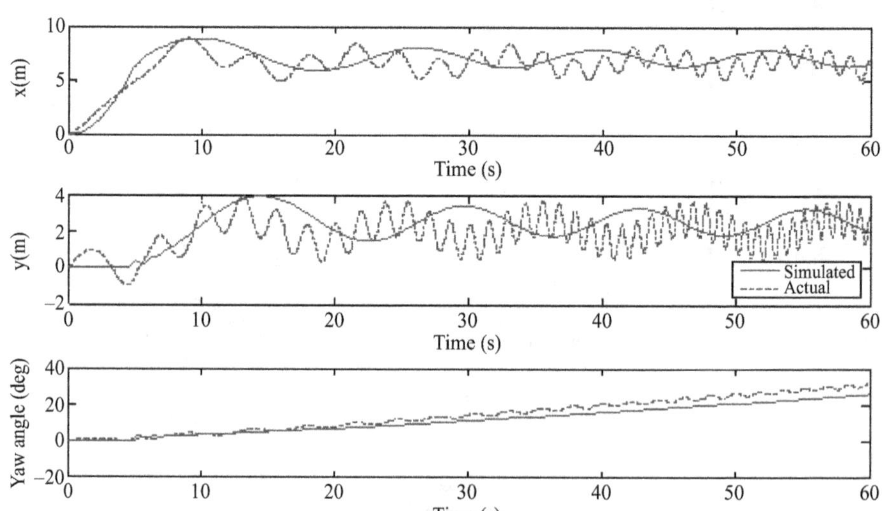

Figure 4.94 Comparison between simulated vs. actual in the water tank at 0.95 m/s [1]

4.14.5 Controller design

To reduce the oscillation response of the top control surface as seen in the open-loop test, a closed-loop control system is required. The surge position is controlled by the propeller speed (using the PID controller) while the heading angle is controlled by the top control surface angle (denoted by φ). There have been various efforts to develop the controller for the underwater robotic vehicle, which include both the (conventional) linear and the (modern) intelligent control schemes. Given the complexities of is control requirements, it is clear that the linear controller is unable to control the vehicle satisfactorily [25]. Intelligent control methods which include NN, sliding mode (SLC), and FLC are more robust and able to adapt to the vehicle's hydrodynamic uncertainties. In addition, they exhibit excellent immunity to disturbances. The SMC [6] is chosen. By using xPC-Target blocksets, the complex SMC scheme can be easily implemented on the underwater robotic vehicle platform. For notational simplicity, define a scalar measure of tracking

$$s = \dot{\tilde{\varphi}} + \lambda\tilde{\varphi} \tag{4.67}$$

where $\tilde{\varphi} = \varphi - \varphi_d$ is the tracking angle error, i.e., the error of the top control surface angle. $\lambda > 0$ is the control bandwidth. For $s = 0$, this expression describes a sliding surface with exponential dynamics

$$\tilde{\varphi}(t) = \exp[-\lambda(t - t_0)]\tilde{\varphi}(t_0) \tag{4.68}$$

which ensures that the tracking error $\tilde{\varphi}(t)$converges to zero in infinite time when $s = 0$ (sliding mode). The error trajectory reaches the time-varying sliding surface in finite time for any initial condition $\tilde{\varphi}(t_0)$, and then slides along the surface toward $\tilde{\varphi}(t) = 0$ exponentially. Hence, the control objective is reduced to finding a control law which ensures that $\lim_{t\to\infty} s(t) = 0$.

In the design of the sliding control law, it is convenient to define a virtual reference φ_r satisfying

$$\dot{\varphi}_r = \dot{\varphi}_d - \lambda\tilde{\varphi} \Rightarrow s = \dot{\varphi} - \dot{\varphi}_r \tag{4.69}$$

Hence, the following expression for $\tau_m\dot{s}$ is obtained.

$$\begin{aligned} \tau_m\dot{s} &= \tau_m(\ddot{\varphi} - \ddot{\varphi}_r) \\ &= (K_m V - \dot{\varphi}) - \tau_m\ddot{\varphi}_r \\ &= -|\dot{\varphi}|s + (K_m V - \tau_m\ddot{\varphi}_r - |\dot{\varphi}|\dot{\varphi}_r) \end{aligned} \tag{4.70}$$

Consider the scalar Lyapunov-like function candidate

$$V_L(s,t) = \frac{1}{2}\tau_m s^2, \tau_m > 0 \tag{4.71}$$

Differentiating V_L with respect to time (as $\dot{\tau}_m = 0$) yields

$$\dot{V}_L = \tau_m s\dot{s} = -|\dot{\varphi}|s^2 + s(K_m\tau_\psi - \tau_m\ddot{\varphi}_r - |\dot{\varphi}|\dot{\varphi}_r) \tag{4.72}$$

Taking the control law for the top fin control surface to be

$$\tau_\psi = -1/K_m(K_d s + K\mathrm{sgn}(s)) \tag{4.73}$$

where

$$\mathrm{sgn}(s) = \begin{cases} 1 & \text{if } s > 0 \\ 0 & \text{if } s = 0 \\ -1 & \text{otherwise} \end{cases} \tag{4.74}$$

yields

$$\dot{V}_L = -|\dot{\varphi}|s^2 + s(-K_d s - K\mathrm{sgn}(s) - \tau_m\ddot{\varphi}_r - |\dot{\varphi}|\dot{\varphi}_r) \tag{4.75}$$

Conditions on the switching gain K_d and K are found by ensuring that $\dot{V}_L \leq 0$. The particular choice

$$K_d + K \geq \tau_m\ddot{\varphi}_r + |\dot{\varphi}|\dot{\varphi}_r \tag{4.76}$$

where $\tau_m = 0.01$, and as the motor is intended to operate slowly, $\dot{\varphi}$ and $\dot{\varphi}_r$ becomes small. This implies

$$\dot{V}_L \leq -(|\dot{\varphi}| + K_d)s^2 - K\mathrm{sgn}(s) \leq 0 \tag{4.77}$$

Note that, $\dot{V}_L \leq 0$ implies that $V_L(t) \leq V_L(0)$, and therefore s is bounded. This in turn implies that $\ddot{V}_L$ is bounded. Hence, $\dot{V}_L$ must be uniformly continuous. Finally, application of Barbalat's lemma, then shows that $s \to 0$ and thus $\tilde{\dot{\phi}} \to 0$ as $t \to \infty$.

The SMC law in (4.73) is implemented as shown in Figure 4.95. When the tracking error is negative (positive), indicating that the position is too small, the control input is increased (decrease) in order to increase the output. The "energy" of the system will diminish slowly as it reaches the sliding surface (near to zero tracking error). To test the robustness of the controller against the external disturbance, a low-frequency (2 Hz and having a peak-to-peak amplitude less than 5) sine wave disturbance was injected into the system. The value of the external disturbance can be observed from the open-loop test where the top control plane oscillates no more than 5° from its initial position. The passive approaches for the SMC require it to be designed off-line in order to achieve the robustness against this disturbance. The control gains were adjusted to achieve the desired output during the simulation stage.

Once the robust controller is designed and simulated to accommodate for the disturbance and the 10% model uncertainties (by decreasing and increasing the hydrodynamic coefficients by 10%) due to the sensitivity [18] of the model parameters on the vehicle response, it was implemented in xPC-Target block diagram as shown in Figure 4.97. But, if other disturbances are not accounted for during the (off-line) design of the robust controller, then they might insert adverse effects on the system. However, as shown in Figure 4.96 or later the pool test, the time response is less oscillatory. The robustness property of SMC can thus compensate

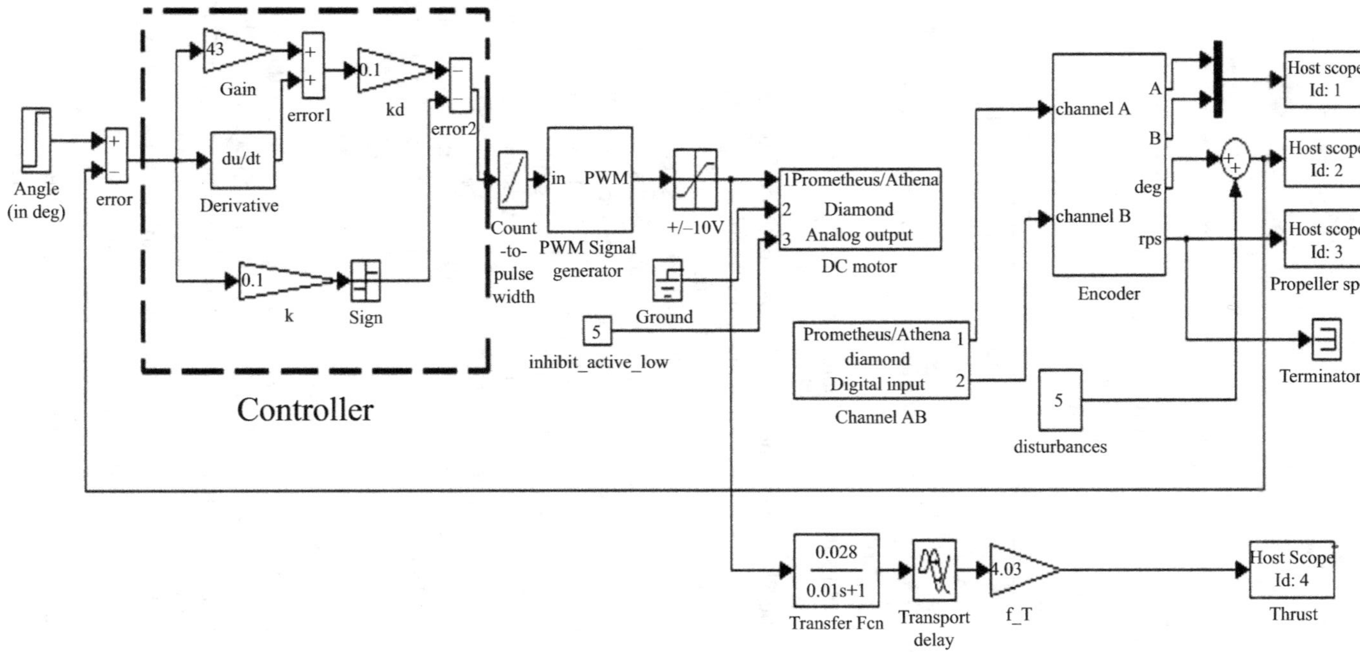

Figure 4.95 SMC system using xPC-Target block diagrams

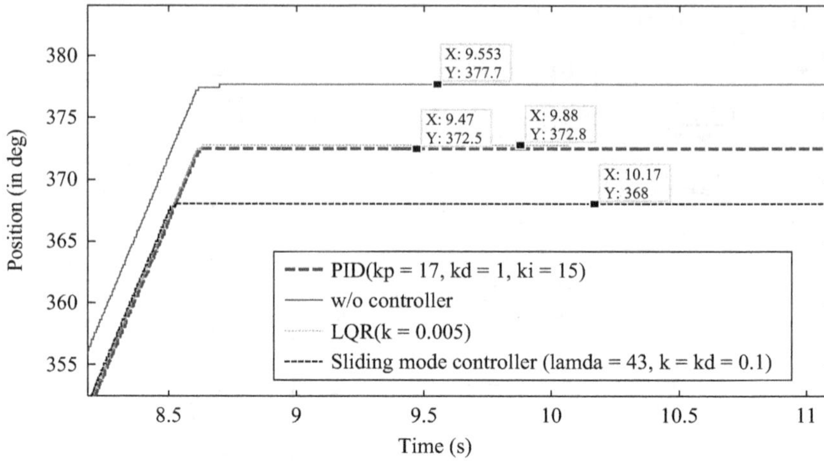

Figure 4.96 Time response of various controllers for certain angle [1]

the disturbances, in this case, the oscillatory behavior of the top control plane and the model uncertainties.

To compare the SMC with other control scheme, few controllers are compared. Besides using PID controller to control the top control surface angular position in water, a LQR was used. In this method, it uses a set of weighting matrices to penalize the states and control inputs of the system. This requirement is set in the cost function to be minimized. The weighting matrices associated with the cost function were applied to the states and inputs, respectively. They were tuned iteratively until the desired performance was achieved. The controller gain, F, for the LQR was calculated and implemented as a feedback gain ($F = 0.05$). This gain was positioned along the feedback loop for the speed. As shown in Figure 4.96, the oscillatory motion has decreased. The steady-state error using the SMC seems to be smaller as compared to the PID and LQR controller. With the results obtained in the hardware-in-the-loop testing, the proposed SMC will be used in subsequent pool test.

4.14.6 Implementation and testing

To verify the proposed rapid control systems prototyping using the xPC-Target systems, the underwater robotic vehicle was tested in the pool. Since the top and bottom control surface are connected and aligned to each other, the bottom control surface is used effectively to change the heading of the underwater robotic vehicle while the top control surface to show the angular position of the fins in water. With the present sensors and system limitation, the application on an obstacle avoidance task seems to be appropriate for the underwater robotic vehicle to perform.

The block diagram used for the pool test can be seen in Figure 4.97. The propeller's speed was controlled by a PID controller. The propeller's motor was connected to the channel 2 of the Analog Output and the fins' motors were

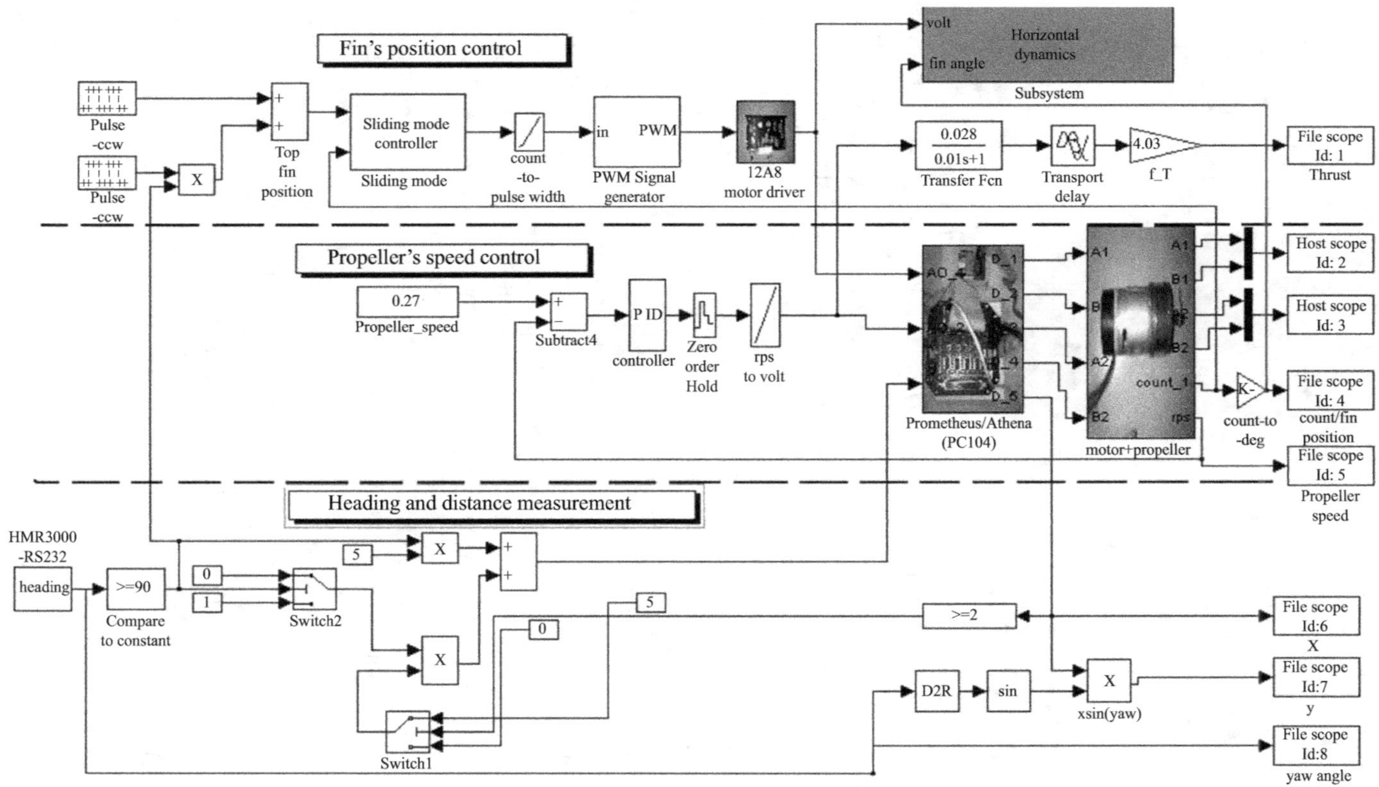

Figure 4.97 xPC-Target block diagram for obstacle avoidance [1]

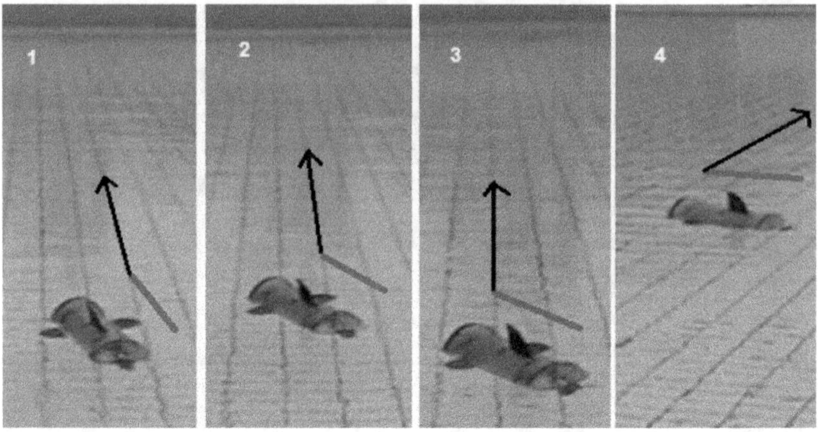

Figure 4.98 Actual motion sequence of underwater robotic vehicle during obstacle avoidance in pool [1]

connected to the channel 1 of the Athena Analog Output. The SMC was selected to control the angular position of the control surface. After the obstacle ("high" TTL signal equivalent of 2 V) was detected, it sent a 5 V to enable the fins' motor to rotate CCW. When the heading angle reads 90° or more, the fins' motor rotates CW to its original position till the vehicle registers the 90° heading angle.

With the control algorithm finalized, the underwater robotic vehicle was tested in the swimming pool as shown in Figure 4.98. As seen in the pool test, the motion of the underwater robotic vehicle is now emulated in the pool environment. Initially, the underwater robotic vehicle was positioned facing the wall of the pool as indicated by the arrow. The front wall facing the underwater robotic vehicle was the obstacle to be avoided. The distance from the underwater robotic vehicle to the pool's wall was around 2 m. Once the 2 m was reached, it began to make a 90° turn to avoid the obstacle (as shown in the sequence 2 onwards in Figure 4.98).

The actual and simulated responses of the underwater robotic vehicle in the pool are shown in Figures 4.99 and 4.100. In Figure 4.99, the switching signal for both altimeter sensor and compass was shown. The obstacle was first detected at around 18 s while maneuvering at a slow propeller's speed of 0.27 rps. The top control surface started to rotate CCW until the heading angle registered 90° to rotate CW. As shown, the top control surface angular position exhibits less oscillation as compared to the open-loop case. As seen in Figure 4.100, the simulated responses of the x, y, and yaw angle from the horizontal plane model were compared with the actual measurement obtained through the sensors. The simulated transient behavior is found to be different as compared to the experiment results. It may be due to the initial water resistance that the underwater robotic vehicle needs to overcome. However, it is quite negligible for this case.

On the other hand, the x, y positions response are quite oscillatory with some overshoot. However, with the SMC, it managed to achieve the steady-state value at around 20 s. The resistance force could increase if the angular position of the top

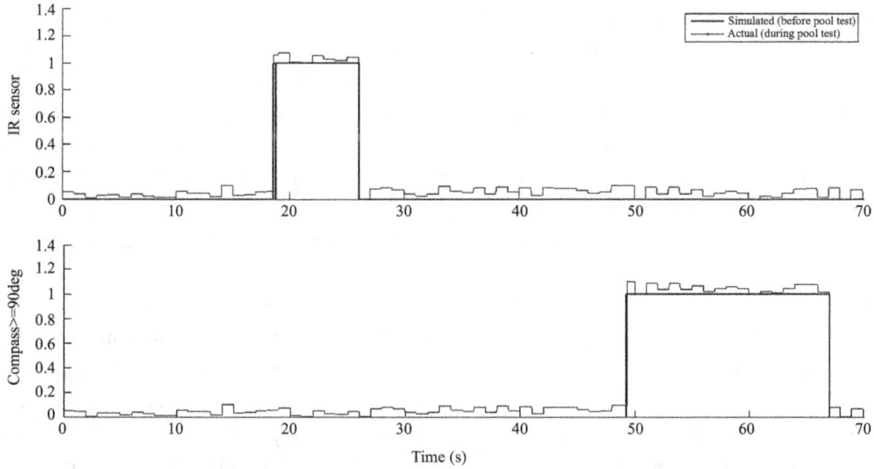

Figure 4.99 Switching signal of underwater robotic vehicle during obstacle avoidance in the pool [1]

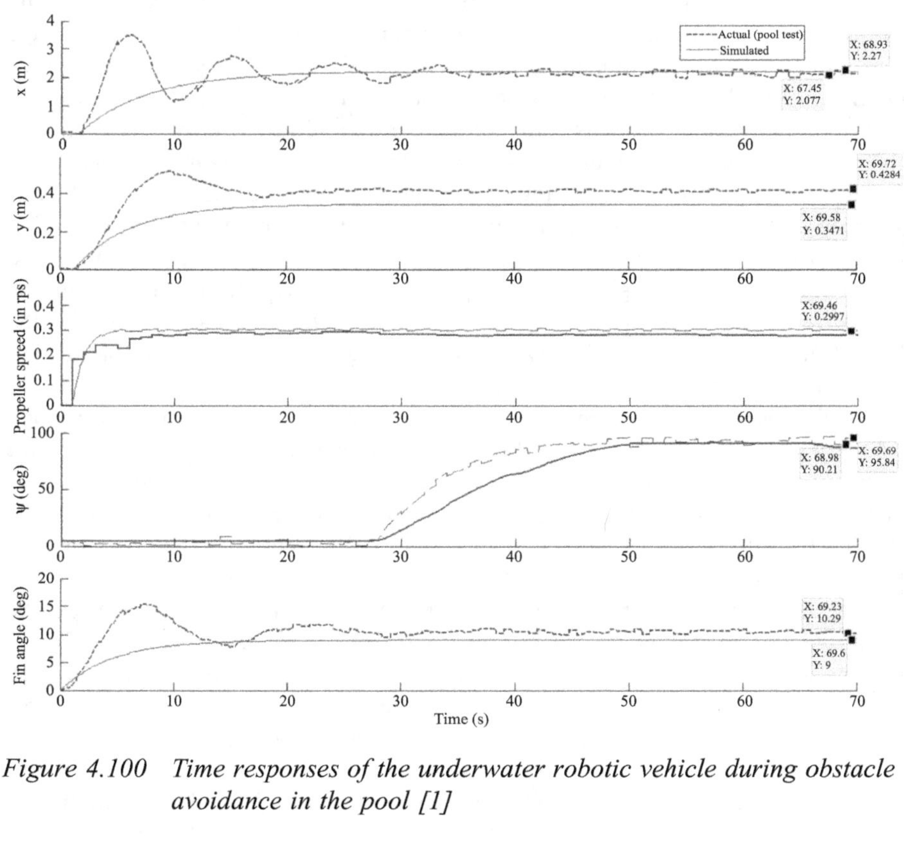

Figure 4.100 Time responses of the underwater robotic vehicle during obstacle avoidance in the pool [1]

control surface increases. Therefore, to decrease the resistance, the top control surface angles were set to 10°, to achieve the desired underwater robotic vehicle's heading direction with least resistance. In summary, the angular position of the top control surface was successfully controlled with least oscillation and the simulated results in both the positions and the heading angle match the actual experiment results in the pool. As observed in the pool test, despite the model uncertainties such as the hydrodynamic parameters and some spouting water, the underwater robotic vehicle was able to perform and behave quite close to the simulated model.

A real-time control systems platform for underwater robotic vehicle control was implemented using the software package (MATLAB®, Simulink®, xPC-Target, Real-Time Workshop, and Microsoft Visual C++ compiler) and commercially available hardware. The steady-state thruster's model was obtained to establish the relationship between the voltages, speeds, and thrusts output. The horizontal plane dynamic including the nonlinear hydrodynamic damping of the underwater robotic vehicle was determined through the water-tank test using xPC-Target simulation and hardware and later verified with ANSYS CFX numerically. As shown in the tests, the relationships between the hydrodynamic loads and the translational and rotational velocity of the vehicle are inherently nonlinear, and the key nonlinear damping coefficients have been obtained.

Using CFD method, it is possible to obtain the nonlinear (or linear) damping coefficients, and thus the results of the simulation are not restricted to a small range of motions. All of these modelings offer a reference point for determining which coefficients need to be investigated when considering an underwater robotic vehicle that has a different shape as compared to the streamlined semi-empirical model. This is very important to reduce costs when hydrodynamic parameters are determined from physical model testing. After the dynamic modeling, the SMC was designed and selected to control the top control surface angle during the heading maneuvering. The pool test results obtained from the sensors match the simulated results obtained from the simulated horizontal dynamic equations. In addition, the angular position of the top control surface exhibits the least oscillation with the SMC. Hence, the proposed xPC-Target from the rapid prototyping to final testing in the pool was performed successfully.

In the hardware and control architecture aspect of the underwater robotic vehicle, the human–machine interface and communication can be used. Since control systems platform run in the host PC running Windows, the graphical human–machine interface can easily be applied to perform client functions. The MATLAB® GUIDE development tool can be used to setup the Ethernet communication, execution of the xPC-Target, changing of various model parameters while running and, plotting of the parameters in real-time. Besides, the sea trial can be conducted and an altimeter for depth measurement can be implemented. Together with the global positioning system, it provides a real-time position of the vehicle.

As seen in the simplicity of the underwater robotic vehicle design, the proposed xPC-Target can also be used in the computer-aided engineering education. Since the underwater robotic vehicle systems is accessible through a TCP/IP network, it could act as a network-share laboratory equipment for both off-campus and on-campus students to verify their control idea. Finally, implementation of network-based control of the controllers, actuators, and sensors using campus-wide network, that is, a wide-area network containing both Ethernet and asynchronous transfer mode at the physical layer can be used. Fundamental issues in network control systems such as network-induced delay when transmitting sensor data and control data can be analyzed. The completed prototype and the testing can be seen in the following video links.

- xPC-Target (setup): http://www.youtube.com/watch?v=-hGg2qFxxK8&layer_token=5a3fb89994 7904c1
- xPC-Target (Open-loop Speed Regulation): http://www.youtube.com/watch?v=sL9S2IkltQ0&layer_token=7f65dcae8a32 f8b1
- xPC-Target (Closed-loop Speed Control): http://www.youtube.com/watch?v=spDnHL1Z4TA&layer_token=961fceec5 d3be913
- xPC-Target (Closed-loop Position Control): http://www.youtube.com/watch?v=SQSSJFyo3gI&layer_token=7f49587925 10bca2
- DC Motor Identification using xPC-Target: http://www.youtube.com/watch?v=ldQn0oFxz9I&layer_token=a5e7e4612d4 a024
- Testing inside the pool: http://www.youtube.com/watch?v=vY2-TvYHKEY
- Testing for x and yaw degree of freedom: http://www.youtube.com/watch?v=XLybvSuJkUg

References

[1] Chin, C. and Lum, S.H., 2011. Rapid modeling and control systems prototyping of a marine robotic vehicle with model uncertainties using xPC Target System. **Ocean Engineering**, 38(17–18), 2128–2141.

[2] Chin, C.S., Lau, M.W., Tan, Y.J., Chee, K.F., and Wong, Y.C., 2009. Development and testing of an autonomous underwater vehicle using industrial xPC-Target platform. **IEEE/ASME International Conference of Advanced Intelligent Mechatronics**, Singapore.

[3] Fehrani, H., Heidari, N., Zakeri, M., Ghaisari, Y., and Jafar, G. 2010. Development, depth control and stability analysis of an underwater remotely-operated underwater vehicle, **8th International Conference on Control and Automation**, Xiamen, China, 814–819.

[4] Byron, J. and Tyce, R., 2007. Designing a vertical/horizontal AUV for deep ocean sampling. **MTS/IEEE OCEAN 2007**, Vancouver, Canada, 1–10.

[5] Ura, T., Nakatani, T., and Nose, Y., 2006. Terrain-based localization method for wreck observation AUV. **MTS/IEEE OCEANS 2006**, Boston, USA, 1–6.

[6] Utkin, V. I., 1977. Variable structure system with sliding modes. **IEEE Transactions on Automatic Control**, 22(2), 212–222.

[7] Fossen, T. I., 1994. **Guidance and Control of Ocean Vehicles**, John Wiley & Sons Ltd., Chichester, New York.

[8] Gertler, M. and Hagen, G., 1967. Standard equations of motion for submarine simulation. **David Taylor Naval Ship Research and Development Center, Technical Report**, Defense Technical Information Center Document, 653 861.

[9] Lamb, S.H., 1932. **Hydrodynamics**. Cambridge University Press, Cambridge.

[10] Aage, C., and Wagner, L.S., 1994. Hydrodynamic manoeuvrability data of a flatfish type AUV. **MTS/IEEE OCEANS 1994**, 3, 425–430.

[11] Yoon, H.K., Son, N.S., and Lee, C.M., 2007. Estimation of the roll hydrodynamic moment model of a ship by using the system identification method and a free running model test. **IEEE Journal of Ocean Engineering**, 32(4),798–806.

[12] Yoon, H.K., and Son, N.S., 2004. Estimation of roll-related coefficients of a ship by using the system identification method (in Korean). **Journal of the Society of Naval Architects of Korea**, 41(4), 53–58.

[13] Jones, D.A., Clarke, D.B., and Brayshaw, I.B, 2002. The calculation of hydrodynamic coefficients for underwater vehicles. **DSTO Platforms Sciences Laboratory**, Fishermans Bend, Australia, Report. DSTO-TR-1329.

[14] Nahon, M., 1993. Determination of undersea vehicle hydrodynamic derivatives using the USAF DATCOM. **MTS/IEEE OCEANS 1993**, 2, **Victoria**, 283–288.

[15] Humphreys, D., 1981. Dynamics and hydrodynamics of ocean vehicles, **MTS/IEEE OCEANS 1981**, 13, 88–91.

[16] Kim, K., Kim, J., and Choi, H.S., 2002. Estimation of hydrodynamic coefficients of a test-bed AUV-SNUUV1 by motion test, **MTS/IEEE OCEANS 2002**, 1, 29–31, 186–190.

[17] Kim, J., Kim, K., and Choi, H.S., 2002. Estimation of hydrodynamic coefficients for an AUV using nonlinear observers. **IEEE Journal of Ocean Engineering**, 27(4), 830–840.

[18] Sen, D., 2000. A study on sensitivity of manoeuvrability performance on the hydrodynamic coefficients for submerged bodies. **Journal of Ship Research**, 45(3), 186–196.

[19] Whitcomb, L.L. and Yoerger, D.R., 1999. Development, comparison, and preliminary experimental validation of nonlinear dynamic thruster models. **IEEE Journal of Ocean Engineering**, 24(4), 481–494.

[20] Yoerger, D.R., Cooke, J.G., and Slotine, J.J.E., 1990. The influence of thruster dynamics on underwater vehicle behavior and their incorporation into control systems design. **IEEE Journal of Ocean Engineering**, 15(3), 167–178.

[21] Tyagi, A. and Sen, D., 2006. Calculation of transverse hydrodynamic coefficients using computational fluid dynamic approach. **Ocean Engineering**, 33(5–6), 798–809.

[22] White, N.M., 1977. A comparison between a single drag formula and experimental drag data for bodies of revolution, **DTNSRDC Report** 77-0028.

[23] Tang, S., Ura, T., Nakatani, T., Thornton, B., and Jiang, T., 2009. Estimation of the hydrodynamic coefficients of the complex-shaped autonomous underwater vehicle TUNA-SAND. **Journal of Marine Science Technology**, 14, 373–386.

[24] Roderick, Barr, A., 1993. Review and comparison of ship maneuvering simulation methods. **SNAME Transaction**, 101, 609–635.

[25] Yoerger, D.R., Cooke, J.G., and Slotine, J.J.E., 1990. The influence of thruster dynamics on underwater vehicle behavior and their incorporation into control systems design. **IEEE Journal of Ocean Engineering**, 15(3), 167–178.

Chapter 5

PIC embedded system design

5.1 Overview of MPLAB IDE

MPLAB IDE is a Windows® OS-based IDE for the PICmicro microcontroller (MCU) families. It allows you to write, debug, and optimize PICmicro MCU applications for firmware product designs. MPLAB IDE includes a built-in text editor, simulator, and project manager. MPLAB IDE also supports the MPLAB ICD 2 in-circuit debugger, PICkit2, PICSTART® Plus, and PROMATE® II programmers, and other microchip or third-party development system tools. The latest MPLAB IDE can be downloaded from www.microchip.com.

The organization of MPLAB IDE tools (see Figure 5.1) by pull-down menus and customizable quick keys provide tools to allow you to:

- Create source code using the built-in editor;
- Assemble, compile, and link source code using various language tools;

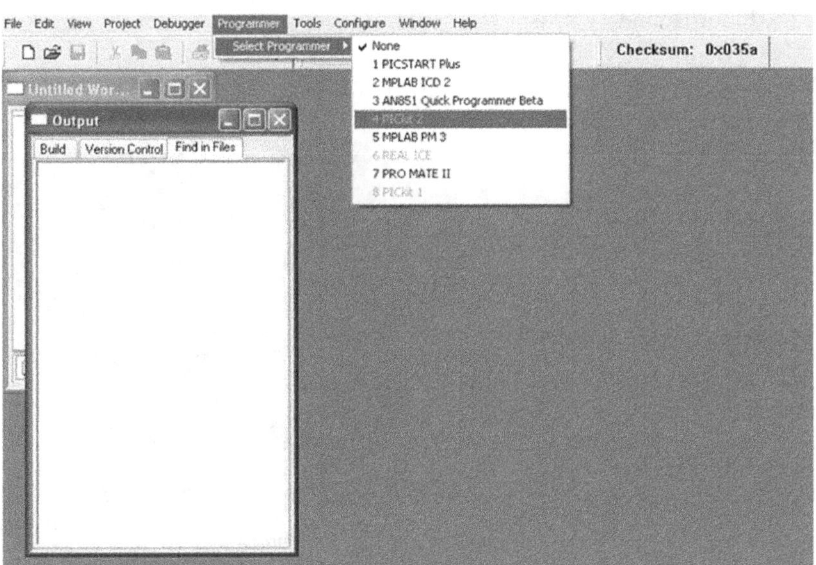

Figure 5.1 MPLAB® IDE Desktop

- Debug the executable logic by watching program flow with the built-in simulator, or in real time with the MPLAB ICD 2 in-circuit debugger or PICkit2;
- Make timing measurements with the simulator or emulator;
- View variables in watch windows;
- Find quick answers to questions from the MPLAB IDE on-line help.

5.2 Intelligent vacuum robot system design

5.2.1 System design and architecture

The target board on the intelligent vacuum robot system can be used as a stand-alone board with a programmed device, with a PICkit2. The details of the intelligent vacuum robot system design used in this section can be seen in the following references [1,2]. Figure 5.2(a) shows the target board on the vacuum robot. The schematic diagram for the robot can be seen in Figure 5.2(b). It consists of the information of the ports used in the microprocessor, in this case, is the PIC18F4620. Note that you can use PIC16 and PIC32 (https://www.microchip.com/). However, the schematic diagram in Figure 5.2(b) will be different.

The location of the components on the board can be seen in Table 5.1. It has the following hardware features:

1. 2 × battery holder AA × 4 cells;
2. PICkit2;
3. 2 × RC servo motor 360° rotation;
4. 12 V DC motor;
5. Object sensor (Ultrasonic sensor);
6. 2 × Floor/Boundary sensor (IR sensor);

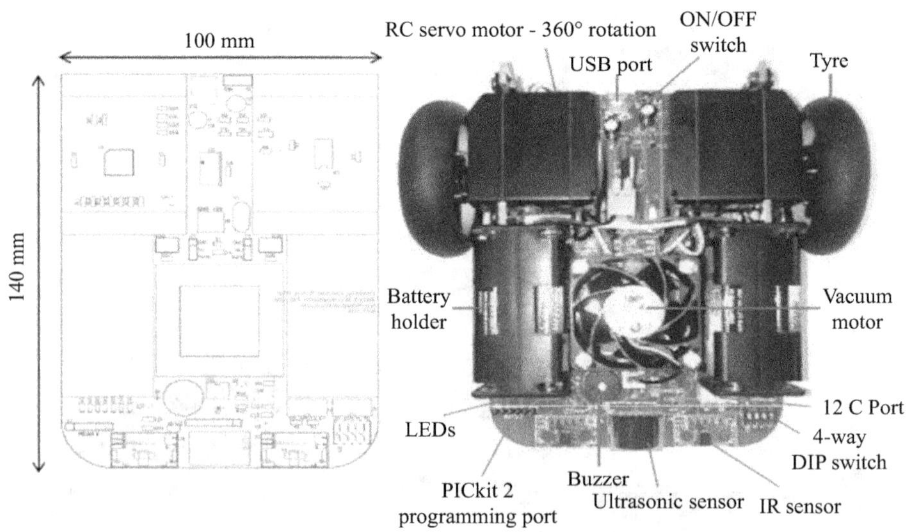

Figure 5.2(a) Target board with components mounted

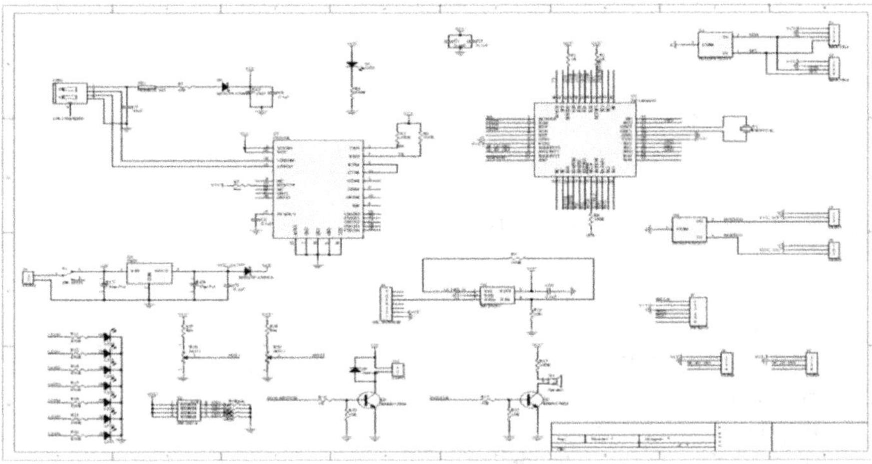

Figure 5.2(b) Target board schematic diagram

Table 5.1. Components used on target board

Main controller		Main controller		IR Sensor	
Location	**Description**	**Location**	**Description**	**Location**	**Description**
B1	Buzzer	Q1, Q2	MMBT3904	C1	0.1 μF
C1, C2, C4, C6, C9, C10	0.1 μF	R1, R2, R16	1K	C2	NL
C3	10 μF	R3	10R	D1	LED
C5	10 nF	R4, R11, R12, R14, R15, R18, R25, R26	430R	J1	CON4
C7, C8	Cap Pol	R5, R6, R8, R9, R19, R20, R21, R24	100K	R1	1M
CN1	CN-USB-MINI	R7, R27, R28	Res	R2	10K
D1, D4, D5, D6, D8, D9, D10, D11	LED	R10, R17, R22, R23	10K	R3	75R
D2, D3	MSS1P5-E3/89A	R13	100R	R4	820K
D7	1N4148	R29, R30	POT1	R5	1K
FB1	FERRITE BD	S1	SW-SPDT	R6	100K
J1, J2	MOLEX4	S2	SW DIP-4	U1	TCND5000_SENSOR
J3, J4	CON3	U1, U4	PESD5V2S2UT	U2	MCP6232
J5	CON2	U2	PIC18f4620		
J6	BK_SENSOR	U3	FT232R		
J7	PICKIT2	U5	5V Regulator -7805		
J8, J9	CON4	U6	MCP6001		
J10	CON2	Y1	CRYSTAL		

7. Buzzer;
8. 2 × Potentiometer;
9. 7 × LED;
10. 4-Way DIP switch;
11. 2 × Ball bearing;
12. 2 × Tyre.

It has the following controller features:

1. Input voltage—12 V;
2. Interface with PC—1 USB;
3. 12C Bus—2;
4. Processor—PIC18F4620;
5. Data memory—64K;
6. CPU speed—40 MHz;
7. PWM Signal ports—2 channels;
8. Digital I/O Ports—16 used by 2-IR Sensors, 7LED, 4-WAY DIP Switch, 1 Buzzer;
9. Analog input ports—Ultrasonic sensor;
10. Programming language—C.

Before computer programming of the control algorithm, working principles or the program flowchart (see Figure 5.3) of the robot was developed to illustrate the flow of the computer programming. The operations of the robot are as follows. The movement of the robot begins upon power-up of the robot. The motors control the forward movement. As the driving motors start to rotate, the robot moves forward. While moving forward, the robot continually tracks the input signal of sensors that are positioned in front (ultrasonic sensor) and bottom of the robot (infrared (IR) sensor). In a typical operation scenario, the ultrasonic sensor (or IR sensor) is activated when an obstacle (or table's edge) is met. The MCU inputs the appropriate signal to the H-bridge circuit that reverses the direction of the driving motor in counterclockwise direction. It primarily drives the robot in reverse direction to avoid the obstacle. On the other hand, the IR sensors are activated when the edges of the table are detected. The MCU inputs the appropriate signal to the H-bridge circuit that reverses the direction of the driving motor in the counterclockwise, and then steers the robot to avoid the edges of the table. It helps to prevent the robot from falling off the table.

Next, to facilitate the prototype testing, analysis, and improvements, the completed program in C code was implemented by downloading the program into the embedded processor as shown in Figure 5.4. The C program was used in the intelligent vacuum robot. The program has several functions (such as FAN MOTOR, LED) to simplify the programming and enable a more systematic application of the functions required to control the robot. The PICkit 2 (see Figure 5.4(a) and Figure 5.4(b)) Development Programmer/Debugger (PG164120) is also a low-cost development tool with an easy-to-use interface for programming and debugging Microchip's Flash families of MCUs. The full-featured Windows® programming interface supports baseline (PIC10F, PIC12F5xx, PIC16F5xx), midrange (PIC12F6xx, PIC16F), PIC18F, PIC24, dsPIC30, dsPIC33, and PIC32 families of 8-bit, 16-bit, and 32-bit MCUs, and many Microchip Serial EEPROM products. In-Circuit-Debugging runs

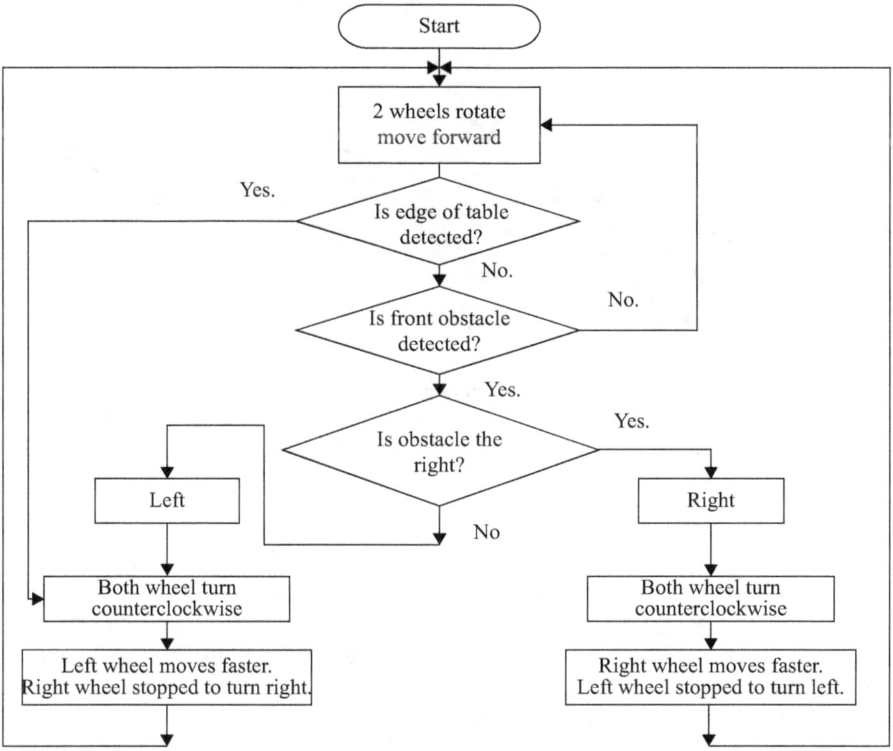

Figure 5.3 Flowchart of control architecture of intelligent vacuum robot cleaner

Figure 5.4(a) Intelligent vacuum robot using PICkit2

halts and single steps the program while the PIC MCU is embedded in the application. When halted at a breakpoint, the file registers can be examined and modified.

After connecting the PICkit 2 to target device, turn on the device power. Double click on the MPLAB IDE icon to select the Programmer from MENU and click on the PICkit 2.

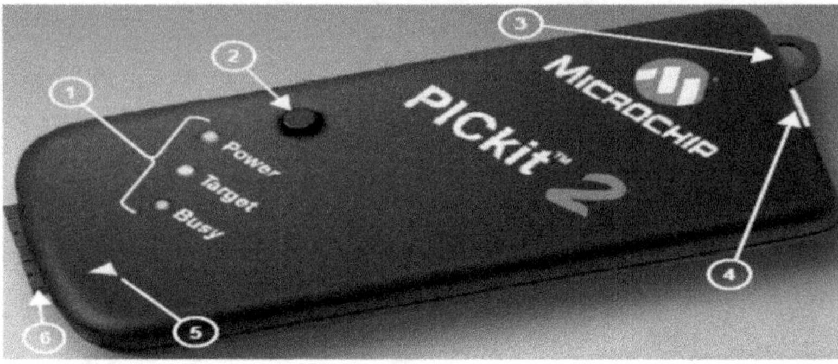

Legend:
1 – Status LEDS
2 – Push button
3 – Lanyard connection
4 – USB port connection
5 – Pin 1 marker
6 – Programming connector

Figure 5.4(b) MPLAB PICkit2

5.2.2 Programming and system implementation

The C program that is used in the intelligent vacuum robot can be seen in the following codes. As shown, the program has several functions. It helps to simplify the programming for the users. The program codes in the "USER Code Section" can be seen in Figure 5.5.

Loading Program – Quick Guide

Step 1. Connect to PICkit2 as shown in Figure 5.4(a).

Step 2. Build the program first; make sure there is no error in the program code in Figure 5.6.

Step 3. Re-establish the connection with PICkit 2 as shown in Figure 5.7.

Step 4. Program the device as shown in Figure 5.8.

This section is intended to be a quick introduction to the MPLAB-IDE v8.xx user interface. It is not intended to discuss all the details of MPLAB IDE but to provide a basic understanding so that you can use the MPLAB IDE right away. The latest version of MPLAB-X IDE and details can be dowloaded from http://www.microchip.com/mplab/mplab-x-ide.

After exploring the MPLAB-IDE Environment, this section will provide a guide to create a source file for prototyping. The following will summarize the steps.

- Setting up the Development Mode;
- Creating a New Project;
- Setting Language Tool Location;
- Selecting the Toolsuite;
- Creating a Simple New Source File;
- Entering Source Code.

To understand the MPLAB IDE environment, you need to invoke the program by double-clicking on the MPLAB icon. The MPLAB IDE desktop as shown in Figure 5.9 will appear.

```
/*********************** User Code Section ***************
*********************************************************/
LATD = 0x00; // off all LEDs
_BUZZER = 0;
PWM_ct = 0;
SERVO = 0;
BUZZ();
SENSE = FALSE;
 while(1)
        {
                if(_DS0 == 1 )
                {
                        _FAN_MOTOR = 1;
                        _LED1 = 1;
                }
                else
                {
                        _FAN_MOTOR = 0;
                        _LED1 = 0;
                }
                if((_IR_R == 0) || (_IR_L ==0) || (Ult_Val < 200))
                {
                        Stop();
                        BUZZ();
                        Reverse();
                        Delay10KTCYx(255);
                        Delay10KTCYx(255);
                        Delay10KTCYx(255);
                        Delay10KTCYx(255);
                        Turn_Right();
                        Delay10KTCYx(255);
                        Delay10KTCYx(255);
                }
                else
                {
                        PWM0_VAL =-2000;
                        PWM1_VAL =2000;
                }
                Calculate_Pulse();
        }
/*********************************************************
```

Figure 5.5 C Program codes

The MPLAB IDE desktop contains the following major elements in Figure 5.10.

1. A menu across the top line;
2. A toolbar below the menu;
3. Workspace in which various files, windows, and dialogs can be displayed.

To understand the function of each icon, just place the cursor over it. Its function will appear below the cursor.

Step 1: Setting up the Development Mode

Before starting your development, you need to **select the device** and decide which **development mode** to use. In this course, we will only be using a **Text editor, Simulator,** and **Debugger**. See the snapshot as shown in Figure 5.11.

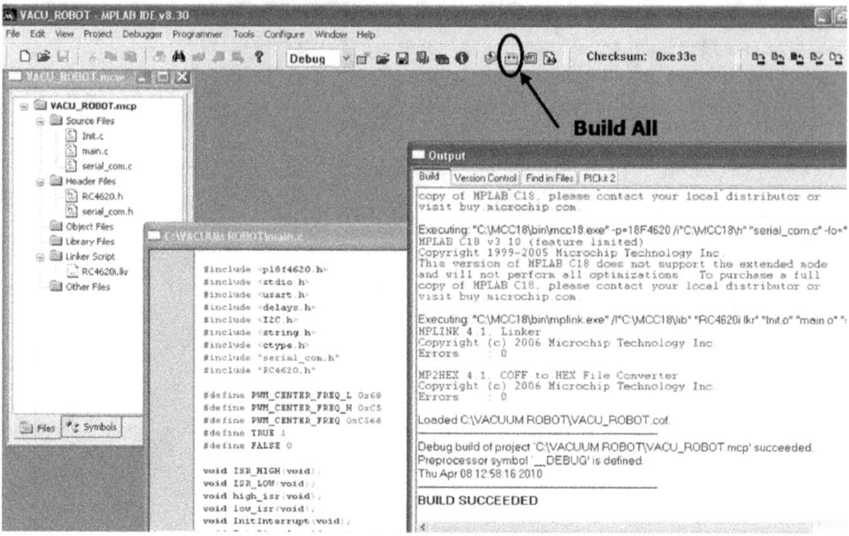

Figure 5.6 Build program

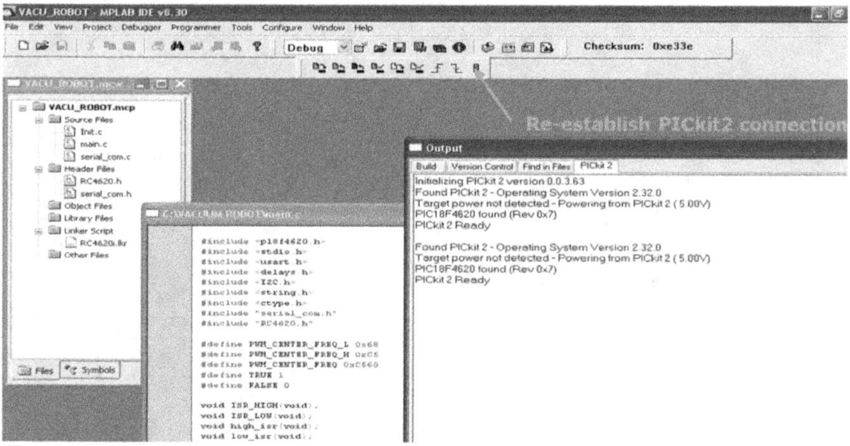

Figure 5.7 Connect to PICkit2

- To select the device, select ***Configure>Select Device***. A dialog box as shown below will be displayed. Choose the **PIC18F4620** from the pull-down list of available **processors** supported and click **OK**.
- The **MPLAB SIM simulator** is used as the development mode when you want to simulate/run the program on MPLAB **without using the hardware.**
- The **PICkit2 debugger** is used as the development mode when you want to run the program **on the hardware**.

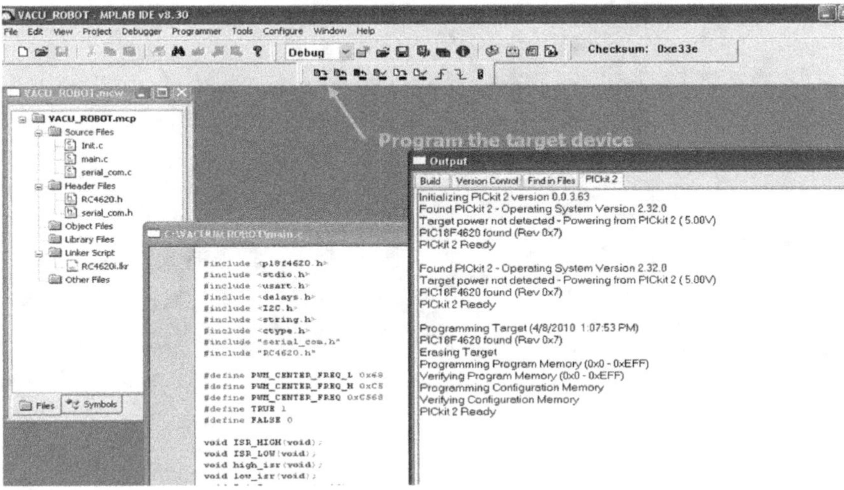

Figure 5.8 Program device

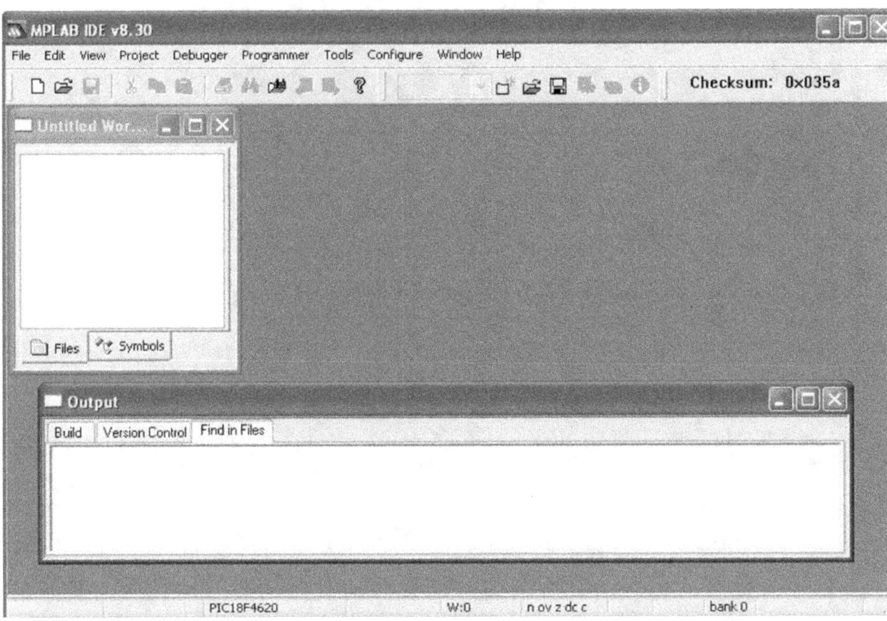

Figure 5.9 MPLAB-IDE v8.30 Desktop

We will use MPLAB SIM, the software simulator. You may switch to any mode at a later time by clicking the ***Debugger>Select Tool>***. See the snapshot as shown in Figure 5.12.

- To select MPLAB SIM simulator as your development mode, select the ***Debugger>Select Tool>MPLAB SIM***. Refer to the display below.

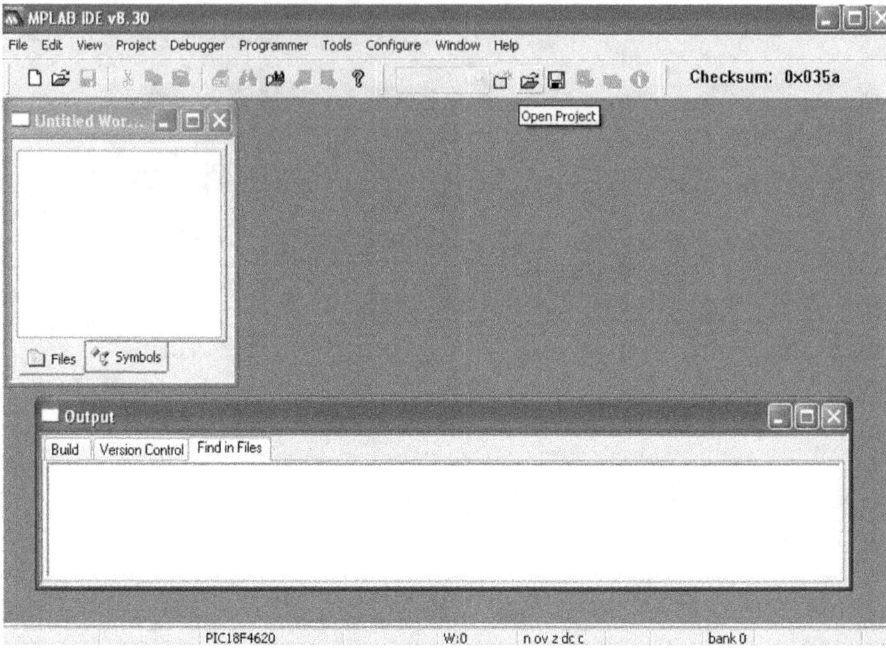

Figure 5.10 Mouse over toolbar icon

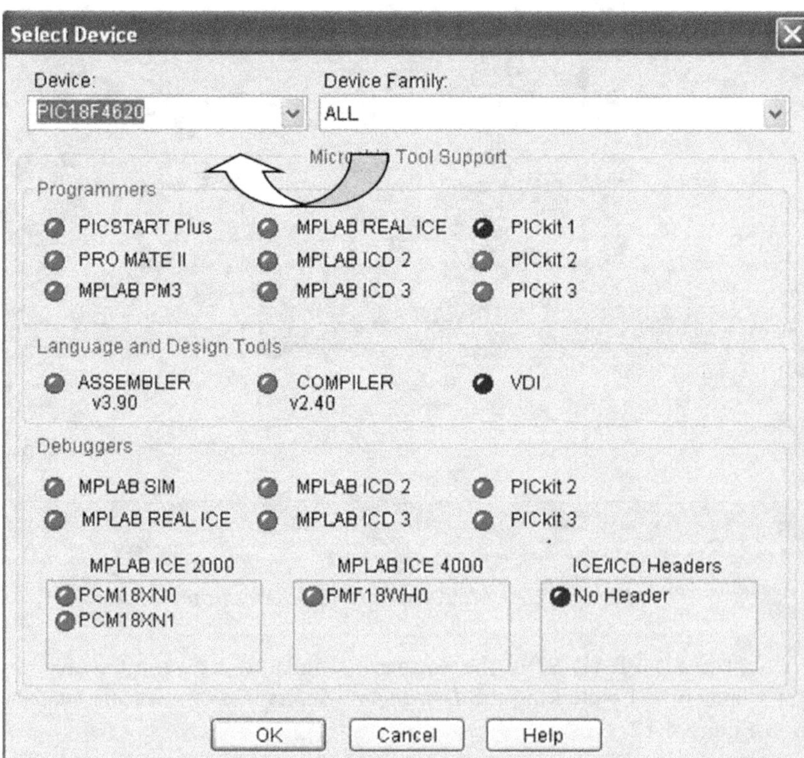

Figure 5.11 Select device dialog

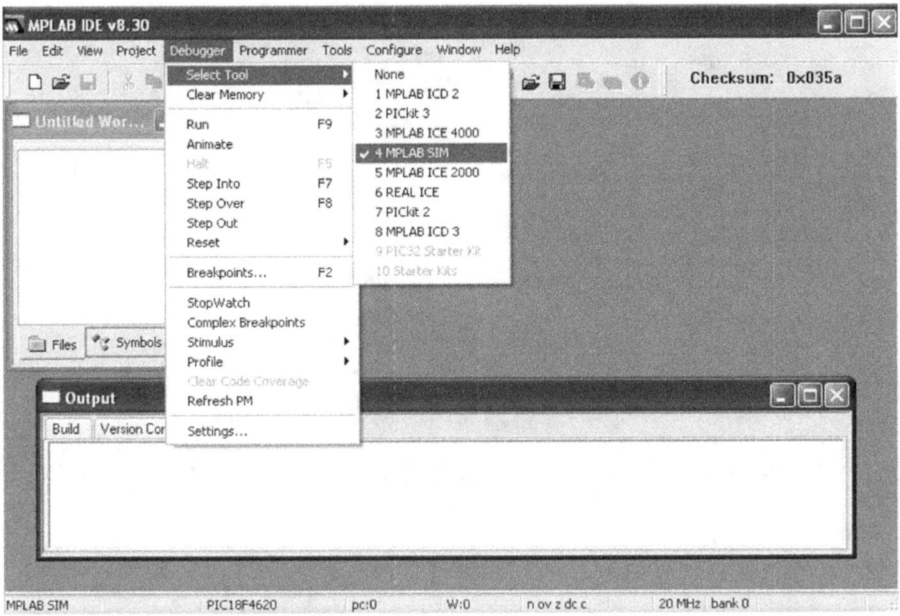

Figure 5.12 Select simulator mode

- After you select MPLAB SIM, you should see the following changes:
  - The status bar on the bottom of the MPLAB IDE window should change to "MPLAB SIM."
  - Additional menu items should now appear in the Debugger menu, and additional toolbar icons should appear in the Debug Tool Bar.

Step 2: Create a New Project

Once the development mode is set, you are ready to simulate your program. However for the simulator to run, you must first create a **source code file** (e.g., **main.c**) and successfully compile the source code. The compiler produces, among other things, a HEX file. This file has the file extension.HEX. In this lab, the file will be named **main.hex**.

- First, use the Windows Explorer to check that the directory C:\MCT\Myproj exists on your computer. If it does not, create it. This will be the default project directory in which you will create your projects.
- To create a new project, select ***Project>New*** from the menu and you will see a dialog box as shown below. Enter a name for your new project (example: lab1) and use the **Browse**. button to select the project directory. Click **OK** when everything is entered. See the snapshot as shown in Figure 5.13.

Step 3: Set Language Tool Location

- Select ***Project>Set Language Tool Locations*** to confirm the location of the Microchip C18 Toolsuite. Click on the "MPLAB C18 C Compiler (mcc18.exe)". The full path to the MPLAB C18 C Compiler will appear in the **Location** text box

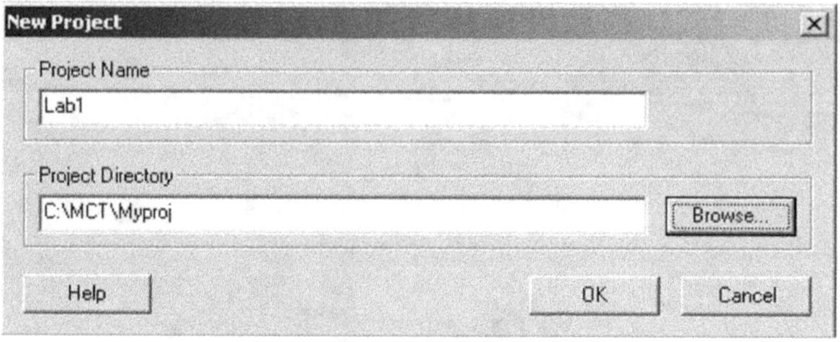

Figure 5.13 New Project dialog

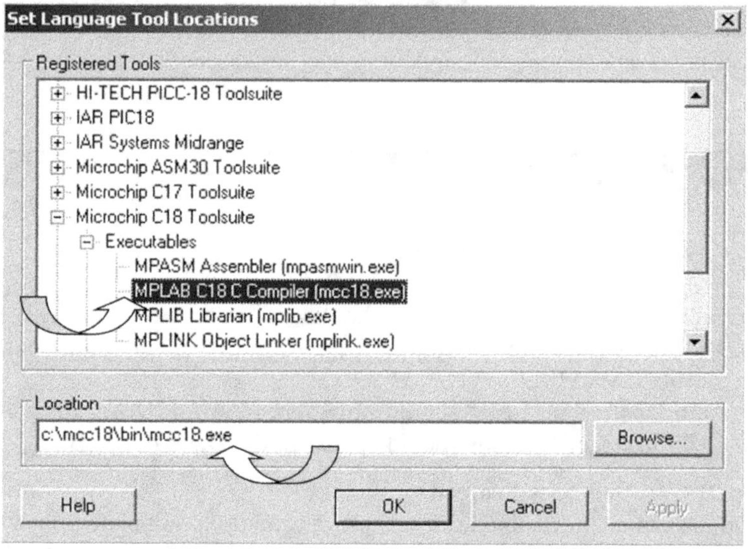

Figure 5.14 Set Language Tool Location dialog

as shown. If it is incorrect or empty, click **Browse** button to locate *mcc18.exe* as shown. Click **OK** after setting the necessary tools. See the snapshot as shown in Figure 5.14.

Step 4: Select Toolsuite

Before you start editing your code, you need to set the language tool suite used. This allows the MPLAB IDE to tailor its operation more accurately about context-sensitive editing and file extensions.

- Select ***Project>Select Language Toolsuite***. For "Active Toolsuite," select "Microchip C18 Toolsuite." The PICmicro language tools will appear under "Toolsuite Contents" as shown below. Make sure the correct locations are selected for all the **four** items as shown in the respective snapshots in Figure 5.15. Click **OK** after selection.

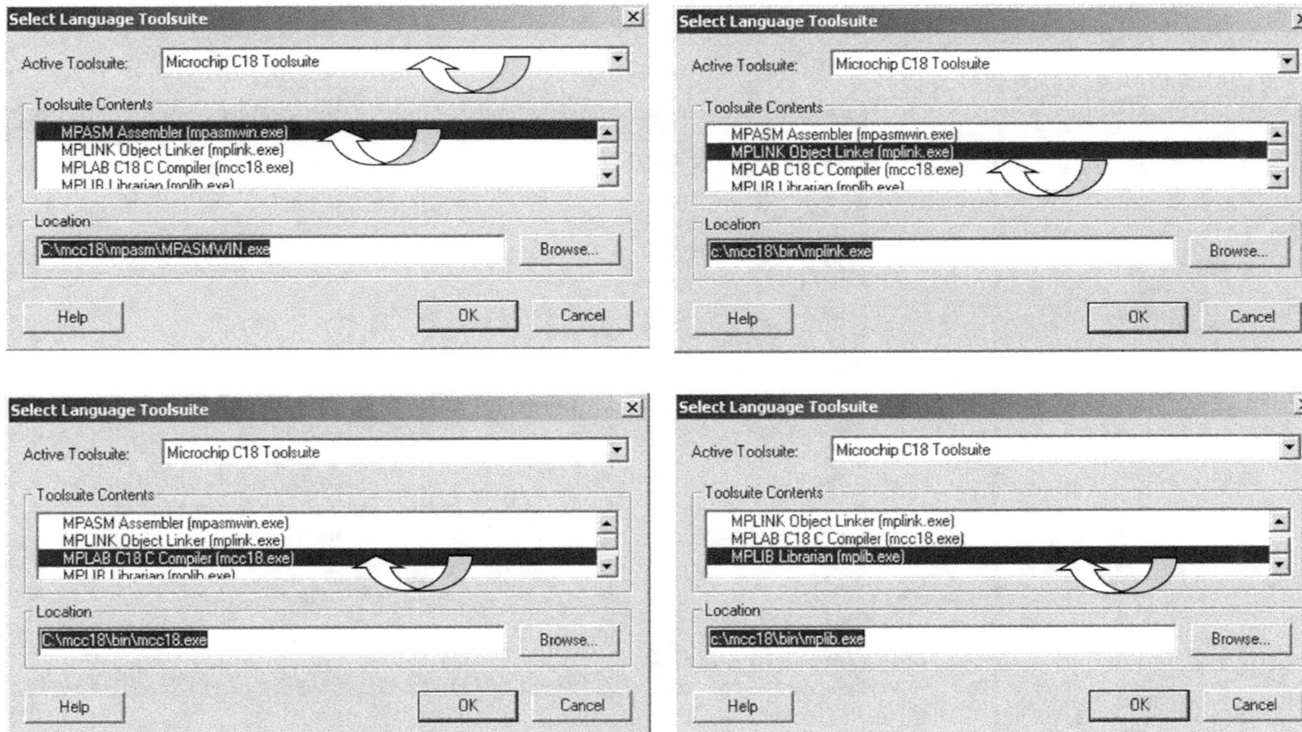

Figure 5.15 Select Language Toolsuite dialog

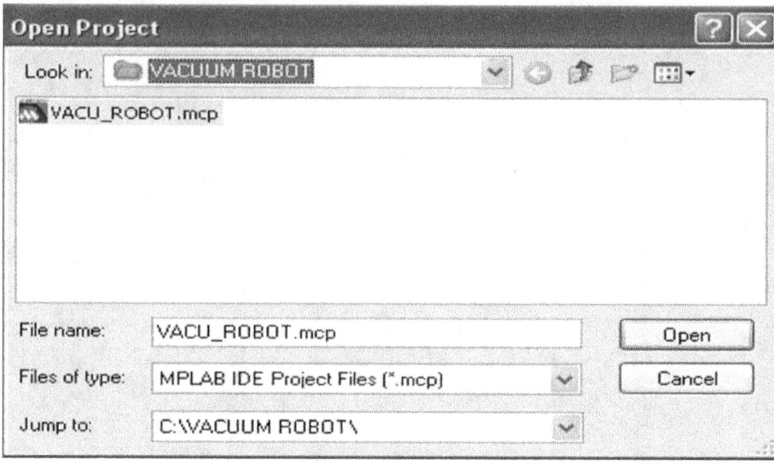

Figure 5.16 Select Open Project dialog

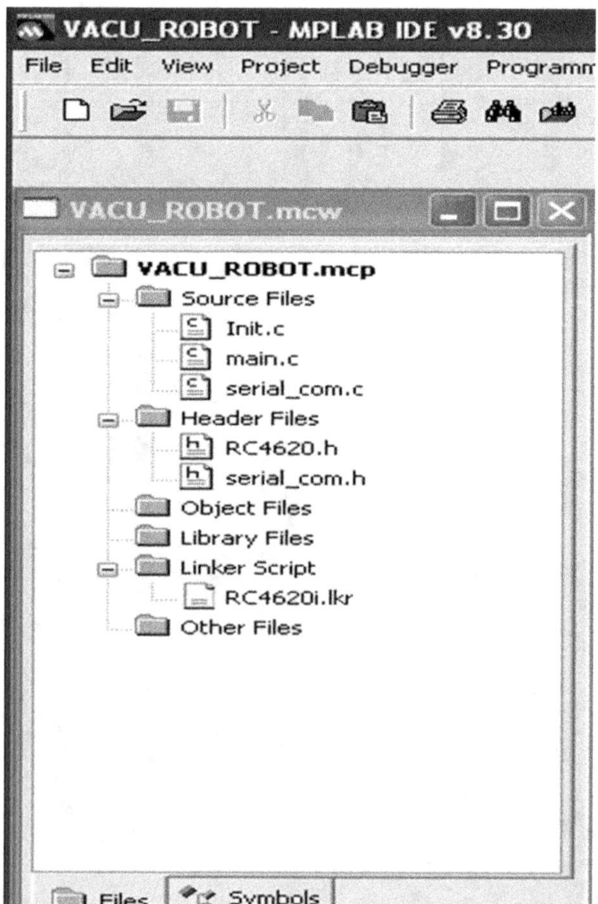

Figure 5.17 Project Pane Window

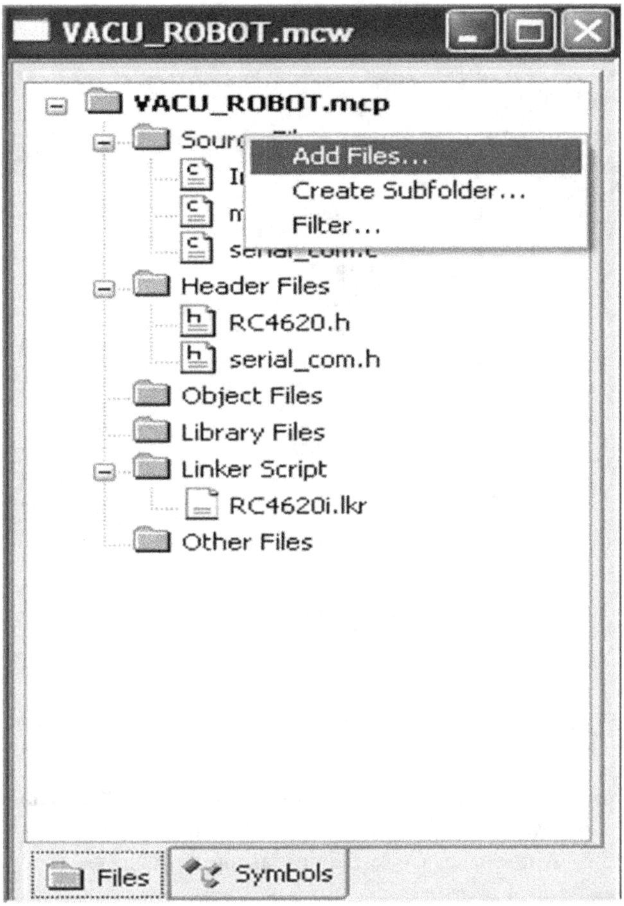

Figure 5.18 Example of Add Files in Project Pane Window

Step 5: Open Project File and Configure Project Pane Window

- Select ***Project>Open*** and select the project file **VACU_ROBOT** and click **Open.** See the snapshot as shown in Figure 5.16.
- Save the project by selecting ***Project>Save to project.***

The right side of the screen is called Project Pane Window. It should contain the following files as shown in Figure 5.17.

If there are missing files, right-click on the folder and Select> Add Files. See the snapshot as shown in Figure 5.18.

Step 6: Entering Source Code

Under Project Pane Window, Select>main.c and right-click to Select>Edit. See the snapshot as shown in Figure 5.19.

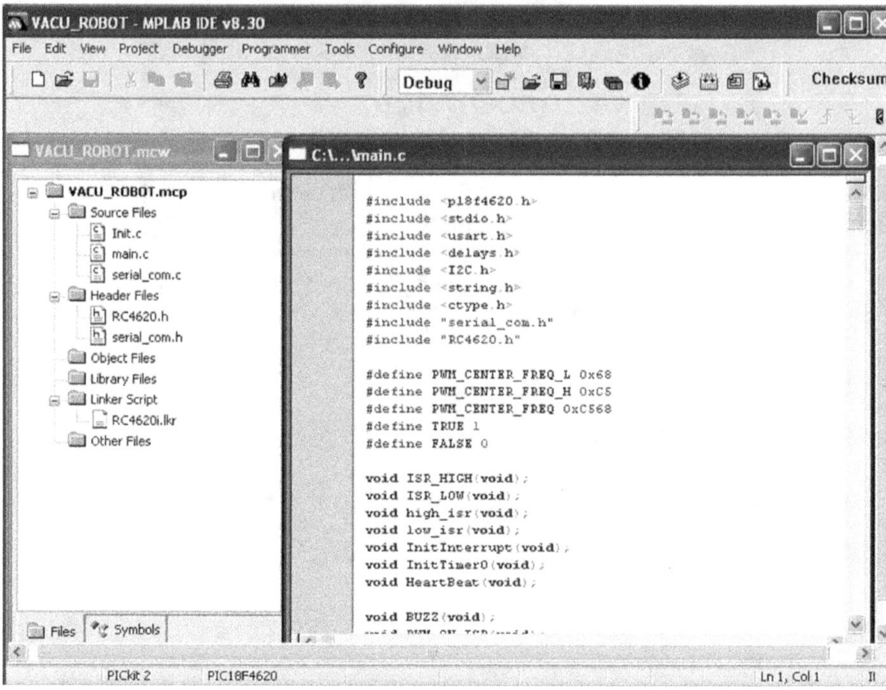

Figure 5.19 Example of editing the source file-main.c

Use the mouse to locate the cursor at the beginning of the main.c. In the User Code Section window, enter the following text exactly. Take note that the codes have to be entered in the User Code Section as shown below.

```
/*********************** User Code Section **************
*********************************************************/
LATD = 0x00; // off all LEDs
_BUZZER = 0;
PWM_ct = 0;
SERVO = 0;
BUZZ();
SENSE = FALSE;

 while(1)
        {
                if(_DS0 == 1 )
                {
                        _FAN_MOTOR = 1;
                        _LED1 = 1;
                }
                else
                {
                        _FAN_MOTOR = 0;
                        _LED1 = 0;
                }
```

```
                if((_IR_R == 0) || (_IR_L ==0) || (Ult_Val < 200))
                {
                        Stop();
                        BUZZ();
                        Reverse();
                        Delay10KTCYx(255);
                        Delay10KTCYx(255);
                        Delay10KTCYx(255);
                        Delay10KTCYx(255);
                        Turn_Right();
                        Delay10KTCYx(255);
                        Delay10KTCYx(255);
                }
                else
                {
                        PWM0_VAL =-2000;
                        PWM1_VAL =2000;
                }
                Calculate_Pulse();
        }
/**********************************************************
****** End of User Code Section ********
**********************************************************/
```

Complete program code for intelligent vacuum robot system design using PIC

```
#include <p18f4620.h>
#include <stdio.h>
#include <usart.h>
#include <delays.h>
#include <I2C.h>
#include <string.h>
#include <ctype.h>
#include "serial_com.h"
#include "RC4620.h"

#define PWM_CENTER_FREQ_L 0x68
#define PWM_CENTER_FREQ_H 0xC5
#define PWM_CENTER_FREQ 0xC568
#define TRUE 1
#define FALSE 0

void ISR_HIGH(void);
void ISR_LOW(void);
void high_isr(void);
void low_isr(void);
void InitInterrupt(void);
void InitTimer0(void);
void HeartBeat(void);

void BUZZ(void);
void PWM_ON_ISR(void);
void PWM_OFF_ISR(void);
void InitTimer1(void);
void ReadAnalogInput(char ch, far int* data);
void Calculate_Pulse(void);
void Reverse(void);
void Turn_Right(void);
void Stop(void);
```

```
#pragma udata MAIN_DATA
int HeartCount;
float f;
int buzz_ct;
int PWM_ct;

int  PWM0_Counter;
unsigned char* ptChar;
char SERVO;
int *POT0,p0;

int *POT1,p1;
int *U_Sensor, Ult_Val;
int PWM0,PWM1;
float pV0,pV1;
float p0_diff,p1_diff;
unsigned char SENSE;
char enter;
int  PWM0_VAL,PWM1_VAL;
#pragma udata

void main (void)
{
        char i;
/******* initialization ************/
        RCONbits.IPEN = 1;
        INTCONbits.GIEH = 0;
        INTCONbits.GIEL = 0;
        Init();
        InitTimer0();
        InitTimer1();
        PWM0_VAL =0;
        PWM1_VAL =0;
        PWM0 = 0;
        PWM1 = 0;
        InitInterrupt();
        Init_Serial_COM(255); //9.6K
        OpenI2C(MASTER, SLEW_ON);
        SSPADD = 0x63;


/**************end of Initialization****/

/**********************************************************
****** User Code Section ***************
**********************************************************/
LATD = 0x00; // off all LEDs
_BUZZER = 0;
PWM_ct = 0;
SERVO = 0;
BUZZ();
SENSE = FALSE;
```

```
while(1)
        {

                if(_DS0 == 1 )
                {
                        _FAN_MOTOR = 1;
                        _LED1 = 1;
                }
                else
                {
                        _FAN_MOTOR = 0;
                        _LED1 = 0;
                }

                if((_IR_R == 0) || (_IR_L ==0) || (Ult_Val < 200))
                {
                        Stop();
                        BUZZ();
                        Reverse();
                        Delay10KTCYx(255);
                        Delay10KTCYx(255);
                        Delay10KTCYx(255);
                        Delay10KTCYx(255);
                        Turn_Right();
                        Delay10KTCYx(255);
                        Delay10KTCYx(255);
                }
                else
                {
                        PWM0_VAL =-2000;
                        PWM1_VAL =2000;
                }


                Calculate_Pulse();
        }

/************************************************************
****** End of User Code Section ********
************************************************************/
        while(1);
}
```

```
/*************************************************************
Interrupt Serive Routines
*************************************************************/
//#pragma code HIGH_INTERRUPT_VECTOR = 0x08
//void high_isr(void)
//{
//	_asm
//	//	goto	ISR_Read_UART
//	_endasm
//}
//#pragma code

#pragma code LOW_INTERRUPT_VECTOR = 0x18
void low_isr(void)
{
	_asm
		goto	ISR_LOW
	_endasm
}
#pragma code

#pragma interruptlow ISR_LOW
void ISR_LOW(void)
{
	ReadAnalogInput(0,U_Sensor);
	Ult_Val = *U_Sensor;
	ReadAnalogInput(1,POT0);
	p0 = *POT0;
	ReadAnalogInput(2,POT1);
	p1 = *POT1;

	if(INTCONbits.TMR0IF){
		InitTimer0();
		if(PWM_ct++ > 8)
		{
			PWM_ON_ISR();
			PWM_ct =0;


/*************************************************************
End of Interrupt Serive Routines
*************************************************************/
```

```
void InitInterrupt(void)
{
        RCONbits.IPEN = 1;
        INTCONbits.GIEH = 1;
        INTCONbits.GIEL = 1;

        INTCONbits.TMR0IF = 0;
        INTCON2bits.TMR0IP = 0;
        INTCONbits.TMR0IE = 1;

        PIR1bits.TXIF = 0;
        PIR1bits.RCIF = 0;
        IPR1bits.TXIP = 1;
        IPR1bits.RCIP = 1;
        PIE1bits.TXIE = 0;
        PIE1bits.RCIE = 1;


}


void InitTimer0(void)
{
        INTCONbits.TMR0IF = 0;
        T0CON = 0x01;        //prescaler 4 over 10Mhz, result in 2.5Mhz
                              // timer value = 2.5ms * 2.5Mhz = 6250 = 0x186A
                              // timer register = 0x10000- 0x186a = 0xE796
        //2.5ms interrupt
        TMR0H = 0xE7;
        TMR0L = 0x96;
        T0CONbits.TMR0ON = 1;
}

void InitTimer1(void)
{
        PIR1bits.TMR1IF = 0;
        PIE1bits.TMR1IE = 0;
        T1CON = 0; //internal 10MHz clock
        TMR1L = PWM_CENTER_FREQ_L ;
        TMR1H = PWM_CENTER_FREQ_H; //Center pulse width of 1.5ms

}
```

```
void HeartBeat(void)
{


        if(++HeartCount >= 400){
                HeartCount = 0;
                _LED0 = 1;
        }else{
                _LED0 = 0;
        }

        Serial_TimeOut++;
}

void BUZZ(void)
{
        _BUZZER = 1;
        Delay10KTCYx(100);
        _BUZZER = 0;
        Delay10KTCYx(50);
        _BUZZER = 1;
        Delay10KTCYx(80);
        _BUZZER = 0;
        Delay10KTCYx(50);
        _BUZZER = 1;
        Delay10KTCYx(40);
        _BUZZER = 0;

}

void PWM_ON_ISR(void)
{
        T1CON = 0;
        PIR1bits.TMR1IF = 0;

        if(SERVO == 0)
        {
                PWM0_Counter =  PWM0 + PWM_CENTER_FREQ + PWM0_VAL;
                ptChar = (unsigned char*)&PWM0_Counter;
                TMR1L = ptChar[0];
                TMR1H = ptChar[1];

                _SERVO0 = 1;
                SERVO = 1;
        }
```

```
        else{
                PWM0_Counter =  PWM1 + PWM_CENTER_FREQ + PWM1_VAL;
                ptChar = (unsigned char*)&PWM0_Counter;
                TMR1L = ptChar[0];
                TMR1H = ptChar[1];

                _SERVO1 = 1;

                SERVO = 0;
        }
        PIE1bits.TMR1IE = 1;
        T1CONbits.TMR1ON = 1;

}

void PWM_OFF_ISR(void)
{
        PIR1bits.TMR1IF = 0;
        PIE1bits.TMR1IE = 0;
        T1CONbits.TMR1ON = 0;
        _SERVO0 = 0;
        _SERVO1 = 0;
}

void ReadAnalogInput(char ch, far int* data)
{
        switch (ch){
                case 0: ADCON0=0b00000011;
                                break;
                case 1: ADCON0=0b00000111;
                                break;
                case 2: ADCON0=0b00001011;
                                break;
                default: ADCON0=0b00001011;
                                break;
        }
        while(ADCON0bits.GO);
        *data = ADRES;
}
```

```
void Calculate_Pulse(void)
{
        pV0 = p0 * 5/1023;
        p0_diff = 2.5 - pV0;
        PWM0 = p0_diff * 500;

        pV1 = p1 * 5/1023;
        p1_diff = 2.5 - pV1;
        PWM1 = p1_diff * 500;

}

void Reverse(void)
{
        PWM0_VAL = 2000;
        PWM1_VAL = -2000;
}

void Turn_Right(void)
{
        PWM0_VAL = 2000;
        PWM1_VAL = 2000;
}

void Stop(void)
{
        PWM0_VAL = 0;
        PWM1_VAL = 0;
}
```

5.2.3 Testing

The vacuum robot was tested on different locations such as carpet floor and tabletop as seen in Figure 5.20. The hardware-in-the-loop testing was performed on the robot. It provided a robust approach to fine-tune the system working in an

Figure 5.20 Intelligent vacuum robot system during hardware-in-the-loop testing (left) and moving on floor (right)

uncertain environment. A further test can be performed to improve the sensing capabilities. For example, an additional sensor can be used to detect dirt level. The dirtbag in the vacuum system can be installed to collect the dirt followed by an enclosure to cover and protect the electronics and motors. The actual prototype and the testing under different environment can be seen in the video links below.

- On Carpet Mat: http://www.youtube.com/watch?v=3elL6LefJso
- On tabletop: http://www.youtube.com/watch?v=faeAPOLqZug

5.3 Remote temperature-sensing system design for patients

The contents used in this section can be found in the following reference [3]. During the severe acute respiratory syndrome (SARS) outbreak in 2003, hospitals became treatment centers in most countries. As patient's core body temperature is one vital parameter for monitoring the progress of the patient's health, it is often measured manually at a frequency ranging from once every few hours to once a day. However, such manual measurement of the temperature of patients requires the efforts of many staff members. Also, when the patients suffer from conditions that result in abrupt changes of the core body temperature, e.g., due to infection at a surgical site after surgery, the staff on duty will not know such a temperature change occurred until the next temperature measurement. Such a delay may lead to patients being unnoticed while their health conditions worsen, which is dangerous because a difference of 1.5 °C can result in adverse outcomes [4]. Furthermore, there is always a need to have a monitoring system to improve the quality of health care [5], such as temperature monitoring of elderly and challenged persons using a wireless remote temperature-monitoring system.

Body temperature can be used to monitor the pain level of a patient following an operation [6] or after shoulder endoprosthesis [7]. In some cases, the tissue temperature was monitored during microwave liver ablation [8] for the treatment of liver metastases. Instead of using a temperature sensor, pulse-echo ultrasound [9] was used to visualize changes in the temperature of the patient's body. In addition, a noncontact temperature-measuring device, such as a thermal imaging camera [10], was successfully used to detect human body temperature during the SARS outbreak. However, it can be quite expensive to equip each patient room with a thermal imaging camera. In addition, there are a few wireless temperature-monitoring systems (e.g., CADI, Primex, and TempTrak) on the market that are used to monitor and store a patient's temperature for medical research by using body sensor networks [11]. Most of these systems consist of an electronic module and a temperature-sensing device. The systems include a stand-alone electronic module with a display screen that allows the temperature sensor data to be transmitted over a secure wireless network.

However, the cost to purchase and maintain these systems can be quite high. It may be difficult to reconfigure these systems to suit the current database system used in the hospital. The current systems using short message service-based telemedicine [12] systems with hardware equipment were developed to monitor the mobility of patients. However, proper hardware and software to manage the messages and the patient's temperature for display on mobile phones are not widely available.

Hence, a medical device to continuously measure the body temperature of patients using a wireless temperature sensor [13,14] is required. With such a wireless temperature sensor system, nurses will no longer have to manually measure the temperature of patients, which will free their time for other tasks and also reduce the risk associated with coming into contact with patients with contagious diseases, such as SARS. The readings will be transmitted wirelessly to the central nurse station, where they can be monitored by the staff on duty. In addition, the current and history of the body temperature measurements can be stored in an online database, which allows the medical staff to access the database when they are not in the hospital.

The objective is to develop and implement a temperature-monitoring system for patients using a wireless remote connection to the nurse's station for frequent real-time monitoring. To the best of our knowledge, a MiWi wireless temperature-monitoring system using a patient and coordinator set design that provides remote internet access to the temperature database has not been reported in a single publication. The temperature-monitoring system was designed based on a proposed patient and coordinator set design approach. The proposed temperature-monitoring system will increase labor efficiency and improve the quality of health care through continuous monitoring of the temperature of patients.

5.3.1 System design and architecture

The core body temperature of a human is one of the vital statistics used to determine individual health. Normal human body temperature (normothermia) is approximately 37 °C, with variations depending on factors such as age and geological factor [15]. When the body fails to regulate itself by dissipating heat, it increases the core body temperature above the normal condition. This condition is called hyperthermia. The highest recorded surviving core body temperature is 45 °C [16]. There are four locations [16] that can be used to measure human core body temperature, rectal (anus), oral (mouth), tympanic (ear), and axillary (armpit), with the rectal measurement considered to be the most accurate in measuring the core body temperature. Each of these locations has its advantages and disadvantages. Rectal measurement of body temperature is the preferred method to perform a core body temperature measurement. Because the aim is to track a patient's core body temperature continuously and wirelessly, the axillary measurement seems to be a viable option. Importantly, the measurement system should measure within the clinically relevant temperature deviation [17,18] of 0.3–0.5 °C and detect changes in the human body temperature (at a resolution of 0.1 °C).

To accomplish the axillary temperature measurement, the wireless remote temperature-monitoring system design (see Figure 5.21) includes a MiWi wireless transceiver (MRF24J40MA), a PIC microprocessor (PIC18LF4620)/MCU, a temperature-sensing circuit, a patient set (on the patients), and a coordinator set (at the nurse's station) for the data coordination between the patients and the central nurse station. The temperature sensor will produce an output voltage with a magnitude equal to the body temperature. The analog-to-digital converter (ADC) has 13 analog channels and a 10-bit resolution in the microprocessor to convert the output

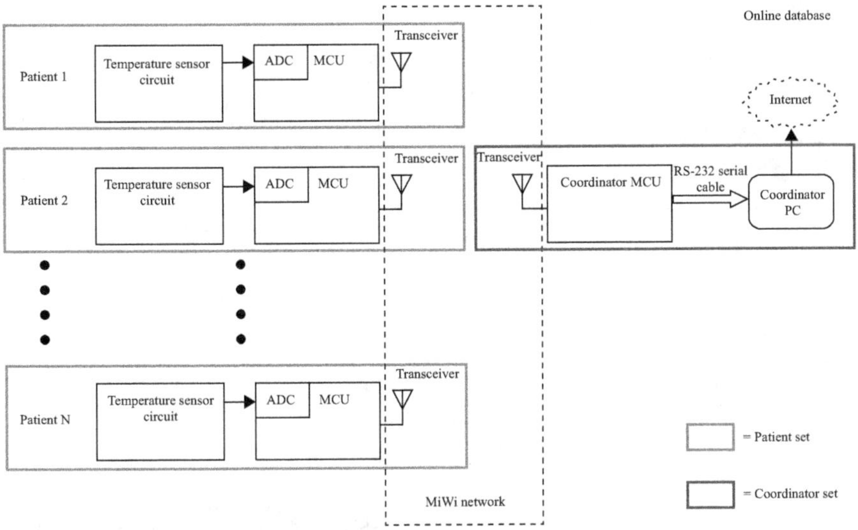

Figure 5.21 An overall block diagram of the multiple patients—coordinator set design approach

voltage into digital values. The patient's transceiver will transmit this digital value via the MiWi wireless protocol stack to the coordinator's transceiver. The coordinator MCU will send the temperature reading to the computer through RS232 that records and stores the temperature measurements in the database. The proposed wireless temperature-monitoring device can support multiple patients when they are connected to the same wireless network.

There are a few wireless and wired temperature sensors available to monitor the temperature of patients. However, the direct application of wireless sensors for web-based remote body temperature monitoring is not widely available. Typically, the temperature resolution is approximately ±0.2 °C for the range 25–60 °C. However, a 1.5 °C loss [4] in body temperature, e.g., due to surgical site infection after surgery, may lead to adverse outcomes. Hence, a resolution of ±0.1 °C is required. Because the highest recorded surviving core body temperature is 45 °C, the temperature range required for the temperature sensor is reduced. For initial prototyping, a nonmedical-grade LM35CAZ (from http://www.ti.com/lit/ds/symlink/lm35.pdf) temperature sensor (see Figure 5.22) with a maximum range of 45 °C and network coverage of 35 m (minimum) on the same floor was used. The temperature sensor used draws only 60 μA from its supply. Thus, the only advantage compared to the medical-grade sensor is the low power consumption. Nevertheless, the sensor can be replaced with any medical-grade temperature probe (such as a tympanic probe or a standard stick-on skin probe) with minor changes in the hardware and software.

The proposed temperature sensor had an output of 10 mV/°C and required 3.3 V. Because the output of the temperature sensor is 10 mV/°C, direct connection of the sensor to the ADC will result in a quantization error of approximately 0.32 °C.

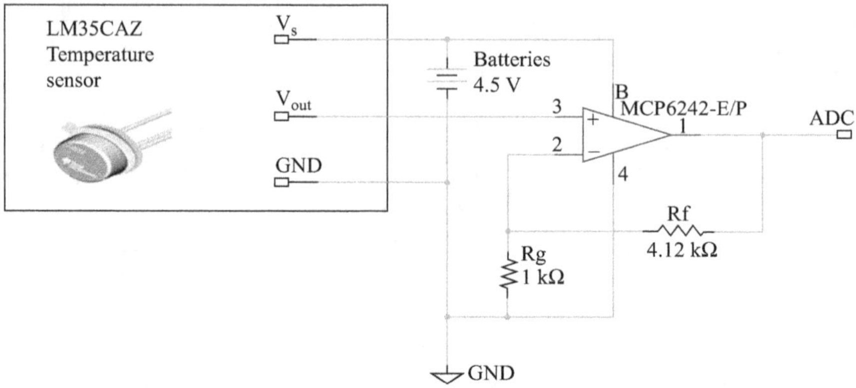

Figure 5.22 Amplifier circuit design

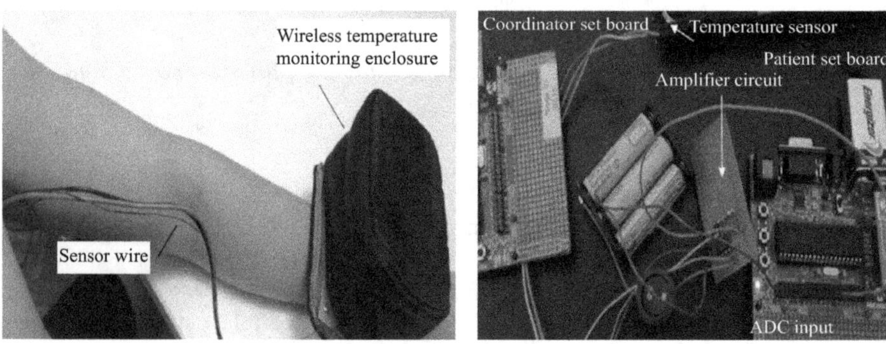

Figure 5.23 Prototype temperature-monitoring system strapped on Patient Set 1 (left) and circuit board (right)

To reduce this quantization error, an amplifier will be introduced. However, the full range of the sensor's output voltage was not used. The ADC has a maximum input voltage of 3.3 V, and the sensor has a maximum theoretical output of 0.45 V (or 45 °C). In this case, amplification of 5 (instead of 7.3) was chosen to prevent the ADC input voltage to have a value higher than 3.3 V. With this gain, the input voltage to the ADC became 50 mV/°C. As a result, the quantization error was reduced to 0.064 °C (rounded up to 0.1 °C). This level of quantization error gives the least significant digit or a resolution of 0.1 °C for the temperature measurement. Figure 5.22 shows the noninverting amplifier circuit design to amplify the output signal of the temperature sensor to achieve the required resolution.

Patient set design: The patient set consists of three main components: temperature sensor circuit, MCU, and transceiver. The electronic components were mounted onto a small PCB board and connected to the patient's MCU. As shown in Figure 5.23, the setup (within the enclosure) was strapped to the patient's arm during temperature monitoring.

Initially, only two patient sets were used. When the patient MCU is turned on, it will establish a connection with the coordinator set using a wireless protocol stack. In the initial phase, the easy-to-use alternative for simple wireless communication of MiWi (by MicroChip) was used to implement the low-cost and low-power wireless control networks (with a battery usage of 95–115 μW compared with ZigBee at approximately 30 mW).

A MiWi wireless transceiver module (MRF24J40MA-2.5GHz IEEE 802.15.4) was mounted onto a carrier board. To allow easy configuration and to monitor packet transactions, a ZENA wireless network analyzer (DM183023) and software were used. Figure 5.24 (top) shows an example of a transmitted message viewed on the ZENA network analyzer. The first line indicates the patient node attempting to join a network, while the second line indicates the coordinator recognizing the patient node as a member of the network.

Because the ADC has 10 bits, the output can be separated into two registers (two bytes). The 2-byte digital output values are sent to the coordinator set with an

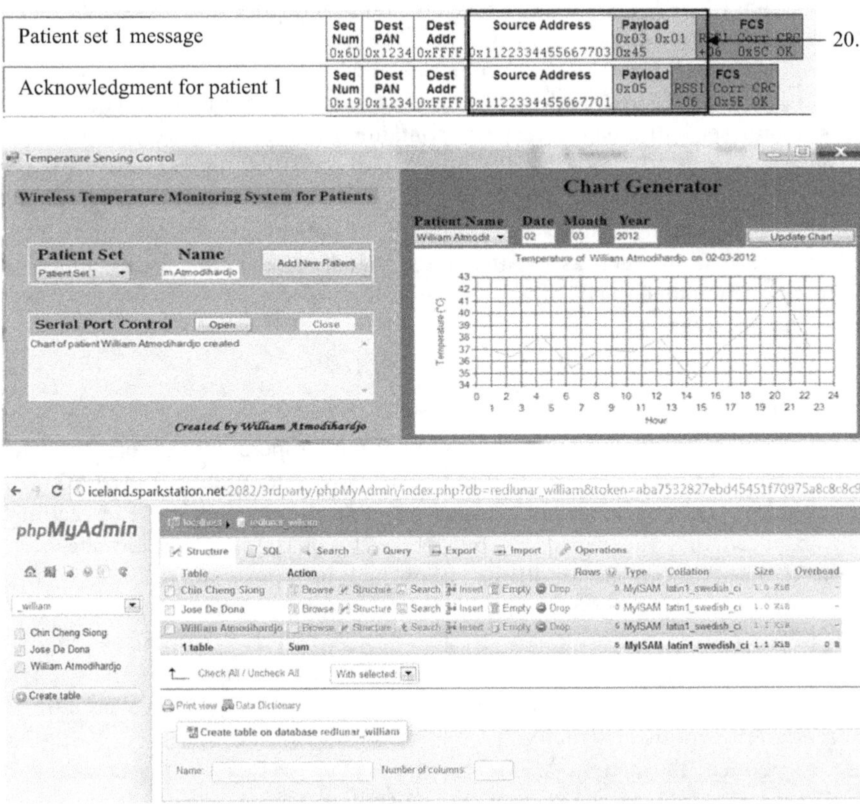

Figure 5.24 ZENA wireless network analyzer output (top), GUI for the coordinator set (middle), and online database at the coordinator (bottom)

additional byte to indicate which patient set is sending the data. Sometimes, there may be a disruption or glitch in the MiWi wireless network that prevents the coordinator from receiving the data from the patient. To overcome this issue, a cyclic redundancy check (CRC) of 16 bits (or 2 bytes in length) was used to determine whether the packet transmitted is valid during the transmission. For example, if the entire set of bits is correct, the CRC value will match.

The patient set temperature conversion process from the temperature sensor is as follows. The temperature sensor of Patient Set 1 will produce an output voltage of 0.365 V or 36.5 °C. This voltage is amplified 5.12 times to 1.868 V. The ADC then converts the voltage into two registers of 1 byte each (conversion to hexadecimal) to obtain 02 46. Subsequently, the MCU adds byte to identify the patient set. If Patient Set 1 is used, the extra byte is 01. Subsequently, the transceiver will transmit the 3 bytes to the coordinator MCU and display the temperature on the computer with the GUI designed using Microsoft Visual Studio C#.

Coordinator set design: As shown in Figure 5.24 (center), the coordinator set requires a similar MCU and transceiver in the patient set. Each patient set was designed to send a message of 3 bytes to the coordinator MCU. The coordinator's MCU performs a few tasks, as shown in the flowchart. The tasks are to establish a connection with each patient set, send an acknowledgment byte to the patient set after receiving the message, and send the message to the coordinator. The GUI of the coordinator set will create a text file containing the temperature reading for a day. This file is stored in the local PC folder with all of the temperature data, the patients' name (with a serial or patient number), and the time stamps. The graphical representation of the temperature distribution of the patient is shown in Figure 5.24. Statistical information, such as the mean and standard deviation for certain time frame, is then computed. The information is displayed on the staff computer near the coordinator patient set.

5.3.2 Programming and system implementation

The hardware setup in the laboratory consists of the temperature sensor, amplifier circuit, coordinator MCU, and patient MCU, as shown in Figure 5.23 (right). The output voltage was calibrated and compared with a digital thermometer reading. It was observed that the sensor output voltage would yield a tolerance level of ±0.1 °C. In addition, the amplifier circuit had an input voltage of 307.4 mV that was amplified 5.12 times to 1.573 V and displays the result of the ADC process in the ZENA network analyzer. The ADC converted the input voltage to the hexadecimal corresponding to 20.3 °C, as shown in Figure 5.24 (top). The test indicated that the ADC conversion process in the patient's MCU was able to measure the temperature correctly.

As mentioned, the patient MCU sent the 3-byte message to the coordinator MCU. To test this feature, the patient set transmitted the message every few seconds to the coordinator to process the message. The coordinator MCU will not send an acknowledgment message until a button is pressed and the CRC is positive. As shown in Figure 5.24 (top), an additional patient has the first byte of its message changed to differentiate between different patients. Figure 5.24 also shows that the coordinator (source address ending with 01 at the payload section) was able to receive messages from multiple patients.

The network coverage range of the MRF24J40MA transceiver was conducted in a school campus to test the network coverage range. The coordinator MCU was placed in a room, while a patient MCU was along the corridor. The coordinator MCU sent a message every few seconds. The signal could be transmitted with a clear line of sight to a distance of approximately 50 m. In addition to testing the indoor network coverage, the range of 120 m provided in the datasheet was validated in a test outdoor at an ambient temperature of 26 °C.

5.3.3 Testing

The proposed wireless remote temperature-monitoring device was tested on three patients. The authors had received the ethics approvals and informed consent from the three subjects before the tests were conducted. The GUI of the coordinator set (for the *nurse's station*) in Figure 5.24 (center) was used. The coordinator set was connected to the coordinator computer via the RS-232 serial cable. Each time the coordinator MCU received a temperature-reading message, it sent the message to the coordinator's computer. The serial port was opened by clicking the "Open" button to allow the coordinator computer to start receiving data from the coordinator MCU. Upon clicking on the "Update Chart" button, the GUI reads and generates the temperature record chart for that particular patient and date. Once the message is received, the GUI converts the inputs to an actual temperature reading.

The user can allocate a patient name tagged with his/her admission serial number. Once allocated, the patient name with the serial number is applied to the local folder in the database. Figure 5.24 (bottom) shows the local temperature record of a subject. The temperature record was saved in the text file, and the data were formatted as "Temperature-Time."

In the nurse's station, the two main components used in the implementation of the online database are the MySQL database and the PHP scripts. An online hosting server called sparkstation.net was used. Each set of patient information includes the temperature, time, and date created by the coordinator. To operate the system correctly, the nurses were trained to retrieve and provide the correct patient information. There is an added feature of using different colors to indicate the patients with high fever; also, in the event of a significant standard deviation (or equivalent to a sharp slope), an alarm will be triggered.

Statistical information, such as the mean and the standard deviation, for evaluating the data variation and the reliability was determined. For an initial trial, 20 min of recording and sampling at an interval of 2 min was conducted to test the proposed system. The subsequent temperature recording was sampled or taken in every hour. The mean of the patient temperature was 36.642 (95% confidence intervals of 35.913 and 37.371). The standard deviation was 1.019 (95% confidence intervals of 0.701 and 1.860), and the median was 36.99 (95% confidence intervals of 36.054 and 37.267).

Also, the wireless remote temperature-monitoring system was tested for another three days (named D1, D2, and D3) on three different male subjects (named P1, P2, and P3) with ages ranging from 28 to 30 years. The interval between each measurement was set to 1 h, starting at 1 AM. The body temperatures of the three subjects were recorded and plotted on a time series plot, as shown in Figure 5.25. The estimated mean temperatures for P1, P2, and P3 on different days were 36.84, 36.75, and 37.81 °C, respectively, and the standard deviations for P1, P2, and P3 were 0.368, 0.364, and 0.352 °C, respectively.

As shown in Figure 5.25, the wireless remote temperature-monitoring system can measure the temperatures of the subjects satisfactorily (i.e., the highest standard deviation was approximately 0.368 °C (within the range of 0.3 to 0.5 °C) [17,18] and less than 1.5 °C in an enclosed building environment throughout the test period. In summary, the proposed system can measure body temperature (at a resolution of 0.1 °C and with a small standard deviation of 0.368 °C) and identify the patient with the highest temperature.

The web-based wireless remote temperature-monitoring device using temperature sensors for the axillary measurement was implemented. The temperature data were sent via the MiWi wireless protocol to the nurse's station connected to the internet. This approach was facilitated by a patient and coordinator set design that connects the process of monitoring, transmitting, alerting, and recording of the temperature. The experimental tests on the proposed device indicated that it provides long-distance coverage, a small temperature resolution of 0.1 °C, and a standard deviation of 0.37 °C. Importantly, the system can increase the labor efficiency and improve the quality of care through 24-h continuous monitoring of the temperature of patients. The use of different wireless network protocols, such as ZigBee and WiFi, can be compared. A standard medical temperature probe, such as a tympanic probe, and standard disposable tape, can be used for final implementation in the hospital.

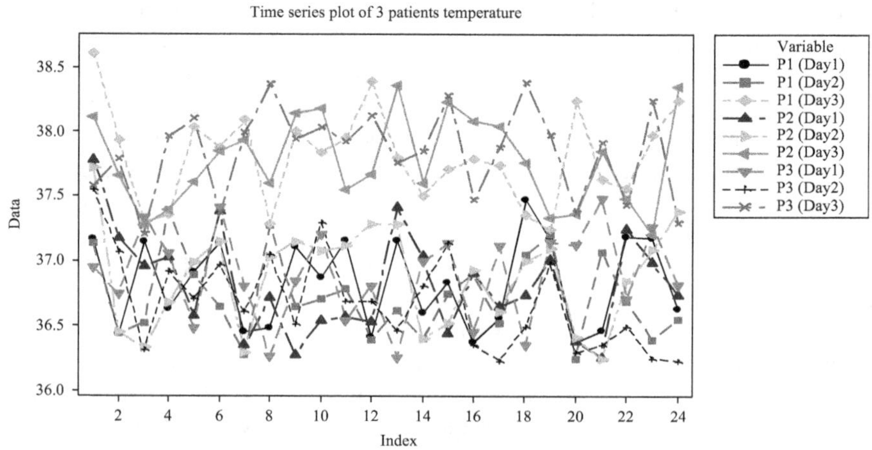

Figure 5.25 Time series plot of three different subjects over three days

5.4 Wall-climbing robot system design

A climbing robot that utilizes dry adhesives as a climbing mechanism is quite useful due to the biomimetic adhesive for different applications such as surveillance, nuclear reactor cleaning, and maintenance. In this section, a tank-like climbing robot is developed to climb most of the wall surfaces. Biomimetic adhesives with nano size hairs are made into tank tracks for climbing. A climbing robot is a type of robot that can transit from the ground to any wall surface. There are a lot of different types of climbing robot in the market. Some utilize the magnetic mechanism to climb metal walls while others utilize an air-suction mechanism to climb. Scientists and engineers have always been intrigued by the way the lizards climb the walls of any surface. For example, the climbing robots will be useful during disasters as it can scour burning high-rise buildings to give a birds' eye view and the location of causalities without endangering the fire-fighters' life.

Vacuum suction is one of the climbing mechanisms used commonly in most climbing robot design [19]. It utilizes the vacuum suction power to achieve strong adhesion force. The disadvantage of the using vacuum suction is the need of powerful motor to achieve a substantial vacuum suction power. It can be quite noisy when the motor is running. Electrostatic adhesion creates an electrically controllable adhesion technology [20] that induces electrostatic charges on a wall using a power supply connected to the compliant pads on the moving robot. It can be electronically turn off and on by remotely electrostatic charges. The disadvantage of the electrostatic adhesion is the needs for an external power source to power the electrostatic circuit. Although the power to induce the charges is small, it still requires an external power source.

In this section, the climbing mechanism uses a gecko rubber pad or biomimetic adhesion. It has very tiny little rubber hairs like mushroom caps of the size of a human hair. These bumps allowed the robot to climb the surfaces of the walls using "van der Waals forces." The van der Waals forces are the theory of attractions between molecules, atoms, and surfaces, as well as many other intermolecular forces [21]. The gecko pad used for the climbing robot can be seen in Figure 5.26.

The electromagnetic adhesive is also used in various climbing robot designs [20]. They have a dominant adhesive force depending on the specifications of the electromagnetic circuit. It can be switched off and on with a flick of a switch while moving on different surfaces. Permanent magnets can also be utilized in the climbing mechanism. They do not rely on an external power source for the magnetic adhesive force. They give strong adhesive force depending on the size of the magnet used. Special magnets called rare-earth magnet are the most durable permanent magnets made from the alloys of rare earth elements. They are stronger than other types of magnets such as ferrite or alnico magnets. There are mainly two types of rare-earth magnets: neodymium and samarium-cobalt magnets. Rare-earth magnets are extremely brittle and prone to corrosion. Therefore, they are usually coated to protect them. A neodymium magnet was chosen as they are easily available in the consumer markets. A small neodymium magnet with a measurement of 2.54 cm $\times$ 2.54 cm $\times$ 0.635 cm has a pull force of approximately 12.52 kg.

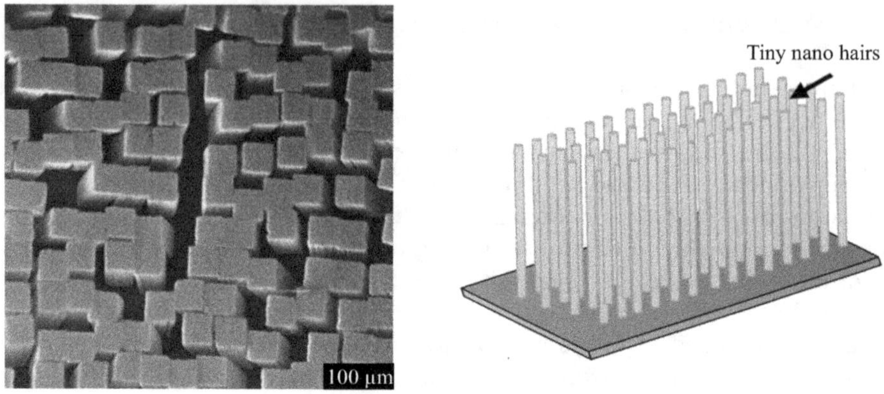

Figure 5.26 Nanotubes are mimicking lizard feet. (see http://www.nisenet.org/scientific-images/nanotubes_mimicking_gecko_feet)

Table 5.2. Comparison between different types of adhesion

Climbing mechanism	External power source	External component needed for climbing	Other factors
Electro adhesive	Low power consumption	Special materials required to generate electrostatic charges	The large surface contact area is needed to carry heavier payloads
Vacuum suction	Required power	Motor required to generate the vacuum suction power	Loud noises are generated
Biomimetic adhesive	Not required	Not required	None encountered
Electromagnetic adhesion	Required power	Excellent magnetic materials are required to induce magnetic forces generated from electricity	Can only climb metal walls
Permanent magnetic adhesion	Not required	Strong earth magnets are needed for strong adhesive force	Can only climb metal walls

The vacuum suction, electrostatic adhesion, and electromagnetic adhesion need an external power source to climb the wall. An external power source would increase the payload. However, the biomimetic adhesive does not need an external power source that reduces the overall payload. The electro-adhesion has a disadvantage as it needs to carry heavier objects that require a large contact surface area to increase its adhesive force. For the biomimetic adhesive, the only disadvantage is to create tiny hairs that are even smaller than the nano scales in order to increase its adhesive force. Hence, both the biomimetic and permanent magnetic adhesion are chosen. The biomimetic adhesion can climb most surfaces including rough or smooth surfaces. The permanent magnet can give extra adhesion for climbing the metallic wall. A comparison of the different type of adhesion methods is tabulated in Table 5.2.

5.4.1 System design and architecture

The prototype designs are presented. During the initial design phase, the robot had a four-wheel design. The small magnetic wheels are the contact point for climbing the walls. The initial design did not have sufficient contact points for the robot to carry heavy components. It also had limited mobility. Thus, few improvements are needed on the initial Prototype Design 1 as seen in Figure 5.27.

The improved Prototype Design 2 as seen in Figure 5.28 has more contact points. It can carry more electronic components, and the mobility is better than the initial design. It also has a lower center of gravity to reduce the gravitational effect on the robot while climbing the vertical walls. The magnetic wheels were removed from the initial design. Instead, a gecko pad tracks were used for climbing.

This Prototype Design 3 as shown in Figure 5.29 consists of two front and back body. With only one body, the robot can climb vertical walls and overcome the

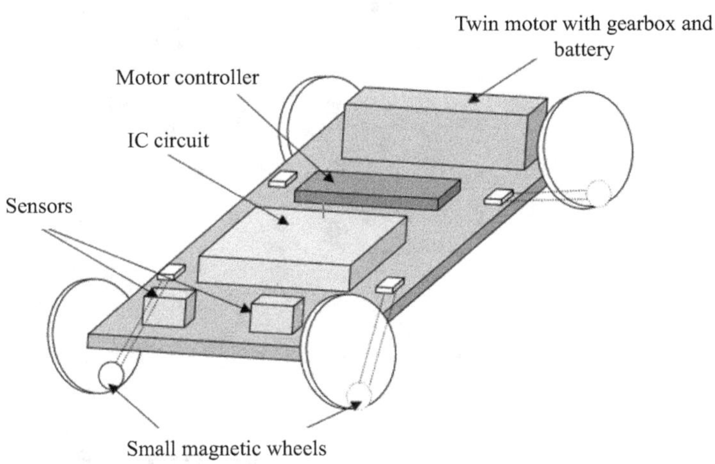

Figure 5.27 Isometric view of Prototype Design 1

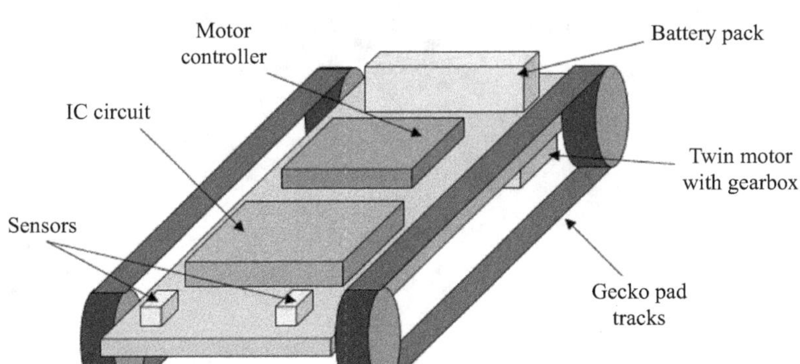

Figure 5.28 Isometric view of Prototype Design 2

corner of the walls. The function of the back body is to support the front body while transiting around the corner. The linkages are flexible enough to make a 90° turn.

Further modifications were made to improve the climbing mechanism of the robot. Magnets were mounted near the bottom of the robot to aid in the climbing. The placement of the electronics was modified. Prototype Design 4 as shown in Figure 5.30 was the final design for the climbing robot.

The neodymium magnets as shown in Figures 5.31–5.33 were mounted on the front bottom of both the front and back of the climbing robot. The battery was

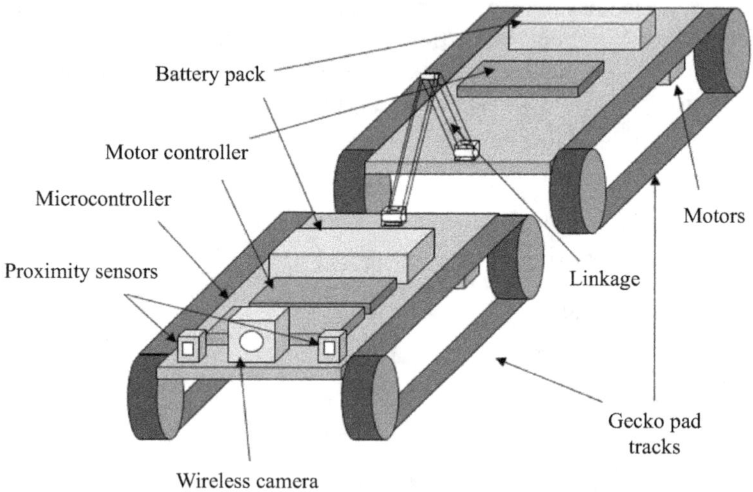

Figure 5.29 Isometric view of Prototype Design 3

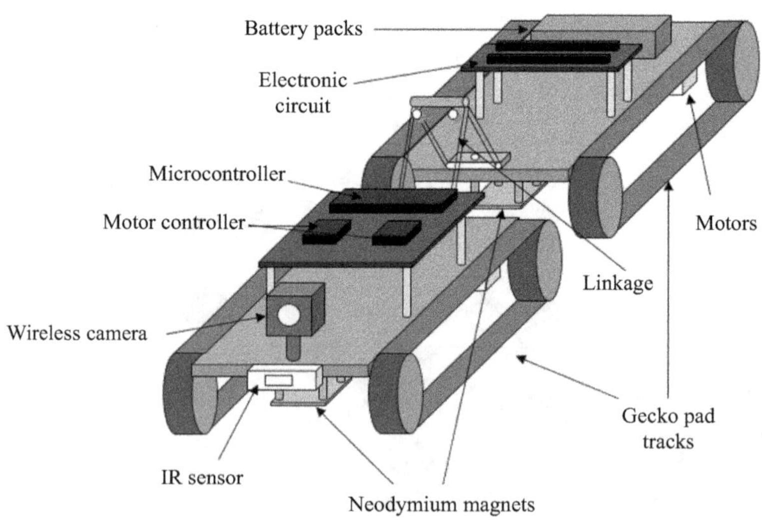

Figure 5.30 Isometric view of Prototype Design 4

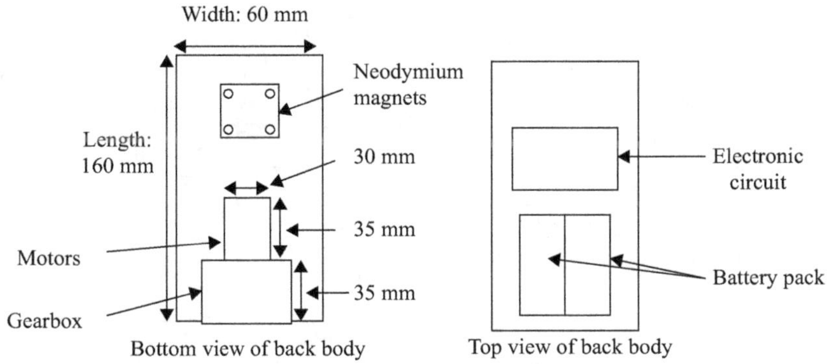

Figure 5.31 Specifications of the back body

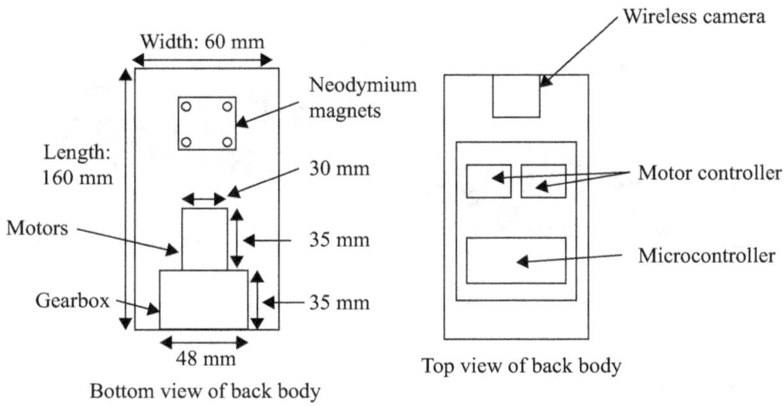

Figure 5.32 Specifications of the front view

placed on the back body to provide the necessary power. The motors and gearbox were fastened at the back to give addition pushing force instead on the front. The overall design became a compact and lightweight climbing robot.

The completed design of the climbing robot can be seen in Figures 5.33 and 5.34. The robot was intended to travel from the ground floor to the vertical wall as shown in Figure 5.35. The link enables the robot to transit smoothly. However, to achieve that, the following calculations are required.

From the design specifications, the estimated speed needed was around 0.1 m/s. The acceleration was around 0.1 m/s^2. With a wheel radius of 30 mm, the wheel angular velocity is as follows.

$$v = \omega r = 0.1\,\text{m/s}$$
$$\omega = \frac{v}{r} = \frac{0.1}{0.03} = 3.33\,\frac{\text{rad}}{\text{s}} = 31.8\,\text{rpm} \tag{5.1}$$

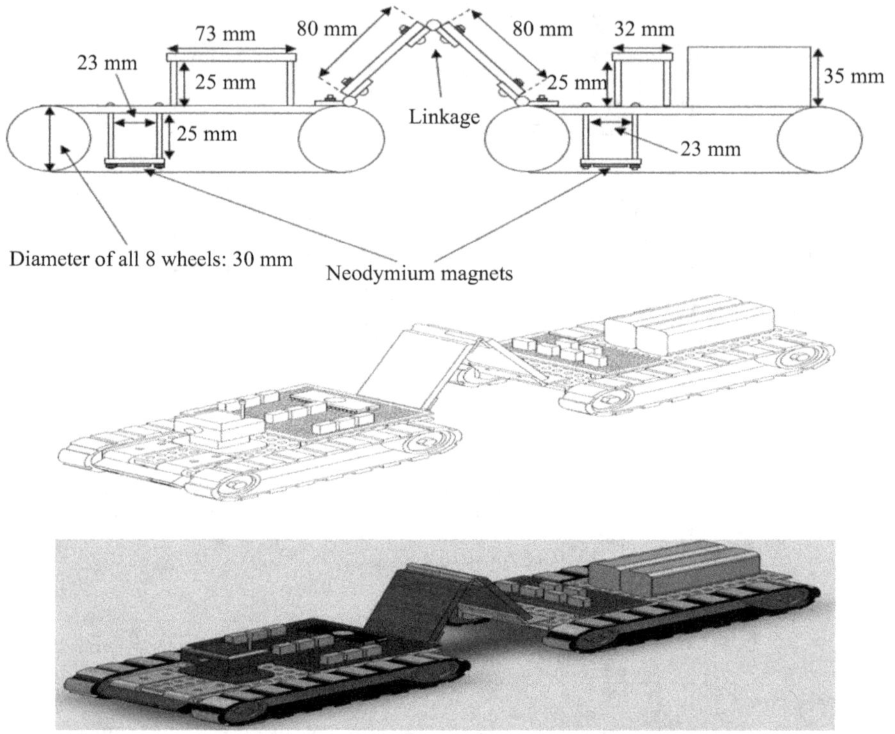

Figure 5.33 Side view of the climbing robot

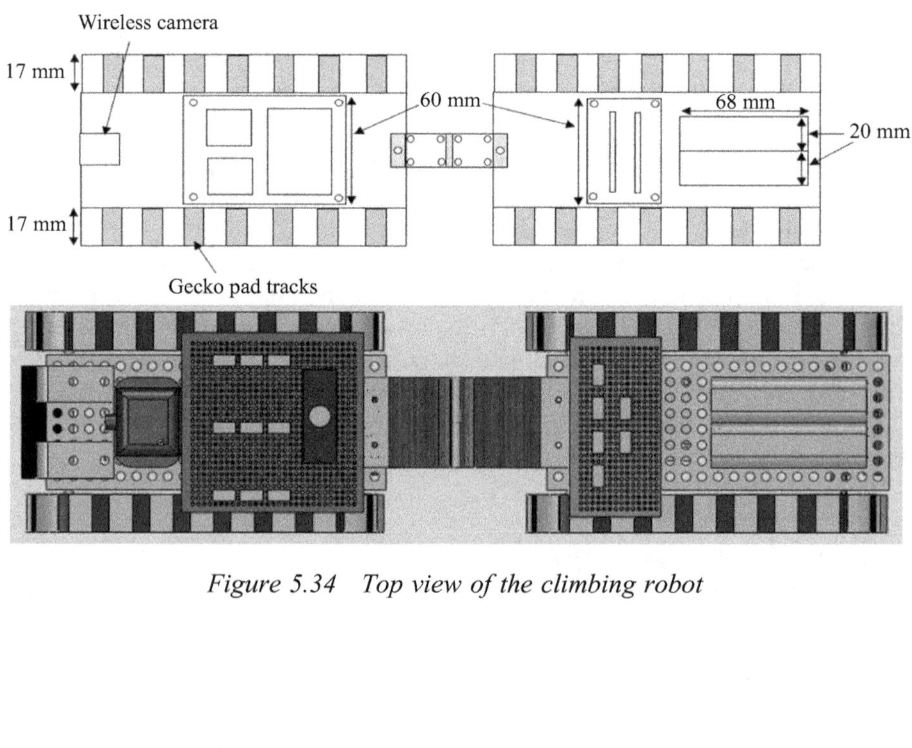

Figure 5.34 Top view of the climbing robot

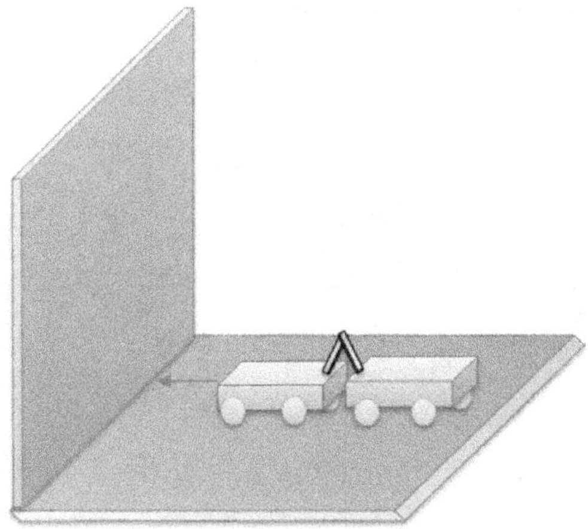

Figure 5.35 Transition from horizontal surface to vertical surface

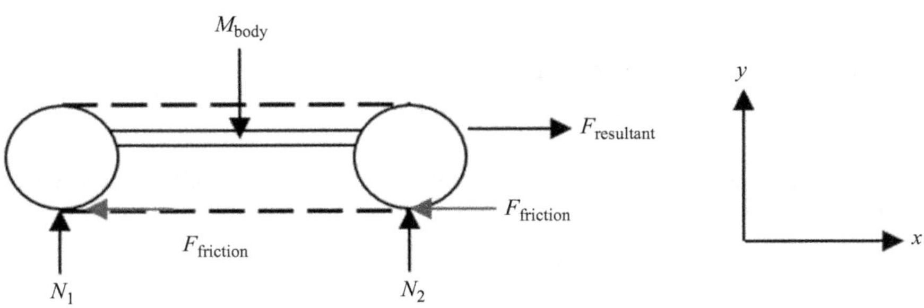

Figure 5.36 Free body diagram of the climbing robot front body

The estimated mass of robot's front body is estimated as

$$M_{body} + M_{battery} + M_{motors+gearbox} + M_{electronics} = 250\ \text{g} + 150\ \text{g} + 200\ \text{g} + 200\ \text{g} = 800\ \text{g} \tag{5.2}$$

Assuming coefficient of friction $\mu = 0.2$ in Figure 5.36, the friction force becomes

$$F_{friction} = \mu\,\text{N} = \mu \times (N_1 + N_2) \tag{5.3}$$

where N is the normal contact force of the climbing robot that is $N_1 + N_2 = 800$ g

From Newton's second law of motion,

$$\sum F_x = ma \tag{5.4}$$

By substituting (5.3) in (5.4), we get

$$F_{\text{resultant}} - \left(F_{friction} \times 2\right) = ma \tag{5.5}$$

$$F_{\text{resultant}} = ma + \left(F_{friction} \times 2\right) \tag{5.6}$$

The resultant force, $F_{\text{resultant}}$ becomes

$$(0.8 \times 0.1) + (0.2 \times 0.8 \times 9.8 \times 2) = 3.216\,\text{N} \tag{5.7}$$

The required maximum torque is approximately

$$T_{\max} = F_{\text{resultant}} \times r \tag{5.8}$$

$$T_{\max} = 3.216 \times 30 \times 10^{-3} = 0.096\,\text{N/m} \tag{5.9}$$

The power required is as follows:

$$P_{\text{required}} = \omega T_{\max} \tag{5.10}$$

$$P_{\text{required}} = 3.33 \times 0.096 = 0.319\,\text{W} \tag{5.11}$$

A safety factor of 2 is added to oversize the motor:

$$P_{\text{required}} = 2 \times 0.319 = 0.638\,\text{W} \tag{5.12}$$

Since the mass of the back body is smaller than the front body, the torque and power required are as follows.

From Figure 5.37,

$$\begin{aligned} F_{\text{friction}} &= \mu\,\text{N} = \mu \times N_2 \\ N_2 &= 400\,\text{g} \end{aligned} \tag{5.13}$$

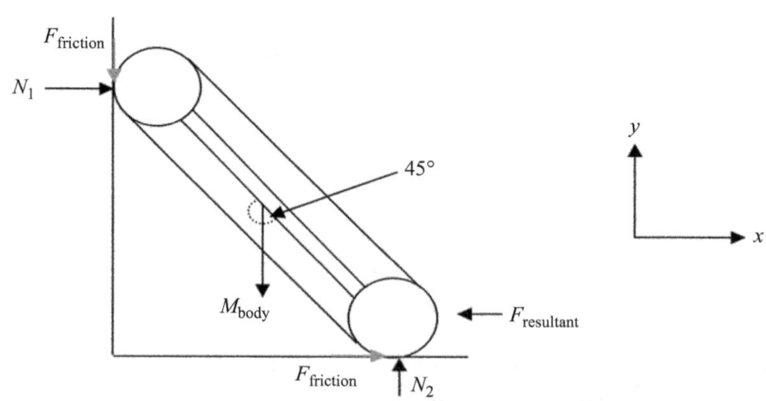

Figure 5.37 Free body diagram of robot transiting from horizontal surface to vertical surface

From Newton's second law of motion,

$$F_{\text{resultant}} - (F_{\text{friction}} \times 2) = ma \tag{5.14}$$

$$\begin{aligned} F_{\text{resultant}} &= ma + F_{\text{friction}} = (0.8 \sin 45^\circ \times 0.1) + (0.2 \times 0.4 \times 9.8 \times 2) \\ &= 1.624\ \text{N} \end{aligned} \tag{5.15}$$

Therefore, the required maximum torque becomes

$$T_{\max} = F_{\text{resultant}} \times r = 1.624 \times 30 \times 10^{-3} = 0.048\ \text{N/m} \tag{5.16}$$

The power required is as follows:

$$P_{\text{required}} = \omega T_{\max} \tag{5.17}$$

$$P_{\text{required}} = 3.33 \times 0.048 = 0.159\ \text{W} \tag{5.18}$$

A safety factor of 2 was added to oversize the motor as shown.

$$P_{\text{required}} = 2 \times 0.159 = 0.319\ \text{W} \tag{5.19}$$

With the calculated torque and power, two motors were compared. The necessary specifications of each motor can be seen in Figure 5.38.

The Mabuchi motor FF-130RA-18100 consumes 0.56 A of current at maximum efficiency. As more current was needed for the motor, it means that a powerful battery would be needed. Thus, it would affect the overall weight of the robot. Therefore, the Mabuchi motor FF-130SH-14230 was a better choice as it consumes 0.23 A of current at maximum efficiency. The speed of FF-130RA-18100 is not as fast as FF-130RA-18100, for a climbing robot. The gearbox Tamiya 70097 Twin-Motor Gearbox Kit was used. The gearbox has 2 gear ratios, namely, 58:1 or 204:1. For a climbing robot, a high torque is important for the robot to transit from the ground to the wall surface. Thus, the 204:1 gear ratio was used to give a higher torque. Based on the previous calculations, the torque required is 0.096 N/m and the power required with a safety factor 2 is 0.638 W. Hence, with a gear ratio of 204:1, the torque that the Twin-motor gearbox kit could produce is: $1.20 \times 10^{-3} \times 204 = 0.2448\ \text{N/m}$.

As shown, the torque from the Twin-motor gearbox kit is higher than required. From the Mabuchi motor specs table, the output power of each motor is 0.76 W.

MODEL	VOLTAGE		NO LOAD		AT MAXIMUM EFFICIENCY					STALL		
	OPERATING RANGE	NOMINAL	SPEED	CURRENT	SPEED	CURRENT	TORQUE		OUTPUT	TORQUE		CURRENT
		V	r/min	4	r/min	A	mN.m	g.cm	W	mN.m	g.cm	A
FF-130SH 14230	2.5–10	6	7,400	0.05	6,060	0.23	1.20	12.3	0.76	6.67	68	1.03

Figure 5.38 Snapshot of Mabuchi motor specifications (cite: https://product.mabuchi-motor.com/detail.html?id=25)

Table 5.3. Gecko pad adhesion test results

	First trial	Second trial	Third trial	Average
Wood surface	1.03 kg	1.04 kg	1.05 kg	1.07 kg
Metal surface	1.16 kg	1.15 kg	1.18 kg	1.16 kg
Concrete surface	1.02 kg	1.01 kg	0.9 kg	0.97 kg
Glass surface	1.4 kg	1.37 kg	1.38 kg	1.38 kg

Since two motors are used for each body, the output power of the two motors is 1.52 W. It is higher than the calculated output power of 0.638 W. To verify the gecko pad can carry the climbing robot's weight, a few simple experimental tests were performed. The tests were conducted on different wall surfaces as shown in Table 5.3.

From the results tabulated in Table 5.3, the gecko pad could withstand the weight of the climbing robot. It was estimated to be around 1 kg onto various wall surfaces. As the robot was designed to be autonomous, few sensors were implemented for navigations. One of the sensors chosen was the magnetic proximity sensors that utilized an electromagnetic field to detect objects. The reason for considering proximity sensors is the high reliability and long functional life without any mechanical parts. However, the shortest sensing distance is 15 mm. The sound wave is transmitted through the transmitter of the sensor and reflected back from the object. The disadvantage of the ultrasonic sensor is the field of vision. The ultrasonic sensor has a field of vision of around 60°. If two ultrasonic sensors are placed near each other, the field of vision will overlap affecting the sensor's range. Hall Effect sensor uses the induced current from the magnetic field to produce an output voltage proportional to the strength of the magnetic field. The Hall Effect sensor can detect objects that are passing through its magnetic field. IR sensor detects the proximity of objects based on the intensity of reflected light from IR LED. The LED transmits IR light and the reflected light from the object is then received by the opto-resistive transistor (receiver). Although it works quite well in the proximity detection, the amount of ambiance lighting will affect the opto-resistive transistor's accuracy. The sensor will be affected by the amount of ambient lighting. The Hall Effect sensor is quite accurate in terms of resolution. The magnetic sensors have a good reliability of accuracy but the sensing distance is too short. Although the laser sensors have the highest accuracy, the price is too expensive. The IR sensor named Sharp sensor 2Y0A41SK was then used. A comparison of different sensors is tabulated in Table 5.4.

The main body of the robot used the universal plate set from Tamiya. The holes on the plate are useful as it can mount various components on the robot's body quickly. It is also lightweight and durable to hold the weight of the components. It is made of acrylonitrile butadiene styrene plastic to withstand the components weight.

Table 5.4. Comparison of sensors

Sensor	Cost	Ease of use	Accuracy
Laser	Very high	Easy	Very high
Ultrasonic	Medium	Medium	High
Infrared	Low	Easy	Medium–High
Hall effect	Low	Medium	Low
Magnetic proximity	Medium	Medium	Low

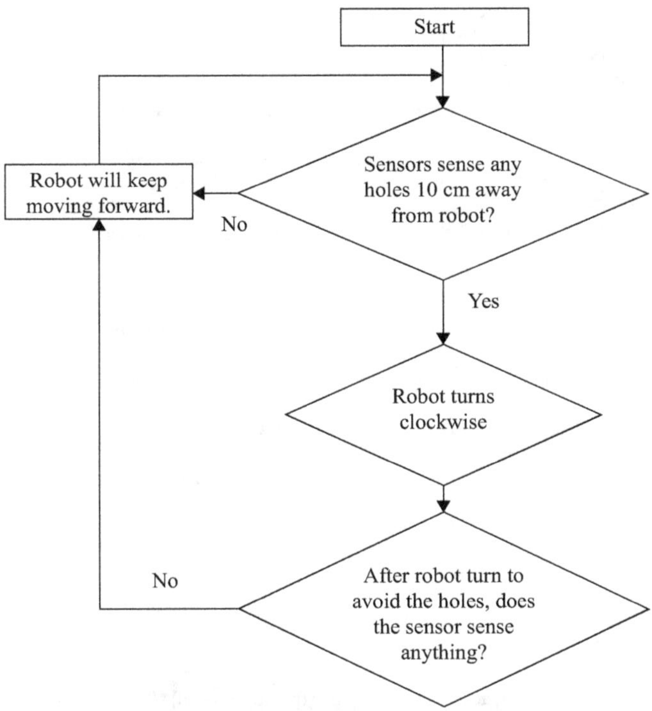

Figure 5.39 Flowchart for the climbing robot

5.4.2 Programming and system implementation

The PIC18F4520 was used as a MCU for the climbing robot. The same MPLAB-IDE was used. The flowchart of the program is shown in Figure 5.39. The climbing robot will move away (counterclockwise) from any holes or gaps below it. The gaps will be detected by IR sensor. The details of the programming were written and downloaded into the PIC18F4520 via MPLAB-IDE and compiler. The use of MPLAB-IDE can be found in Section 5.1.

PIC program codes for the climbing robot

```
#include <p18f4520.h>
#include <delays.h>
int result;

void main(void)
{
        TRISA = 0x01;                //RA0 as sensor input
        TRISB = 0x00;                //Initialize Port B as outputs
        TRISC = 0x00;                //Initialize port C as outputs
        TRISD = 0x00;                //Initialize port D as outputs

        /*ADC Module setup*/
        ADCON1bits.PCFG = 1110;              //Select AN0
        ADCON1bits.VCFG0 = 0;                //Vref +Source
        ADCON1bits.VCFG1 = 0;                //Vref -Source
        ADCON0bits.CHS = 0000;   //Select analog channel 0, RA0

        ADCON2bits.ADCS = 010;   //ADC conversion clock FOSC/32
        ADCON2bits.ACQT = 001;   //ADC Conversion clock 2TAD
        ADCON2bits.ADFM = 1;     //ADC Result format select bit Right justified
        ADCON0bits.ADON = 1;     //ADC conversion module enabled
        while(1)
        {
                /*Sensor program*/
                ADCON0bits.GO = 1 ;                  //ADC conversion in progress
                while(ADCON0bits.GO == 1);           //wait for conversion to finish
                //ADCON0bits.ADON = 0;   //Turn off ADC module
                result = ADRES;
                if(result >= 100)        //Move forward
                {
                        /*Front motor controller (positive speed)*/
                        PORTDbits.RD0 = 1; //Input 1
                        PORTDbits.RD1 = 1; //Enable A
                        PORTDbits.RD2 = 0; //Input 2

                        PORTDbits.RD3 = 1; //Input 3
                        PORTCbits.RC5 = 1; //Enable B
                        PORTCbits.RC4 = 0; //Input 4
                        /*======================================*/
                        /*Back motor controller (positive speed)*/
                        PORTBbits.RB4 =     1;       //Input 1
                        PORTBbits.RB3 =     1;       //Enable A
                        PORTBbits.RB2 = 0; //Input 2

                        PORTCbits.RC7 = 0; //Input 3
                        PORTDbits.RD4 = 1; //Enable B
                        PORTCbits.RC6 = 1; //Input 4
                        /*======================================*/
                }
```

```
            else if(result < 100)    //Move backwards and turn clockwise
            {
                    /*Front motor controller (negative speed)*/
                    PORTDbits.RD0 = 0; //Input 1
                    PORTDbits.RD1 = 1; //Enable A
                    PORTDbits.RD2 = 1; //Input 2

                    PORTDbits.RD3 = 0; //Input 3
                    PORTCbits.RC5 = 1; //Enable B
                    PORTCbits.RC4 = 1; //Input 4
                    /*========================================*/


                    /*Back motor controller (negative speed)*/
                    PORTBbits.RB4 =     0;      //Input 1
                    PORTBbits.RB3 =     1;      //Enable A
                    PORTBbits.RB2 = 1; //Input 2

                    PORTCbits.RC7 = 1; //Input 3
                    PORTDbits.RD4 = 1; //Enable B
                    PORTCbits.RC6 = 0; //Input 4
                    /*========================================*/
                    Delay1KTCYx(250);

                    /*Front motor controller (Turn clockwise)*/
                    PORTDbits.RD0 = 1; //Input 1
                    PORTDbits.RD1 = 1; //Enable A
                    PORTDbits.RD2 = 0; //Input 2

                    PORTDbits.RD3 = 0; //Input 3
                    PORTCbits.RC5 = 1; //Enable B
                    PORTCbits.RC4 = 1; //Input 4
                    /*========================================*/
                            /*Back motor controller (Stop back body from moving)*/
                    PORTBbits.RB4 =     1;      //Input 1
                    PORTBbits.RB3 =     0;      //Enable A
                    PORTBbits.RB2 = 0; //Input 2

                    PORTCbits.RC7 = 1; //Input 3
                    PORTDbits.RD4 = 0; //Enable B
                    PORTCbits.RC6 = 0; //Input 4
                    /*========================================*/
            }
        }
}
```

A motor controller is a device to control the performance of an electric motor. A motor controller can start and stop the motor, select forward or reverse rotation, and regulate the speed. The L298 is an integrated circuit with high-voltage and high-current dual full-bridge driver. It can drive inductive loads such as solenoids, relays, DCs, and stepping motors. One L298 can drive two motors with an output current of 2 A at voltages from 4.5 to 36 V. The L293 is a quadruple high-current half-H driver to drive inductive loads such as solenoids, relays and bipolar stepping motors. It can

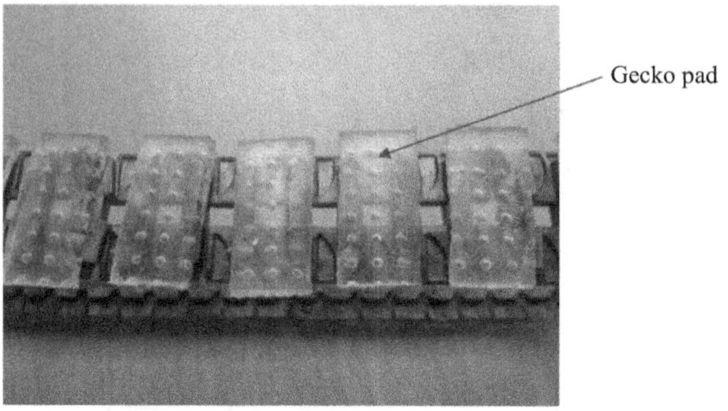

Figure 5.40 Tank tracks with gecko pad

Figure 5.41 Overview of the electronic circuit board

produce an output current of 1 A at voltages from 4.5 to 36 V. The L293 can give an output current of 1.2 A while the L298 can produce an output current of 2 A. From the Mabuchi motor datasheet, each of the motors needs to draw 0.66 A. It implies that the L298 can produce enough current to drive the motors. As shown in Figure 5.40, the gecko pad was cut into small rectangular pieces and glued onto the tracks as shown. Note that the gecko pad will undergo wear and tear.

A simple circuit was constructed as shown in Figures 5.41 and 5.42. The wireless camera was mounted on the front body of the robot. When the main battery runs out, the wireless camera can still operate. The IR sensor was placed below the wireless camera to detect any front objects that are near to the sensor. The schematic diagram of the electronic circuit is shown in Figure 5.43.

The sharp IR sensor needs to be calibrated for its input–output, namely, voltage vs. distance detected. A simple sensor calibration test was set up as shown in Figure 5.44. The results are tabulated as shown in Figures 5.45 and 5.46.

The output voltage of the sensor can be seen in Figure 5.45. As the object is further away from the sensor, the voltage drops. For the MCU to read the output voltage of the sensor, a simple relationship was derived from the analog voltage

Figure 5.42 Front view of the robot

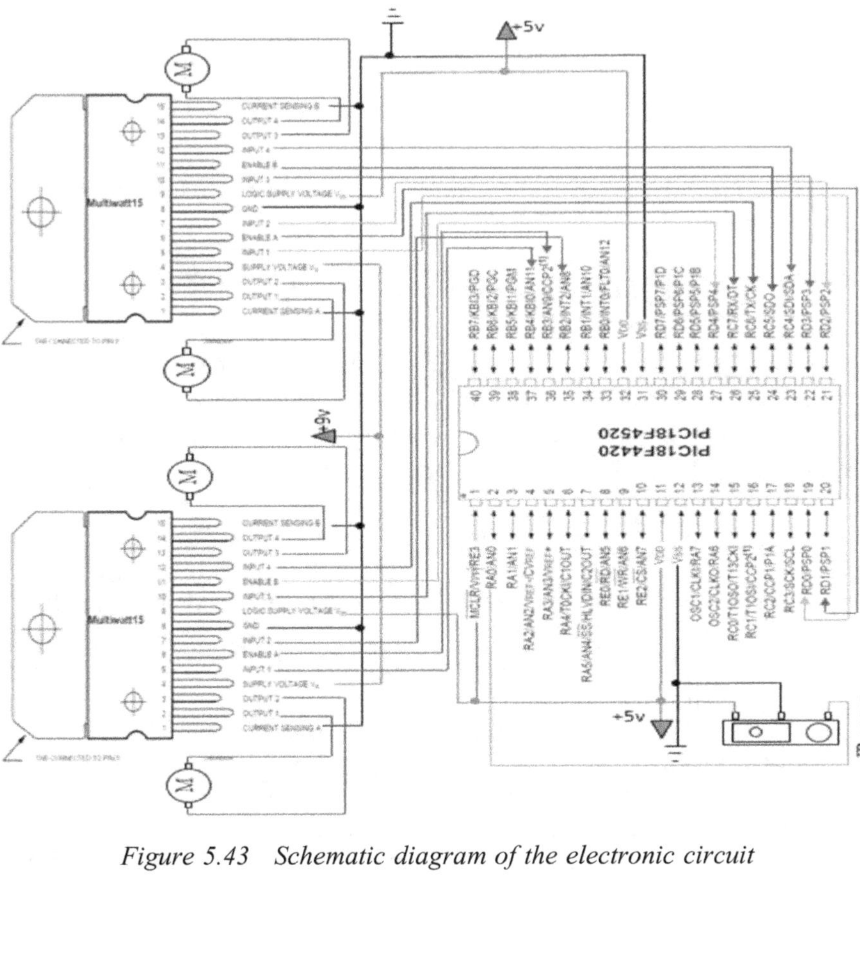

Figure 5.43 Schematic diagram of the electronic circuit

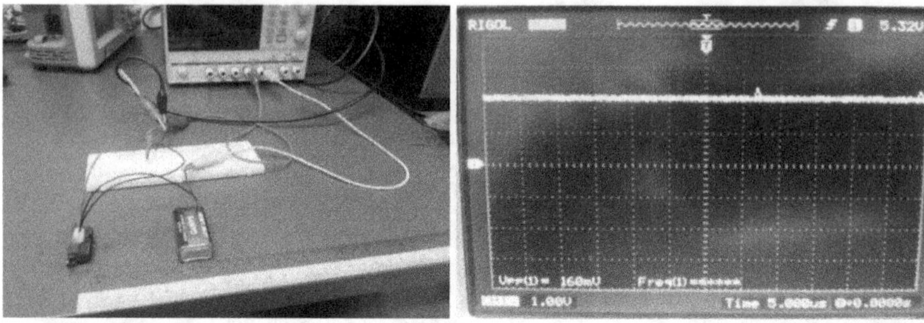

Figure 5.44 Sensor calibration test setup (left) and sensor output voltage waveform at 0.5 cm (right)

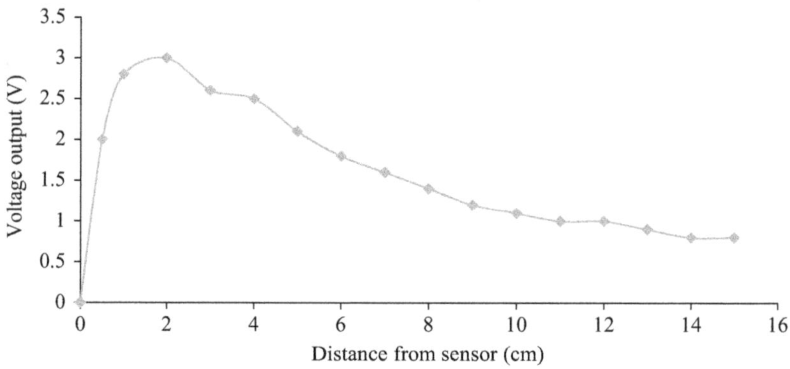

Figure 5.45 Sensor characteristics

output to a digital voltage output. The MCU will need an ADC. The PIC18F4520 MCU has a 10-bit ADC. The MCU requires 5 V for the analog to a digital reference voltage. The same +5 V is equivalent to 1,024 bits (in binary: 11111111) via a simple relationship.

$$X = \frac{\text{Sensor's voltage}}{\text{Reference voltage}} \times 1,024 \tag{5.20}$$

where X is the digital value for the MCU. For example, the MCU will detect an analog voltage of +3 V equivalent to

$$X = \frac{3}{5} \times 1,024 = 614.4 \approx 614 \tag{5.21}$$

The MCU gives 614 as an interpretation of +3 V that is equivalent to 2 cm. With this calibration test, the Sharp IR sensor can communicate with the MCU. As shown in Figure 5.46, the neodymium magnets were mounted on the bottom front of the robot. They were mounted just 0.7 cm above the ground to give sufficient

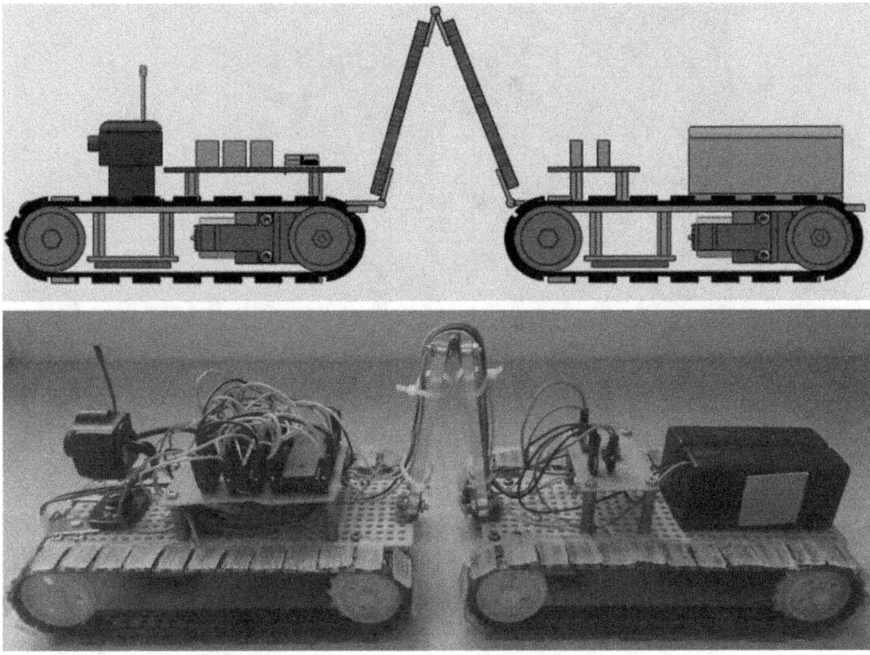

Figure 5.46 Overview of the climbing robot

attraction force to hold the robot to the metallic wall. It could not be placed too near to the wall as it would attract to the surface and remain motionless. Conversely, it could not place too far away from the ground as the attraction force will be weak to hold the robot firmly on the wall.

5.4.3 Testing

The intelligent climbing robot was designed and tested experimentally as shown in Figure 5.47. Several tests were carried in indoor environment. The biomimetic material used was not as good as the one that was fabricated. The motors and gearboxes chosen were sufficient to drive the robot from the horizontal to the vertical surface. The electronic components work with the motors without overheating of electronic components. The robot could achieve the required speed and torque to transit from the horizontal surface to the vertical surface. The uniqueness of the intelligent climbing robot making was the use of biomimetic material to enable the robot to climb most surfaces without using an external power source. As a result, the robot can adhere to the vertical surface for long hours. The design using the biomimetic material is non hazardous without affecting the wall surfaces. The main disadvantage of using the biomimetic material is poor adhesiveness after numerous trials. It needs to be cleaned and changed.

As the climbing robot was only equipped with one IR sensor, additional sensors can be implemented on the climbing robot to make it more intelligent.

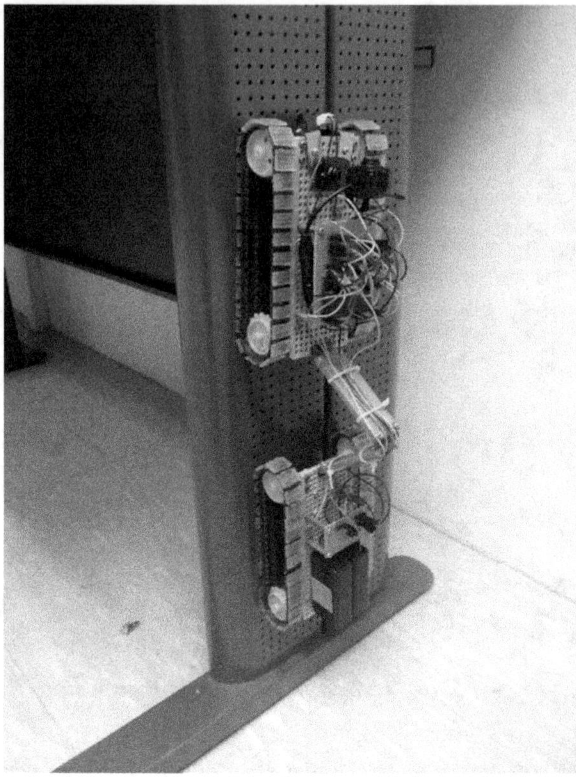

Figure 5.47 Intelligent climbing robot climbing table leg

For example, temperature sensors can be implemented to measure the surrounding temperatures and a smarter wireless camera sensor for object recognition. The pressure sensor can be used to gauge the pressure exerted against the surface. The robot can be equipped with lightweight protective casings to protect the electronic components. The electronic components can be upgraded to military grades for more robust applications. The completed prototype and its testing can be seen in the following video links.

- https://youtu.be/IM6cNa6kYwA
- https://youtu.be/ABbXBgfEqYA

5.5 Magnetic conveyor system design

The details of the magnetic conveyor system design used in this section can be found in the following reference [22]. High-speed trains are one application of a transport system that uses the magnetic levitation technology. The TransRapid in Germany [23], for example, uses a combination of a synchronous linear drive and an electromagnetic levitation system. They began development about 50 years ago of what is referred to as an electromagnetic system. The electromagnetic system can be divided into the electrodynamics suspension system, electromagnet suspension system (EMS), and most recent hybrid forms using a permanent magnet. Besides the TransRapid in Germany, the applications of these magnetic suspension systems were also developed on other high-speed train systems such as HSST [24] and MLX [23,25] in Japan, UTM [26] in Korea, and MagLev in China [27].

As shown in the above, the magnetic levitation systems have practical importance in the railway industry. Another critical application of the EMS system is in transport system [28–32] such as semiconductor assembly line, biochemical manufacturing, industry tools production, food processing, and baggage conveyance, of which further studies have not done much development, despite the general interest in these systems. The recent studies have been made in the magnetic levitation technology to substitute traditional conveyor or assembly line technology. High operational availability combined with low operating costs are therefore the main demands of modern conveyor systems. Conventional conveyor vehicles are often based on the pulley, and chain concepts have suffered from friction and high maintenance. As the demand for expensive maintenance increases exponentially with the conveyor speed, its upper limit is often restricted.

In recent years, magnetic levitation techniques have been used for eliminating friction due to mechanical contact, decreasing maintainable cost and achieving proper precision positioning at higher speed. A conveyor system using the magnetic levitation principles is proposed to tap into these tangible benefits of a magnetic levitation principle. To improve the levitation gaps, a hybrid approach of using EMS with repulsive levitation is used jointly with contactless sensing and an array of magnets in the LSM.

However, there are some disadvantages in these EMS systems. The magnetic levitation force is inherently unstable [31–34] and highly nonlinear, which requires a closed-loop controller to stabilize the system. For the past few decades, the magnetic levitation systems used many linear and nonlinear control strategies based on both model-based and model-free method [33]. A sliding-mode controller [34] is implemented for the conveyance system to provide a systematic approach to the problem of maintaining stability and performance in the face of modeling imprecision. To the best of our knowledge, the details of implementing the sliding-mode control system involving the MCU for the electromagnetic actuated conveyance system have not been seen in the literature. The potential application for the proposed prototype could be used for material transportation in biochemical and semiconductor plants as they need to be completely free from dust and slight noise.

5.5.1 System design and architecture

The novel magnetic conveyor system is shown in Figure 5.48. The system has a dimension of 0.5 m × 0.5 m × 0.3 m and weight 3.5 kg (excluding the electronics circuits). The prototype is designed to carry a load of 2 kg and a maximum speed of 0.6 m/s with an acceleration of 1.2 m/s^2. The three main parts are a base assembly

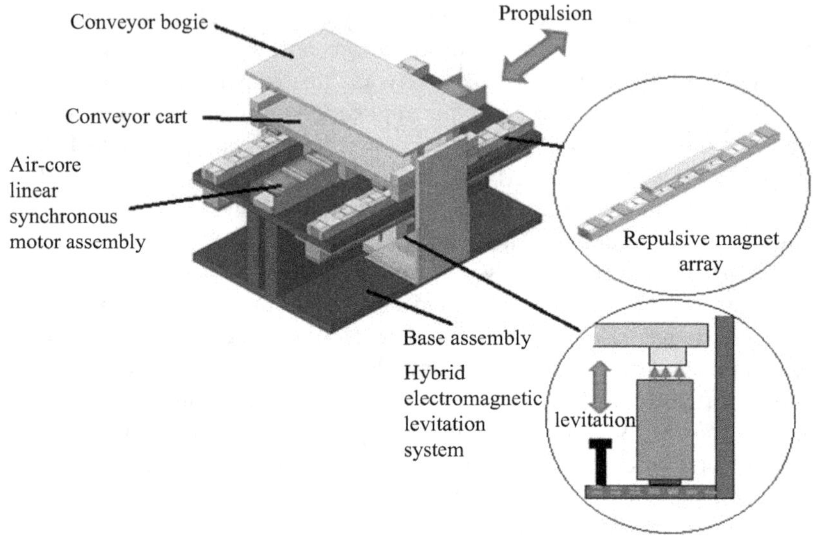

Figure 5.48 3D view of magnetic conveyance system

(0.20 kg), a conveyor bogie (3 kg), and a conveyor cart (0.30 kg). In the magnetic conveyance system, it has a hybrid electromagnetic levitation system, consisting of a linear synchronous motor (LSM) assisted by contactless sensing for a propulsion drive that keeps track of the position of the conveyor cart. To ensure that the system levitates appropriately, an array of permanent magnets is placed on the top and bottom surface of the base assembly.

This arrangement of the repulsive magnet array helps to avoid misalignment in the lateral direction and to maintain the initial levitation gap and propulsion. On the other hand, a hybrid EMS system using repulsive approach is used to provide additional magnetic levitation without consuming additional electrical energy. During the propulsion, a MCU and IR sensors are used to maintain the reference gap between the magnetic coil and the guideway. Figure 5.49 shows the front view of the final design and its components within the conveyor system.

The magnetic levitation is accomplished by the hybrid electromagnetic levitation system. The hybrid electromagnetic levitation system uses two permanent magnets and electromagnets (see Figure 5.50). The former compensates the primary levitation force, while the latter produces magnetic forces to control the air gap to the desired reference point. The material used for the electromagnetic core was laminated silicon steel of 0.0005 m thick with layers of Neodymium magnets on top.

As shown in Figure 5.50, the two IR sensors measure the air gaps on both sides of the conveyor cart. The sensor signal goes into the PIC18F4520 MCU to actuate the two electromagnets mounted on the left and right sides of the cart. The control algorithm is implemented in MPLAB environment that interfaces the MCU to the user's computer (installed with the MPLAB software) via a PICkit2 programmer/debugger. The other side of the compiler is connected to the MCU through

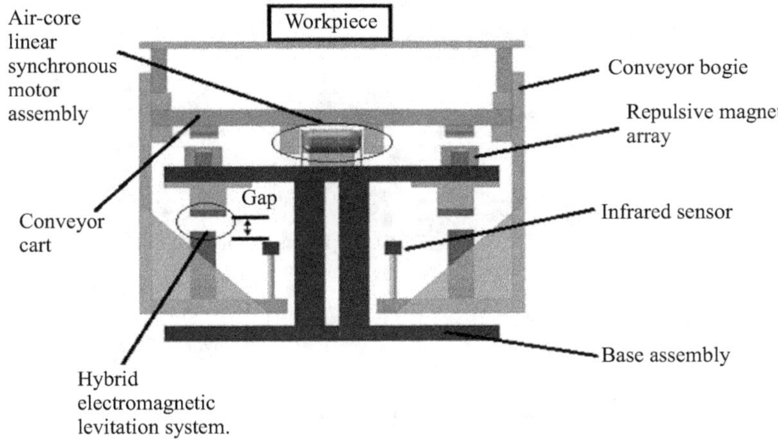

Figure 5.49 Front view of magnetic conveyance system

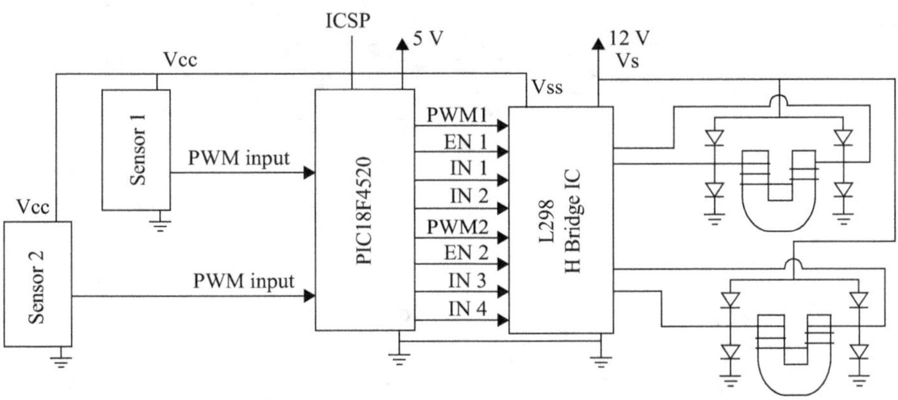

Figure 5.50 Simplified electrical schematic for hybrid EMS system

in-circuit serial programming that uses two input/output pins to serially input and output the data. L298 Dual H-Bridge integrated chip is used to toggle the current's direction of the two leads connected to the electromagnet coils for levitation. As noticed, a set of diodes (IN4936) is used to protect the power source whenever the power is turned off, or an inductive load has its power source removed.

The LSM (see Figure 5.51) consists of a rotor being the permanent magnets (mounted on the conveyor cart), a stator with winding coils (mounted on the top of the base assembly), and contactless sensing to produce the forward or reverse propulsion. By alternating the direction of current in the stator coil (using the three-phase supply), propulsion of the conveyor cart is achieved at a specific frequency. This propulsion is obtained from the thrust force generated by the interactions of magnetic fields between the permanent magnets and the stator windings as shown in Figure 5.52.

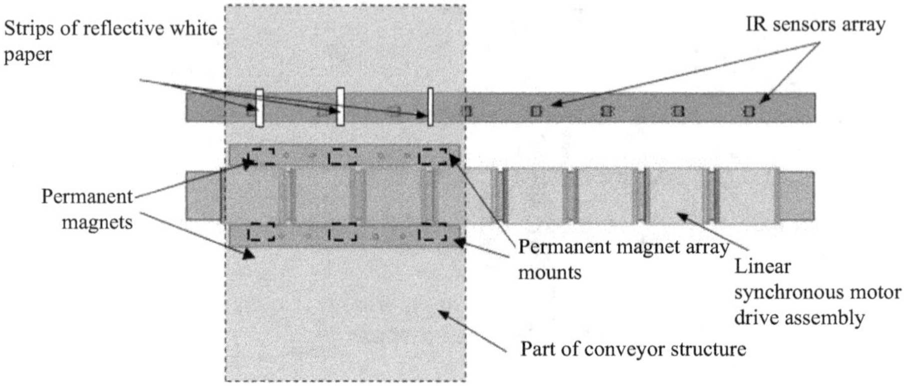

Figure 5.51 Design of LSM using contactless sensing

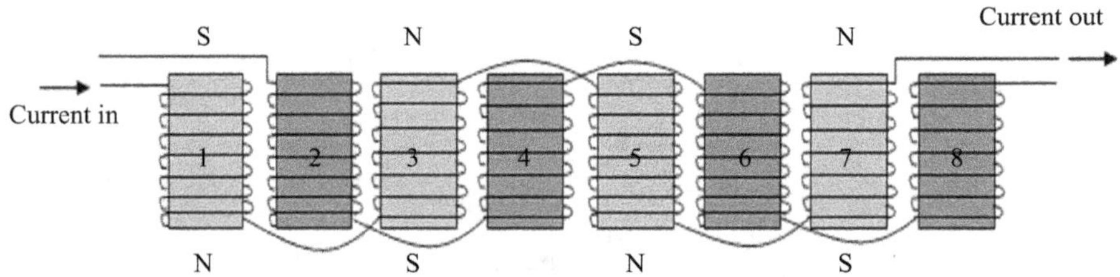

Figure 5.52 Magnetic polarity when Phase A (or Phase B) is energized in the forwarding direction

Besides using the permanent magnets and stator with the coils, the LSM drive requires some form of commutation during the propulsion. The array of IR sensors in Figure 5.53 is used. The LSM requires an array of IR sensors to continually keep track of the position of the conveyor cart. Strips of reflective white paper are attached to the underside of the conveyor cart facing the IR sensors. They are aligned to the center line of the three magnets in the magnet array to provide the displacement. On the other hand, the extra arrays of permanent magnets are also used to provide extra repulsive forces to maintain the initial levitation gap and to prevent any lateral misalignment during propulsion.

The details of the circuit design and operating principle of the LSM are presented. The driving circuit for the stator coils consists of another two H-Bridge controlled by the PIC18F4520 MCU and the commutation signals from the IR sensors. To reduce the wiring and control complexity, the stator coils are separated into two different phases denoted by Phase A (or Phase B) in Figure 5.53. For example, Coils 1, 3, 5, and 7 are interconnected to form Phase A, and Coils 2, 4, 6, and 8 are interconnected to form Phase B (not shown). The magnetic polarity of the stator coils at any instance can be reversed by changing the current direction for the Phase A (or Phase B). With this design, there are coils that would be energized when the magnet

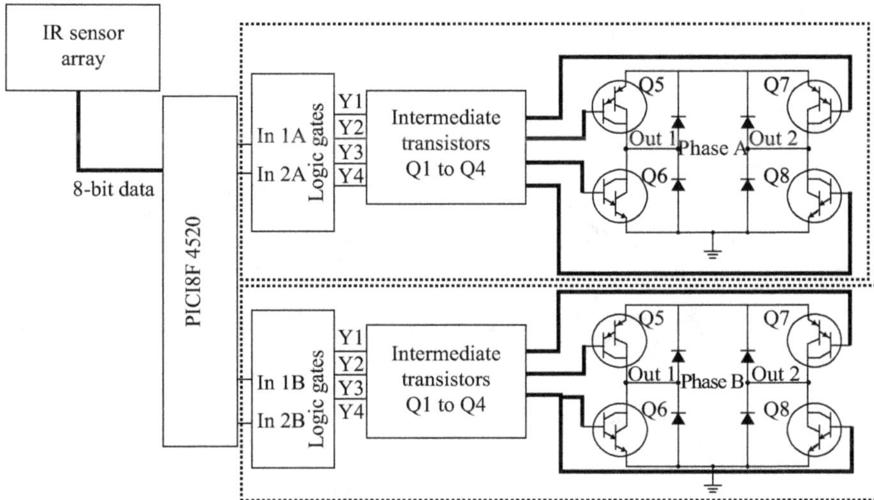

Figure 5.53 Simplified schematics of LSM system

array is not in its vicinity; however, it is the result of the design trade-off for more straightforward wiring, control, and reduction of circuit components.

To determine the position of the conveyor cart during the forward or reverse travel, an array of IR sensors is used. As shown in Figure 5.54, the sensors provide the signal to the MCU via a logic gate that activates the Phase A or B of the coils. The logic gate blocks consist of NAND (74STD), and NOR (7400N) gates manipulate the inputs and the four outputs to the intermediate transistors, Q1 to Q4. For clarity in the diagram, the details of the outputs from the logic gate blocks to the base of the transistors Q1 to Q4 (BJT NPN 2N2222) and then Q5 to Q8 (Darlington NPN TIP127) are not shown in the schematic diagram.

To achieve the forward or reverse motion of the conveyor cart, the corresponding outputs for Phase A or B (see Table 5.5) based on the locations' feedback from the sensors are required. The transistors Q5 to Q8 will be in its "On" state or "Off" state depending on the state of Q1 to Q4. The current flows through the load in one direction when Q5 and Q8 are "On" with Q6 and Q7 are "Off" or vice versa. It ensures the current flows in either forward or reverse direction to activate the Phase A (or Phase B) of the coils.

As the transistor outputs depend on the sensor inputs, the operations of the sensor array are explained. The sensors act as proximity sensors to detect the presence of the conveyor cart or the reflective white strips attached. If the reflective white strip is aligned with a sensor, the corresponding sensor activates and sends a signal to the PIC18F4520 MCU to activate the set of coils in Table 5.6. As shown in Table 5.6, the sensors are numbered from 1 to 8 (left to right). For example, the current flows from left to right in Phase A stator coils are defined as A+ (or A− in the opposite direction). With the position of the eight sensors and the layout of the three reflective white strips, there are 14 possible positions at which the sensors

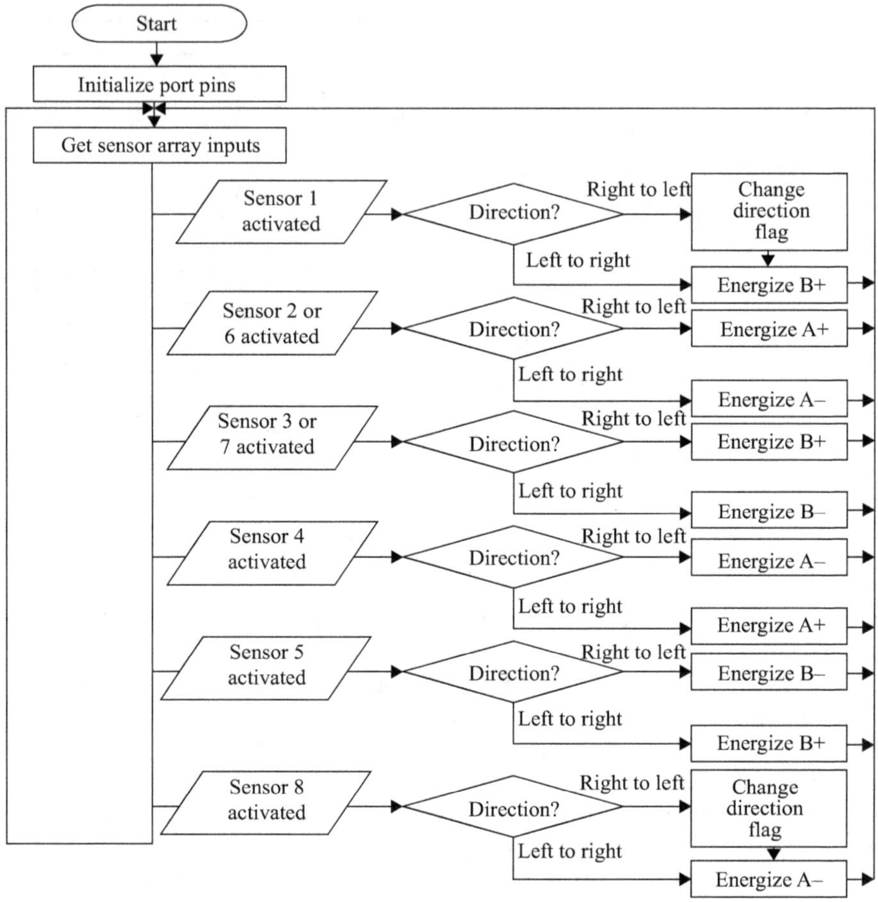

Figure 5.54 Programming flowchart for air-core linear drive

Table 5.5. Truth table of logic gate block

Sensor inputs		**Inputs to Q1 to Q4**			
In 2	**In 1**	**Y1**	**Y2**	**Y3**	**Y4**
0	0	0	1	0	1
0	1	1	1	0	0
1	0	0	0	1	1
1	1	1	1	1	1

could detect the reflective white strips. Depending on the position of the cart, the sensors that are activated would change the direction of travel through the A+/A– or B+/B– coils. The commutation sequence of the air-core linear drive movement (from left to right and vice versa) is shown in Figure 5.54.

Table 5.6. Air core linear drive commutation sequence

Position number	Movement from left to right		Movement from right to left	
	Voltage across coils	**Detecting sensor**	**Voltage across coils**	**Detecting sensor**
1	B+	Sensor 1	A−	Sensor 8
2	A+	Sensor 4	B−	Sensor 5
3	B−	Sensor 3	A+	Sensor 6
4	A−	Sensor 2	B+	Sensor 7
5	B+	Sensor 5	A−	Sensor 4
6	A+	Sensor 4	B−	Sensor 5
7	B−	Sensor 3	A+	Sensor 6
8	A−	Sensor 6	B+	Sensor 3
9	B+	Sensor 5	A−	Sensor 4
10	A+	Sensor 4	B−	Sensor 5
11	B−	Sensor 7	A+	Sensor 2
12	A−	Sensor 6	B+	Sensor 3
13	B+	Sensor 5	A−	Sensor 4
14	A−	Sensor 8	B+	Sensor 1

Table 5.7. Required specification and design constraints of EMS

Required specification			
Levitation force	>30 N	Nominal air gap Maximum current	0.006 m <5 A
Design constraints			
Magnet depth (d)	0.020 m	Magnet width (w)	0.020 m
Magnet height (h)	0.065 m	Magnet width (p)	0.020 m
Magnetic field intensity of the magnet (H_c)	920,000 At/m	Resistance of the coil (R_{coils})	8.1 Ω
Length of the magnet in the magnetization direction (l)	0.003m	Relative permeability of the magnet (μ_m/μ_o)	1.05
Equivalent mass of the conveyor cart (m)	3 kg	Relative permeability of the track (μ_T/μ_o)	1.00

The required specifications and design constraints of EMS are shown in Table 5.7. According to the required specifications and design constraints in Table 5.7, pole width $p = 0.02$ m in Figure 5.55 is then adjusted. After winding area is calculated, winding number, $N = 1{,}300$ depending on the rated current I of coil diameter could be yielded. The calculated winding number, rate current of coil diameter, and pole area $A_g = 0.0004\ \text{m}^2$ are used to compute the final magnetic levitation force, F_m.

The maximum current (see Table 5.7) required was obtained by slowly increasing the current from zero, and recording the current that enables the conveyor cart to levitate. The coercive force of the permanent magnet was measured using the

DGM-202 gauss meter in a magnetic circuit that consists of two permanent magnet and electromagnetic cores with a 0.006 m air gap. With the reading of the magnetic field intensity taken to be 706 Gauss, it could be calculated using fundamental magnetic circuit analysis to obtain the coercive force of approximately 920,000 At/m.

The magnetic levitation system using the hybrid electromagnets could be described mathematically. In this design, there are two electromagnets working together to achieve levitation. The magnetic flux path for the physical arrangement of the electromagnet and track (in Figure 5.55) is modeled by the equivalent circuit as seen in Figure 5.56.

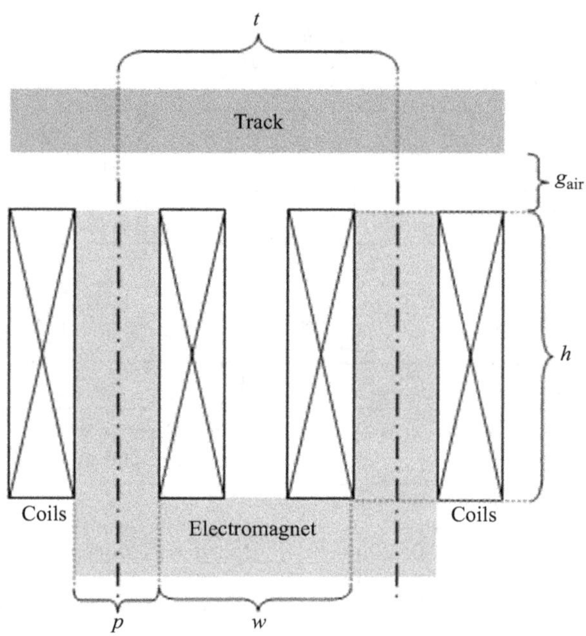

Figure 5.55 Physical arrangements of an electromagnet with track

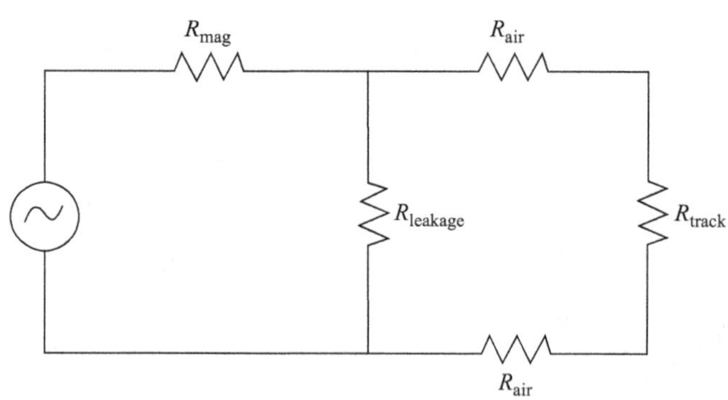

Figure 5.56 Magnetic flux path circuit

The attractive force and the input current for the electromagnet are determined as shown. First, the air gap reluctance with the presence of lateral fringe flux is given by:

$$\Re_{airgap} = \frac{z}{\mu_0 l_m \left(p + \frac{2z}{\pi}\right)} \tag{5.22}$$

where permeability of free space is $\mu_0 = 4\pi \times 10^{-7}$, l_m is the length of the magnet, z is the air gap distance, and p is the width of the pole.

The reluctances for the track, magnet, and leakage flux are computed as:

$$\Re_{track} = \frac{(t + 4p)}{\mu_0 \mu_T \times l_m \times p} \tag{5.23}$$

$$\Re_{magnet} = \frac{(2h + w + 2p)}{\mu_0 \mu_M \times l_m \times p} \tag{5.24}$$

$$\Re_{leakage} = \frac{2w}{\mu_0 l_m h} \tag{5.25}$$

where μ_T and μ_M are the relative permeabilities of the track and electromagnet, respectively.

The magnetic levitation force produced by the EMS system could be written as:

$$\begin{aligned} F_m &= \frac{\phi^2}{\mu_0 A_g} = \frac{MMF_{total}^2}{\mu_0 A_g R_{total}^2} \\ &= \frac{(2H_c l + NI)^2}{\mu_0 A_g R_{total}^2} = \frac{\mu_0 A_g (2H_c l + NI)^2}{4z^2} \end{aligned} \tag{5.26}$$

where N is the number of turns, H_c is the magnetic field intensity of magnet, A_g is the effective cross-sectional area of the airgap (considering the fringing effect), z is the airgap distance, I is the current input to the coils, MMF_{total} is the total magnetomotive force from the permanent magnets and coils, ϕ is the magnetic flux of the airgap, and R_{total} is the total reluctance computed by:

$$R_{total} = \frac{(\Re_{track} + 2\Re_{airgap})(\Re_{leakage} + \Re_{magnet}) + \Re_{magnet}\Re_{leakage}}{\Re_{leakage} + 2\Re_{airgap} + \Re_{track}} \tag{5.27}$$

With two electromagnets used for the levitation, the levitation force in (5.26) could be simplified to:

$$F_{m_1} = G_1 I_1{}^2 \tag{5.28a}$$

$$F_{m_2} = G_2 I_2{}^2 \tag{5.28b}$$

where F_{m_1}, G_1, and I_1 are the levitating force, levitating force gain, and air gap length of a left side electromagnet, respectively; F_{m_2}, G_2, and I_2 are the levitating force, levitating force gain, and air gap length of a right-side electromagnet, respectively.

In (5.28), the actual magnetizing currents I_1 and I_2 are produced by electromagnet driving circuits; therefore, the dynamic relation between the actual voltage and the magnetizing current has to be considered in control system design. The input current for the EMS system can be determined. The relationship between electromagnet core flux $\phi_{\max}$, magnetizing current (I that is I_1 or I_2) and terminal voltage (V) is given by:

$$V = N\frac{d\phi_{magnet}}{dt} + IR_{coils} \tag{5.29}$$

where R_{coils} is the coil resistance and $\phi_{magnet} = \mu_0 A_g NI/2z$ is the electromagnet flux.

Rearranging it, the actual magnetizing current can be obtained as:

$$\frac{dI}{dt} = -\frac{R_{coils}}{R_{coils}}I + \frac{V}{L_{coils}} \tag{5.30}$$

where $L_{coils} = \mu_0 N^2 A_g/2z$ is the inductance of an electromagnet.

The control voltage in (5.29) is regulated according to

$$V = V_{\text{dc}} \times d \tag{5.31}$$

where d represents the control duty cycle of the driving circuit and V_{dc} is the voltage supply to the driving circuit. The output voltage V can be controlled in a designed range $V \in [0,\ V_{dc}]$ by tuning the duty cycle command from the PWM controller.

According to the Newtonian law, the actual dynamic of the conveyor cart could be written as:

$$m\ddot{z}(t) = F_{m2} + F_{m1} + F_d = G_2 I_2^2 + G_1 I_1^2 + F_d \tag{5.32}$$

where m (=3 kg) is the mass of the conveyor cart, $\ddot{z}$ is the acceleration of the conveyor cart, $F_{m_1}(t)$ and $F_{m_2}(t)$ are the attractive force from the electromagnets, and $F_d(t)$ is the disturbance force. The actual force was measured using a vernier force sensor, model DFS-BTA. It is assumed that the force reading from the force sensor equals to the frictional force. The average force obtained from the experiment was used to determine the coefficient of friction between the conveyor cart's and its base (=0.30 kg) during the initial levitation and propulsion. The coefficient of friction ($\mu = 0.12$) was computed using the force obtained ($F_d = 0.38$ N) and divided it by the mass of the conveyor cart.

With the friction force obtained, the thrust force produced by the LSM system could be computed using equation (5.32). As shown in Figure 5.57, at the different current inputs to the two electromagnets (on the left and right side), the resulting thrust forces were measured. It gives a good estimate of the thrust force produces for given input currents. For example, the estimated total thrust produced with an input current of 5 A to the two electromagnets are approximately equal to 11.5 N.

Also, a nonlinear characteristic of the magnetic levitation force is influenced not only by the air gap distance and input current, but also by the variation of the

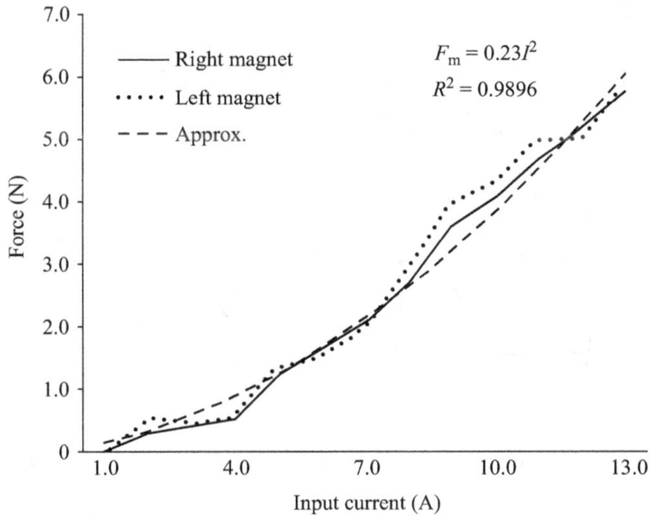

Figure 5.57 Estimated thrust force vs. input current

permeability in different operating points. As the permeability is not easy to be obtained precisely, the variation in electromagnet gain (G_1, G_2) due to permeability changed with different operating points and temperature variation within the coil that result in modeling errors is considered into the lumped uncertainty. Therefore, the model includes these variations and the controller developed should be robust to tolerate these uncertainties and unmodeled dynamics.

Considering the parameter variations and external disturbance, the actual dynamic model can be represented realistically as:

$$(m + \Delta m)\ddot{z}(t) \\ = (G_f + \Delta G_2)I_2^2 + (G_f + \Delta G_1)I_1^2 + F_d \tag{5.33}$$

where Δm, ΔG_2, and ΔG_1 denote the uncertainties introduced by system parameters and unmodeled dynamics.

By defining a lumped uncertainty vector:

$$L_u = -\Delta m\ddot{z}(t) + \Delta G_1 I_1^2 + \Delta G_2 I_2^2 + F_d \tag{5.34}$$

Equation (5.34) can be rewritten as:

$$\ddot{z}(t) = \frac{G_f}{m}u + L_u \tag{5.35}$$

where $u = I_2^2 - I_1^2$; the band of the lumped uncertainty is assumed to be given by $|L_u| < \rho$ and ρ is a given positive constant.

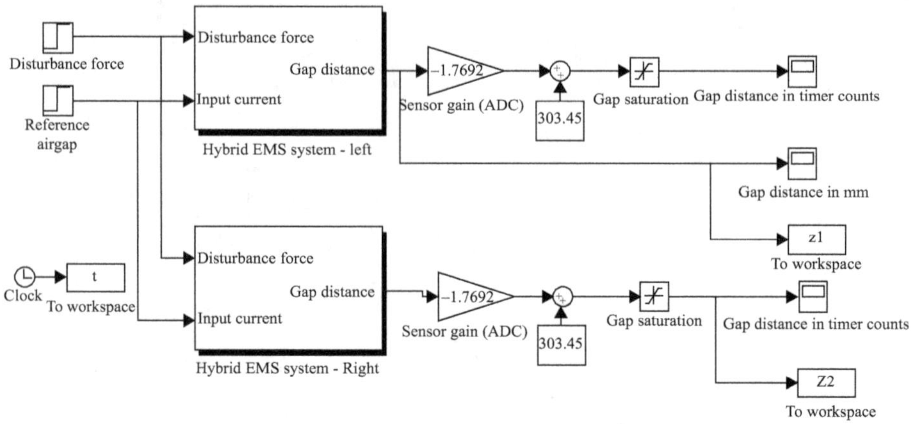

Figure 5.58 Overall block diagram of nonlinear EMS system model

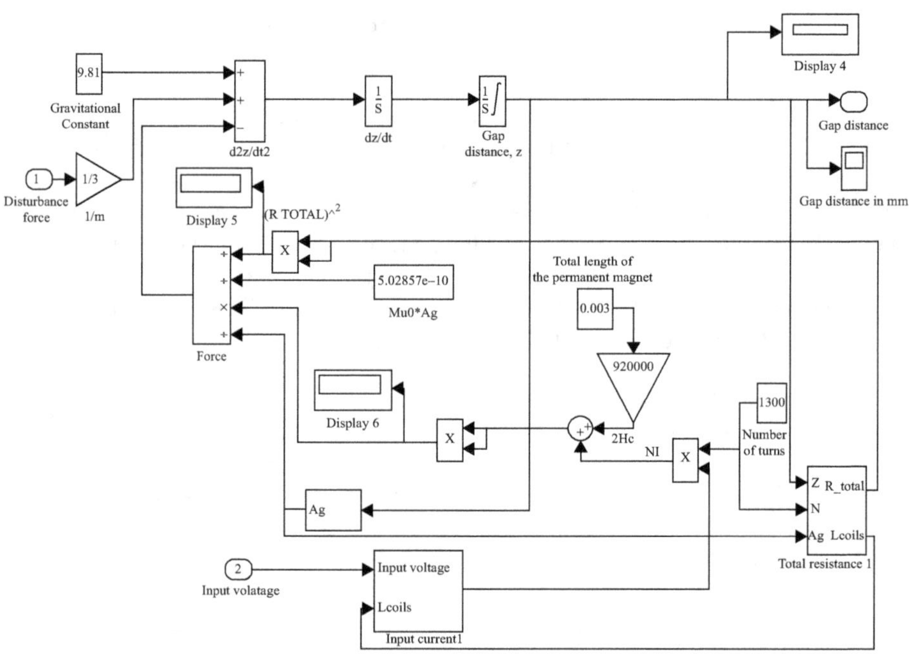

Figure 5.59 Detailed block diagram of hybrid EMS system

As shown in Figures 5.58 and 5.59, the air gaps were measured by the two sensors. These sensors were mounted deliberately to offset the inoperative range of 0 to 0.04 m. The distance between the hybrid electromagnet and the guideway was measured to be approximately 0.060 m. The desired gap distance is set as 0.006 m (or 291 counts) in the simulation. To obtain the actual gap distance, the digital-to-analog

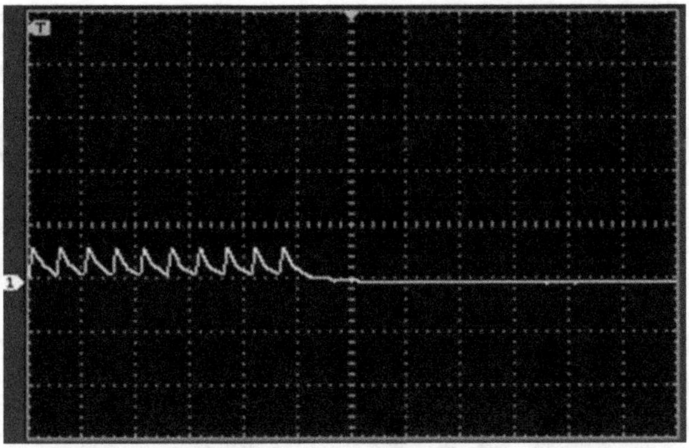

Figure 5.60 Image of the overall oscilloscope output for open-loop system (0.002 m/div and 0.02 s/div)

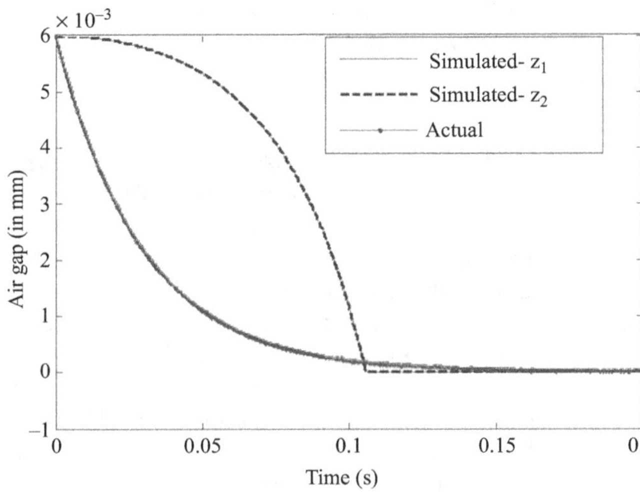

Figure 5.61 Open-loop time response of the air gap

conversion is required. During the test, the conveyor cart was then raised by an increment of 0.001 m to 0.007 m to determine the average trend line between the digital count (y) and the air gaps (z). It was found that the average relationship between the two air gaps is approximately given by $y = -1.769z + 303.45$. The measured air gaps on both sides of the conveyor cart are quite close to each other. Hence, only one electromagnet was used in the subsequent controllers' comparisons.

The transient response of the actual air gap was captured as shown in Figure 5.60. As compared with Figure 5.61, there are some differences in the

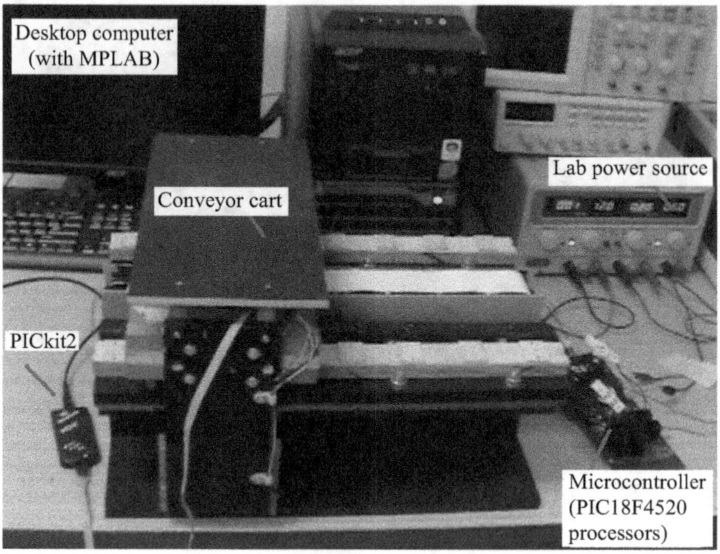

Figure 5.62 Experimental setup of magnetic levitation system for conveyance test

transient response due to the model uncertainties. However, both the responses show that it reaches steady state within 0.1 s. As the response indicates the plant behavior in open-loop condition, it shows that the conveyor cart falls onto the electromagnets due to its weight. Without the controller, the EMS system is unable to levitate the conveyor cart to maintain the desired levitation. The current excitation to the magnets is therefore ineffectual to rise fast enough to create the force required.

As observed in the open-loop response, the EMS control system requires some feedback control for the conveyance system. Instead of using the PID controller, it is recognized that H-Infinity control is an effective method for attenuating such disturbances and that the sliding-mode control scheme is able to compensate for the unknown disturbances. The purposes of this section are to examine the proposed sliding-mode control as compared to other controllers (such as PID and H-Infinity controller) used for magnetic levitation systems. To achieve this, a magnetic levitation system experiment is constructed as shown in Figure 5.62. The controllers were designed, and their performance for set-point regulation was experimentally evaluated and compared with the simulated results.

5.5.2 Programming and system implementation

Different controllers such as PID and sliding-mode controller were used in the magnetic conveyor system. The programs were written in MPLAB-IDE and compiled into the PIC18f4520 during the actual implementation. Note that the parameters used in the controllers can be changed to suit the uncertainties in the actual working environment. The details of the coding and use of MPLAB-IDE can be seen in Section 5.1.

(Programming for levitation using PID controller

```
#include <p18f4520.h>
#include <adc.h>
#include <pwm.h>
#include <stdlib.h>
#include <delays.h>

int GapDistance1=0,GapDistance2=0,PWMout1=0,PWMout2=0;
int RefHeight1=470,RefHeight2=470;              //Sensor count value when at
reference air gap.
int Error1=0,Error2=0;
int ErrorSum1=0,ErrorSum2=0;
int Sum1=0,Sum2=0;
int Previous1=0,Previous2=0;
int ErrorPrevious1=0,ErrorPrevious2=0;
int PID_Out1=0,PID_Out2=0;
float Kp = 0.03;                    //Error gain
float Ki = 0.01;                    // Gain for PID controller's integral action
float Kd = 0.02;                    //Gain for PID controller's differential action
float Ts = 0.005;                   // Sampling time
//The optimum gain values are in red.

void SenseDistance1 (void);
void SenseDistance2 (void);
void Do_PID1 (void);
void Do_PID2 (void);
void main( void )
{
        TRISA = 0x00;                   //Port A as Output (to control the direction of
current)
        TRISD = 0x00;                   //Port D as Output
        TRISC = 0x00;                   //Port C as Output
        PORTC=0xFF;                     //Turn off PortC output
        ADCON0bits.ADON = 0x01;//Enable A/D  module
//PWM period =[(period ) + 1] x 4 x TOSC x TMR2 prescaler
        OpenPWM1(0xFF);
        OpenPWM2(0xFF);
```

```
while(1)                          //Loop forever
{
// Controlling flow of current to produce repulsive magnetic force when coil is too the
guide way.
Current is being controlled according to Air gap error. When there is a negative Error,
there will be a change in current flow.

if ((Error1 <= 0) & (Error2 <= 0))
{
PORTA=0x05;                    //0101
}
if ((Error1 > 0) & (Error2 > 0))
{
PORTA=0x0A;                    //1010
}
if ((Error1 <= 0) & (Error2 > 0))
{
PORTA=0x09;                    //1001
}
if ((Error1 > 0) & (Error2 <= 0))
{
PORTA=0x06;                    //0110
}

SenseDistance1 ();             //Loop to convert Sensor 1 analog output to digital value
SenseDistance2 ();             //Loop to convert Sensor 2 analog output to digital value
Do_PID1();                     //Loop to do PID calculation for sensor 1 air gap error to
                               produce appropriate  amount of current to system.
Do_PID2();                     //Loop to do PID calculation for sensor 2 air gap error to
                               produce appropriate  amount of current to system.
SetDCPWM1(PID_Out1);           //Output PWM from CPP1 to produce Analog output to coil
1. SetDCPWM2(PID_Out2); // Output PWM from CPP2 to produce Analog output to coil
2.
}
}

void SenseDistance1 ()                      //Loop to capture air gap for coil 1.
{
OpenADC( ADC_FOSC_4 &ADC_RIGHT_JUST &ADC_4_TAD,ADC_CH11
&ADC_INT_OFF & ADC_VREFPLUS_VDD & ADC_VREFMINUS_VSS, 0 );
Delay10TCYx( 5 );                  // Delay for 50 instruction cycle
ConvertADC();                               // Start analog to digital conversion for
sensor1 input.
while( BusyADC() );                // Wait for completion
GapDistance1 = ReadADC();          // Read result
}
void SenseDistance2 ()                      //Loop to capture air gap for coil 2.
{
```

```
OpenADC( ADC_FOSC_4 &ADC_RIGHT_JUST &ADC_4_TAD,ADC_CH12
&ADC_INT_OFF & ADC_VREFPLUS_VDD & ADC_VREFMINUS_VSS, 0 );
Delay10TCYx( 5 );                // Delay for 50 instruction cycle
ConvertADC();                            // Start analog to digital conversion for
sensor2 input.
while( BusyADC() );              // Wait for completion
GapDistance2 = ReadADC();        // Read result
}

void Do_PID1 ()                          //PID to control current into coil 1.
{
Error1 = RefHeight1 - GapDistance1;      //Calculating error of air gap 1.
ErrorSum1 = ErrorSum1 + Error1;  //Sum up error for integral action
PID_Out1 = (Kp*Error1) + (Ki*ErrorSum1)*Ts + (Kd*(Error1-ErrorPrevious1))/Ts;
        //PID Equation
if(PID_Out1 >= 1023 || PID_Out1 <= -1023)//Anti-windup condition
{
ErrorSum1 = (PID_Out1 - (Kp*Error1) - ((Kd*(Error1-ErrorPrevious1))/Ts))/(Ki*Ts);
}
        if(PID_Out1 < 0)                 //Setting lower saturation limit for output.
        {
        PID_Out1=PID_Out1*(-1);
        }
        if(PID_Out1 > 1023)      //Setting upper saturation limit for output.
        {
        PID_Out1 = 1023;
        }
ErrorPrevious1 = Error1;         //Storing air gap error of coil 1.
}
void Do_PID2 ()                          //PID to control current into coil 2.
{
Error2 = RefHeight2 - GapDistance2;      //Calculating error of air gap 2.
ErrorSum2 = ErrorSum2 + Error2;  //Sum up error for integral action
PID_Out2 = (Kp*Error2) + (Ki*ErrorSum2)*Ts + (Kd*(Error2-ErrorPrevious2))/Ts;
        //PID Equation
if(PID_Out2 >= 1023 || PID_Out2 <= -1023)//Anti-windup condition
{
ErrorSum2 = (PID_Out2 - (Kp*Error2) - ((Kd*(Error2-ErrorPrevious2))/Ts))/(Ki*Ts);
}
        if(PID_Out2 < 0)                 //Setting lower saturation limit for output.
        {
        PID_Out2 = PID_Out2*(-1);
        }
        if(PID_Out2 > 1023)      //Setting upper saturation limit for output.
        {
        PID_Out2 = 1023;
        }
ErrorPrevious2 = Error2;         //Storing air gap error of coil 2.
}
```

Programming for levitation using Sliding Mode controller

```
#include <p18f4520.h>
#include <adc.h>
#include <pwm.h>
#include <stdlib.h>
#include <delays.h>

int GapDistance1=0,GapDistance2=0,PWMout1=0,PWMout2=0;
int RefHeight1=460,RefHeight2=460;          //Sensor count value when at
reference air gap.
int Error1=0,Error2=0;
int SlideMode1_out=0,SlideMode2_out=0;
int Sign1=0,Sign2=0;
int DError1=0,DError2=0;
int Function1=0,Function2=0;
float Kd = 4.0;          //Constant to compensate for disturbance
float K = 5.5;           //Gain to amplify error of air gap
float K1 = 5.23;                //Gain to amplify function for sliding mode
controller.
float K2 = 4.72;                //Gain to amplify sign for sliding mode controller.
float Ts = 0.5;          //Sampling time
//The optimum gain values are in red.


void SenseDistance1 (void);
void SenseDistance2 (void);
void SlideMode1 (void);
void SlideMode2 (void);

void main( void )
{
        TRISA = 0x00;                   //Port A as Output (to control the direction
of current)
        TRISD = 0x00;                   //Port D as Output
        TRISC = 0x00;                   //Port C as Output
        PORTC=0xFF;                     //Turn off PortC output
        ADCON0bits.ADON = 0x01;//Enable A/D  module

//PWM period =[(period ) + 1] x 4 x TOSC x TMR2 prescaler
        OpenPWM1(0xFF);
        OpenPWM2(0xFF);

while(1)                //Loop forever
{
SenseDistance1 ();      //Loop to convert Sensor 1 analog output to digital value
SenseDistance2 ();      //Loop to convert Sensor 2 analog output to digital value
SlideMode1();           //Loop to do Sliding Mode calculation for sensor 1 air gap
                        error to produce appropriate  amount of current to system.
SlideMode2();           //Loop to do Sliding Mode calculation for sensor 2 air gap
```

```
// Controlling flow of current to produce repulsive magnetic force when coil is too the
guide way.
Current is being controlled according to Air gap error. When there is a negative Error,
there will be a change in current flow.

if ((Error1 > 0) & (Error2 > 0))
{
PORTA=0x0A;        //1010
}
if ((Error1 < 0) & (Error2 < 0))
{
PORTA=0x05;        //0101
}
if ((Error1 < 0) & (Error2 > 0))
{
PORTA=0x09;        //1001
}
if ((Error1 > 0) & (Error2 < 0))
{
PORTA=0x06;        //0110
}

SetDCPWM1(SlideMode1_out);          //Output PWM from CPP1 to produce Analog
output to coil 1. SetDCPWM2(SlideMode2_out);          //Output PWM from CPP2 to
produce Analog output to coil
        }
}

void SenseDistance1 ()                    //Loop to capture air gap for coil 1.
{
OpenADC( ADC_FOSC_4 &ADC_RIGHT_JUST &ADC_4_TAD,ADC_CH11
&ADC_INT_OFF & ADC_VREFPLUS_VDD & ADC_VREFMINUS_VSS, 0 );
Delay10TCYx( 5 );                   // Delay for 50 instruction cycle
ConvertADC();                             // Start analog to digital conversion for
sensor1 input.
while( BusyADC() );                 // Wait for completion
GapDistance1 = ReadADC();           // Read result
//Setting upper and lower saturation limit for sensor 1 ADC value to eliminate inaccurate
reading due to sudden spike of output voltage from sensor.
        if(GapDistance1 > 540)
        {
        GapDistance1=540;
        }
        if(GapDistance1 < 460)
        {
        GapDistance1=460;
        }
}

void SenseDistance2 ()                    //Loop to capture air gap for coil 2.
```

```
{
OpenADC( ADC_FOSC_4 &ADC_RIGHT_JUST &ADC_4_TAD,ADC_CH11
&ADC_INT_OFF & ADC_VREFPLUS_VDD & ADC_VREFMINUS_VSS, 0 );
Delay10TCYx( 5 );                // Delay for 50 instruction cycle
ConvertADC();                            // Start analog to digital conversion for
sensor2 input.
while( BusyADC() );              // Wait for completion
GapDistance1 = ReadADC();        // Read result
//Setting upper and lower saturation limit for sensor 2 ADC value to eliminate inaccurate
reading due to sudden spike of output voltage from sensor.

        if(GapDistance2 > 520)
        {
        GapDistance2=520;
        }
        if(GapDistance2 < 460)
        {
        GapDistance2=460;
        }
}

void SlideMode1()                //Sliding Mode controller to control current into
coil 1.
{
Error1 = (RefHeight1 - GapDistance1);            //Calculating error of air gap 1.
DError1 = Error1/Ts;                     //Differentiate Gap Error
Function1 = DError1+(K*Error1);          //Switching function for sliding mode
controller.
        if(Function1 > 0)
        {
        Sign1 = 1;
        }
        if(Function1 == 0)
        {
        Sign1 = 0;
        }
        else if(Function1 < 0)
        {
        Sign1 = -1;
        }
SlideMode1_out = (K1*Function1) - (K2*Sign1) - Kd;       //Total Sliding Mode
equation
//Setting upper and lower saturation limit for Sliding mode control output.
        if(SlideMode1_out < 0)
        {
        SlideMode1_out = -1*SlideMode1_out;
        }
        if(SlideMode1_out > 1023)
        {
        SlideMode1_out = 1023;
```

```
        }
}

void SlideMode2()                    //Sliding Mode controller to control current into
coil 2.
{
Error2 = (RefHeight2  - GapDistance2);               //Calculating error of air gap 2 .
DError2 = Error2/Ts;                         //Differentiate Gap Error
Function2 = DError2 + (K*Error2);            //Switching function for sliding mode
controller.
        if(Function2 > 0)
        {
        Sign2 = 1;
        }
        if(Function2 = = 0)
        {
        Sign2 = 0;
        }
        else  if(Function2 < 0)
        {
        Sign2 =  -1;
        }
SlideMode2_out = (K1*Function2)   - (K2*Sign2)  - Kd;
//Setting  upper  and lower  saturation limit  for Sliding mode control output.
        if(SlideMode2_out < 0)
        {
        SlideMode2_out =   -1*SlideMode2_out;
        }
        if(SlideMode2_out  > 1023)
        {
        SlideMode2_out = 1023;
        }
}
```

The numerical simulations of these controllers were implemented via MATLAB®/ Simulink® software. Throughout the simulation, a variable-type solver - ODE45 was used. In order to mimic the actual motion, the maximum gap was set at 0.008m and the reference air gap was targeted as 0.006m. Due to the hardware limitations, the corresponding sampling time is determined to be around 2ms. The control algorithms were written in C language using the MPLAB environment. The C programs were downloaded into the PIC18F4520 via the PICkit2 debugger. The air gap signal was subsequently measured by the sensors using analog-to-digital converter and its signal were later captured using the laboratory oscilloscope for comparisons with the simulated results.

As shown in the experimental test using the PID controller (using controller's parameters as ($k_p = 8$, $k_i = 2$, $k_d = 3$), the controller is trying to maintain at the reference air gap for a period of time before it slides to its steady-state condition in Figure 19. It is observed that during forward propulsion, the conveyor tends to oscillate before attracted to the electromagnet. To circumvent this problem, the electromagnets need to be demagnetize when the air gap becomes small. This helps to prevent the track from getting attracted to the electromagnets.

To accomplish this, a robust control method such as the sliding-mode controller is used to compensate for the errors and uncertainties in the model. The following shows how the sliding-mode controller was implemented. To achieve the control objective, the tracking error and its derivative are defined as: $e_z = z_d - z$ and $\dot{e}_z = \dot{z}_d - \dot{z}$, respectively. The sliding surface is designed as follows:

$$S = \dot{e}_z + \lambda e_z \tag{5.36}$$

where λ is a positive constant. By taking the derivative of the sliding surface with respect to time and use $\ddot{z} = G_f u/m + L_u$, then

$$\dot{S} = \ddot{z}_d - \frac{G_f}{m} u - L_u + \lambda \dot{e}_z \tag{5.37}$$

The sliding-mode control law is designed as:

$$u_{\mathrm{smc}} = \frac{m}{G_f}\left(k_1 S + k_2 \operatorname{sgn}(s) + \lambda \dot{e}_z + \ddot{z}_d\right) \tag{5.38}$$

where m and G_f are both positive system parameters; k_1 and k_2 are positive control parameters.

The ideal current command I^* is calculated as:

$$I^* = \sqrt{u_{\mathrm{smc}}},\ \text{if } \mathrm{u_{smc}} \geq 0 \tag{5.39a}$$

$$I^* = \sqrt{-u_{\mathrm{smc}}},\ \text{if } \mathrm{u_{smc}} < 0 \tag{5.39b}$$

By taking the overall control system into consideration, the control voltage produced by current modules is designed as:

$$V = \dot{I}^* L_{\mathrm{coils}} + I^* R_{\mathrm{coils}} \tag{5.40}$$

Finally, the actual control duty is obtained as:

$$d_{\mathrm{smc}} = V / V_{\mathrm{dc}} \tag{5.41}$$

Define a Lyapunov function candidate as: $V_L = S^2/2$. The derivative of V_L with respect to time is expressed as follows:

$$\begin{aligned}\dot{V}_L &= S\dot{S} \\ &= s\left(\ddot{z}_d - \frac{G_f}{m}u - L_u + \lambda\dot{e}_z\right) \\ &= -k_1S^2 - k_2S\,\mathrm{sgn}(s) - SL_u \\ &= -k_1S^2 - |s|(k_2 - |L_u|) \\ &\leq -k_1S^2\end{aligned} \tag{5.42}$$

where $k_2 > |L_u|$ and k_1 are positive constants to be designed to achieve the desired levitation gap and to ensure $\dot{V}_L \leq 0$. Since $\dot{V}_L \leq 0$ implies that $V_L(t) \leq V_L(0)$, and therefore S is bounded. This, in turn, implies that $\ddot{V}_L$ is bounded. Hence $\dot{V}_L$ must be uniformly continuous. Finally, application of Barbalat's lemma shows that $S \to 0$ and thus $e_z \to 0$ as $t \to \infty$.

As the sigmoid function *sgn*(.) gives a chattering control signal, it could be eliminated by smoothing out the control law discontinuity inside a boundary layer by replacing the function with *sat*(.). This substitution assigns a low-pass filter structure to the dynamics of the sliding surface inside the boundary layer as seen below.

$$u_{\text{smc}} = \frac{m}{G_f}(k_1S + k_2sat(S) + \lambda\dot{e}_z + \ddot{z}_d) \tag{5.43}$$

where

$$\mathrm{sat}(s/\Phi) = \begin{Bmatrix} \mathrm{sgn}(s), & \text{if } |S/\Phi| > 1 \\ S/\Phi, & \text{if otherwise} \end{Bmatrix} \tag{5.44}$$

moreover, Φ is the boundary layer thickness greater than zero.

Figure 5.63 shows the final control system design for the sliding-mode controller. The controller calculates the amount of current to be applied to the EMS system using

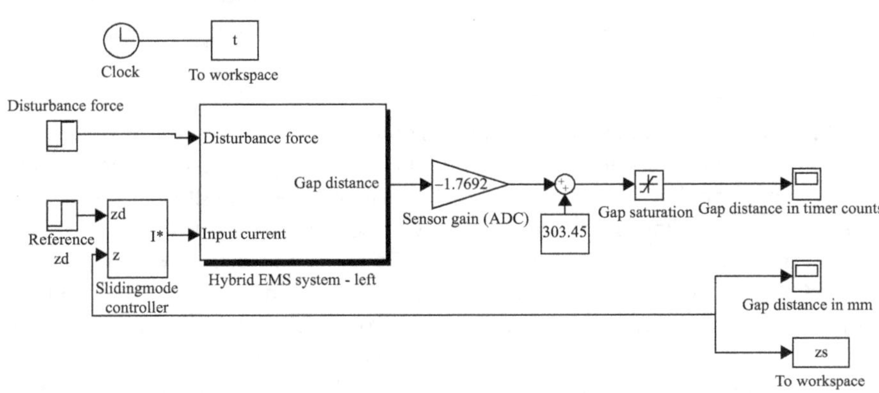

Figure 5.63 Block diagram of sliding-mode controller for hybrid EMS system

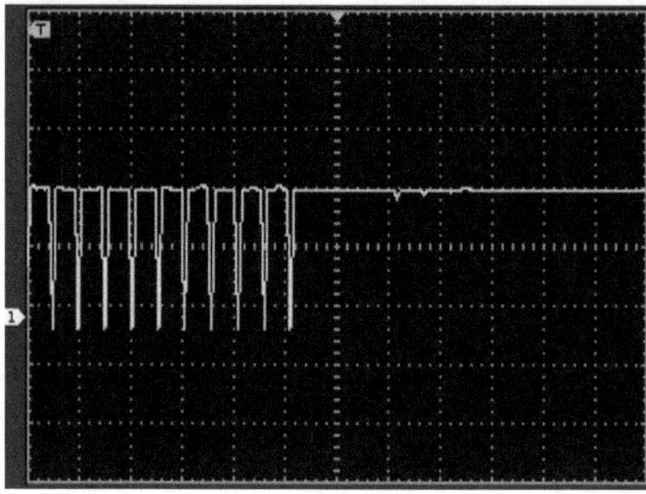

Figure 5.64 Oscilloscope signal output for sliding-mode control system (0.001 m/div and 0.02 s/div)

the control law in (5.43). The amount of supply current is controlled via the PWM. The following parameters that satisfy the constraints are chosen as $\lambda = 9$, $k_1 = 18$, $k_2 = 1{,}100$, $\Phi = 1.5$. The analog output voltage from the sensor is converted to a digital value through the A/D converter in the PIC18F4520. The reference gap is compared with the actual value before generating the control action for the EMS system. When the tracking error is negative (positive), indicating that the air gap is too small, the control input is increased (decrease) to increase the current. The "energy" of the system will diminish slowly as it reaches the sliding surface (near to zero air gap error).

Figure 5.64 shows the oscilloscope images obtained from one of the electromagnetic coils using the sliding-mode controller. The image in Figure 5.65 shows the controller trying to maintain the reference air gap for a period. However, there is still some slight chattering as seen in the actual responses. However, the responses of the simulated and actual results exhibit the same steady value after 0.4 s.

5.5.3 Testing

Table 5.8 shows the comparisons of various controllers on the hybrid EMS system. A predefined uncertainty level of 30% for the Δm and ΔG was used for the comparison. The simulated results for each controller are plotted as shown in Figure 5.66. The sliding-mode controller gives the smallest error as observed from the maximum or infinity norm measured of the air gap as compared with its peer such as the H-Infinity controller. However, both the sliding-mode controller and H-Infinity controller can maintain a distance from the guideway, which prevents the electromagnets to be attracted to the guide way in the case of the PID controller. With the parametric uncertainty presence in the EMS's model, the sliding-mode controller seems to perform better than the PID and H-Infinity controller in terms of the steady-state error.

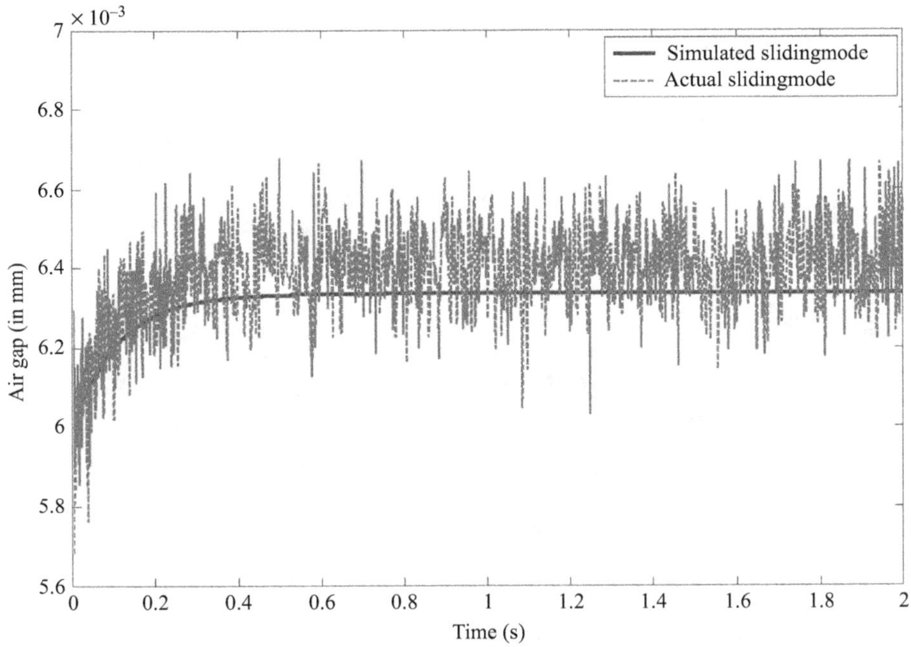

Figure 5.65 Sliding-mode closed-loop time response of air gap

Table 5.8. Summary of various controllers' responses vs. compared to sliding-mode controller (with or without uncertainties)

	(H Infinity)		**(PID)**		**Sliding mode**	
	without	**with**	**without**	**with**	**without**	**with**
Maximum air gap	7×10^{-3}	7×10^{-3}	7×10^{-3}	7×10^{-3}	6×10^{-3}	6×10^{-3}
Minimum air gap	6×10^{-3}	6×10^{-3}	0	0	6×10^{-3}	6×10^{-3}
Infinity norm	7×10^{-3}	7×10^{-3}	7×10^{-3}	7×10^{-3}	6×10^{-3}	6×10^{-3}
Sum of squared error	5×10^{-5}	5×10^{-5}	28×10^{-2}	28×10^{-2}	1×10^{-4}	8×10^{-5}
Steady-state error	1×10^{-3}	1×10^{-3}	6×10^{-3}	6×10^{-3}	5×10^{-5}	3×10^{-4}

As seen in the table, the sum-of-squared error and infinity norm demonstrates that the H-Infinity controller achieves a better performance in disturbance attenuation as compared to the others. However, the steady-state performance of the sliding-mode controller is superior to the H-Infinity controller. From the experimental results, we can also conclude that these robust controllers are useful for designing nonlinear control systems with variations in the models. In the test, we can observe that the performance of the both H-Infinity and sliding-mode controller in the disturbance attenuation is much better as compared to the PID controller. Finally, the experimental results show that the proposed sliding-mode

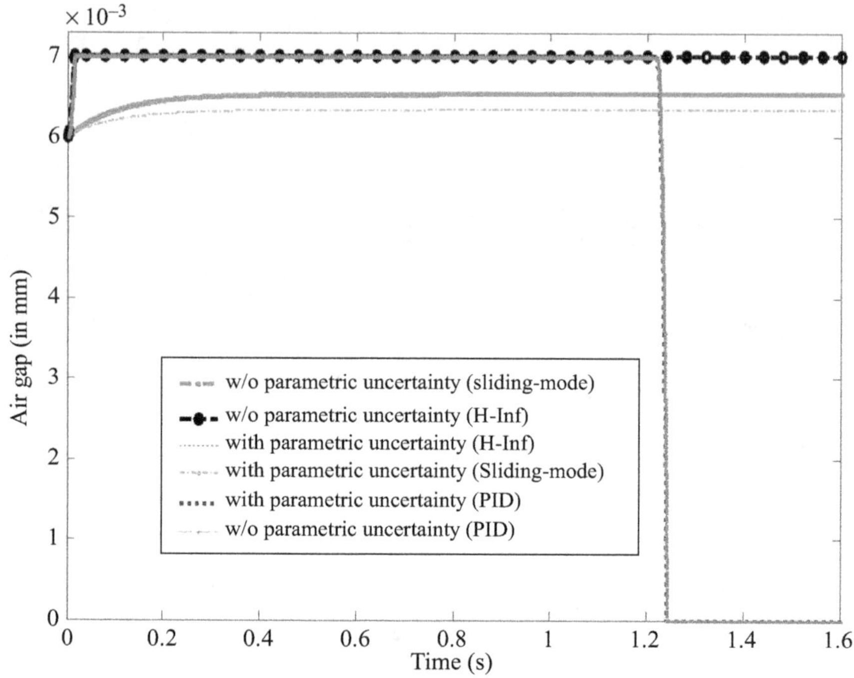

Figure 5.66 Comparisons of controllers' closed-loop time response

control scheme improves the response when compared to the classically used feed-forward controller such as the PID and the H-Infinity controller.

The prototype MCU EMS system and LSM for conveyance purposes was designed and implemented for possible applications in the biochemical and semi-conductors manufacturing plants. The LSM for propulsion was built using the noncontact digital commutation via the electronics, sensors, and the PIC18F4520 MCU. The hybrid electromagnetic levitation system uses two permanent magnets and two electromagnets for vertical levitation to reduce the power consumption. The sliding-mode feedback controller was then designed to control the air gaps while moving. As compared to other controllers, the proposed conveyance system using sliding-mode control scheme can perform the conveyance function under the disturbance and 30% of parameters uncertainties. Further research will focus on increasing the performance of the controller when the fault occurred in the hybrid EMS system and applied it on a larger scale of the actual manufacturing environment. The actual videos tested on various environments can be seen below.

- Magnetic levitation in action: http://www.youtube.com/watch?v=T6hSpV1j6V8
- Magnetic propulsion and levitation: http://www.youtube.com/watch?v=KZEO4sYLoSA&feature=channel&list=UL

References

[1] Chin, C. Application of an intelligent table top vacuum robot cleaner in mechatronics system design education. **Journal of Robotics and Mechatronics**, 2011, 23(5): 645–657.

[2] Chin, CS and Yue, KM. Vertical stream curricula integration of problem-based learning (PBL) using autonomous vacuum robot in mechatronics course. **European Journal of Engineering Education**, 2011, 36(5): 485–504.

[3] Chin, CS, Atmodihardjo, W, Woo, WL, and Mesbahi, E. Remote temperature monitoring device using a multiple patients-coordinator set design approach. **ROBOMECH Journal**, 2015, 2: 4.

[4] Mahoney, C and Odom, J. Monitoring intraoperative normothermia: A meta-analysis of outcomes with costs. **AANA Journal**, 1999, 67(2): 155–164.

[5] Yamakoshi, KI. In the spotlight: Bioinstrumentation. **IEEE Reviews in Biomedical Engineering**, 2011, 4:6–8.

[6] Jiann, SS, Chun, YD, Yeong, RW, and Wei, ZS. A novel fuzzy pain demand index derived from patient-controlled analgesia for postoperative pain. **IEEE Transactions on Biomedical Engineering**, 2007, 54(12):2123–2132.

[7] Graichen, F, Arnold, R, Rohlmann, A, and Bergmann G. Implantable 9-channel telemetry system for in vivo load measurements with orthopedic implants. **IEEE Transactions on Biomedical Engineering**, 2007, 54 (2):253–261.

[8] Yang, D, Converse, MC, Mahvi, DM, and Webster, JG. Measurement and analysis of tissue temperature during microwave liver ablation. **IEEE Transactions on Biomedical Engineering**, 2007, 54(1):150–155.

[9] Liu, DL and Ebbini, ES. Real-time 2-D temperature imaging using ultrasound. **IEEE Transactions on Biomedical Engineering**, 2010, 57(1);12–16.

[10] Seffrin, RJ. Thermal imaging for detecting potential SARS infection, National Conference on Thermal Imagers for Fever Screening – Selection, Usage and Testing 2003;1–8, Sheraton Towers, Singapore, see https://www.irinfo.org/06-01-2003-seffrin/.

[11] Bae, J, Song, K, Lee, HW, Cho, HW, and Yoo, HJ. A low-energy crystal-less double-FSK sensor node transceiver for wireless body-area network. **IEEE Journal of Solid-State Circuits**, 2012, 47(11):2678–2692.

[12] Scanail, I, Cliodhna, N, Ahearne, B, and Lyons, GM. Long-term telemonitoring of mobility trends of elderly people using SMS messaging. **IEEE Transactions on Information Technology in Biomedicine**, 2006, 10 (2):412–413.

[13] Malhi, K, Mukhopadhyay, SC, Schnepper, J, Haefke, M, and Ewald, H. A zigbee-based wearable physiological parameters monitoring system. **IEEE Journal Sensors**, 2012, 12(3):423–430.

[14] Shahriyar, R, Bari, MF, Kundu, G, Ahamed, SI, and Akbar, MM. Intelligent mobile health monitoring system, **International Journal of Control and Automation**, 2009, 2(3):13–27.

[15] Ivy, AC. Comment: What is normal body temperature? **Gastroenterology**, 1945, 5:326–329.

[16] Jeffrey, RS. Recovery from severe hyperthermia (45°C) and rhabdomyolysis induced by methamphetamine body-stuffing. **Journal of Emergency Medicine**, 2007, 8(3):93–95.

[17] Bridges, E and Thomas, K. Noninvasive measurement of body temperature in critically ill patients. **Critical Care Nurse**, 2009, 29(3):94–97.

[18] Langham, GE, Maheshwari, A, Contrera, K, You, J, Mascha, E, and Sessler, DI. Noninvasive temperature monitoring in postanesthesia care units. **Anesthesiology**, 2009, 111(1):90–96.

[19] Xiao, J and Sadegh, A. City-Climber: A New Generation Wall-climbing Robots. 2008, The City College, City University of New York USA.

[20] Harsha Prahlad, RP. Electroadhesive robots-wall climbing robots enabled by a novel, robust, and electrically controllable adhesion technology. ***IEEE International Conference on Robotics and Automation***, 19–23 May 2008, Pasadena, CA, USA. DOI: https://doi.org/10.1109/ROBOT.2008.4543670.

[21] Clark, J. Intermolecular bonding - Van der Waals forces, 2000. Retrieved January 28th, 2012, http://www.chemguide.co.uk/atoms/bonding/vdw.html

[22] Chin, C, and Wheeler, C. Sliding-mode control of an electromagnetic actuated conveyance system using contactless sensing. **IEEE Transactions on Industrial Electronics**, 2013, 60(11):5315–5324.

[23] Cassat, A and Jufer, M. MAGLEV projects technology aspects and choices. **IEEE Transactions on Applied Superconductivity**, 2001, 12(1):915–925.

[24] Kusagawa, S, Baba, J, Shutoh, K, and Masada, E. Multipurpose design optimization of EMS-type magnetically levitated vehicle based on genetic algorithm. **IEEE Transactions on Applied Superconductivity**, 2004, 14(2): 1922–1925.

[25] Yoshida, K, Takami, H, and Fujii, A. Smooth section crossing of controlled-repulsive PM LSM vehicle by DTC method based on new concept of fictitious section. **IEEE Transactions on Industrial Electronics**, 2004, 51(4):821–826.

[26] Han, SH, Yun, S, Kim, HK, Kwak, YH, Park, HK, and Lee, SH. Analyzing schedule delay of mega project: Lessons learned from Korea Train Express. **IEEE Transactions on Engineering Management**, 2009, 56(2):243–256.

[27] Wang, J, Wang, S, and Zheng, J. Recent development of high-temperature superconducting Maglev system in China. **IEEE Transactions on Applied Superconductivity**, 2009, 19, 3(2):2142–2147.

[28] Banucu, R, Albert, J, Reinauer, V, *et al.* Automated optimization in the design process of a magnetically levitated table for machine tool applications. **IEEE Transactions on Magnetics**, 2010, 46(8):2787–2790.

[29] Appunn, R, Schmulling, B, and Hameyer, K. Electromagnetic guiding of vertical transportation vehicles: Experimental evaluation. **IEEE Transactions on Industrial Electronics**, 2010, 57(1):335–343.

[30] Warberger, B, Kaelin, R, Nussbaumer, T, and Kolar, JW. Bearingless motor for high-purity pharmaceutical mixing. **IEEE Transactions on Industrial Electronics**, 2012, 59(5):2236–2247.

[31] Yang, SM. Electromagnetic actuator implementation and control for resonance vibration reduction in miniature magnetically levitated rotating machines. **IEEE Transactions on Industrial Electronics**, 2011, 58(2):611–617.

[32] Chin, CS, Wheeler, C, Quah, SL, and Low, TY. Design, modeling and experimental testing of magnetic levitation system for conveyance applications. **Proceedings of the 2nd IEEE International Conference on Computing, Control and Industrial Engineering**, Wuhan, China, August 2011, 174–179.

[33] Wai, RJ and Lee, JD. Adaptive fuzzy-neural-network control for Maglev transportation system. **IEEE Transactions on Neural Networks**, 2008, 19 (1):54–70.

[34] Lee, JD and Duan, RY. Cascade modeling and intelligent control design for an electromagnetic guiding system. **IEEE/ASME Transactions on Mechatronics**, 2011, 16(3):470–479.

Chapter 6
ARDUINO® embedded system design

6.1 Remotely operated vehicle system design

ROV's development has gained increasing significance in the 1980s when a significant amount of offshore evolution has gone beyond the capabilities of human beings. ROVs have been used in deep ocean exploration and construction. However, the ROV has been dominated by expensive components. In addition, they involved skilled technicians and specially designed craft to launch and recover the vehicle. The high cost has led to research works on designing a low-cost and functional ROV. This section will show a low-cost ROV design for shallow water applications that provides a visual display via smartphone. Despite operating with Wi-Fi that incurs a large decrease in the functional range of the ROV, it is cheaper to implement than acoustic systems. With built-in Wi-Fi module in most smartphones, it enables command to be sent easily to the mobile device for remote control and inspection purpose.

6.1.1 System design and architecture

Figure 6.1 shows a completed ROV prototype. The ROV has an integrated structure that measures 44 cm long, 20 cm wide, and 20 cm high. It weighs approximately 2.3 kg. It consists of necessary components including the primary hull, two horizontal propellers, and two vertical and electronics inside the main hull. It has a webcam housed in the waterproof camera housing for navigation and inspection purpose.

The final design was performed using CAD drawings as shown in Figure 6.2. The designs were drawn using Design Spark Mechanical. This CAD software allows users to modify to design model effortlessly and hence accelerates the process of prototyping. The ROV was designed differently as compared to those designs available in the market. The typical ROVs have the frame of a cubic-like structure made of rods and elbows. The ROV was designed to resemble a catamaran style submarine. The ROV hull comprises of a Tupperware box and a horizontal frame section made of polyvinyl chloride (PVC) pipes with camera. The ROV is watertight such that no water leaks into the components to affect the ROV's buoyancy or damage the electronics. It operates as the cheapest way of floatation. The ROV is made slightly positively buoyant for easy recovery as no tethering was involved in recovering the ROV. The buoyancy was counteracted by the weight of the thrusters, electronics, and some ballast weights.

Figure 6.1 Completed ROV prototype

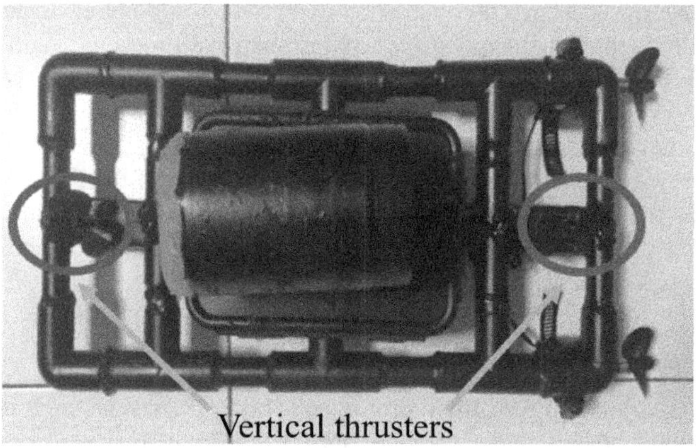

Figure 6.2 Plan view of prototype

The motors used for the vertical thrusters are two 12 V DC brushless motor submersible water pumps. The motor was combined with a Traxxas Propeller Right 4.0 mm Villain EX to act as a thruster. The thrusters were fastened onto a DIY plastic bracket and then secured onto the PVC pipes in an upright position. These thrusters provide the heave (or vertical) motion of the ROV. Both propellers were faced to provide the downwards thrust. When the vertical thrusters were switched off, the positive buoyant ROV will ascend in a controlled manner. As shown in Figure 6.3, the vertical thrusters were attached near to the bow and the stern of ROV to stabilize the ROV during heave motion.

The design incorporates another two similar thrusters placed horizontally on both sides of the rear to provide the surge and yaw motion. They were mounted on the PVC pipes with metal pipe clamps. To maneuver the ROV in the yaw motion, only one of the horizontal thrusters was activated. All the four thrusters were

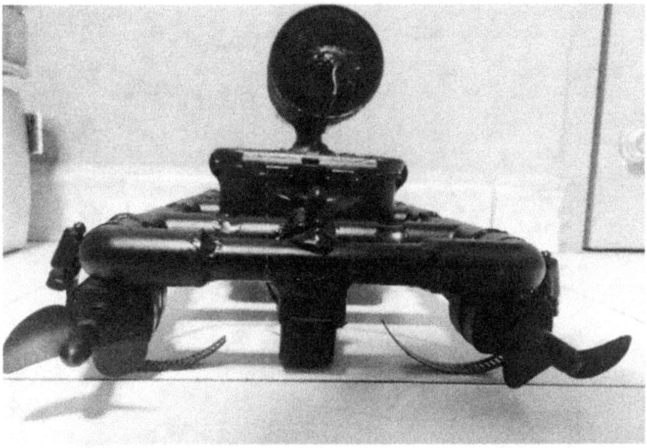

Figure 6.3 Rear view of a prototype

connected to the ARDUINO® Yun™. The ROV was designed for shallow water application, having a few DOF. These motions include moving in the surge, heave, and yaw. There are several methods to provide the command signals to the ROV. Wired cables are used to provide a reliable and fast data transmission. Bluetooth is a wireless communication standard for communicating with Bluetooth devices. Bluetooth module can be found in most modern devices (mobile phones, smartphones, GPS devices). However, it does not provide comparable transfer speed than the wired cable. Additionally, the drawback is the short-range transmission. Bluetooth operates at a frequency lower than 2.4 MHz. A final decision was made on Wi-Fi technology. It can be accessed from the built-in Wi-Fi support from the microprocessor and command signals can be transmitted from the phone to the ROV. The only disadvantage is the limited range underwater as water absorbs most of the radiofrequency waves.

The ROV can be made neutrally buoyant or positively buoyant. A neutrally buoyant ROV allows it to maintain at cerain depth with less power consumption. If the ROV is positively buoyant, it needs thrusters to submerge the vehicle. Thus, more power is required to maintain the vehicle at a particular operating depth. A positively buoyant ROV is easier to recover. The proposed ROV has no tethering cable as Wi-Fi communication was used. A propulsion system designed for the ROV has at least two or more thrusters. The placement of the thrusters is critical to allow maximum maneuverability and controllability. The thrusters are positioned to produce thrusts and moments to move the ROV.

Brushless electric motors provide a higher thermal efficiency over brushed motors. A water jet thruster consists of a small diameter pump with an inlet and outlet duct. The outlet duct provides the thrust. There exist different types of micropump system such as an impeller, diaphragm, and peristaltic pump configurations. Peristatic pumps provide bidirectional flow to force water through a tube in the housing. A motor syringe system consists of a syringe and a plunger that is

Figure 6.4 ARDUINO® Yun (retired model, see: https://store.arduino.cc/usa/arduino-yun)

controlled by a linear stepper motor. This system is used to change the buoyancy of the vehicle. Propellers are typically driven by small electric motors with excellent maneuverability. The brushless DC motor water pumps were chosen for the ROV. The motors used for the vertical thrusters are two 12 V DC brushless motor-submersible water pumps with propellers.

Cameras are a vital component to provide the visual capability of the ROV. It needs to be compact, reliable, and low cost to meet the requirements. A USB Logitech Webcam C170 was selected to stream live videos and provide color video. The images captured by the webcam are converted into JPEG files in the micro-SD card attached to the ARDUINO® Yun board. These images can be easily read by the SD card reader from a computer. A microprocessor communicates with the controller of the ROV and sends the command signals to the thrusters and camera.

ARDUINO® Yun as shown in Figure 6.4 has a microcontroller called ATmega32U4 microcontroller running Linux® and the OpenWrt wireless stack. The ARDUINO® Yun is easy to code in C-Programing language. The ARDUINO® Yun has USB host capabilities through OpenWrt-Yun Linux® distribution. The USB webcam is connected to the Yun and shell scripts are run through Ubuntu terminal on the laptop to control the webcam. The ARDUINO® Yun has seven PWM output pins. Three of the pins were connected to the thrusters.

A power source is required to supply electrical energy to the ARDUINO® board, the thrusters, and the webcam. Eight AA rechargeable batteries were used to provide the ROV with 12 V power supply. The batteries were placed onboard the ROV. As the Wi-Fi communication linkage was chosen, a router is needed. A router is a networking device that can interconnect two or more computers. The computer, iPhone, and ARDUINO® Yun will connect to the router's network. The locked tight container provides the ROV with a reliable watertight structure to house the electronics. The PVC was chosen as part of the ROV's frame due to its

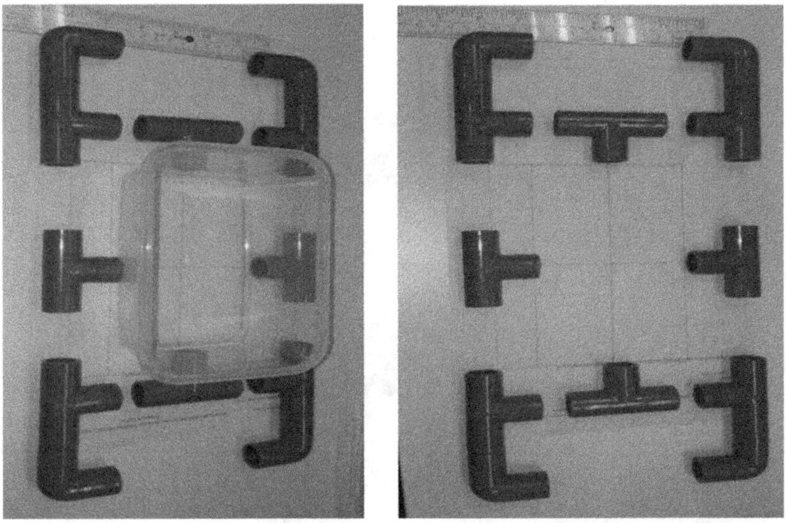

Figure 6.5 Outline of Tupperware exterior with PVC fittings

Figure 6.6 Hole saw drilling a hole in Tupperware

inexpensive cost and lightweight. They are easy to cut into different lengths and have internal buoyancy when sealed. The fabrication process can be discussed in two parts. The primary hull of the ROV was fabricated followed by the camera housing.

The framework of the hull is shown in Figure 6.5. After the length of the PVC pipes was measured, they were cut by a tube cutter. The centroid of the holes was marked on masking tape for the accuracy of drilling the holes to the right dimensions. As shown in Figure 6.6, the holes were drilled using a hole saw. The pieces of PVC pipes were fitted using 90° elbows and "T" joints. The horizontal PVC frame was assembled with the Tupperware as seen in Figure 6.7.

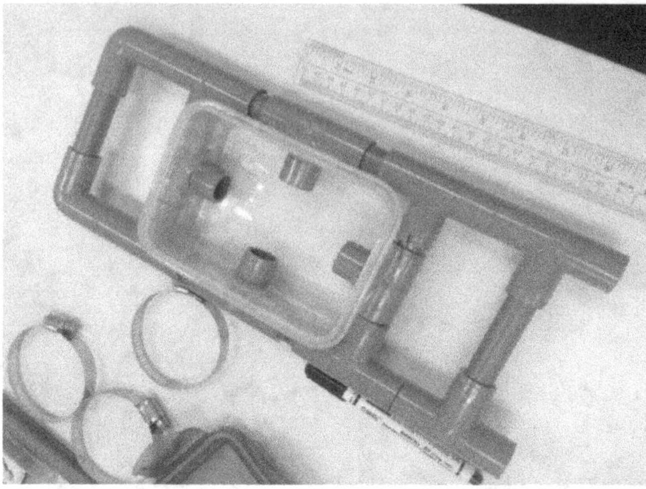

Figure 6.7 Horizontal PVC frame assembled with the Tupperware

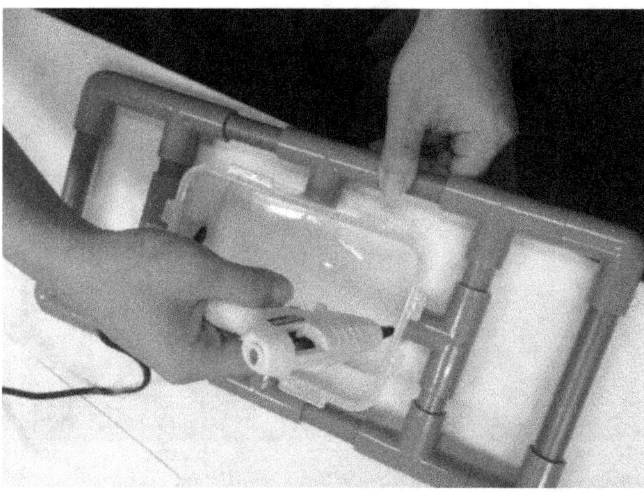

Figure 6.8 Gluing the PVC frame onto holes of Tupperware with hot glue gun

The frame was checked to be symmetrical before applying glue on the PVC tubes and Tupperware using a hot glue gun as shown in Figure 6.8. The fitting between all the PVCs was glued and sealed to ensure watertight integrity. At this stage, the fabrication process was halted to design the electrical circuits. It is followed by dry testing of the motors via the ARDUINO® Yun and the "Yun buddy" application.

As shown in Figure 6.9, the thruster assemblies were made of 12 V DC brushless motor submersible water pumps and Traxxas Propeller Right 4.0 mm

Figure 6.9 Thruster assembly

Villain EX. The propeller was mounted on the water pump. The side exhaust of the water pump was removed and sealed with hot glue. Then it was attached to the propeller. Next, the completed set was attached to the permanent magnet of the brushless DC water pump. A few holes were drilled on the PVC tubes with a hand drill for the electrical wires of the motors in the Tupperware. The wires of the motors as shown in Figure 6.10 will penetrate the holes of PVC frame and the Tupperware. The hot glue gun was used to waterproof the Tupperware to prevent short circuit.

A few motor mounts were made using readily available plastic. After checking the alignment of both the motor mounts, hot glue gun was used to seal any remaining gaps of the holes. The vertical thrusters were then mounted onto the holders. The picture of the ROV after the thrusters and electronics were installed can be seen in Figure 6.10.

A glass jar was used to house the webcam. The bottom of the jar was removed as it has no flat surface for good vision. An acrylic block was cut into a circular shape and mounted onto the glass jar using hot glue as shown in Figure 6.11. Next, a hole was drilled through the cover of the jar for the USB cable to pass through.

Two short PVC pipes were used as holders for the camera housing as shown in Figure 6.11. The hot glue gun was used to glue the holders onto the top of the

Figure 6.10 ROV prototype with electronics and thrusters

Figure 6.11 PVC holders for the mounting of the camera housing

Tupperware and the bottom of the glass jar. Figure 6.11 shows the prototype after mounting the camera housing. Hence, the entire fabrication process of the ROV was completed. As seen in Figures 6.12 and 6.13, a black coat of paint was applied onto the ROV for esthetics purpose.

The completed schematic diagram can be seen in Figure 6.14. The circuit layout consists of the thrusters' connection and the motor drivers. As shown in Figure 6.15, the black wire was connected to the ground of the ARDUINO® Yun. As shown in Figure 6.16, the negative terminal of the batteries was hooked onto the ground. The positive terminal of the batteries was connected to the positive end.

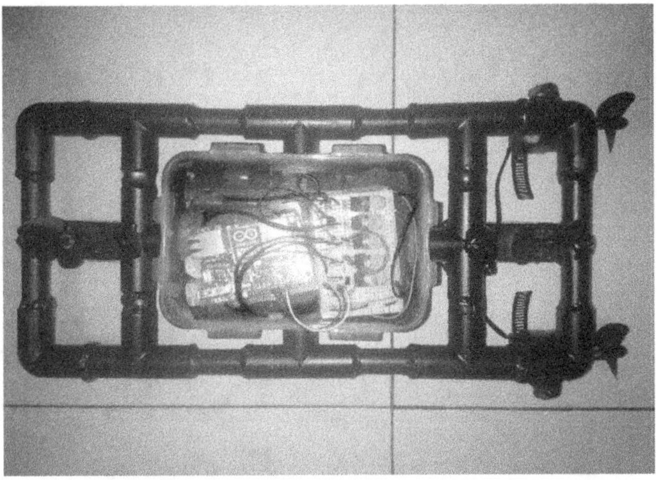

Figure 6.12 Plan view of ROV (camera housing removed)

Figure 6.13 Side view of the ROV

As seen in Figure 6.17, wires were hooked onto the positive–negative columns of the breadboard so that both sides of the breadboard can be used.

As shown in Figure 6.18, Pin 2, Pin 3, and Pin 4 were hooked to the gate pins of each metal-oxide semiconductor field-effect transistor (MOSFET). As shown in Figure 6.19, a yellow jumper wire was connected on the third and fourth MOSFETs (from the left), so that Pin 2 can control the two thrusters concurrently. The four 10k Ω pull-down resistor were placed between the signal and ground pins of the four MOSFETs. The positive terminals of the motors were connected to the positive column of the breadboard. Three 10k Ω resistors were connected in series and white wire was hooked onto the analog pin (A0) to measure the voltage of the

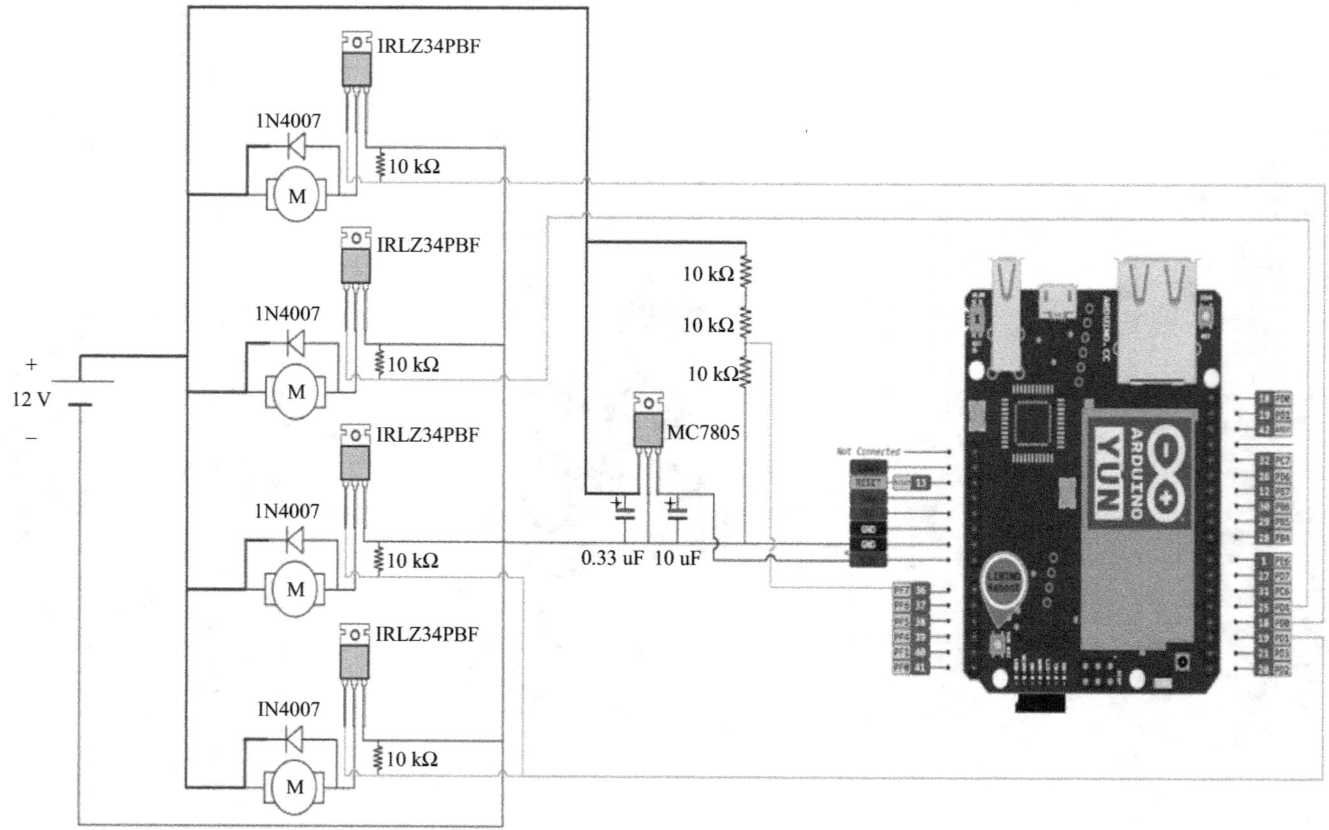

Figure 6.14 Schematic diagram of circuit connection

Figure 6.15 Wire connection 1

Figure 6.16 Wire connection 2

Figure 6.17 Wire connection 3

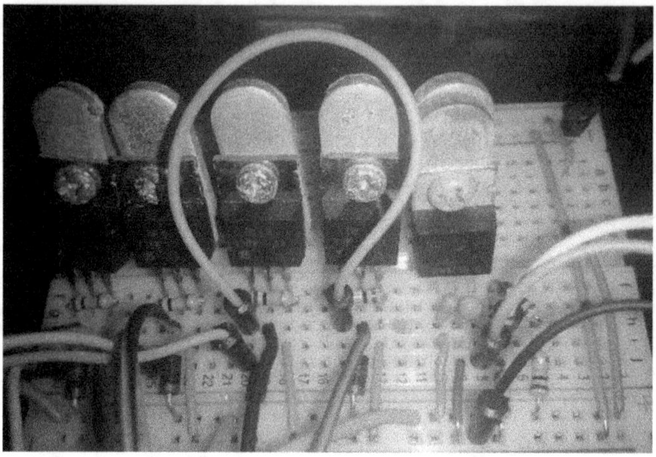

Figure 6.18 Wire connection 4

Figure 6.19 Wire connection 5

power supply. The diodes were placed on the positive terminal of the breadboard and the drain pin of the MOSFETs. The input pin of the voltage regulator was connected to the capacitor and the positive terminal of the batteries. The ground of the voltage regulator was connected to the ground of the breadboard.

Based on the schematic diagram of the circuit connection (see Figure 6.14), the design of the electrical circuit was made of the ARDUINO® Yun, four 12 V DC brushless motor-submersible water pumps, 12 V of battery, a voltage regulator, MOSFETs, diodes, capacitors, and resistors. A user cannot drive the motors directly from the ARDUINO® board. Hence, an intermediate switching device is needed. In the circuit, the power MOSFETs (IRL34PBF) are used as a switch to control the motors. These are *n*-channel MOSFETs as the switching device.

The pinout of the device is G (Gate), D (Drain), and S (Source) for this particular model. Four MOSFETs were driven from Pins 2, 3, and 4. Two of the MOSFETs was connected to Pin 2 concurrently. The wires were plugged into these three pins connected to the individual gate terminal of the MOSFETs. The PWM will deliver a HIGH signal to the transistor's gate. When the MOSFET receives a gate voltage above the threshold voltage, the motor will switch on. When the switch is turned off, the motor will stop. The ARDUINO® Yun contains no onboard voltage regulator. In the circuit, the positive terminal of the 12 V batteries is connected through a voltage regulator. The negative terminal of the battery was then connected to the ground of the ARDUINO®. The heat sink was used to prevent any overheating caused by the voltage regulator.

6.1.2 Programming and system implementation

The ARDUINO® software IDE is a program that allows users to write program codes for the ARDUINO® board in a simple programming language. The code that is written in ARDUINO® programming language is translated into C++ programing language via AVRGCC compiler. ARDUINO® IDE allows users to write sketches to interact with the pins on the board. ARDUINO® Yun provides a library called "Bridge" that allows the Linux® processor and the microcontroller to communicate with each other. It provides the possibility to enable ARDUINO® sketches to run shell scripts and collect data from the AR9331 processor. The Yun also works with Wi-Fi, enabling wireless connection between the board and the laptop.

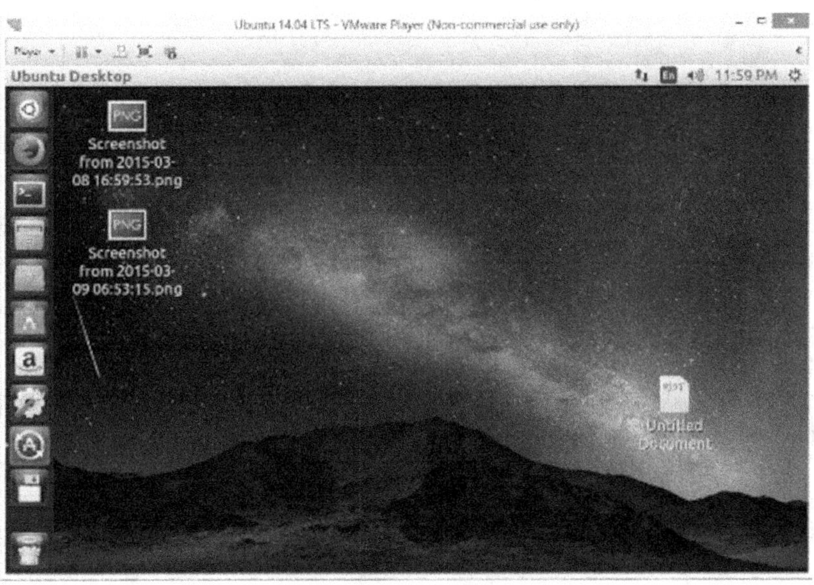

Figure 6.20 Ubuntu operating system on VMware player

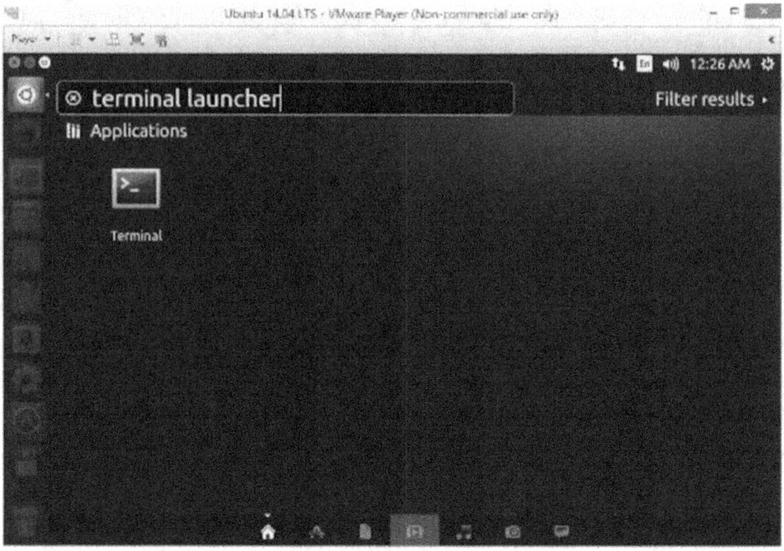

Figure 6.21 Opening the terminal application on Ubuntu Linux®

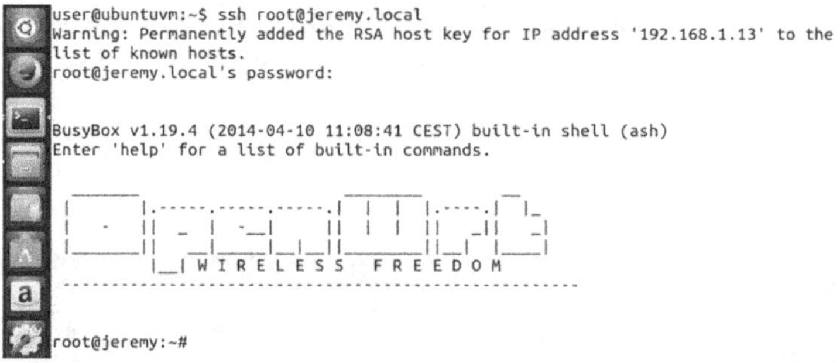

Figure 6.22 Connection to Yun via SSH

VMware player is a virtual machine that allows two operating systems to work simultaneously (see Figures 6.21 to 6.22). The Ubuntu Linux® as shown in Figure 6.21 was used. See the details in Chapter 2 for the Ubuntu Linux®.

The Yun operates in a Linux® distribution named OpenWrt-Yun. After connecting with the OpenWrt-Yun system, the user needs to work with a command line via SSH to communicate with the ARDUINO® Yun Linux® machine. To go to the command line interval (CLI), the terminal application is launched on Ubuntu Linux® to use CLI as shown in Figure 6.22. The terminal creates a connection between the Yun board Linux® processor and Ubuntu Linux® to enable the user to send and receive commands through the CLI and operate the USB webcam. CLI was used to install the software required for the Yun. SSH is an abbreviation for Secure Shell.

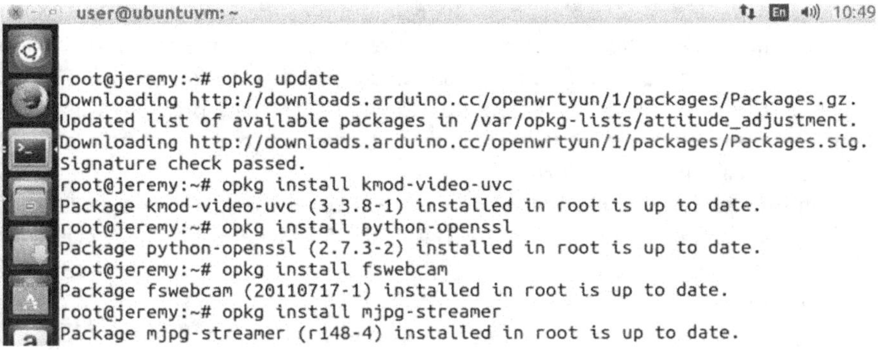

Figure 6.23 OKPG update and installation of drivers

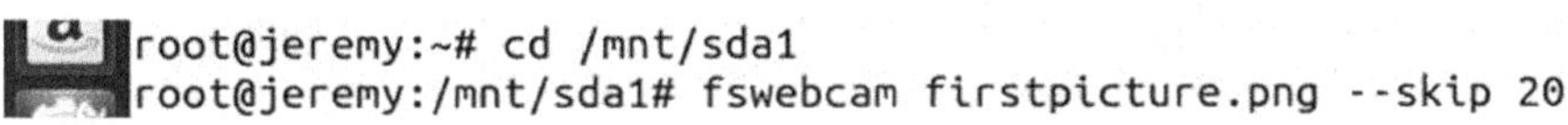

Figure 6.24 Command line for SD card to store images and test webcam

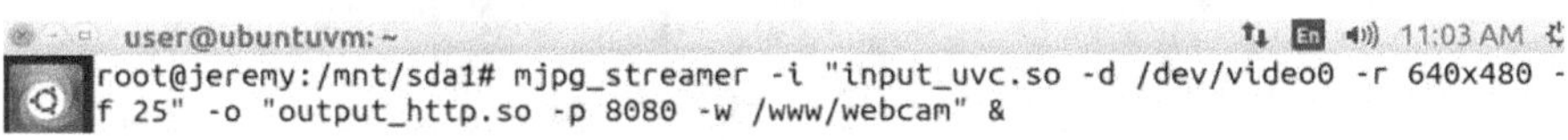

Figure 6.25 Command line for video streaming

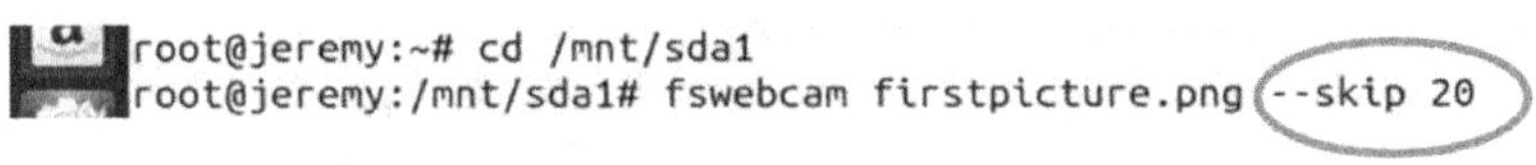

Figure 6.26 Skip argument on CLI

It is a network protocol using a set of guidelines to inform the computer how to send data from one place to another securely. SSH can be used for transmitting data, commands, and files. The commands used on the CLI are as follows.

As shown in Figure 6.22, the connection to the ARDUINO® Yun was established through SSH after configuring the ARDUINO® Yun to the local area network.

The package manager (opkg update command) was updated to activate the latest packages to install the desired webcam application as seen in Figure 6.23. To stream live video, the UVC driver (see Section 2.5.3 Image video unit) for the webcam application and the mjpeg-streamer program. As shown in Figure 6.24, the command line "cd/mnt/sda1" saved the images into the micro-SD card. The webcam was tested by entering the next command line as shown in Figure 6.24. Note that the "-skip-20" argument was included before processing the image.

The images were read from the SD card. Figure 6.25 shows the images captured using the command lines. Figure 6.26 shows the command line that allows the streaming of the video. The "–r" sets the resolution of the video to 648×480, and "–p" argument sets the port for video streaming. By entering http://xxxxx.local:8080/stream.html, the video was streamed live from the webcam. The pictures appeared to be black or distorted on the computer. The problem was solved using the "–skip n option" that command the fswebcam to skip the first n images from the webcam before capturing next.

Yun Buddy is an iPhone application from the Apple Store (see Figure 6.27). The interface sends command signals to the ARDUINO® Yun wirelessly. The ARDUINO® Yun will send the voltage through the PWM outputs to the designated MOSFET to activate the motor. Yun buddy eases the difficulty for the writing of sketches. It requires uploading the "Bridge" example to the Yun through ARDUINO® Software IDE.

As the Bridge example has already been uploaded to the Yun, the Wi-Fi interface of the AR9331 allows communication with the ATmega32U4. Any digital pins can be used for control of the MOSFETs as long as they were defined with pinMode() as OUTPUT in the ARDUINO® program. Digital Pin 0 and 1 were used to connect to the Linux® chip. The iPhone, laptop, and ARDUINO® Yun were

Figure 6.27 Yun Buddy application

connected to the same WLAN. The Yun Buddy app was used to send the control signal to the ARDUINO® Yun wirelessly. The ATmega32U4 receives the signal and sends a HIGH to the PWM outputs to activate the thrusters remotely. It can be established by triggering the on/off button for the respective digital pins on the Yun Buddy interface as shown in Figure 6.28. It shows the control interface of Yun Buddy application. For instance, when Pin 2 is switched to high, both the vertical thrusters will turn on.

6.1.3 Testing

As shown in Table 6.1, the ROV operated at a forward speed of 22.5 cm/s and a yaw rate of 0.285 rad/s in freshwater. The maximum operating depth of the ROV is 22 cm, and the maximum operation duration is approximately 3 h. Before the tests were performed, the electrical circuit was validated. The following underwater tests were performed in an inflated swimming pool that has a diameter of 150 cm and a depth of 35 cm as shown in Figures 6.29 and 6.30.

The buoyancy of the ROV was tested after the electronics and hardware were installed. The ROV was too positively buoyant due to the excessive air trapped in the Tupperware and camera housing. The vertical thrusters were inadequate to

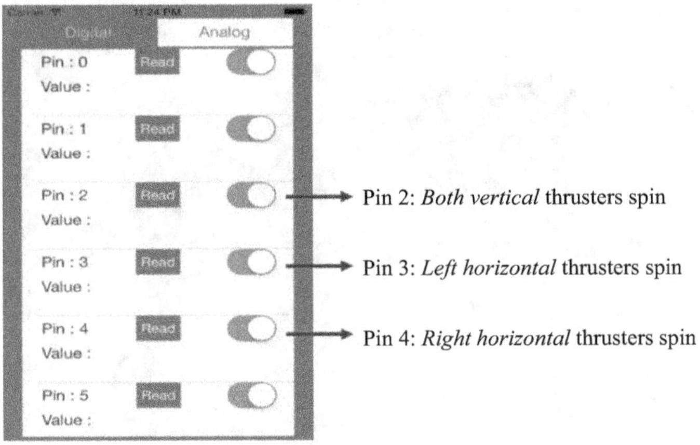

Figure 6.28 Yun Buddy control interface

Table 6.1 Experimental test results of the ROV

Experimental tests	Results
Forward speed	22.5 cm/s
Yaw rate	0.285 rad/s
Maximum operating depth	22 cm
Maximum operating time	3 h

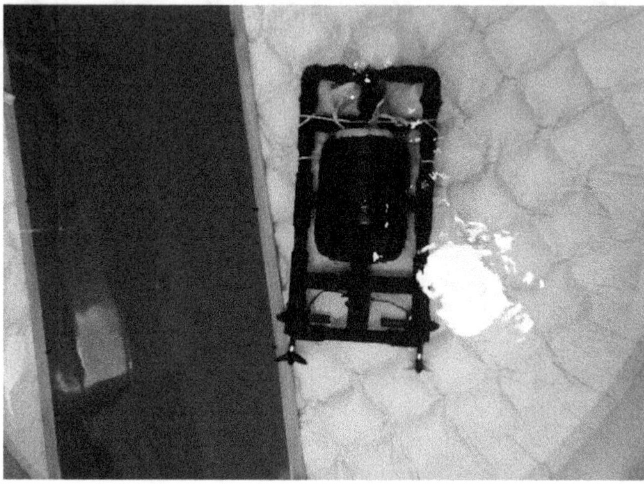

Figure 6.29 ROV tested underwater 1

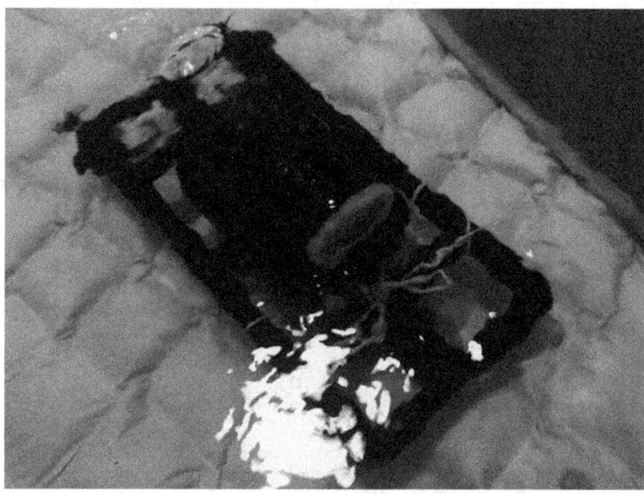

Figure 6.30 ROV tested underwater 2

overcome the buoyant force. Hence, an additional ballast weight was added to the bottom of the ROV. To keep the ROV maneuvering on the surface, weights such as screws were added to the Tupperware to trim ballast. Thus, it increased the stability of the ROV.

When the ROV submerged to 30 cm depth, the control of the ROV was lost. The ROV was tested at 22 cm depth. By adjusting the buoyancy of the ROV to allow the Tupperware (that housed the ARDUINO® Yun) closer to the water surface when the vertical thrusters were operating, the propagation losses could be minimized. This maximum operating depth of the ROV was determined by the

Figure 6.31 ROV maneuvering forward

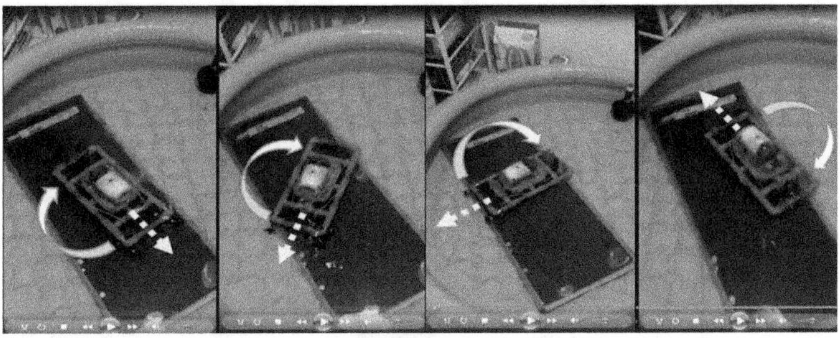

Figure 6.32 ROV moving in the yaw direction

most efficient range of radiowaves between the router and the controller board. As water has a high permittivity, it will absorb most of the radio signals transmitted. Despite the high attenuation of water, the ROV was able to move in different trajectories. The live streaming of the video aided the navigation of the ROV. The ROV was able to maintain in water for long period. However, the ROV was unable to perform the reverse motion due to the brushless DC motor. The user's control of the "Yun Buddy" app on the iPhone was used.

Figure 6.31 shows a few screenshots taken while the ROV was moving. To find the forward speed of the ROV, a starting point was defined for the ROV. The distance was divided by the time taken for the ROV to travel between these two points to determine the forward speed. This test was performed with the video footage. The forward speed of the ROV was approximately 22.5 cm/s. As shown in Figure 6.32, the ROV took approximately 11 s to perform the 180° rotation. Hence the yaw rate was approximately 0.285 rad/s. The ROV's maneuverability in the surge, heave, yaw, and video images capturing were performed satisfactorily. The ROV operated at a forward speed of about 22.5 cm/s and then yaw rate of 0.285 rad/s. The maximum operating depth of the ROV was 22 cm for approximately 6 h. The use of Wi-Fi communication limited the operating depth of the ROV. The program codes for testing the ROV are shown below.

Program codes for ROV testing

Bridge

```
#include <Bridge.h>
#include <YunServer.h>
#include <YunClient.h>

// Listen on default port 5555, the webserver on the Yún
// will forward there all the HTTP requests for us.
YunServer server;

void setup() {
  // Bridge startup
  pinMode(13, OUTPUT);
  digitalWrite(13, LOW);
  Bridge.begin();
  digitalWrite(13, HIGH);

  // Listen for incoming connection only from localhost
  // (no one from the external network could connect)
  server.listenOnLocalhost();
  server.begin();
}

void loop() {
  // Get clients coming from server
  YunClient client = server.accept();

  // There is a new client?

  if (client) {
    // Process request
    process(client);

    // Close connection and free resources.
    client.stop();
  }

  delay(50); // Poll every 50ms
}
```

```
void process(YunClient client) {
  // read the command
  String command = client.readStringUntil('/');

  // is "digital" command?
  if (command == "digital") {
    digitalCommand(client);
  }

  // is "analog" command?
  if (command == "analog") {
    analogCommand(client);
  }

  // is "mode" command?
  if (command == "mode") {
    modeCommand(client);
  }
}

void digitalCommand(YunClient client) {
  int pin, value;

  // Read pin number
  pin = client.parseInt();

  // If the next character is a '/' it means we have an URL
  // with a value like: "/digital/13/1"
  if (client.read() == '/') {
    value = client.parseInt();
    digitalWrite(pin, value);
  }
  else {
    value = digitalRead(pin);
  }

  // Send feedback to client
  client.print(F("Pin D"));
  client.print(pin);
  client.print(F(" set to "));
  client.println(value);

  // Update datastore key with the current pin value
  String key = "D";
  key += pin;
  Bridge.put(key, String(value));
}

void analogCommand(YunClient client) {
  int pin, value;

  // Read pin number
  pin = client.parseInt();
```

```
  // If the next character is a '/' it means we have an URL
  // with a value like: "/analog/5/120"
  if (client.read() == '/') {
    // Read value and execute command
    value = client.parseInt();
    analogWrite(pin, value);

    // Send feedback to client
    client.print(F("Pin D"));
    client.print(pin);
    client.print(F(" set to analog "));
    client.println(value);

    // Update datastore key with the current pin value
    String key = "D";
    key += pin;
    Bridge.put(key, String(value));
  }
  else {
    // Read analog pin
    value = analogRead(pin);

    // Send feedback to client
    client.print(F("Pin A"));
    client.print(pin);
    client.print(F(" reads analog "));
    client.println(value);

    // Update datastore key with the current pin value
    String key = "A";
    key += pin;
    Bridge.put(key, String(value));
  }
}

void modeCommand(YunClient client) {
  int pin;

  // Read pin number
  pin = client.parseInt();

// Read pin number
pin = client.parseInt();

// If the next character is not a '/' we have a malformed URL
if (client.read() != '/') {
  client.println(F("error"));
  return;
}

String mode = client.readStringUntil('\r');
```

```
  if (mode == "input") {
    pinMode(pin, INPUT);
    // Send feedback to client
    client.print(F("Pin D"));
    client.print(pin);
    client.print(F(" configured as INPUT!"));
    return;
  }

  if (mode == "output") {
    pinMode(pin, OUTPUT);
    // Send feedback to client
    client.print(F("Pin D"));
    client.print(pin);
    client.print(F(" configured as OUTPUT!"));
    return;
  }
   client.print(F("error: invalid mode "));
   client.print(mode);
}
```

6.2 Smart control of marine-tracked vehicle for surveillance

As seen in the previous section, the use of smartphones became popular due to an available device such as accelerometer/gyroscope, ambient light sensor, camera, proximity sensor, a GPS, heart-rate sensor, Wi-Fi, Bluetooth, and touchscreen. Engineers can use their smartphones to monitor data, view historical data over a period, receive notifications based on the measurements or even remotely control a prototype to run a test. The connection between a remote device and the smartphones is established using wired or wireless technology. There are a lot of operating systems that can incorporate with software to program and control a remote device. These remote devices controlled by smartphone have both industrial and commercial use.

To enable smartphones to perform remote control functions, an application can be designed and installed into the phones. It removes the need for a new physical remote control unit. Furthermore, with the sudden boom in the smartphone industry, research has shown that 56% of adult globally use smartphones. This degree of penetration will facilitate the transfer of this technology among consumers. The application can be disseminated through portals like the Android or Apple application stores and will be easily accessible. As such, using smartphones as a platform for the remote control of vehicles is a highly attractive technology.

Figure 6.33 ARDUINO® Uno (see https://store.arduino.cc/usa/arduino-uno-rev3)

6.2.1 System design and architecture

The objective is to design and build a functional remotely controlled marine-tracked vehicle based on the interaction between the embedded system and the smartphone. With the use of a marine-tracked vehicle, the condition of the hull, compartments, and confined spaces can also be recorded and surveyed. The platform used in the embedded system is ARDUINO® Uno (see Figure 6.33) and Android application for smartphone side. Both platforms are easily found on the market and are inexpensive.

The method to achieve control of an ROV using existing smartphone applications is to establish a connection between the embedded system and smartphone device by Bluetooth protocol and to control the vehicle by direct drive mode (DDM). DDM is a mode of controlling the vehicle by sending directional commands through the touchscreen buttons on the smartphone as shown in Figure 6.34. For example, if the user wants the vehicle to reverse, the user touches the reverse button on the smartphone's screen. The vehicle receives the commands, analyzes it, and then applies the corresponding movement. It gives the user the ability to change the vehicle's direction and to steer it away from obstacles. The smartphone application will be an Android application build by MIT App Inventor. The application will establish the connection of the Smartphone to the ARDUINO® Uno board which is mounted on the Marine Track Vehicle by Bluetooth Protocol. Bluetooth is a digital radio communication protocol that can come and go ad hoc. Compared to the Wi-Fi protocol which operates at the same frequency, it is more suitable for a longer-term wireless infrastructure, as each node is assigned a fixed IP address. Most of the smartphone devices today are integrated with Bluetooth modules, and it is easy to initiate a communication. Devices are linked together via Wireless Personal Area Network. The main reason for using Bluetooth communication is the low power usage as compared to Wi-Fi. A real-time camera viewer on the smartphone can be used for surveillance.

As shown in Figure 6.35, the MIT App Inventor allows users to develop Android applications for Android phones by using the web servers. The app

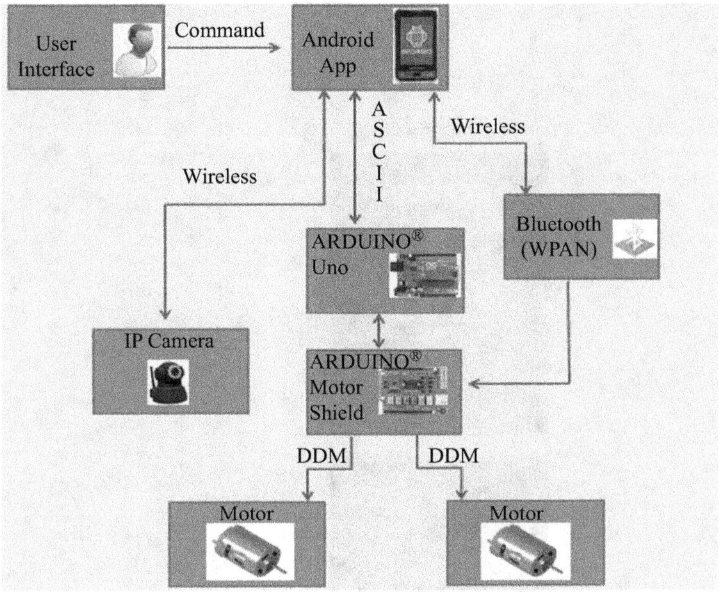

Figure 6.34 Block diagram of the overall system

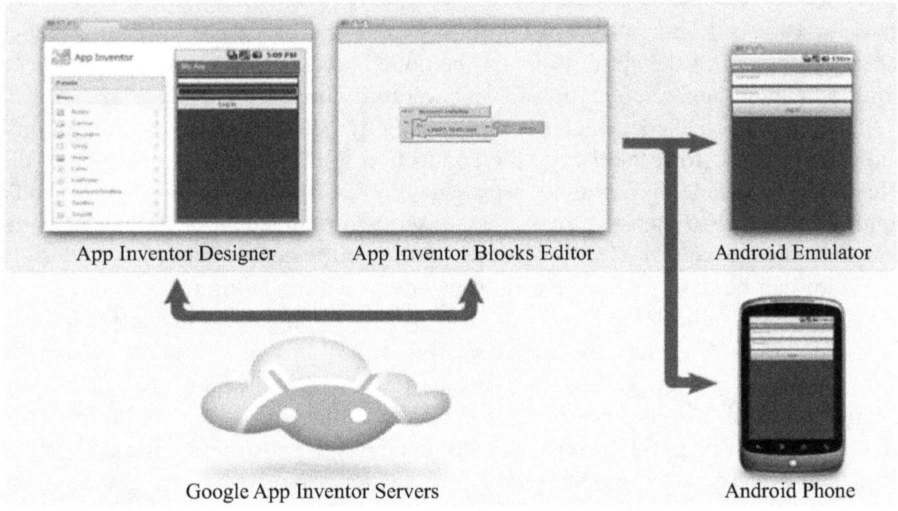

Figure 6.35 MIT App Inventor interface

inventor servers also save your work and help to keep track of the work progress. Applications are built by selecting the components that needed for your application in the App Inventor Designer page. After that, through App Inventor Blocks Editor, program blocks can be assembled with reference to the components chosen, and it specifies how the components behave and respond to human interface. The App

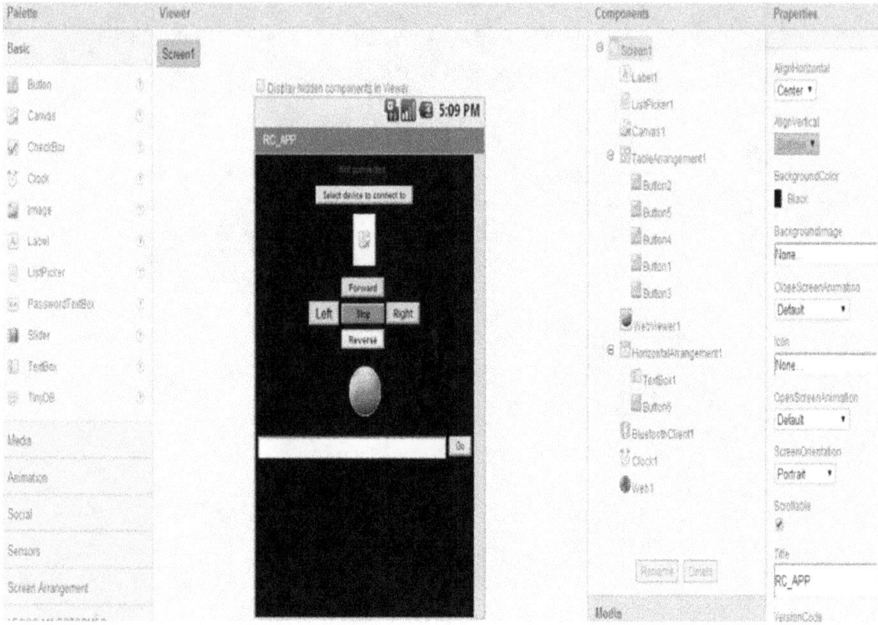

Figure 6.36 MIT App Inventor Designer GUI

Inventor Blocks Editor allows users to assemble the programs by fitting pieces together. Testing of the application can be done by using the Android Emulator to produce a stand-alone application for the Android Phone.

The marine-tracked vehicle communicates by Bluetooth protocol, directions, and a web viewer for establishing the connection to the IP camera. Therefore, the Bluetooth client and five buttons represent the directions (forward, reverse, left, right, and reverse). The web viewer is chosen and dragged from the list of components into the viewer. The properties of the components such as text, design, and color can be defined from the right palette as shown in Figure 6.36.

The Block Editor Interface is defined by C++ language. Blocks for general behaviors are dragged into the Block Viewer. It is followed by blocks of specific components that are dragged into the Block Viewer to join the pieces together. First, a connection is established by choosing the general block of the Bluetooth blocks to join the block (if and else) to set the outcome blocks as shown in Figure 6.37. If the Bluetooth is successfully connected, the text color should be set to green. If the Bluetooth is unsuccessfully connected, then the text color should be set to red as not connected. Second, we set the behavior of the direction buttons by choosing the general block of Button 1, followed by joining the Bluetooth client block (call) to send the bytes number to the ARDUINO® as shown in Figure 6.38. The numbers are directly linked to the codes set up by ARDUINO® codes and converted by using ASCII. The respective Buttons 2, 3, 4, and 5 are set using the same blocks but different bytes number as shown in Figure 6.38. Finally, set up the web view by choosing the general block of button 6 and joining it with block (if and else), then followed by web viewer block (call). The text block can be seen in Figure 6.38.

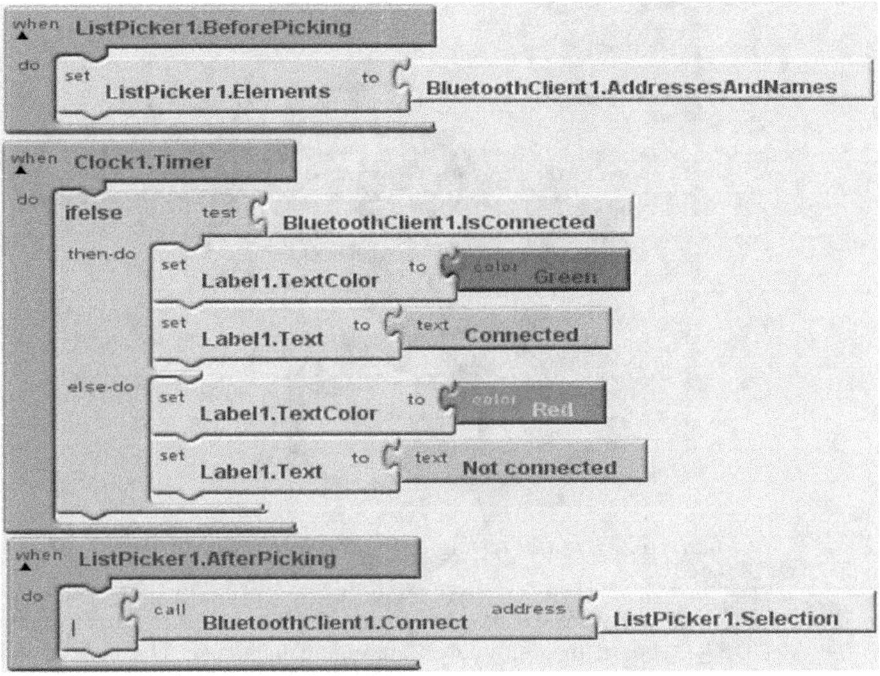

Figure 6.37 MIT App Inventor Block Editor Interface 1

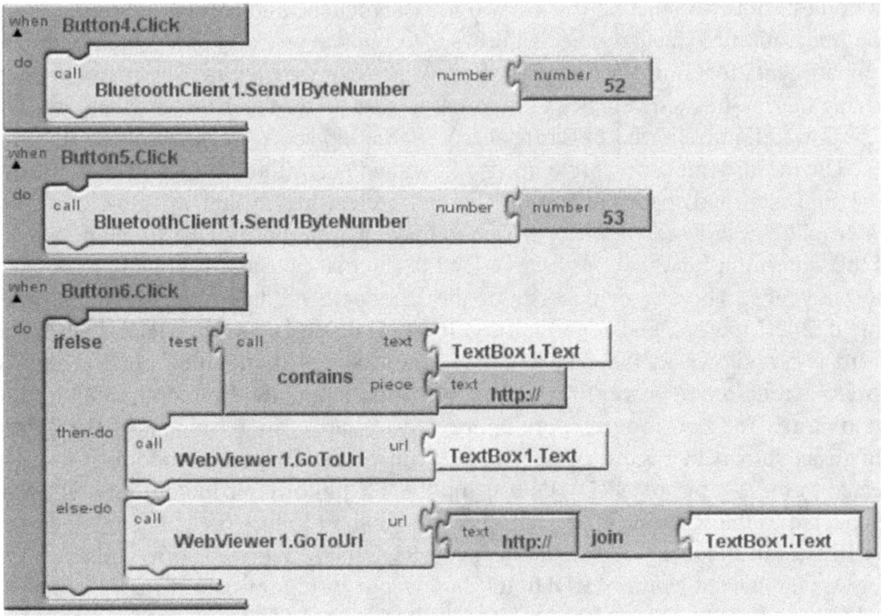

Figure 6.38 MIT App Inventor Block Editor Interface 2

Figure 6.39 Initial testing of marine-tracked vehicle

The ARDUINO® software IDE is a program that allows users to write sketches for the ARDUINO® board in a simple programming language. The code that is written in ARDUINO® programming language is translated into C++ programing language to the AVR-GCC compiler (an essential piece of open-source software that makes the final translation readable by the microcontroller). As shown in Figure 6.39, the Pololu Zumo chassis kit is a robot platform with tracked wheels. The kit comes with two silicone tracks, two idler sprockets, and acrylic mounting plate. The integration of tracked wheels allows it to maneuver freely. The main body has compartments for four AA batteries and two sockets for motors. The motor used in driving marine-tracked vehicle is micro-metal gear motor that has a free-run speed of 625 rpm, stall current of 1,600 mA at 6 V, and a top speed of 100 cm/s at 500 g.

The marine-tracked vehicle needs to conduct surveillance of the hull in the dry dock and confined spaces in water. The entire electrical system was enclosed in a watertight box and sealed with silicone to make it waterproof. Due to the buoyancy of the watertight box, weights were added to the marine-tracked vehicle to counter the buoyancy. The circuit consists of the Bluetooth module, ARDUINO® Motor Shield, 2 DC motors, and a 9 V battery. In the Fritzing software, ARDUINO® Uno board is not shown as the ARDUINO® Motor Shield is mounted on it. The 9 V battery is required to power the ARDUINO® Uno board, the Bluetooth module, and the motors. The two motors were connected to the screw terminals of the Motor Shield as shown in Figure 6.40. The RX pin of the Bluetooth module was connected to the TX pin of ARDUINO®, and the TX pin of the Bluetooth module was connected to the RX pin of ARDUINO® as shown in Figure 6.41. The GND pin of the Bluetooth module was hooked to the ARDUINO®'s ground pin, and the VCC pin was connected to the ARDUINO®'s 5 V pin in Figure 6.41. Finally, the RTS and CTS pin of the Bluetooth module were connected to complete the circuit.

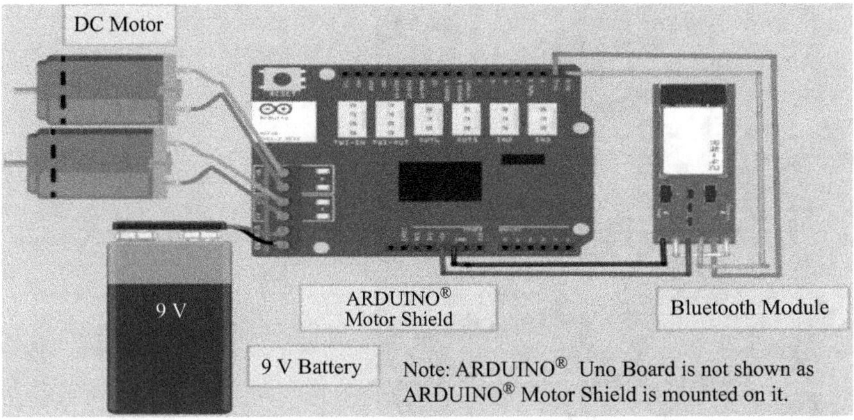

Figure 6.40 Physical electrical layout of the marine vehicle by Fritzing software

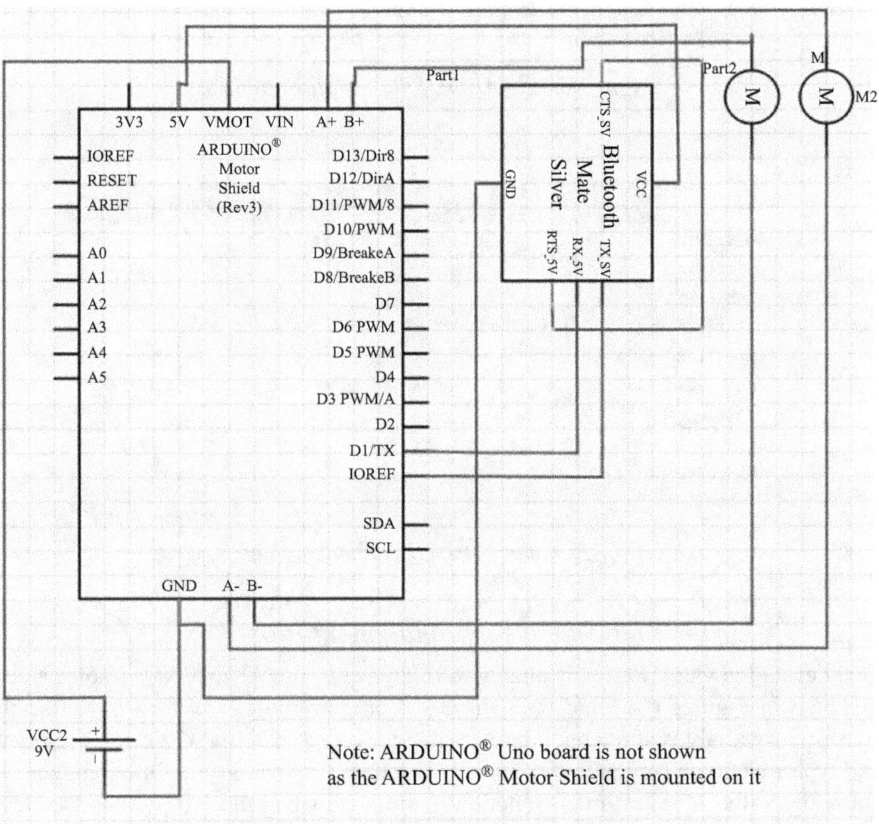

Figure 6.41 Schematic circuit diagram for the marine vehicle by Fritzing software

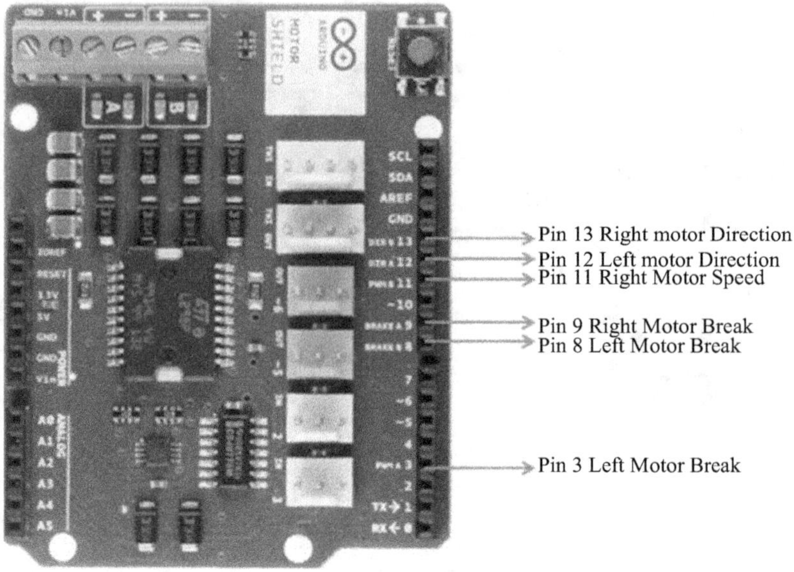

Figure 6.42 ARDUINO® motor shield

```
dc_motor_2 §
int motor1Pin1 = 13; // pin 13 on Motor shield left motor direction
int motor1Pin2 = 8; // pin 8 on Motor shield left motor break
int enable1Pin = 11; // pin 11 on Motor shield controls left motor speed
int motor2Pin1 = 12; // pin 12 on Motor shield right motor direction
int motor2Pin2 = 9; // pin 9 on Motor shield  controls right motor break
int enable2Pin = 3; // pin 3 on Motor shield controls right motor speed
int state;
int flag=0;         //makes sure that the serial only prints once the state
int stateStop=0;
void setup() {
```

Figure 6.43 ARDUINO® Sketches 1

6.2.2 Programming and system implementation

First, define the variable motor pins with reference to the ARDUINO® motor shield as shown in Figures 6.42 and 6.43. In C++ programming, we need to define the types of our variables at the beginning. Second, every ARDUINO® sketch needs two void type functions: [setup()] and [loop()] function. One void type function does not return any value. The [setup ()] function is run once when the ARDUINO® is powered up, and the [loop()] function is executed continuously. The [setup()] function is where the pin mode is initialized at the start while the [loop()] function is where the codes are repeated. In the [setup()] function, pins are set by calling [pinMode()] with

```
    pinMode(motor1Pin1, OUTPUT);
    pinMode(motor1Pin2, OUTPUT);
    pinMode(motor2Pin1, OUTPUT);
    pinMode(motor2Pin2, OUTPUT);
    pinMode(enable1Pin, OUTPUT);
    pinMode(enable2Pin, OUTPUT);
    // sets enable1Pin and enable2Pin high so that motors are turned on:
    // initialize serial communication at 9600 bits per second for bluetooth model HC-06:
    Serial.begin(9600);
}
```

Figure 6.44 ARDUINO® Sketches 2

```
void loop() {
    //if some date is sent, reads it and saves in state
    if(Serial.available() > 0){
      state = Serial.read();
      flag=0;
    }
    // if the state is '1' the ROV will go forward
    if (state == '1') {
        digitalWrite(motor1Pin1, HIGH);
        digitalWrite(motor2Pin1, HIGH);
        analogWrite(enable1Pin, 175);
        analogWrite(enable2Pin, 175);


        if(flag == 0){
          Serial.println("Go Forward!");
          flag=1;
        }
```

Figure 6.45 ARDUINO® Sketches 3

two variables. The first variable is the pin number and the second variable indicates an (INPUT) or (OUTPUT) pins as shown in Figure 6.44. For motor pins, the variable is (OUTPUT), and the sensors used a variable as (INPUT). The HC-06 Transistor-transistor logic (TTL) transceiver module by MDFLY electronics that has a baud rate range of 1,200–115,200 bps. Therefore, serial communication for Bluetooth is set at 9,600 bps using the function [Serial.begin()] as shown in Figure 6.45.

Third, in the [loop()] function, direction movements are defined using the function [digitalWrite()] by calling out the pin number and setting it to (HIGH) or (LOW) state similar to PWM. The speed of the individual motor is defined by the function [analogWrite()] by calling the pin number and setting the [analogWrite()] to a scale of 0–255. The [analogWrite(255)] requests a 100% duty cycle and [analogWrite(127)] is a 50% duty cycle (on half the time). For example, setting the motor Pin 1 and 2 to high and [analogWrite(175)], the ROV will drive forward at 75% speed. All the directions movements are set with [digitalWrite()] and [analogWrite()] function according to the state number defined in the MIT App Inventor Block editor via the ASCII table.

Finally, the sketch was compiled into machine code and uploaded to the ARDUINO® by USB port. When uploading the sketch, the Bluetooth connection is removed to prevent interference to the ARDUINO® Uno board. The complete ARDUINO® Sketch can be seen in this section.

ARDUINO® Sketch

```
int motor1Pin1 = 13; // pin 13 on Motor shield left motor direction
int motor1Pin2 = 8; // pin 8 on Motor shield left motor break
int enable1Pin = 11; // pin 11 on Motor shield controls left motor speed
int motor2Pin1 = 12; // pin 12 on Motor shield right motor direction
int motor2Pin2 = 9; // pin 9 on Motor shield  controls right motor break
int enable2Pin = 3; // pin 3 on Motor shield controls right motor speed
int state;
int flag=0;        //makes sure that the serial only prints once the state
int stateStop=0;
void setup() {
  // sets the pins as outputs:
  pinMode(motor1Pin1, OUTPUT);
  pinMode(motor1Pin2, OUTPUT);
  pinMode(motor2Pin1, OUTPUT);
  pinMode(motor2Pin2, OUTPUT);
  pinMode(enable1Pin, OUTPUT);
  pinMode(enable2Pin, OUTPUT);
  // sets enable1Pin and enable2Pin high so that motors are turned on:
  // initialize serial communication at 9600 bits per second for bluetooth model HC-06:
  Serial.begin(9600); }
void loop() {
  //if some date is sent, reads it and saves in state
  if(Serial.available() > 0){
   state = Serial.read();
   flag=0;
  }
  // if the state is '1' the ROV will go forward
  if (state == '1') {
    digitalWrite(motor1Pin1, HIGH);
    digitalWrite(motor2Pin1, HIGH);
```

```
    analogWrite(enable1Pin, 175);  //Spins the motor on Channel A at 70% speed
forward direction
    analogWrite(enable2Pin, 175);  //Spins the motor on Channel B at 70% speed
forward direction
    if(flag == 0){
      Serial.println("Go Forward!");
      flag=1;
    }
  }
    // if the state is '2' the ROV will turn left
  else if (state == '2') {
    digitalWrite(motor1Pin1, HIGH);
    digitalWrite(motor1Pin2, LOW);
    digitalWrite(motor2Pin1, LOW);
    digitalWrite(motor2Pin2, LOW);
    analogWrite(enable1Pin, 250);  //Spins the motor on Channel A at full speed
forward direction
    analogWrite(enable2Pin, 50);   //Spins the motor on Channel B at 20% speed
reverse direction
    if(flag == 0){
      Serial.println("Turn LEFT");
      flag=1;
    }
    delay(1500);
    state=3;
    stateStop=1;   }
  // if the state is '3' the ROV will Stop
  else if (state == '3' || stateStop == 1) {
    digitalWrite(motor1Pin1, LOW);
    digitalWrite(motor1Pin2, LOW);
    digitalWrite(motor2Pin1, LOW);
    digitalWrite(motor2Pin2, LOW);
```

```
    analogWrite(enable1Pin, 0);
    analogWrite(enable2Pin, 0);
    if(flag == 0){
     Serial.println("STOP!");
     flag=1;
    }
    stateStop=0;
  }
  // if the state is '4' the ROV will turn right
  else if (state == '4') {
    digitalWrite(motor1Pin1, LOW);
    digitalWrite(motor1Pin2, LOW);
    digitalWrite(motor2Pin1, HIGH);
    digitalWrite(motor2Pin2, LOW);
    analogWrite(enable1Pin, 50);   //Spins the motor on Channel A at 20% speed reverse
direction
    analogWrite(enable2Pin, 250);  //Spins the motor on Channel B at full speed forward
direction
    if(flag == 0){
     Serial.println("Turn RIGHT");
     flag=1;
    }
    delay(1500);
    state=3;
    stateStop=1;
  }
  // if the state is '5' the ROV will Reverse
  else if (state == '5') {
    digitalWrite(motor1Pin1, LOW);
    digitalWrite(motor2Pin1, LOW);
    analogWrite(enable1Pin, 175);
```

```
        analogWrite(enable2Pin, 175);  //Spins the motor on Channel B at 70% speed
reverse direction
        if(flag == 0){
          Serial.println("Reverse!");
          flag=1;
        }
    }
    //For debugging purpose
    //Serial.println(state);
}
```

The ARDUINO® Sketch is an open-source initiative to support designers, engineers, researchers, and hobbyist from physical prototyping to developing the actual product. There is a library of electronic parts that provides technical resources for learning electronics with ARDUINO®. It also adopts the electronic design automation (EDA) approach to ARDUINO® [1]. EDA is a category of software tools for designing systems, integrated circuits (ICs), and printed circuit boards (PCBs). Fritzing software is considered to be an EDA as it can create circuit schematic diagrams and PCBs based on the prototype board layout. The ARDUINO® Uno board is an open-source general-purpose microcontroller programming and prototyping platform that can be easily programmed to control things or respond to sensory input. The ARDUINO® Uno Board is based on ATmega328 microcontroller that is an inexpensive small computer. It has 14 digital input/output pins that can be used as PWM outputs, six analog inputs, a 16 MHz ceramic resonator, a USB connection, a power jack, and ICSP header with a reset button. It can be powered by USB cable, AC/DC adapter, or a battery.

The ARDUINO® Motor Shield is a dual full-bridge driver designed to drive inductive loads such as solenoids, relays, DC motors, and stepping motors based on the L298. It allows users to drive two DC motors with the ARDUINO® board and allows them to control the speed and direction of each motor independently. It has two separate channels, A and B that uses four ARDUINO® pins to drive the motor. There are a total of eight pins on the shield that can use each channel separately to drive two DC motors. The ARDUINO® Motor Shield is mounted directly on top of the ARDUINO® Uno board with the respective pins to control multiple motors.

6.2.3 Testing

With the integration of the camera and tracked wheels to the system, the marine-tracked vehicle could maneuver and conduct surveillance underwater as shown in Figure 6.46. The vehicle was controlled from a distance by smartphone to reduce the risk of operator exposing to the hazardous environment.

Figure 6.46 Testing of the marine-tracked vehicle in the water

Table 6.2 Test results of marine-tracked vehicle

Surveillance distance on land	Max. 27 m
Surveillance distance in water	Max. 20 m
Depth in water	Max. 25 cm
Speed	0–50 cm/s
Operating hours	4 h

As tabulated in Table 6.2, the test results have shown that the marine-tracked vehicle could survey from a maximum distance of 27 m on land and 20 m in water. The maximum depth in water is restricted to 25 cm as Bluetooth protocol was used to establish a connection between the smartphone and the marine-tracked vehicle. The speed of the marine-tracked vehicle was reduced to 50 cm/s due to the weights attached to counter the buoyancy. Many current and future embedded systems can be customized and adapted to be monitored by smartphones. This allows users to interact with the embedded devices in the industry and home easily. There are a lot of embedded systems in the market that achieve a similar function. The results show that the marine-tracked vehicle can be used to survey the hull, compartments, hazardous areas, and confined spaces that cannot be accessed easily.

The marine-tracked vehicle integrates a high-performance Smartphone with ARDUINO® and utilizes the Bluetooth technology as an efficient, secure, and reliable connection. With further improvements such as GPS and Wi-Fi, long-distance surveillance can be established. Multiple cameras and sensors can be added to the marine-tracked vehicle for better surveillance. The application can also be modified by adding features to display more data, surrounding temperature, and connection strength. By integrating pressure-balanced closure equipment, pressure-tolerant electronic systems [2], acoustic modems, and transponder, deep water surveillance can be achieved [3].

The completed prototype and testing of the robot can be seen in the following video link.

- Marine-tracked vehicle: https://youtu.be/NrnK8WpaJU4

6.3 A sloth bear-inspired pole-climbing robot

Instead of working on an underwater robot, a land-based robot can be designed. Cable inspection is subjected to high-risk and dangerous environment. With the intensive electric and magnetic fields involved, the climbing robot is used to accomplish the task. The robot will climb up the pole using the gripping and driving systems. There were several designs of the pole climbing robot developed in the past years. The locomotion mechanism revolves around the gripping, climbing, and traveling systems. There are few examples like the Robo-Sloth, Four-Legged Multipurpose Rope Climbing Robot, and Hand-bot. The initial proposed design went through many amendments due to some constraints. However, the objective of the robot is to move along the hanging structure such as pole. The overall complexity of construction should be minimal to reduce the cost of the robot. Instead of using both wheels and gripping arms, the design is simplified using a pair of gripping arms. The newly proposed climbing mechanism includes a pair of gripping arms with rubber pads to provide a firmer grip. The traveling mechanism involves rotation of the arms for climbing.

The development of the robot was a bio-mimicry development and it mimics the movement of the limps of a sloth bear. The design adopts the structure of the bear's limbs as the gripper and the back as a hinge where it straightens and moves forward. The design of the robot consists of four legs and four grippers, each controlled by a single motor. The movement of the robot relies on the synchronization of both the legs and motors. The grippers are used to hold onto the pole steadily so that the robot can move freely. The robot uses its arms to grab the object and maintain the body stability while climbing. The motor driving the arms enables the orientation of the body to grip on the pole while climbing.

6.3.1 System design and architecture

The mechanical model of the entire body of the robot consists of three bar mechanisms (see Figures 6.47 and 6.48), performing the actions to move steadily along the pole. The links will rotate about each joint to achieve the climbing motion. We would need to control the rotation about the joint. Thus, a power source needs to be placed between the joints. A motor could be placed between the links to provide the appropriate speed to move the robot steadily. Moreover, choosing the appropriate power transmission mechanism is crucial as high torque will be acting at these points, especially when the robot carrying a certain payload.

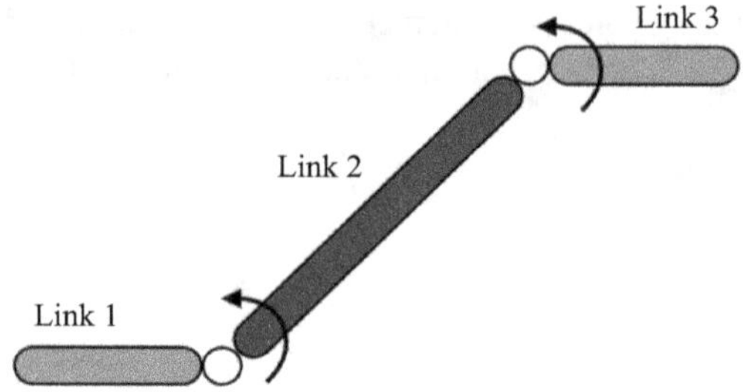

Figure 6.47 Entire body mechanism

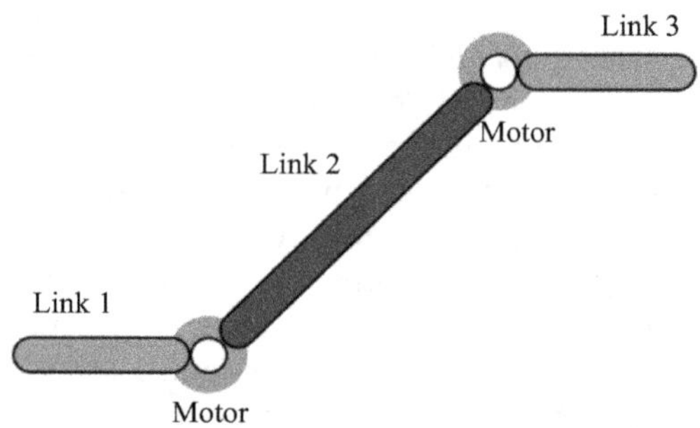

Figure 6.48 Links with the motor in place

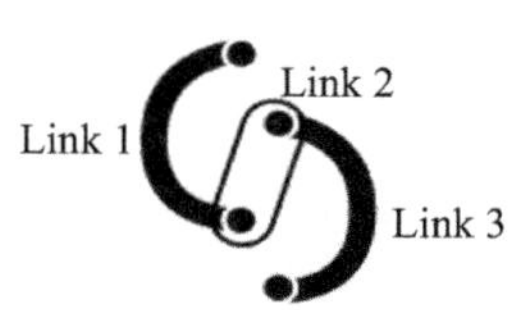

Figure 6.49 Rotational links

The design of the gripper needs to ensure that the robot does not fail on the clamping mechanism. It consists of two curves and one straight bar to enable the rotation of the motor to open and close the gripper as shown in Figures 6.49–6.52.

This section deals with design, fabrication, and testing of an autonomous cable climbing with the ability to move freely from pole to pole. An algorithm to control the motion is proposed. Hence, the requirements are as follows.

1. Type of adhesion and gripping mechanism: It includes a pair of gripping arms with a rubber pad on the inner surface of the grippers. The arms of the gripper are fixed to free rotating joints such that when the joint rotates, the rotation of

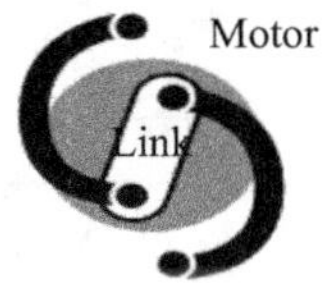

Figure 6.50 Rotational links with motor

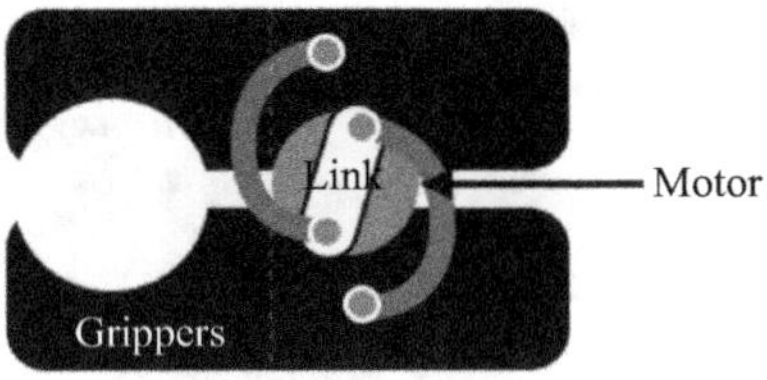

Figure 6.51 Entire gripper

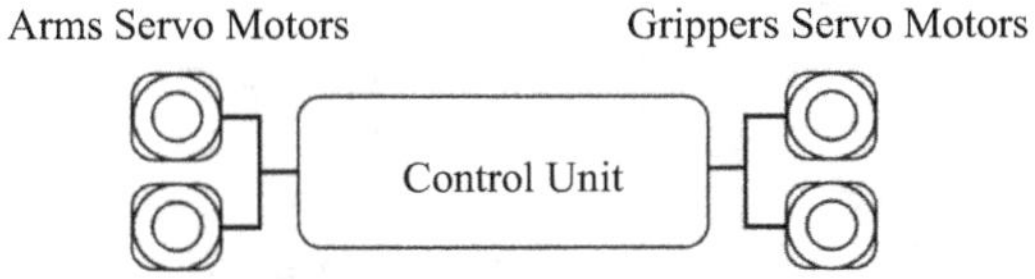

Figure 6.52 Control unit overview

the motor about the joints will create movement to enable the gripper to grip onto the pole. Moreover, the reverse rotation will allow the gripper to release itself from the grip.

2. Rotation and traveling mechanism: The rotational module aims to allow the robot to travel on the pole or suspended pole. The module consists of two motors, each drive one set of the gripping arm. The two arms will take turns to rotate. The respective gripper will release its grip when it rotates. The rotation about the joint should achieve a 180° turn for both the arms.

The entire system consists of two sets of gripping and climbing system with the mirrored mechanism. They are positioned back to back with each other for ascending and descending down the pole. Two linear actuators will be used and positioned between the two sets of gripping system. When the actuators retracted, it will cause the arms to retract and grip the pole. Once the robot is gripped onto the pole, the DC motor mounted on each arm will drive the gears. The rotation of the wheels will rotate clockwise on one side and counterclockwise on the other.

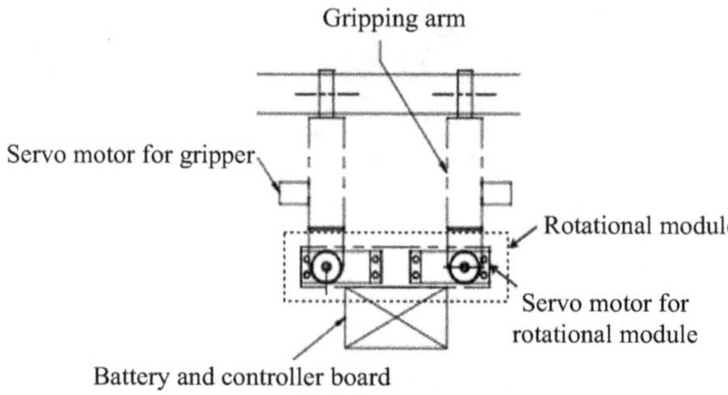

Figure 6.53 Side view of the robot moving horizontally

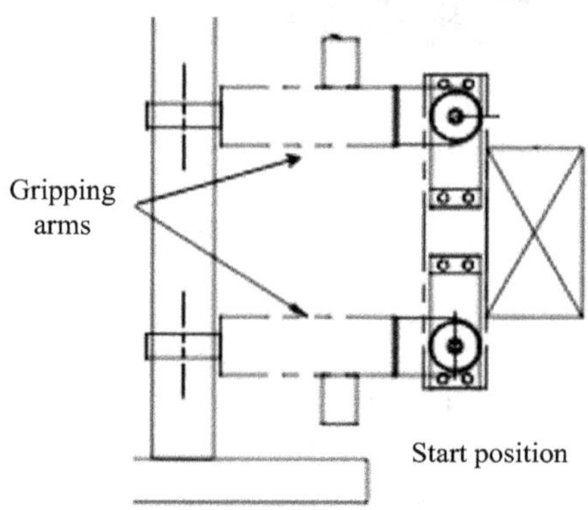

Figure 6.54 Side view of the robot moving vertically

Two extended arms were created at the top part of the gripping and climbing system to enable the robot to travel along the pole.

One set of the gripping system is shown in Figure 6.53. It consists of a position-servo motor that is programmed to rotate the joint at 90°. Once the joint is fixed at 90°, the pulley will stay on the pole. The motor connected the pulley with a set of belt and gear system starts to drive and allow the robot to travel along the pole. Few IR sensors are installed on the top area of the robot to act as the "eye" and to activate the next designated motion. One sensor is installed at the arm to sense the pole. There are sensors at the bottom area as shown in Figure 6.54. The entire system consists of a pair of gripping arms and a rotational module. Each joint is driven by a motor that would create a rotating moment about each arm joint and cause the robot to make a 180° turn to grip the next point. In summary, the locomotion of the pole-climbing robot can be seen in Figure 6.55.

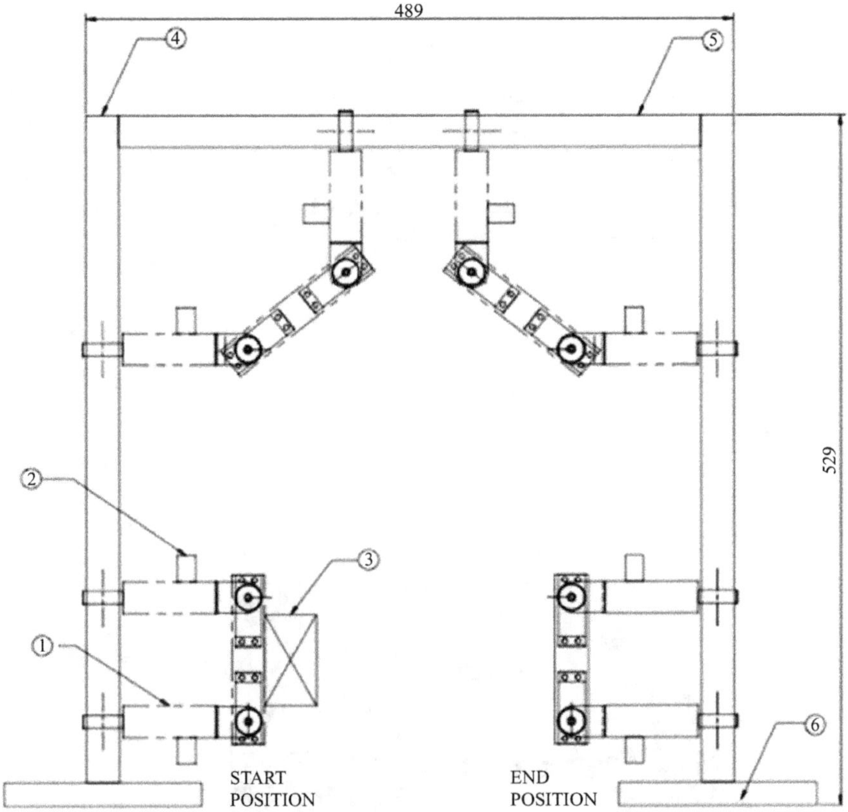

Figure 6.55 Tasks of the pole-climbing robot

As shown in Figures 6.56 and 6.57, the motor rotates to drive the primary and secondary gears to achieve a closing position and allow the robot to grip onto the pole. The applied force will increase to an acceptable level by rotating the motor and gears. To ensure the grip is firmed, a layer of rubber pad is attached to the inner surface of the gripper. The friction between the pole and rubber pad will prevent the robot from slipping while climbing.

The whole mechanism consists of three links where they rotate about the two joints as shown in Figure 6.58. A motor is placed at each joint to control the rotation and climbing motion. It can provide the appropriate speed to make the motion steady. Choosing a correct motor torque is essential as the joint carries the necessary payload. The entire system consists of a pair of gripping arms and a rotational module. Each joint is driven by a motor that would rotate about each arm joint and cause the robot to make a 180° turn to grip the next point. As shown in Figures 6.59 and 6.60, the rotation at the center of the straight bar allows the two curve bars to close the grippers for a good grip. The final design of the sloth bear-inspired pole-climbing robot can be seen in Figure 6.61.

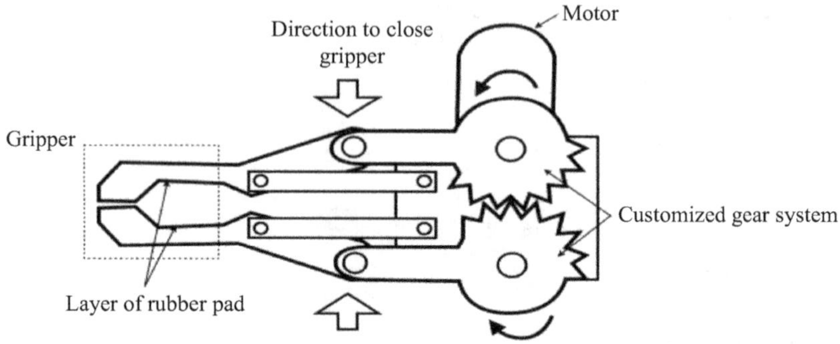

Figure 6.56 Top view of the gripper showing rotation of the motor and its movement in close/open position

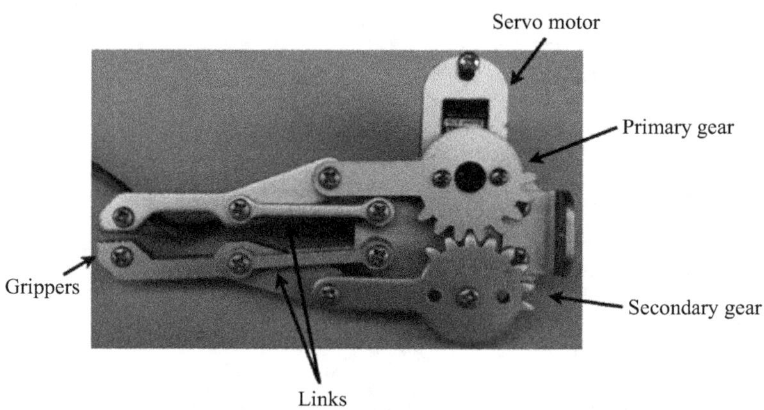

Figure 6.57 Picture of the gripper

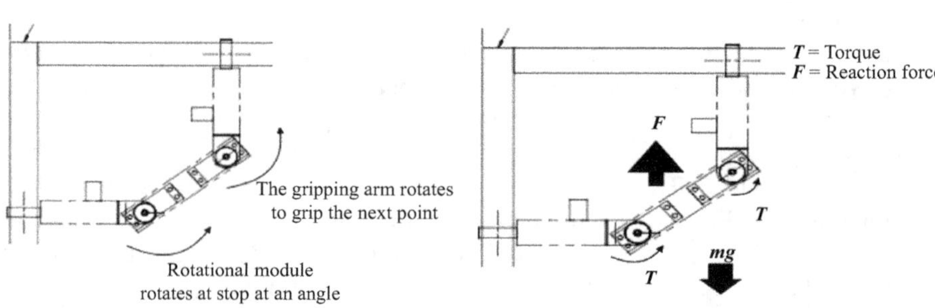

Figure 6.58 Horizontal climbing with rotation of arms

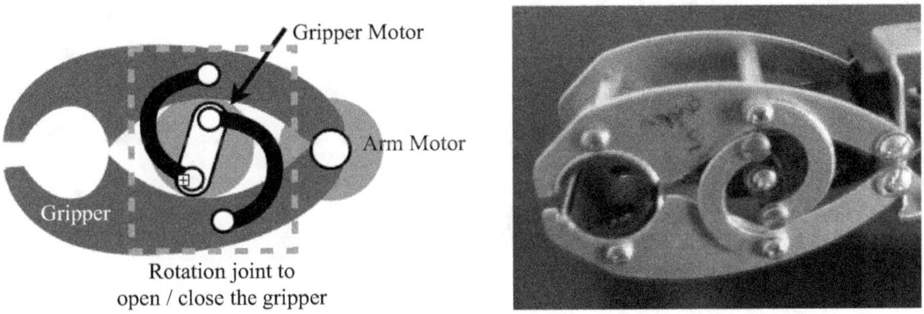

Figure 6.59 Design of the gripper and rotation of the motor to achieve the close position

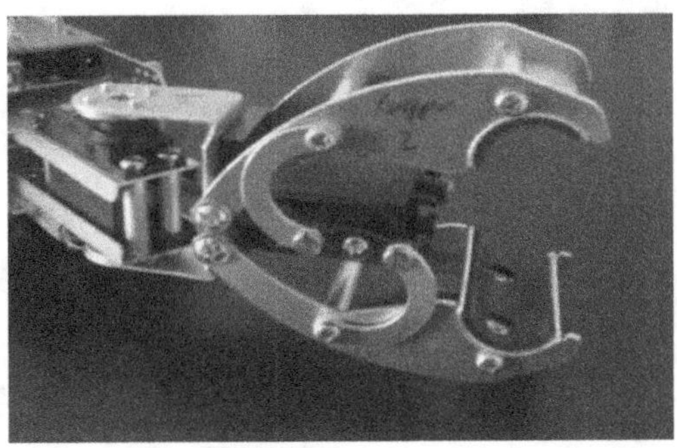

Figure 6.60 Design of the gripper and rotation of motor to achieve open position

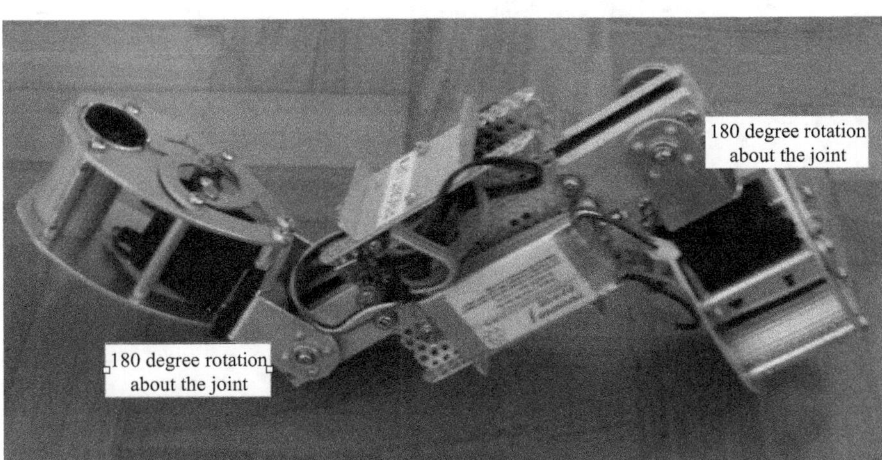

Figure 6.61 Final design of sloth bear-inspired pole-climbing robot

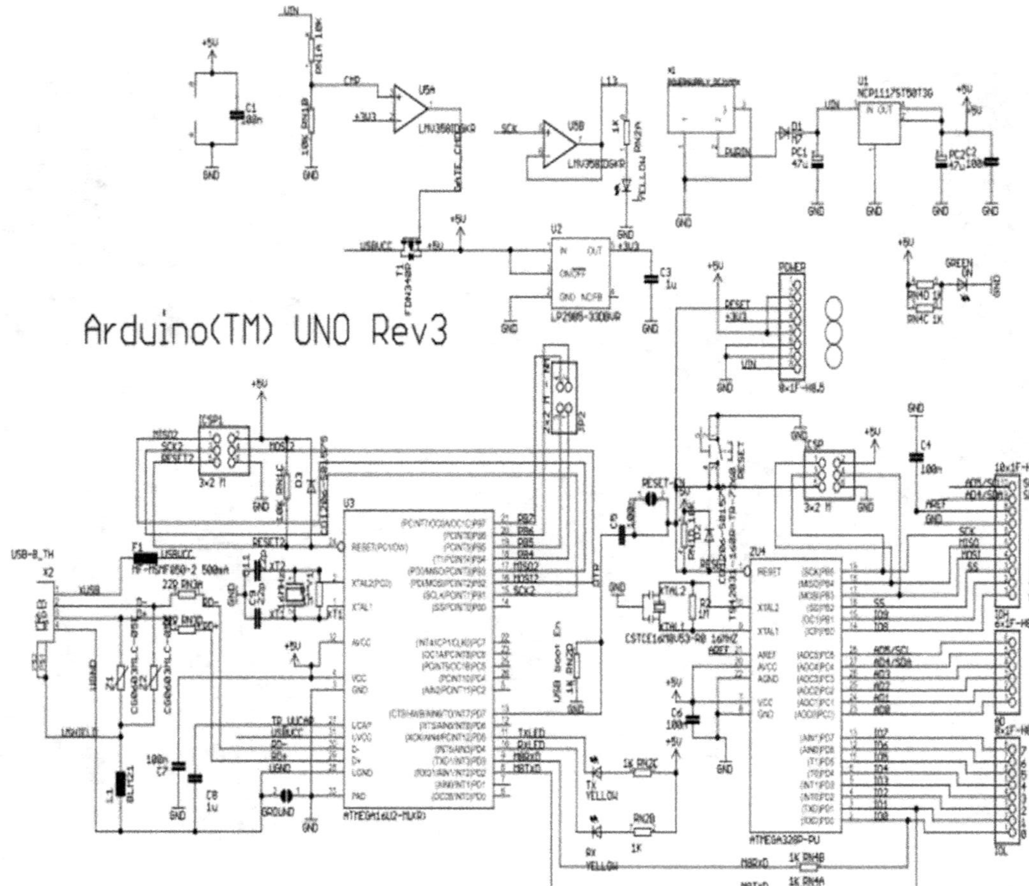

Figure 6.62 Electrical schematic of ARDUINO® Uno Rev3

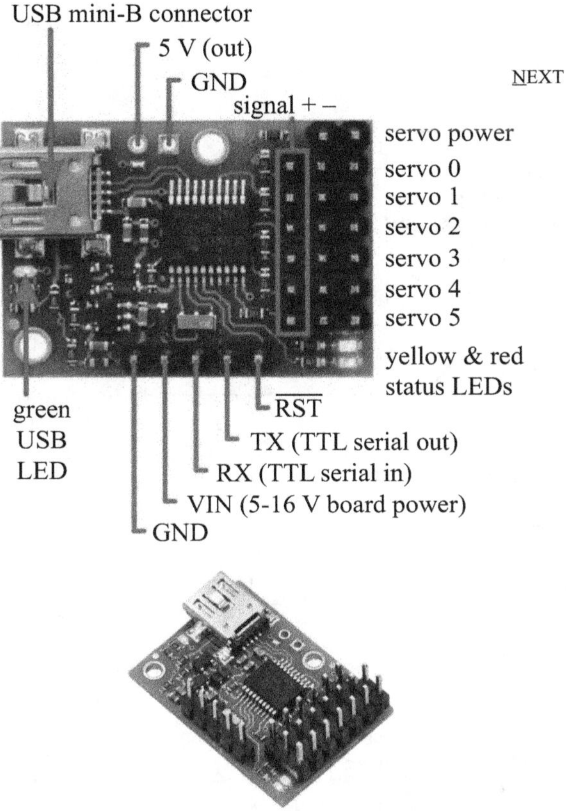

Figure 6.63 Servo controller board used for sloth bear-inspired pole-climbing robot

6.3.2 Programming and system implementation

The IR sensors were installed on the top of the robot to act as the "eye" and activate the next designated move. One sensor is installed on the arm to sense the end of the pole. There are sensors at the bottom area for the robot to "know" when to stop. These IR sensors can detect obstacles. In this section, the ARDUINO® Uno R3 in Figure 6.62 was used. The servo controller board in Figure 6.63 for controlling the angle of the servo motor was utilized. The program was written in sketches before downloading into ARDUINO® Uno.

The entire obstacle with total distance of 1,550 mm requires 30 s. The system algorithm in the flowchart is shown in Figure 6.64.

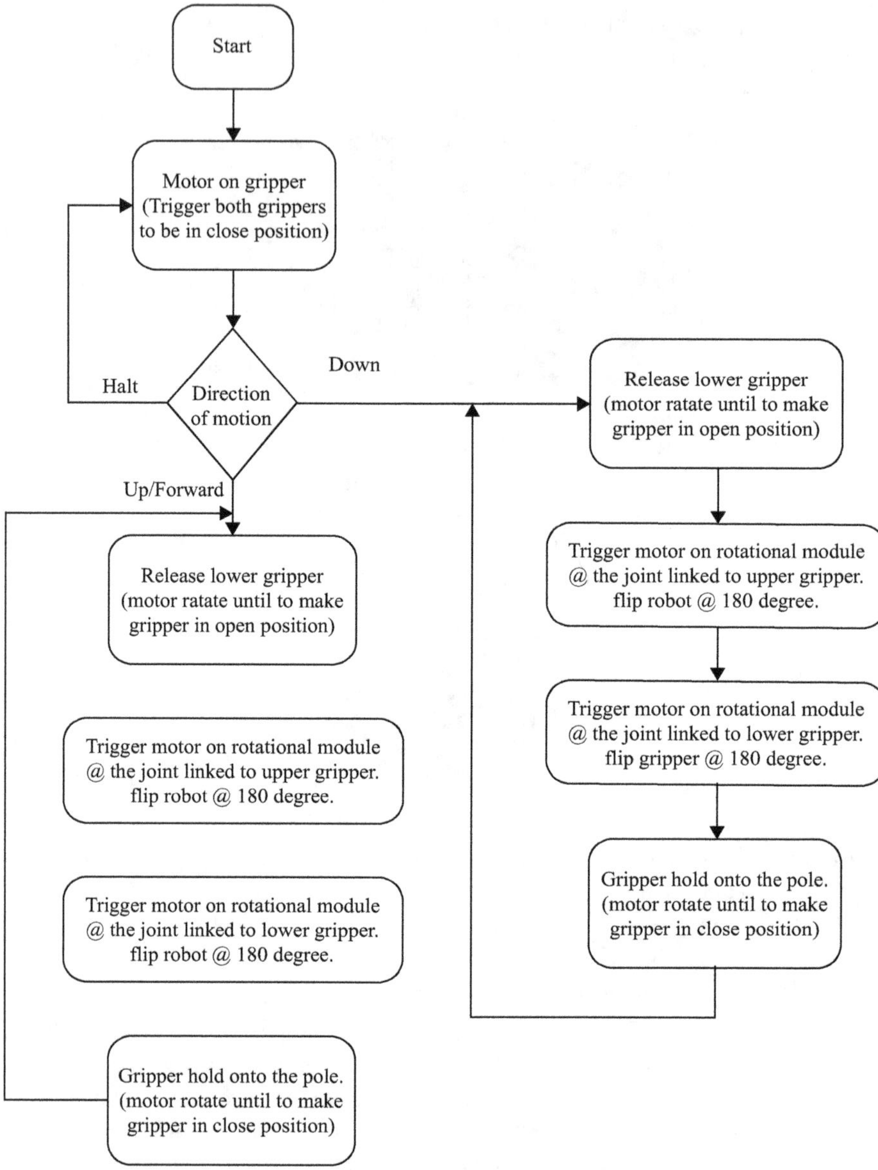

Figure 6.64 Flowchart of control logic for climbing

The program codes used for sloth bear-inspired pole-climbing robot can be seen below.

```
#include <NewSoftSerial.h>

#define txPin 7
#define rxPin 8

#define arm1  1
#define arm2  2
#define gripper1 3
#define gripper2 5

int  buttonPin = A0;
int LEDPin =  12;

/*
unsigned int gripper1_open = 4700;
unsigned int gripper1_close = 7000;
unsigned int gripper2_open = 4700;
unsigned int gripper2_close = 7000;
*/
unsigned int gripper1_close = 4700;
unsigned int gripper1_open = 7000;
unsigned int gripper2_open = 4700;
unsigned int gripper2_close = 7000;

unsigned int arm1_90 = 2800;
unsigned int arm1_270 = 9700;

unsigned int arm2_90 = 2800;
unsigned int arm2_270 = 9500;


NewSoftSerial mySerial(rxPin, txPin);

void setup()// run once, when the sketch starts
{
  pinMode(buttonPin, INPUT);
  pinMode(LEDPin, OUTPUT);

  digitalWrite(LEDPin, HIGH);   // set the LED on
  mySerial.begin(57600);
  delay(1000);
}
```

```
//Send a Set Target command to the Maestro.
//Target is in units of quarter microseconds, so the normal range is 4000 to 8000.
void Maestro_cmd(unsigned char servo, unsigned int target)
{
  mySerial.print(0xAA,BYTE); //start byte
  mySerial.print(0x0C,BYTE); //device id
  mySerial.print(0x04,BYTE); //command number
  mySerial.print(servo,BYTE); //servo number
  mySerial.print(target & 0x7F, BYTE);
  mySerial.print((target >> 7) & 0x7F,BYTE);
}

void Open_Gripper(unsigned char gripperID)
{
  unsigned int pos;


  if (gripperID == gripper1)
  {
   pos = gripper1_open;
  }else if( gripperID== gripper2)
  {
   pos = gripper2_open;
  }
  mySerial.print(0xAA,BYTE); //start byte
  mySerial.print(0x0C,BYTE); //device id
  mySerial.print(0x04,BYTE); //command number
  mySerial.print(gripperID,BYTE); //servo number
  mySerial.print(pos & 0x7F, BYTE);
  mySerial.print((pos >> 7) & 0x7F,BYTE);
}
```

```
void Close_Gripper(unsigned char gripperID)
{
  unsigned int pos;

  if (gripperID == gripper1)
  {
   pos = gripper1_close;
  }else if( gripperID== gripper2)
  {
   pos = gripper2_close;
  }
  mySerial.print(0xAA,BYTE); //start byte
  mySerial.print(0x0C,BYTE); //device id
  mySerial.print(0x04,BYTE); //command number
  mySerial.print(gripperID,BYTE); //servo number
  mySerial.print(pos & 0x7F, BYTE);
  mySerial.print((pos >> 7) & 0x7F,BYTE);
}

void mSetSpeed(unsigned char servoID, unsigned int speedValue)
{
 mySerial.print(0xAA,BYTE); //start byte
 mySerial.print(0x0C,BYTE); //device id
 mySerial.print(0x07,BYTE); //command number
 mySerial.print(servoID,BYTE); //servo number
 mySerial.print(speedValue & 0x7F, BYTE);
 mySerial.print((speedValue  >> 7) & 0x7F,BYTE);
}


void mSetAccel(unsigned char servoID, unsigned int accValue)
{
 mySerial.print(0xAA,BYTE); //start byte
 mySerial.print(0x0C,BYTE); //device id
 mySerial.print(0x09,BYTE); //command number
 mySerial.print(servoID,BYTE); //servo number
 mySerial.print(accValue & 0x7F, BYTE);
 mySerial.print((accValue >> 7) & 0x7F,BYTE);
}
```

```
void moveArm(unsigned char armID, unsigned int armPos)
{

   mySerial.print(0xAA,BYTE); //start byte
   mySerial.print(0x0C,BYTE); //device id
   mySerial.print(0x04,BYTE); //command number
   mySerial.print(armID,BYTE); //servo number
   mySerial.print(armPos & 0x7F, BYTE);
   mySerial.print((armPos >> 7) & 0x7F,BYTE);
}

void blinkLED()
{
  for(int i=0; i<5; i++)
   {
    digitalWrite(LEDPin, LOW);   // set the LED on
    delay(100);
    digitalWrite(LEDPin, HIGH);   // set the LED on
    delay(100);
   }
}
void loop(){
 /*
   while(analogRead(buttonPin) > 250) continue;
   delay(1000);

   while(1)
   {
    digitalWrite(LEDPin, LOW);   // set the LED on
     delay(1000);
     digitalWrite(LEDPin, HIGH);   // set the LED on
     delay(1000);
   }*/
 /*
 while(1)
 {
 delay(1000);
  Open_Gripper(gripper2);
  delay(1000);
  Close_Gripper(gripper2);

  delay(1000);
 }

  Open_Gripper(gripper2);
  delay(1000);
  Close_Gripper(gripper2);
  */
```

```
 // delay(1000);
//   moveArm(arm1, arm1_270);
/*
  delay(4500);
  moveArm(arm2, arm2_90);
  delay(4500);
  moveArm(arm2, arm2_270);
  */
  /*
  Close_Gripper(gripper1);
  delay(1000);
  Close_Gripper(gripper2);
  delay(1000);
  */
  delay(2000); //delay after power on
  Open_Gripper(gripper1);
  delay(500);
  Open_Gripper(gripper2);
  delay(1000);
  moveArm(arm1, arm1_90);
  delay(1000);
  moveArm(arm2, arm2_270);
  delay(5000);

  while(analogRead(buttonPin) > 250) continue;  //wait for start button

  blinkLED();

  digitalWrite(LEDPin, LOW);   // set the LED on
  delay(1000);

  Close_Gripper(gripper1);
  delay(500);
  Close_Gripper(gripper2);
  delay(1000);

  while(analogRead(buttonPin) > 250) continue;  //wait for start button

  blinkLED();

    //start loop #1
    Open_Gripper(gripper2);
  delay(2000);
  moveArm(arm1, arm1_270);
  delay(1000);
  moveArm(arm2, arm2_90);
  delay(5000);
```

```
  Close_Gripper(gripper2);
  delay(5000);


  //start loop #2
  Open_Gripper(gripper1);
  delay(2000);
  mSetAccel(arm2 ,3);
  delay(100);
  moveArm(arm2, arm2_270);
  delay(1000);
  mSetAccel(arm2 ,3);
  delay(100);
  moveArm(arm1, arm1_90);
  delay(5000);
  Close_Gripper(gripper1);

  //start loop #3

  Open_Gripper(gripper2);
  delay(2000);
  moveArm(arm1, arm1_270);
  delay(1000);
  moveArm(arm2, arm2_90);
  delay(5000);
  Close_Gripper(gripper2);
  delay(5000);

  //start loop #4
  Open_Gripper(gripper1);
  delay(2000);
  mSetAccel(arm2 ,3);
  delay(100);
  moveArm(arm2, arm2_270);
  delay(1000);
  mSetAccel(arm2 ,3);
  delay(100);
  moveArm(arm1, arm1_90);
  delay(5000);
  Close_Gripper(gripper1);

  //start loop #5

  Open_Gripper(gripper2);
  delay(2000);
  moveArm(arm1, arm1_270);
  delay(1000);
  moveArm(arm2, arm2_90);
  delay(5000);
  Close_Gripper(gripper2);
  delay(5000);

  while(1);
}
```

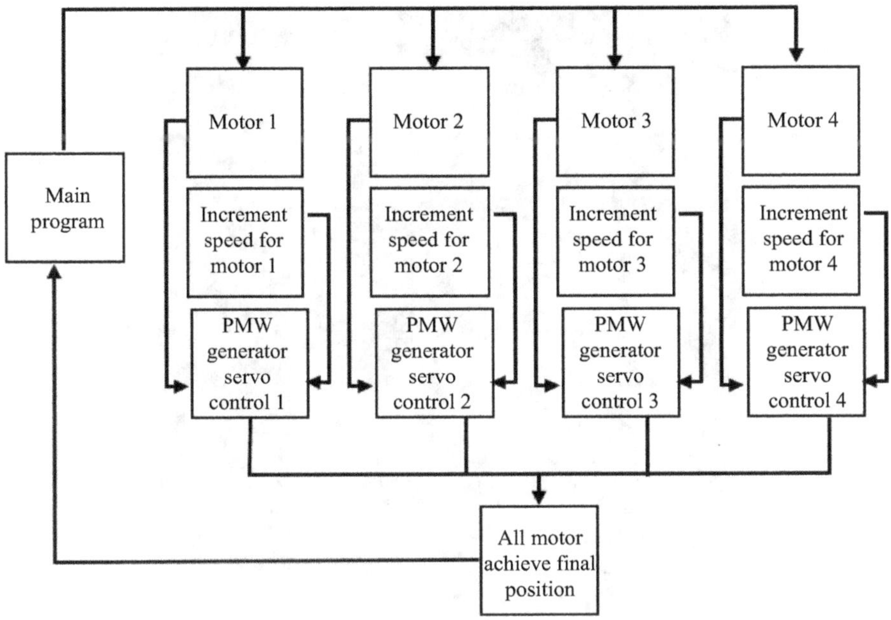

Figure 6.65 Flow of servo control logic

To accomplish the task of climbing, the design of the system includes the control of four motors as shown in Figure 6.65. To coordinate the functionality of these components, an electronic circuit was created. The execution codes were written in C programming language with a series of task-achieving behaviors. With the use of microcontroller and robot programming, the control logic can be easily modified. The control logic starts by sending a signal to the robot to start. The next move involves positioning of the robot onto the pole. Once the robot is on the pole, the timer will count to carry out the next movement to release the lower gripper and rotate the body unit it reaches 180°. Another signal will close the gripper until it reaches the top position. The whole process repeats until it reaches the end of the path as shown in Figure 6.55.

6.3.3 Testing

The robot was tested on the real pole to validate the algorithm and gripping force. The arms produced sufficient force to rotate the body of the robot while climbing the pole. The initial design of the grippers did not provide a firm grip. Hence, it was unable to withstand the robot weight and motion. However, the design of the grippers was modified. There is a need to use more sensors to allow smooth operation of the robot in an uncertain environment. The completed prototype (see Figure 6.66) and testing of the robot can be seen in the following video link.

- Sloth bear-inspired pole-climbing robot: http://youtu.be/4tXWArbzPfw

Figure 6.66 Gripping motion of the robot

References

[1] Wilcher, D. (2012). Learn Electronics with Arduino. Springer, New York.

[2] Virr, L.E. (1987). 'Role of electricity in subsea intervention', Physical Science, Measurement and Instrumentation, Management and Education – Reviews, *IEE Proceedings A*, 134(6), pp. 547–576.

[3] Jiarong, Z., Gang, Q., and Can, W. (2012). Computational Intelligence and Design (ISCID), 2012 Fifth International Symposium on Computational Intelligence and Design, Hangzhou, China, 28–29 October 2012.

Chapter 7
Raspberry Pi-embedded system design

7.1 Fouling detection system

The details in this section can be found in [1]. The colonization of fouling organisms on marine structures has been a perceptual phenomenon in the marine industry. Commonly found fouling organisms include barnacles and algae that tenaciously grip onto marine structures such as the ship's hull, columns of offshore structures, and marine debris. Fouling activity can take place immediately especially when a vessel is stationary. These organisms form a layer of increased surface roughness that would lead to a turbulent flow layer along the surface of the hull, creating a viscous effect, and thereby influencing the tremendous effort by a naval architect to design an ideal hull form. By reducing frictional resistance generated by the hull flow-through seawater, a ship owner can save on additional fuel consumption since frictional resistance constitutes up to 90% of the total drag force. Acquiring the same speed with additional fuel consumption or using the same amount of fuel input that results in a lower ship's speed and undesirable situations that a ship with a fouled hull condition will experience. Slow-speed vessels with a deep draft where greater part of the hull surface gets in contact with seawater tend to attract the habitation of fouling organisms as fast-moving vessels create a hydrodynamic force that affects the capability to attach themselves onto the hull surface [2].

Transportation of cargo via shipping, being the cheaper alternative, accounts for approximately 80% of the global trade [3]. Such demand for sea trade relates to shipping owners planning on cheaper shipping methods to attain higher profitability. Thus, monitoring of fouling conditions is essential to prevent delays. Researcher in [4] conducted a study to analyze fuel efficiency of 44 general cargo ships using Automatic Identification System. Half the fleet was loaded with 7,700 deadweight tonnage while the other half was loaded with 12,700 deadweight tonnage. The result shows that the optimum speed range is present when the ship is propelling at 7.8 to 9.7 knots.

During a voyage, a vessel can be subjected to adverse weather condition that is not part of predetermined factors. On the other hand, proper loading of cargo and fouled hull condition can be monitored to improve the fuel efficiency before a vessel departs. It is important to emphasize on proper drydocking schedule to prevent the accumulation of fouling organism over time. Some ship owners may neglect the importance of such maintenance that causes high fuel consumption and longer voyages and in most cases when the fouling activity is overwhelmed, operational costs will rise.

Fouling rank developed by the researchers in [5] enabled a quantification method for a visual estimate of benthic cover. This method is developed by capturing five random images from the surface to the keel of the yacht and visually judges the percentage of fouling. Table 7.1 shows the fouling rank developed by the researchers.

An experiment was conducted on a Japanese destroyer to determine the consequences of fouling based on various durations out of the dock and varying the ship speed. The result shows that more fouling tonnage is accumulated when a longer duration is spent at sea when the destroyer is traveling at a lower speed. Fouling on the ship's hull had been of great concern, and numerous precautions had been formulated to prevent the occurrence. One of such methods in the prevention of fouling is by using anti-fouling (AF) systems. Researchers [6] used environmental-friendly foul release coatings that successfully reduced fuel consumption on a 33-knot aluminum catamaran by 12%. However, the use of AF systems is governed by regulations, and such applications on the hull are limited.

Monitoring the condition of the hull cannot be performed until a vessel is scheduled for dry-docking. Using image-processing methods [7] one can observe the growth rate of barnacles throughout the year by counting the number of pixels belonging to barnacles on a captured image. The result showed that during warmer temperatures of the year, the percentage of the growth of barnacles increased. An intelligent surveillance system was designed [8] using computer vision (CV) by obtaining the hue-saturation value to identify the types of fouling located on the ship's hull. The method experimented on 40 fouling images extracted from the Internet gave an accuracy of 79.97%. The most commonly found macrofouling organisms are the acorn barnacles. They are mostly hermaphroditic invertebrates capable of multiplying themselves rapidly and usually occur in overcrowding situations encrusted on the surface marine structures. The adhesion strength of barnacles is so strong that expensive hull surface treatments are needed for removal.

This section focuses on the method to identify the fouling organisms and severity of the fouling condition on the ship hull using machine learning. It allows the ship to identify the location and the fouling automatically without the need of hiring divers.

Table 7.1 Fouling rank

Rank	Description	Visual estimate of fouling cover
0	No visible fouling	0%
1	Slime fouling only	0%–1%
2	Light fouling	1%–5%
3	Considerable fouling	6%–15%
4	Extensive fouling	16%–40%
5	Very heavy fouling	41%–100%

7.1.1 System design and architecture

The design of the system architecture has been through many iterations with the aim to design a system that is portable, yet cheap and user-friendly, cardinally eliminating the potential risk taken by hired divers to observe fouling. The proposed system consists of a portable mounting device designed to be mounted on the bulwark of the offshore support vessel with variable extension rods from the gunwale to the design waterline to observe fouling conditions. Observation method used in this system is the Raspberry Pi microcontroller (see https://www.raspberrypi.org) and its camera module. The end of the rod mounts a Raspberry Pi camera enclosed in a watertight enclosure that will use CV to identify the location of fouling. The system will serve the following functions:

1. Only one man is needed to mount and operate the system.
2. Capture an image based on the location of interest.
3. Highlight the area of fouling, calculate fouling percentage, and show fouling rank.
4. The GUI that allows the user to observe fouling and records the data.

The use of Raspberry Pi in this system serves as a low-cost method of tackling the detection and identification of fouling organisms. Python and OpenCV are modules installed in this Linux®-based microcontroller. See the following websites for some information.

- http://www.pyimagesearch.com/2015/10/26/how-to-install-opencv-3-on-raspbian-jessie/
- http://docs.opencv.org/2.4/doc/user_guide/ug_traincascade.html

The fouling information needed by a shipowner can be known through a wireless SSH connection by connecting wireless-enabled device such as a laptop in the same network as the Raspberry Pi. The software and hardware used are provided in Tables 7.2 and 7.3, respectively.

The prototype was designed to accommodate placement anywhere along the bulwark of a vessel. As different parts of the hull are designed to have a different curvature, length variations of the prototype components are necessary. Figure 7.1

Table 7.2 Software components

Software	Purpose
Solidworks 2013	System design
Raspbian OS	Operating system
TightVNC Viewer	Remote control
Adobe Photoshop CS 5	Image adjustment
EasyPHP 3.0	Webserver, MySQL
Notepad ++	Programming tool
Python 2.7	Programming tool
WinSCP	File transfer

Table 7.3 Hardware components

Hardware	Quantity	Purpose
Winch	1	Hoisting/Lifting
Suction cup	2	Support device
Aluminum rod	5	Extension
Prototype design	1	Main structure
Raspberry Pi 2	1	Microcontroller
Camera module	1	Capture device
Wifi USB dongle	1	Provide SSH
8 GB MicroSD card	1	Storage
Portable charger	1	Power

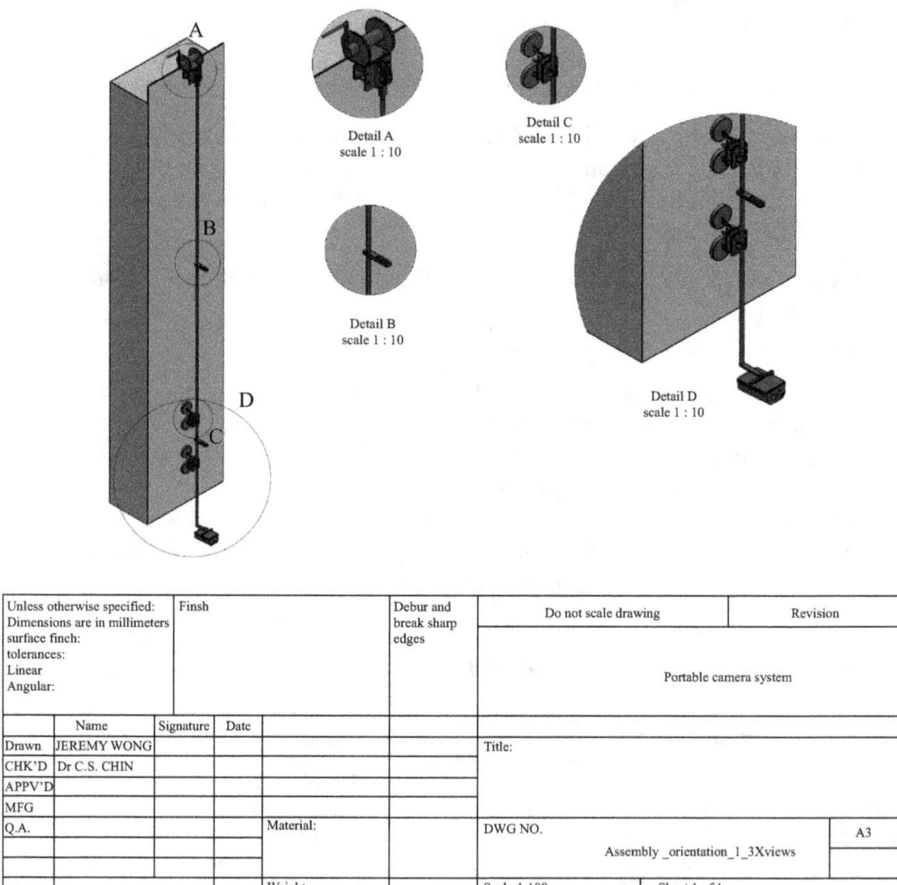

Figure 7.1 Proposed prototype design orientation 1

shows the design ideas of the fouling system. The final design comprises two different orientations for implementation of the system. The proposed prototype design for orientation 1 is designed for mounting on a mid-ship section of a hull or on a position on the hull where hull curvature is minimal. On the contrary, proposed prototype design orientation 2 is designed to suit installation at sections of the hull form with greater curvature. The design selected for prototype testing is orientation 1. The prototype was designed to accommodate placement anywhere along the bulwark of a vessel. As different parts of the hull are designed to have a different curvature, length variations of the prototype components are necessary. Two different orientations were catered for implementation of the system. The proposed prototype design for orientation 1 is designed for mounting on a mid-ship section of a hull or on a position on the hull where hull curvature is minimal. On the contrary, proposed prototype design orientation 2 is designed to suit installation at sections of the hull form with greater curvature. The design selected for prototype testing is orientation 1.

The proposed design for orientation 1 is designed for mounting on a mid-ship section of a hull or on a position on the hull where hull curvature is minimal. The proposed design will be mounted on the bulwark plate of the vessel using a clamping mechanism. A pivot that is attached to the clamping mechanism will connect five 1 m aluminum rods with the camera being connected at the bottom of the last rod. The camera enclosure at the bottom can be adjusted to adapt to a correct position of the hull. Detail A shows a winch mounted on top of the gunwale clamp that is clamped onto the bulwark plating of the hull. A closer view of Detail A can be seen in Figure 7.2.

The top part of this system includes a winch for hoisting up the rods and camera. The winch chosen is a Maxpull GM-1-SI 100 kg capacity manual hand winch with an automatic brake applied when the load is lifted halfway. The handle

Figure 7.2 Close-up of Detail A

Figure 7.3 Maxpull GM-1-SI 100 kg capacity manual hand winch

of the Maxpull GM-1-SI winch will not swing back even if force is not applied. Figure 7.3 shows the winch used in the system.

Detail B shows a clamp secured to a part of the rod that will be attached to the hook of the wire rope connected to the winch. The clamp is called the rod clamp and is designed to accommodate hoisting and lowering of the assembly. Detail C shows the suction cups assembly made up of two Laser DPT-96 Suction Cups capable of securing itself on a metal plate. The assembly is designed to allow allowance due to the curvature of the hull form of the offshore support vessel. The bracket that held on to the rod is designed to restrict movement when the ship is propelling forward, during assembly and dismantling of the system. Figure 7.4 shows the Detail C in the design.

The purpose of the suction cups is to maintain a fixed-fixed beam to allow testing of prototype without damaging the hull. The suction vacuum holds itself in place firmly. It restricts movement and increases portability of the system. However, the use of suction cup may not be usable on all types of vessel.

Proposed prototype design orientation 2 can be seen in Figure 7.5 to suit installation at sections of the hull form with greater curvature. The support arm shown in Detail C allows the extension to support the length of rods. It contains a series of holes to accommodate the extension in accordance to the hull curvature, a screw, and a nut placed in the correct hole to maintain the required length. In comparison with the two designs, design orientation 1 is more reliable for a

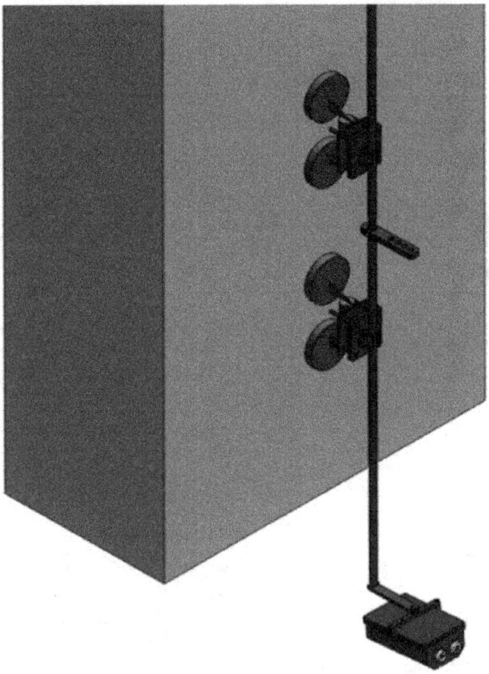

Figure 7.4 Close-up of Detail C

prototype design than design orientation 2 due to the length of the extension. The system is designed to require single man to operate. Preinstallation of the suction cups onto the ship hull is necessary during the voyage. The preinstallation work includes accurate alignment of suction cups where the handles of the suction cups must be perpendicular to the waterline. This installation should be done when the vessel is anchored in port. Installation of the gunwale clamp onto the bulwark plate must be collinear to the position of the suction cups.

The assembly involves lowering of the camera and aluminum rods part by part. Dis-assembly requires hoisting of the camera assembly and removal of the parts. Both procedures will be assisted by the Maxpull GM-1-SI 100 kg capacity manual hand winch with a 12 m length of 5 mm diameter wire. Figures 7.6 and 7.7 show the entire assembly process. The process repeats for the next four lengths of rods. When the last rod with a connector bracket to the gunwale clamp is aligned with the gunwale, the operator will lower the entire length of the rod to the point that connected to the gunwale clamp with an M8 screw and nut. The angle of the rods should be at least 15° away from the hull. The final step is to lower further until the rods meet the pole bracket of the suction cup assembly to hold the entire assembly firmly.

It is recommended to purchase the "Raspberry Pi Model B Starter Kit" as it comes with preinstalled NOOBS. The Wi-Fi adapter is needed to connect to a working terminal using remote client "TightVNC Viewer"(see http://www.penguintutor.com/linux/tightvnc). The items purchased from Adafruit enables GPS

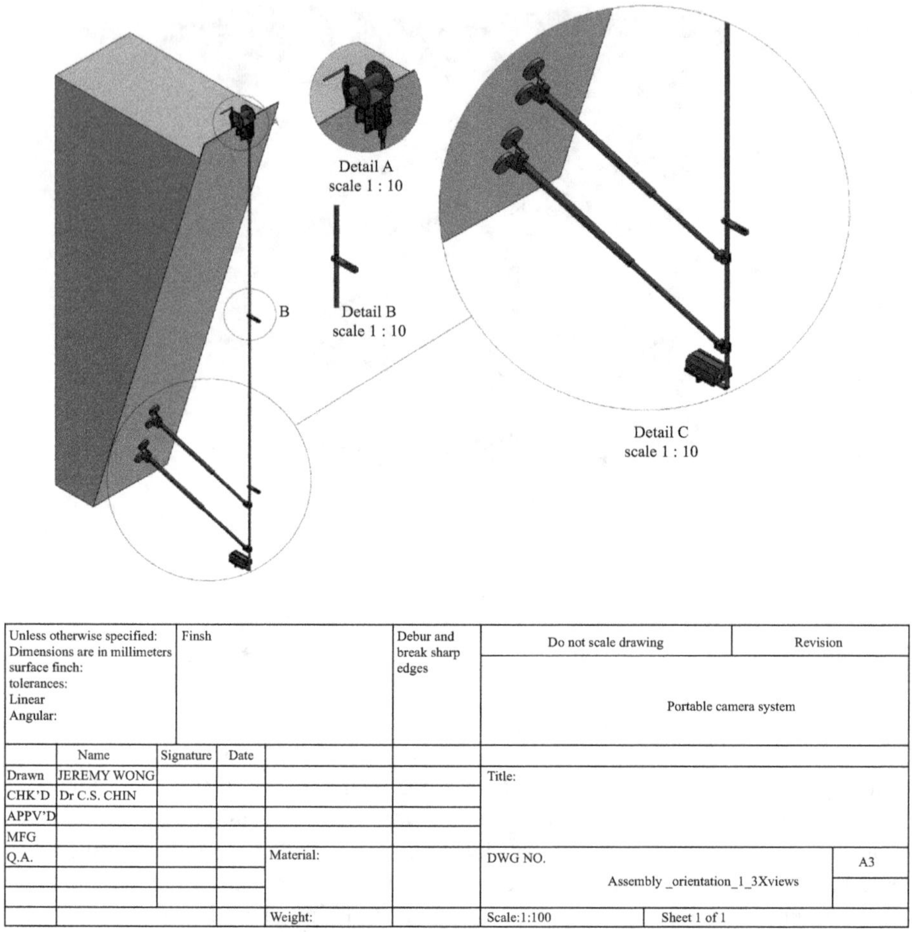

Figure 7.5 Proposed prototype design orientation 2

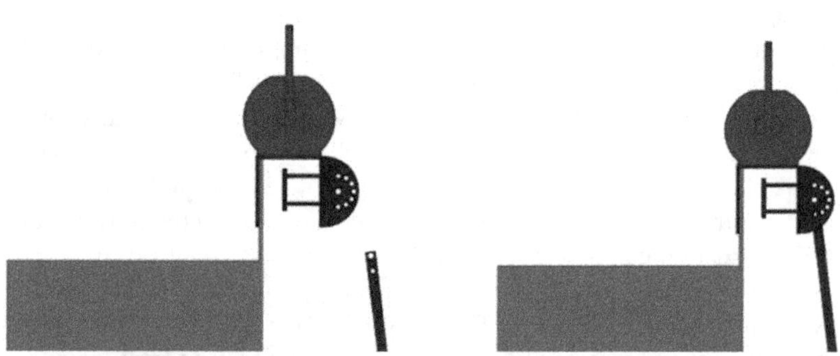

Figure 7.6 Connection of the pole assembly with gunwale clamp

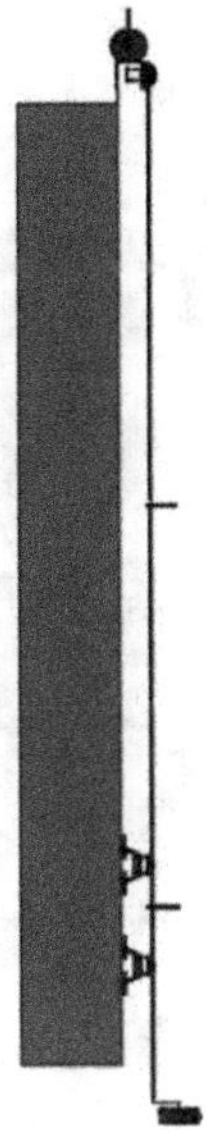

Figure 7.7 Proper alignment of the system

Table 7.4 List of useful sites

Website	Purpose
https://www.raspberrypi.org	Information related to Raspberry Pi
http://www.pyimagesearch.com/	Information on using OpenCV Python
http://www.robopapa.com/	Information on using OpenCV Python
http://www.9lessons.info/2012/09/multiple-file-drag-and-drop-upload.html	Information on PHP Drag and Drop user upload
https://pythonprogramming.net/haar-cascade-object-detection-python-opencv-tutorial/	Information related to Haar cascade
http://code.tutsplus.com/	Information on setting up the database server

connectivity to the Raspberry Pi. The direct power supply will not be present on board the vessel for testing. Thus a 5,000 mAh power bank will be used to supply power to the Raspberry Pi. However, for programming and testing purposes, the direct power supply will be used. Table 7.4 shows a list of useful sites used in this section.

7.1.1.1 First-boot procedures

To begin using the Raspberry Pi (see Figure 7.8), we need to install an operating system. Download NOOBS from the Raspberry Pi website as shown below.

Step 1: Download the NOOBS package from the website https://www.raspberrypi.org/help/noobs-setup/

Figure 7.8 Layout for programming on Raspberry Pi

Figure 7.9 Raspberry Pi MicroSD slot

Step 2: Format the SD card by download the tool from the SD Association's website.

Step 3: Unzip the downloaded NOOBS package into the SD card.

Step 4: Plug in keyboard and mouse into the Raspberry Pi USB socket, connect to a monitor using an HDMI cable. Insert the SD card into the Raspberry Pi as shown in Figure 7.9.

Step 5: The Raspberry Pi will boot with a list of different operating systems to install. Raspbian OS is chosen for the Raspberry Pi. If prompt for user credentials and password, the default credentials are pi and raspberry, respectively.

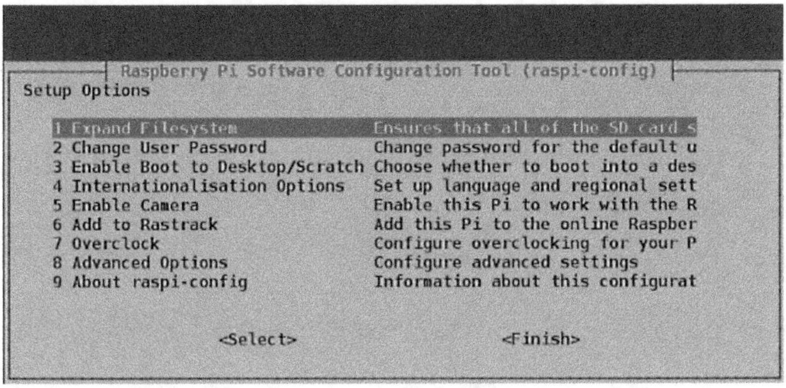

Figure 7.10 Raspberry Pi configuration menu

Step 6: When the installation is done, the monitor should appear as shown in Figure 7.10. The configuration screen can be accessed later in the terminal using:

```
sudo raspi-config
```

Step 7: Choose Option 3: Enable Boot to Desktop/Scratch and hit enter. It will lead to the "Boot Option" window shown in Figure 7.11. Select the second option, that is, "Desktop Log in as user 'pi' at the graphical desktop" and hit the enter button. The Raspberry Pi will boot into a terminal with no GUI options, and the desktop will not be shown during start-up.

Step 8: Select Finish using the keyboard and the system should reboot as shown in Figure 7.12.

Step 9: To update to the latest firmware by entering:

```
sudo rpi-update
```

7.1.1.2 Installing Raspberry Pi cam

To enable the camera, we must physically install the camera module onto the board.

Step 1: Place the initial setup as shown Figure 7.13. The camera module silver ports should be facing up.

Step 2: Slide the cables into position with the silver ports facing the HDMI ports. Snape the bracket back to its original position. The cables should slot firmly as shown in Figures 7.14 and 7.15.

For the software installation, follow the below steps.

Step 1: Boot up Raspberry Pi.

Step 2: Open the Terminal and run

```
sudo raspi-config
```

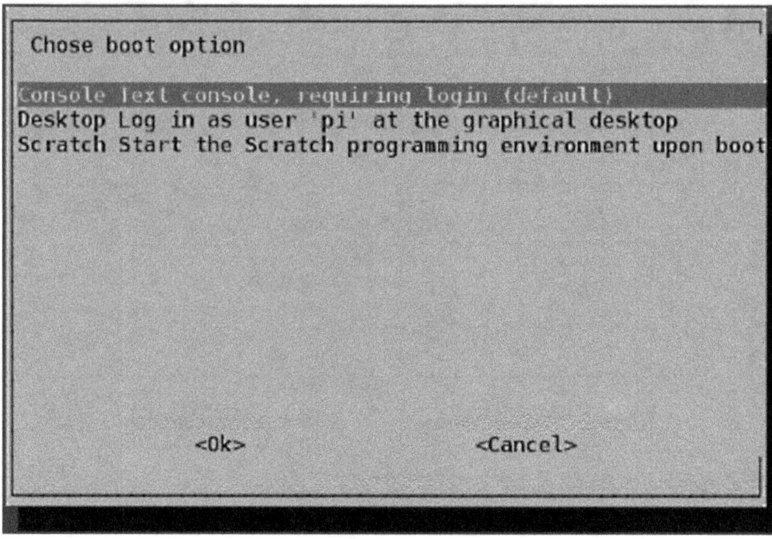

Figure 7.11 Raspberry Pi boot option

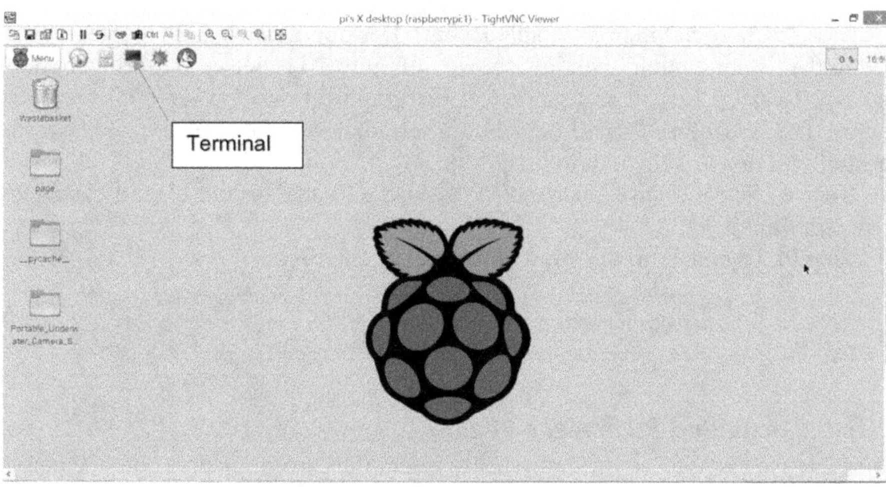

Figure 7.12 Raspberry Pi desktop

If the "camera" option is not listed, run

```
sudo apt-get update
sudo apt-get upgrade
```

Run the configuration command

```
sudo raspi-config
```

You should see the "camera" option now.

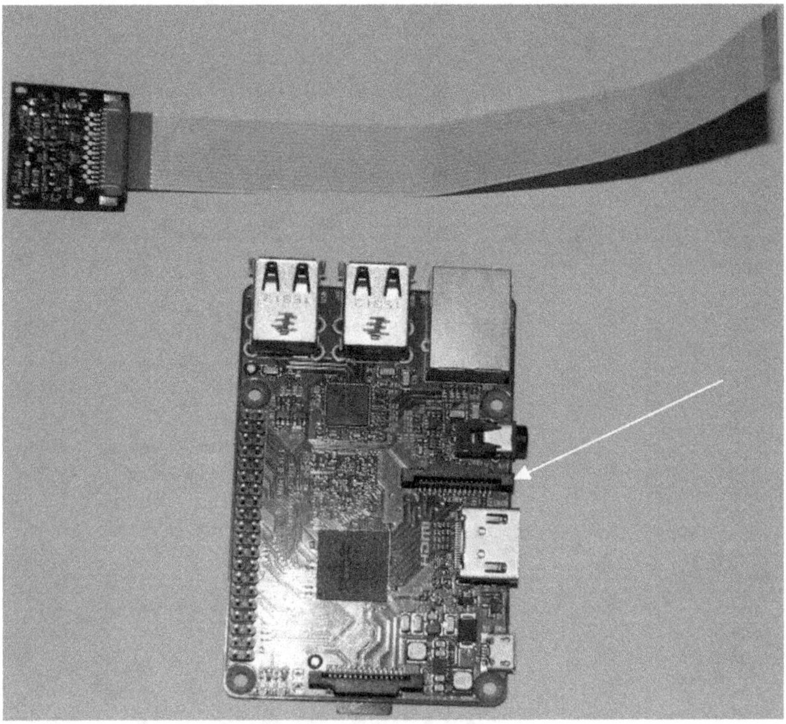

Figure 7.13 Initial setup assembly

Figure 7.14 Raspberry Pi camera module setup

Step 3: Navigate to the “camera” option and enable it (see Figure 7.16). Then select <Finish> and reboot the Raspberry Pi. To capture an image in jpeg format, type the following into the terminal:

```
raspistill -o image.jpg
```

Note that the “image” is the name of the image.

Figure 7.15 Raspberry Pi with Pi camera module

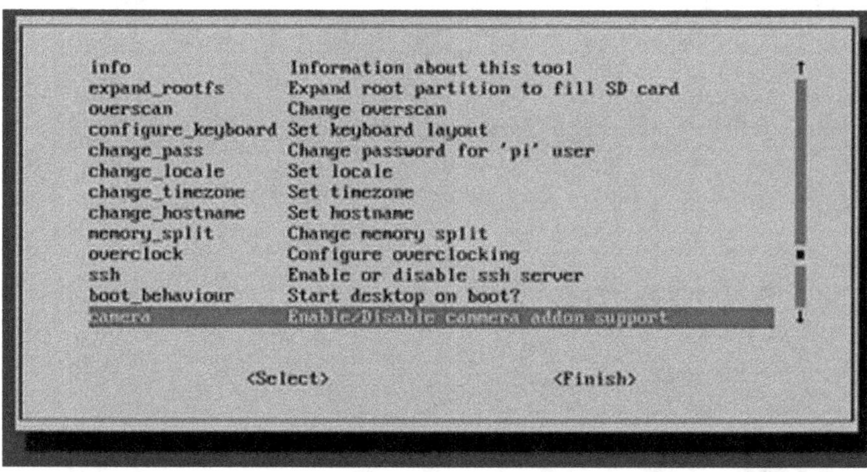

Figure 7.16 Enabling Raspberry Pi camera

7.1.1.3 Setting up Wi-Fi to enable remote control for Raspberry Pi

Virtual network computing (VNC) is a graphical desktop-sharing system that allows remote control of the desktop from any computer. In this setup, the Raspbian OS is accessible to make necessary changes in Raspberry Pi. The following steps are used for the VNC installation if it is missing. Note that the latest Raspberry Pi comes with VNC.

Step 1: Open up the terminal and install the TightVNC package

```
sudo apt-get install tightvncserver
```

Step 2: Run TightVNC Server which will prompt you to enter a password and an optional view-only password: tightvncserver.

Step 3: Type

```
vncserver :1
```

Step 4: Install the VNC client:

```
sudo apt-get install xtightvncviewer
```

Step 5: To find out the IP address of the Raspberry Pi:

```
ifconfig
```

Record down the "inet addr" that will be used as local IP address.

Step 6: On the laptop connected to the Raspberry Pi, download the VNC software from this website:

http://www.tightvnc.com/download.php

Step 7: Install the software.

Step 8: Once setup is completed, open the TightVNC client and type in the IP address onto the "Remote Host" box and add port ":1" behind the IP address. Type in the password chosen in the initial setup and click "connect". Once completed, the Raspbian Desktop should appear.

7.1.1.4 Remotely copy files to and from Raspberry Pi

FileZilla uses SSH File Transfer Protocol to transfer files via Raspberry Pi

Step 1: Download FileZilla from this link

https://filezilla-project.org/

Step 2: Load the program as shown in Figure 7.17.

Step 3: Click on File -> Site Manager looks like Figure 7.18.

Step 4: Input the Host and Port information. The protocol should be changed to SFTP - SSH File Transfer Protocol and login type should be standard. Type the user and password (User: pi, Password: Raspberry) then click connect. At this point, the files in the working desktop and Raspberry Pi directory can be seen. WinSCP can be used instead of FileZilla. It can be downloaded from https://winscp.net/eng/download.php

7.1.1.5 Remotely controlling Raspberry Pi terminal

PuTTy telnet client allows the Raspberry Pi terminal to control remotely by the computer through the same SSH connection. It can be downloaded from: http://www.chiark.greenend.org.uk/~sgtatham/putty/download.html

It can be used if the automatic start up for tightvncserver did not work.

7.1.1.6 Set up database and PHP webpage

Apache is a popular web server application to install on the Raspberry Pi. Apache can serve HTML files over HTTP, and with additional modules, it can serve dynamic web pages using scripting languages such as PHP.

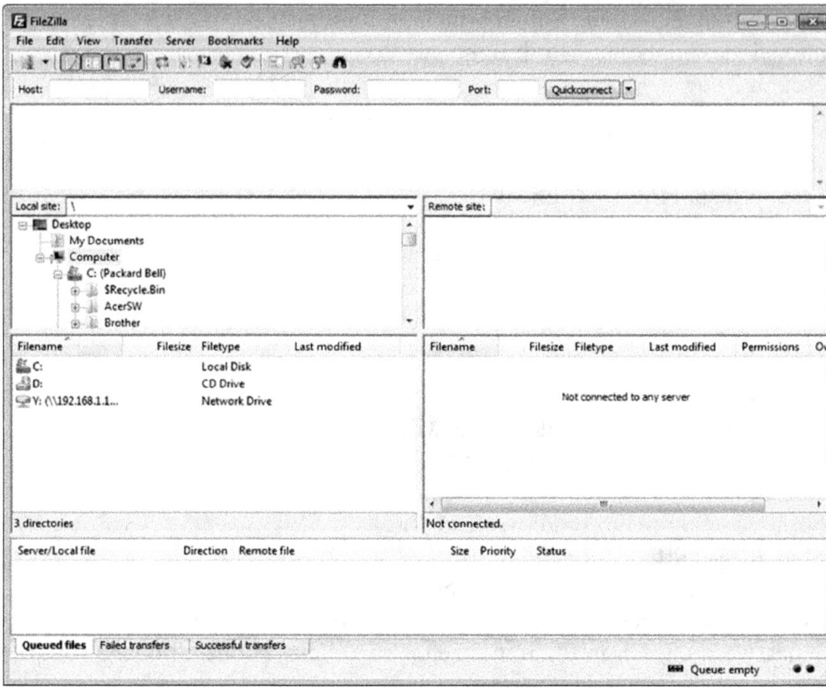

Figure 7.17 FileZilla GUI

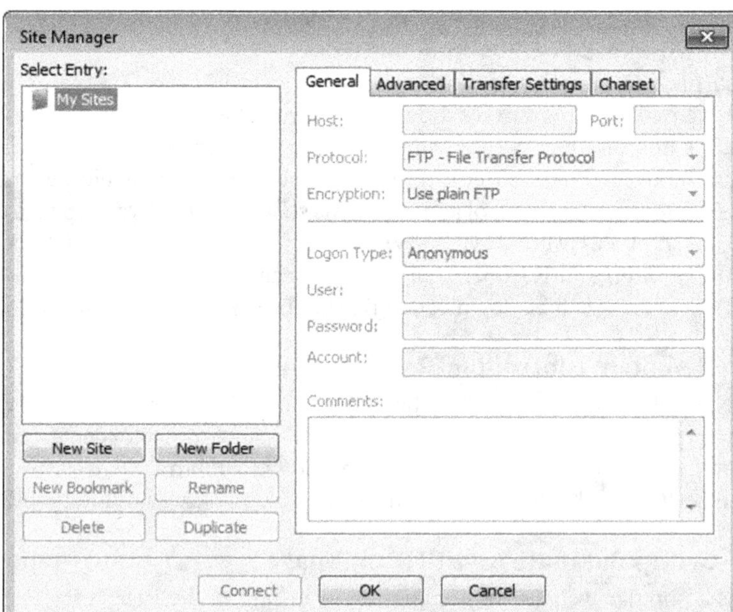

Figure 7.18 Input SSH information for FileZilla

Step 1: Install Apache 2

```
sudo apt-get install apache2 –y
```

Step 2: By default, installing Apache will create a test HTML file in the web folder (www). The default webpage can be accessed by http://localhost/ or http://192.168.1.16 (depending on Pi's IP address)

Step 3: Access either the web page in step 2 via a browser. The Apache2 Debian Default Page should appear.

The HTML file is the file that is being manipulated to change the webpage design. It is located at */var/www/html/index.html* in the Raspberry Pi directory. By default, the HTML directory and index.html are both owned by user "root." Change the owner to get permissions by typing:

```
sudo chown pi: index.html
```

To allow Apache server to process PHP files, install PHP5 and the PHP5 module for Apache. Type the following command to install:

```
sudo apt-get install php5 libapache2-mod-php5 –y
sudo rm index.html
sudo nano index.php
```

Create some PHP content to display:

```
<?php echo "hello world"; ?>
```

Accessing the webpage will show hello world printed on a white page now. PHP5 and Apache2 are now installed. A database management system is set up using MySQL to store and maintain data. Since extensive database of images is required for training to classify images, this web server is created to invite contributors to share images to improve the accuracy of classification. PhpMyAdmin on the contrary provides an easy interface for users to administrate.

- PhpMyAdmin: https://www.siteground.com/tutorials/phpmyadmin/phpmyadmin_create_database.htm
- Database: http://www.onlinebuff.com/article_step-by-step-to-upload-an-image-and-store-in-database-using-php_40.html
- Database: http://www.sitepoint.com/php-amp-mysql-1-installation/

7.1.1.7 Installing MySQL and Easyphp and PhpMyAdmin on computer

Easyphp is a webserver and capable of allowing the user to host whatever the user needs on the website.

Step 1: Download Easyphp application: http://www.easyphp.org/

After Easyphp is installed, Apache server, MySQL, and PhpMyAdmin will be completely configured and ready to work.

Step 2: In the directory "C:\Program Files\EasyPHP 3.0" on the PC, there is a subfolder "www" that contains the PHP pages. The subfolder "MySQL/data" is used to contain MySQL databases.

7.1.1.8 Installing MySQL on Raspberry Pi

Step 1: Install MySQL server package

```
sudo apt-get install MySQL-server
```

Step 2: Prompt for password.

Input password for root user, this field can be left empty so that no password is required.

Step 3: To allow interaction between MySQL and Python, we must install the Python bindings:

```
sudo apt-get install python-mysqldb
```

Step 4: Installation of PHPMyAdmin package to enable an easy interface to work with:

```
sudo apt-get install phpmyadmin
```

Step 5: Select Apache2 if prompt for the type of server to run.

Step 6: Select <Yes> for configuring the database for phpmyadmin. Set a password for PHPMyAdmin.

Step 7: Setup Apache to include PHPMyAdmin installation, enter:

```
sudo nano /etc/apache2/apache2.conf
Include /etc/phpmyadmin/apache.conf
```

Save and exit by pressing CTRL+X and then <Y>.

Step 8: Open a new terminal and restart Apache2 service by entering the following command:

```
sudo /etc/init.d/apache2 restart
```

Step 9: On the computer, access PHPMyAdmin from a browser. Input user and password and the page will appear as shown in Figure 7.19.

To allow external users to contribute, we need to setup port forwarding and disable the firewall. To set up port forwarding, it will vary across different routers. For a Singtel Aztech DSL605EU router, we follow the following steps.

Port to forward:

Name	Port
PHPMyAdmin	80
MYSQL	3306

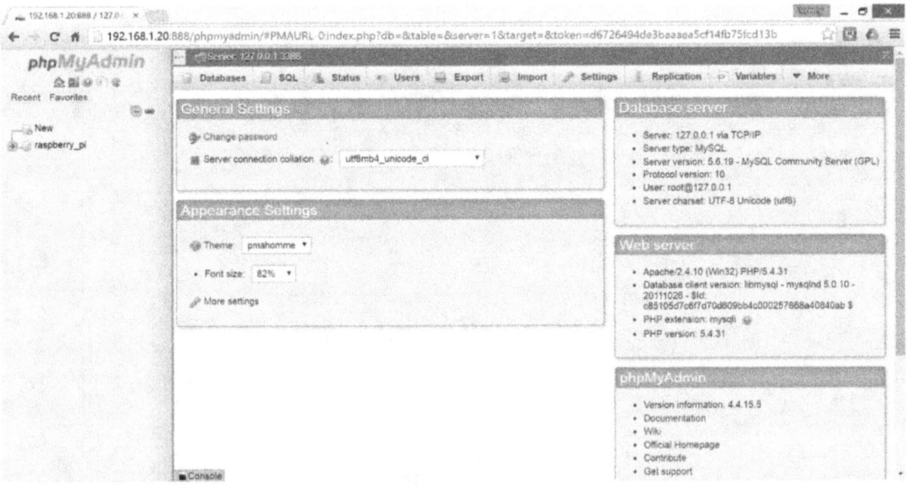

Figure 7.19 PHPMyAdmin homepage

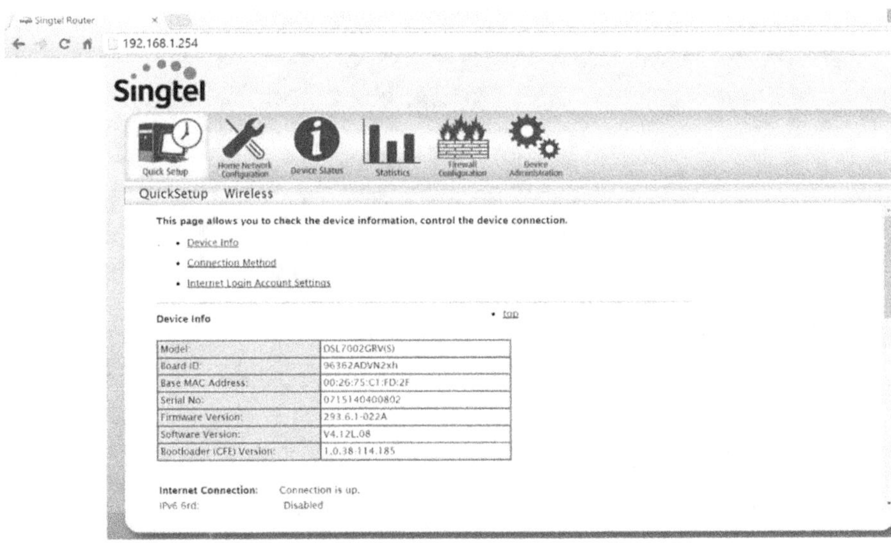

Figure 7.20 Port Forwarding step 1

Step 1: Input the internal IP address of your router on a browser. A page similar to Figure 7.20 will appear.

Step 2: Go to Firewall Configuration -> Port Forwarding. Input the following details as shown in Figure 7.21.

Step 3: Click Save/Apply and port forwarding setup is done.

7.1.1.9 Creation of database

Image database (see Figures 7.22 and 7.23) was created for the user to contribute the fouling images.

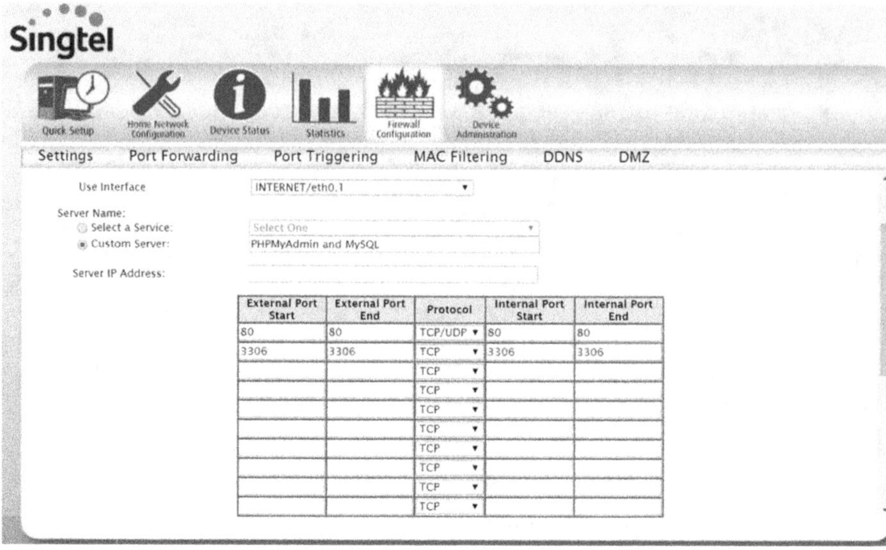

Figure 7.21 Port forwarding Step 2

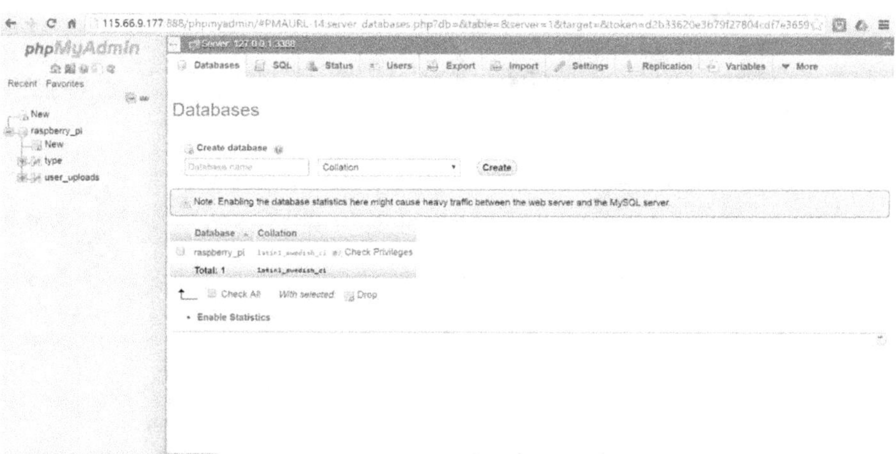

Figure 7.22 Creation of database using PHPMyAdmin

Step 1: Go to http://115.66.9.177:888/phpmyadmin (Personal IP address followed by /phpmyadmin).

Step 2: Select <New> from the control panel on the screen and enter a database name. Click create.

Step 3: A database will consist of tables and columns with different categories.

Step 4: The types of field are shown in Figure 7.24.

Table 7.5 shows the types of fields used in this database.

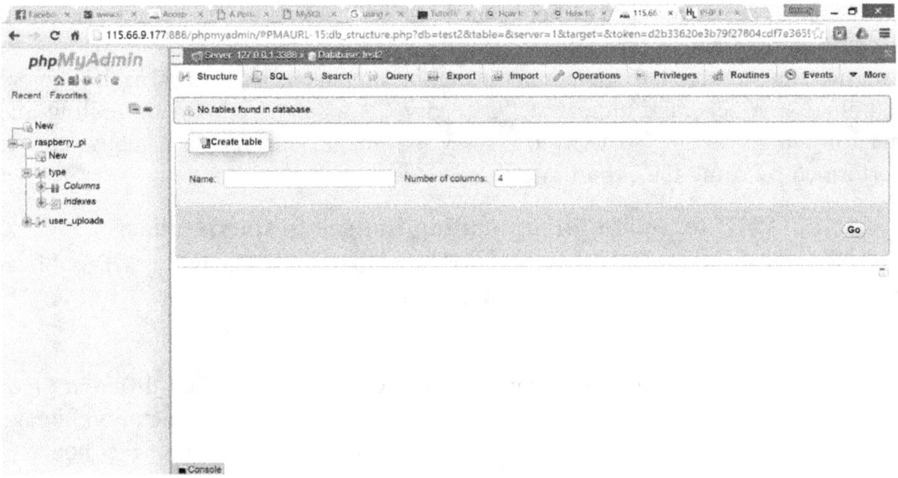

Figure 7.23 Creating tables in database

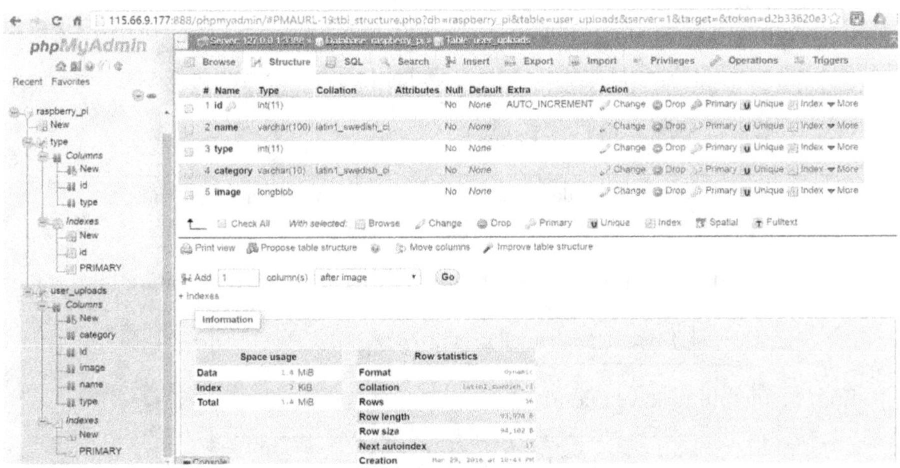

Figure 7.24 Database layout

Table 7.5 Fields used in a database

Int	A 4-byte integer
Varchar	A variable length of the string. Bracket indicates a maximum number of characters allowed.
Longblob	Used for images that are lesser then 4 GiB stored with a 4-byte prefix indicating the length of the value.

Step 5: Name the types of fields needed. "Id" is used to number the images. It is set to autoincrement for more images. "Name" is to indicate the specific type of fouling. "Type" refers to the common fouling organisms such as barnacles, green/red algae, mussels, and tunicates. "Category" is a selection of microfouling and macrofouling. "Image" is used to display the image. Backup the database can be performed by using the export tab.

7.1.1.10 PHP template for uploading images to the database

A PHP template is used as the website for users to upload. The link is accessible at http://115.66.9.177:888/ when the server is up.

7.1.1.11 OpenCV

OpenCV is a CV and machine learning library. It includes 2,500 different algorithms that are capable of performing tasks, namely, face recognition, image stitching, and many more. Installation of OpenCV requires a total of few hours.

Step 1: Install development tools

```
$ sudo apt-get install build-essential git cmake
```

Step 2: Install image I/O packages that allow us to load image file formats such as JPEG and PNG

```
$ sudo apt-get install libjpeg-dev libtiff5-dev libjasper-dev libpng12-dev
```

Step 3: Install Video I/O packages

```
$ sudo apt-get install libavcodec-dev libavformat-dev libswscale-dev libv4l-dev
$ sudo apt-get install libxvidcore-dev libx264-dev
```

Step 4: Install the GTK development library to display images for GUI interfaces

```
$ sudo apt-get install libgtk2.0-dev
```

Step 5: Install added dependencies

```
$ sudo apt-get install libatlas-base-dev gfortran
```

Step 6: Install the Python 2.7 and Python 3 header files to compile OpenCV and Python bindings

```
$ sudo apt-get install python2.7-dev python3-dev
```

Install OpenCV 3.0 source code

Step 1: Install OpenCV 3.0

```
$ cd ~
$ wget -O opencv.zip https://github.com/Itseez/opencv/archive/3.0.0.zip
$ unzip opencv.zip
$ wget -O opencv_contrib.zip https://github.com/Itseez/opencv_contrib/archive/3.0.0.zip
$ unzip opencv_contrib.zip
```

Step 2: Set up Python

```
$ wget https://bootstrap.pypa.io/get-pip.py
$ sudo python get-pip.py
```

Step 3: Create Python virtual environment

```
$ mkvirtualenv cv
```

Step 4: Install NumPy to compute fast arrays in powerful N-dimension

```
$ pip install numpy
```

Compile and install OpenCV
Step 1: Build file

```
$ workon cv
$ cd ~/opencv-3.0.0/
$ mkdir build
$ cd build
$ cmake -D CMAKE_BUILD_TYPE=RELEASE \
```

Step 2: Compile

```
$ make -j4
```

The "-j4" switch stands for the number of cores to use when compiling OpenCV.
Step 3: Install

```
$ sudo make install
$ sudo ldconfig
```

7.1.2 Programming and system implementation

Figure 7.25 describes the flowchart of the Python program. The first step in this program requires the selection of live video or perspective live video. The perspective live video will convert three-dimensional frames to two-dimensional frames to generate a perspective view. The program considers the keel with a curvature that leads to a false detection by the program. The captured image will use Haar cascade algorithm to classify the types of barnacles. The outcomes of the program will allow the user to observe the density of the image covered by barnacles and the percentage of barnacles in a given area.

7.1.2.1 Haar cascade object recognition

Creating a Haar cascade for Image or Video Object Classification requires the following steps.

1. Collection of "Negative" or "Background" images.
   These images contain the object of interest. The higher the number of images, the higher the accuracy. Around 500 images of acorn barnacles, mostly

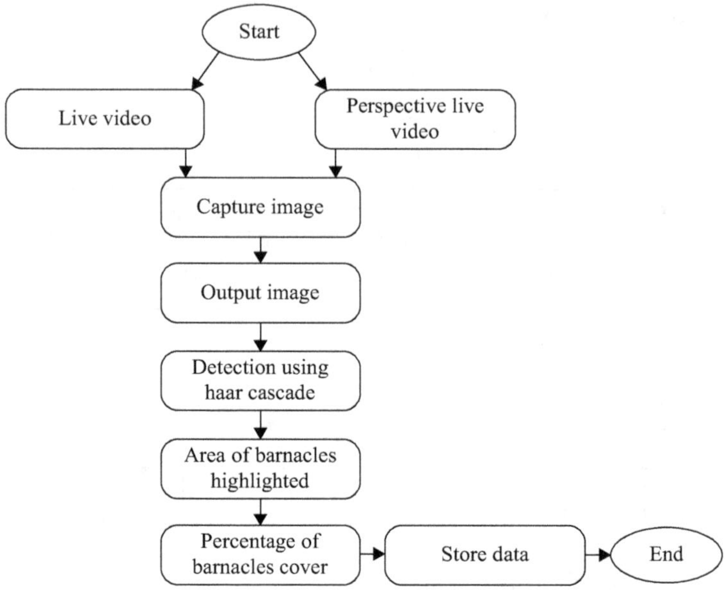

Figure 7.25 Program flowchart

Chthamalus (see http://www.wildsingapore.com/wildfacts/crustacea/other-crust/ciriipedia/euraphia.htm), were taken along the coastline of Singapore using a high-quality digital single-lens reflex (DSLR) camera and 40 best images were manually selected for the training of the classifier.

2. Collection of "Positive" images.
   These images come in different sizes and preferably bigger than a training window size. The greater the number of images, the higher the tendency of accuracy. Haar cascade uses machine-learning algorithms for classification of objects based on the huge bulk of input images provided. It learns from the input patterns and tries to make generalizations based on other images. Around 4,000 negative images were used from the Internet. Negative images are larger than positive images to allow superimposing accurately. Superimposing creates a greater number of positive images by overlaying each cropped image of barnacles on top of each negative image.
3. Creating positive vector files by stitching all "Positives" images together.
   These files contain information on the positive samples used for training. A vector file is generated using OpenCV command line for training. The process for superimposing is repeated for each cropped barnacle.
4. Merging vector files
   A total of 25 vector files were generated. Merging these vector files will generate a final vector file for cascade training.
5. Train the cascade
   The *cascade.xml* will be trained and used to compare input images taken from the Raspberry Pi camera module. The process of the training is illustrated in Figure 7.26.

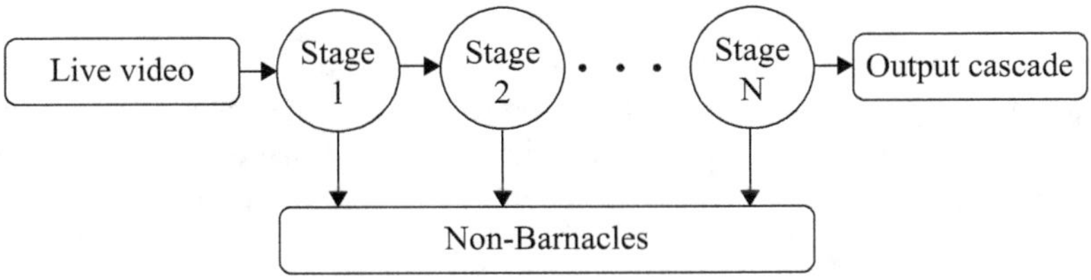

Figure 7.26 Cascade classifier structure

The "positive" and "negative" images need description files to locate the path to each image, along with the number of objects, and their locations (e.g., "bg.txt" file for positive images, "pos.txt" for positive images). Ideally, the path to each image contains rectangular coordinates using exact cropped images of interest. To minimize the training time, training must be conducted on a faster operating system and not on the Raspberry Pi itself. The OpenCV for windows is downloaded from: http://opencv.org/downloads.html

Negative images are downloaded from the Internet using a script for automatic extraction using URLs from image-net.org.

```
import urllib.request
import cv2
import numpy as np
import os

def store_raw_images():
  neg_images_link = '//image-net.org/api/text/imagenet.synset.geturls?wnid=n00523513'
  neg_image_urls = urllib.request.urlopen(neg_images_link).read().decode()
  pic_num = 1

  if not os.path.exists('neg'):
    os.makedirs('neg')

  for i in neg_image_urls.split('\n'):
    try:
      print(i)
      urllib.request.urlretrieve(i, "neg/"+str(pic_num)+".jpg")
      img = cv2.imread("neg/"+str(pic_num)+".jpg",cv2.IMREAD_GRAYSCALE)
      # should be larger than samples / pos pic (so we can place our image on it)
      resized_image = cv2.resize(img, (100, 100))
      cv2.imwrite("neg/"+str(pic_num)+".jpg",resized_image)
      pic_num += 1

    except for Exception as e:
      print(str(e))
```

The script extracts negative images from image-net.org, converts into grayscale, and resizes each image to 100×100 pixels. Each negative image should be larger than positive images to allow superimposing to be performed accurately. A total of 4,000 negative images were used. Some pictures hosted that are not available will show a plain image with nothing on it. Therefore, a script is written to delete these images.

```
def find_uglies():
    match = False
    for file_type in ['neg']:
        for img in os.listdir(file_type):
            for ugly in os.listdir('uglies'):
                try:
                    current_image_path = str(file_type)+'/'+str(img)
                    ugly = cv2.imread('uglies/'+str(ugly))
                    question = cv2.imread(current_image_path)
                    if ugly.shape == question.shape and not(np.bitwise_xor(ugly,question).any()):
                        print('That is one ugly pic! Deleting!')
                        print(current_image_path)
                        os.remove(current_image_path)
                except Exception as e:
                    print(str(e))
```

The next step is to create a descriptor file for negative images. Around 4,000 pictures of flowers and animals were used from image-net.org.

```
def create_pos_n_neg():
    for file_type in ['neg']:


        for img in os.listdir(file_type):


            if file_type == 'pos':
                line = file_type+'/'+img+' 1 0 0 50 50 \n'
                with open('info.dat','a') as f:
                    f.write(line)
            elif file_type == 'neg':
                line = file_type+'/'+img+'\n'
                with open('bg.txt','a') as f:
                    f.write(line)
```

A bg.txt file is created by running the above script. These file contains information on negative images. The next step is to create positive images. Create directory “info” by typing “*mkdir info*” into the terminal. “*opencv_createsamples*” command prepares a training dataset of positive images by superimposing on negative images while manipulating the properties (rotation and intensity) of the positive images to generate more positive images. To create these positive images, the following commands can be used.

```
opencv_createsamples -img test.jpg -bg bg.txt -info info/info.lst -pngoutput info -
maxxangle 0.5 -maxyangle 0.5 -maxzangle 0.5 -num 4000
```

Around 500 images of barnacles were taken along the coastline of Singapore using a high-quality DSLR camera, and 40 best images were manually selected for training the classifier as shown in Figure 7.27.

These images are needed to be converted to bmp format in "opencv_samples()." An open-source software (bulk image converter) is used to convert all samples to bmp format in Figure 7.28 resulting in the following Figure 7.29. The software can be downloaded from: https://sourceforge.net/projects/bulkimageconver/.

Cropping of the region of interest (ROI) was performed using a utility called "Haar training Utility" by a third-party software to ease the cropping process. The utility can be downloaded at: http://www.tectute.com/2011/06/opencv-haartraining.html. Using Opencv "cvtColour" function (see Figure 7.30), these images were converted to grayscale. The "cvtColour" script was placed in the same folder where the 40 barnacles images and other colored images were removed. Figure 7.31 shows the pictures of barnacles converted to grayscale.

The "Objectmarker.exe" is provided in the utility and the images can be cropped. Figure 7.32 shows an image of objectmarker.exe.

Step 1: Draw a rectangular box on the ROI with this utility.

Step 2: Multiple ROI(s) can be selected from 1 image, Hit <Space> when every region is selected and hit <Enter> after completed. Hitting <Enter> will get to the next image and the process will repeat for the 40 barnacles. Around 5 – 10 samples were taken from each image.

Step 3: When the process is completed. "info.txt" will be created in: <Haar-training Stuff/STEPS/step 02/>shown in Figure 7.33.

More "positive" images can be produced by overlaying positive images of cropped barnacles over negative images. Since 25 cropped images of barnacles were used, the process is repeated 25 times to generate 10,000 "positive" images. Examples of "positive" images are shown in Figure 7.34.

A positive vector file is now needed to stitch of these "positive" images together using the following commands.

```
opencv_createsamples -info info/info.lst -num 4000  -vec positives.vec
```

7.1.2.2 Merging the vector files and training the cascade

The last part of creating the Haar cascade is to train the cascade. In general, higher the number of training stages, the more accurate the results. Training more stages requires a longer time to compensate for better accuracy. Around 25 vector files were computed into one vector file.

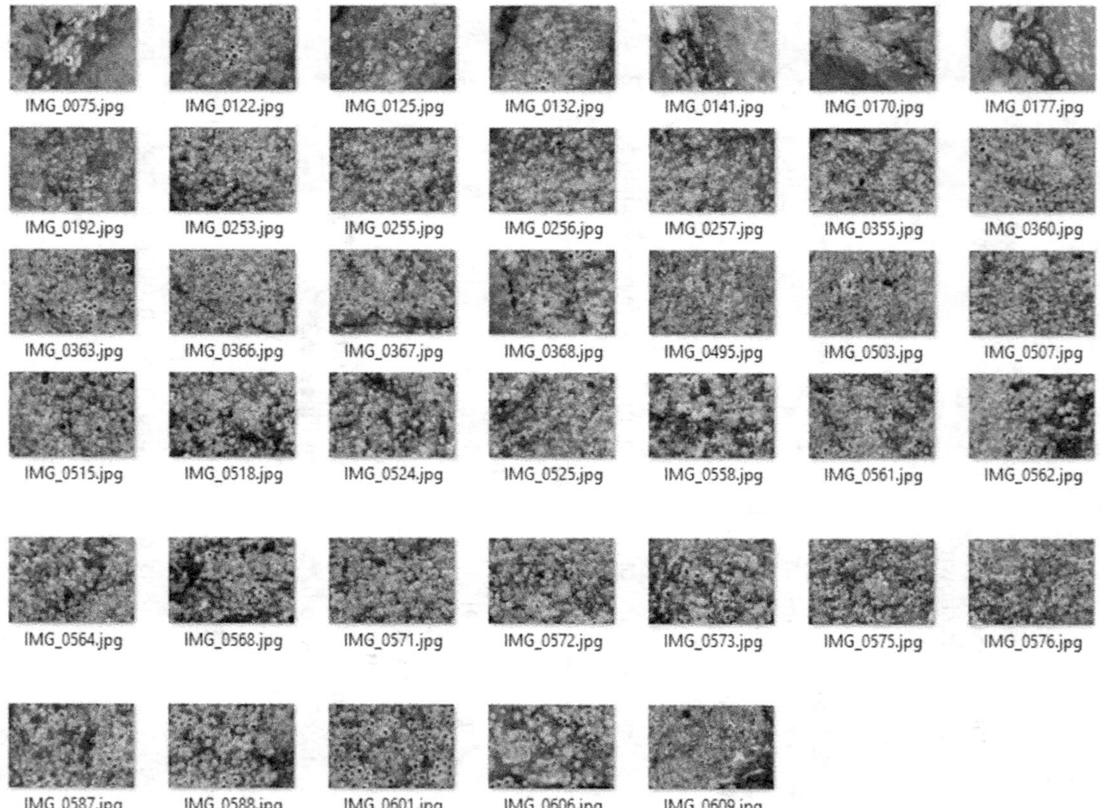

Figure 7.27 Forty best images of barnacles (.jpg)

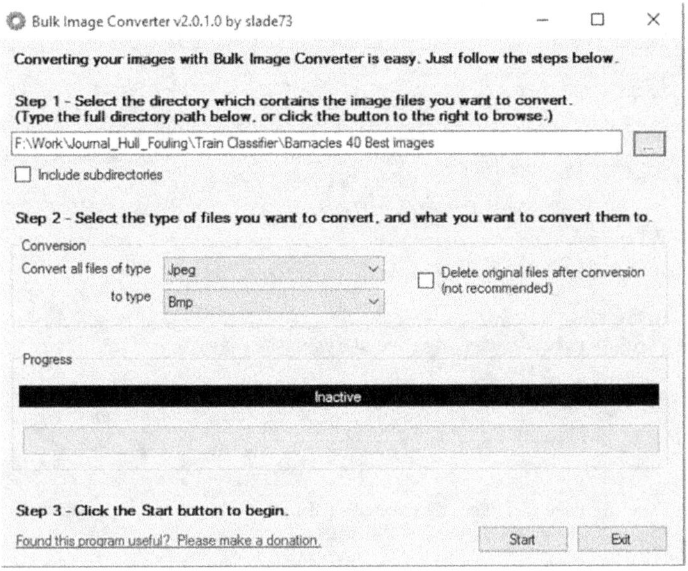

Figure 7.28 Bulk image converter

```
import sys
import glob
import struct
import argparse
import traceback

def exception_response(e):
    exc_type, exc_value, exc_traceback = sys.exc_info()
    lines = traceback.format_exception(exc_type, exc_value, exc_traceback)
    for line in lines:
        print(line)

def get_args():
    parser = argparse.ArgumentParser()
    parser.add_argument('-v', dest='vec_directory')
    parser.add_argument('-o', dest='output_filename')
    args = parser.parse_args()
    return (args.vec_directory, args.output_filename)
```

```
def merge_vec_files(vec_directory, output_vec_file):
# Check that the .vec directory does not end in '/' and if it does, remove it.
        if vec_directory.endswith('/'):
                vec_directory = vec_directory[:-1]
        # Get .vec files
        files = glob.glob('{0}/*.vec'.format(vec_directory))

        # Check to make sure there are .vec files in the directory
        if len(files) <= 0:
                print('Vec files to be merged could not be found from directory:
{0}'.format(vec_directory))
                sys.exit(1)
        # Check to make sure there are more than one .vec files
        if len(files) == 1:
                print('Only 1 vec file was found in directory: {0}. Cannot merge a single
file.'.format(vec_directory))
                sys.exit(1)


        # Get the value for the first image size
        prev_image_size = 0
        try:
                with open(files[0], 'rb') as vecfile:
                        content = ''.join(str(line) for line in vecfile.readlines())
                        val = struct.unpack('<iihh', content[:12])
                        prev_image_size = val[1]
        except IOError as e:
                print('An IO error occurred while processing the file: {0}'.format(f))
                exception_response(e)


        # Get the total number of images
        total_num_images = 0
        for f in files:
                try:
                        with open(f, 'rb') as vecfile:
                                content = ''.join(str(line) for line in vecfile.readlines())
                                val = struct.unpack('<iihh', content[:12])
```

```
                    num_images = val[0]
                    image_size = val[1]
                    if image_size != prev_image_size:
                        err_msg = """The image sizes in the .vec files differ.
These values must be the same. \n The image size of file {0}: {1}\n
                            The image size of previous files: {0}""".format(f,
image_size, prev_image_size)
                        sys.exit(err_msg)

                    total_num_images += num_images
            except IOError as e:
                print('An IO error occurred while processing the file: {0}'.format(f))
                exception_response(e)


        # Iterate through the .vec files, writing their data (not the header) to the output file
        # '<iihh' means 'little endian, int, int, short, short'
        header = struct.pack('<iihh', total_num_images, image_size, 0, 0)
        try:
            with open(output_vec_file, 'wb') as outputfile:
                outputfile.write(header)

                for f in files:
                    with open(f, 'rb') as vecfile:
                        content = ''.join(str(line) for line in vecfile.readlines())
                        data = content[12:]
                        outputfile.write(data)
        except Exception as e:
            exception_response(e)

if __name__ == '__main__':
        vec_directory, output_filename = get_args()
        if not vec_directory:
            sys.exit('mergvec requires a directory of vec files. Call mergevec.py with -v
/your_vec_directory')
        if not output_filename:
            sys.exit('mergevec requires an output filename. Call mergevec.py with -o
your_output_filename')

        merge_vec_files(vec_directory, output_filename)
```

Figure 7.29 Best images of barnacles

```
import cv2
import glob

i=1
for img in glob.glob("*.bmp"):
    cv_img = cv2.imread(img)

    if (type(cv_img) == type(None)):
        break

    gray_image = cv2.cvtColor(cv_img, cv2.COLOR_BGR2GRAY)
    cv2.imshow('detection', gray_image)
    cv2.imwrite('IMG_'+str(i)+'.bmp', gray_image)
    i=i+1

    if (0xFF & cv2.waitKey(10) == 27) or cv_img.size == 0:
        break

cv2.waitKey(0)
cv2.destroyAllWindows()
```

Figure 7.30 Convert to grayscale script

After merging the vectors file into one, ten stages were used for training the final cascade file using the following command.

```
opencv_traincascade -data data -vec positives.vec -bg bg.txt -numPos 3600 -numNeg 2000 -numStages 10
```

7.1.2.3 Artificial neural networks identification

The results taken using Haar cascade from each type of fouling can be combined from ANNs to identify fouling accurately. Further developments aim to develop a complete categorization of the types of fouling. The ANNs make comparisons of the input with the output and generate its output when the target is matched. A vast image database is necessary as images of high-quality fouling species are difficult to find. Figure 7.35 shows a combing Haar cascade with neural networks.

As object recognition requires significant positive and negative images to achieve a higher accuracy, a web interface is created for the user to contribute different fouling images. By using PHP 5.0, the web interface is hosted on the Apache Web Server supported by MySQL to store the fouling images. The database server stores and organizes content in a web browser used by an end-user. Hosted on a web server, the hypertext pre-processor (PHP) is an interpreter embedded within a webpage. It is used to interpret the PHP file when a user request access to a page. Figure 7.36 shows the relationship in setting up

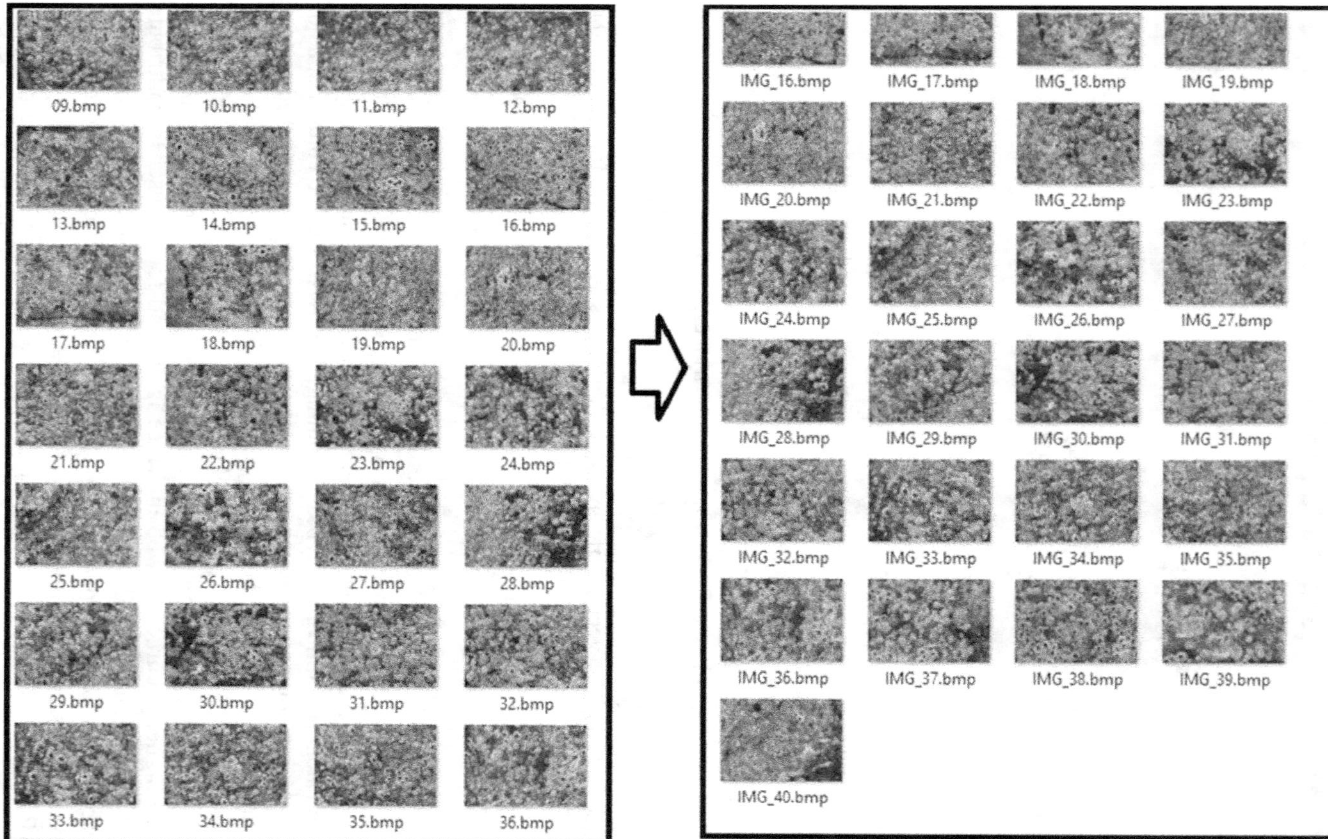

Figure 7.31 Images converted to grayscale

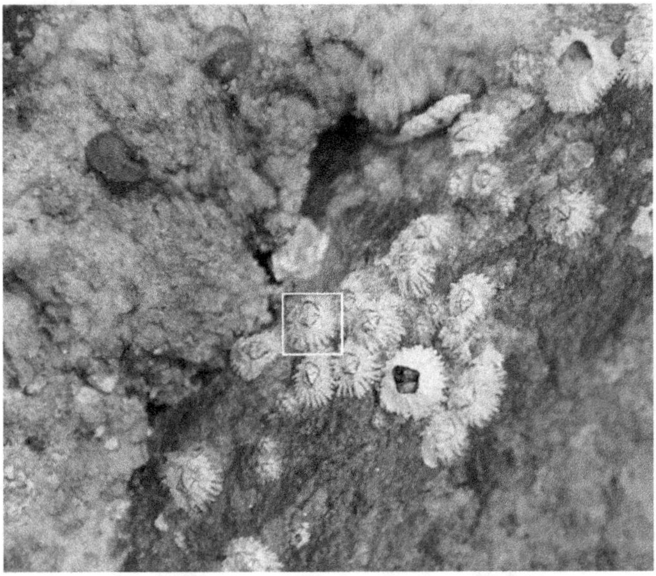

Figure 7.32 Cropping of images using Objectmarker

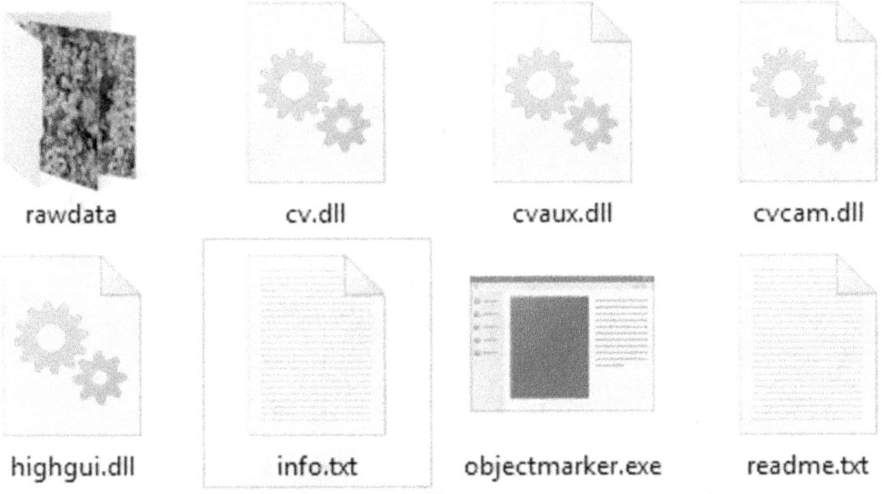

Figure 7.33 Image info file generated

the webpage to allow users to contribute fouling organisms. The users can contribute fouling organism to the database server by choosing image files from their library and uploading them. The image files will be automatically uploaded to the database server, and contributors would browse and download all fouling images stored in the database. Figure 7.37 shows the initial webpage

Figure 7.34 Positive sample created

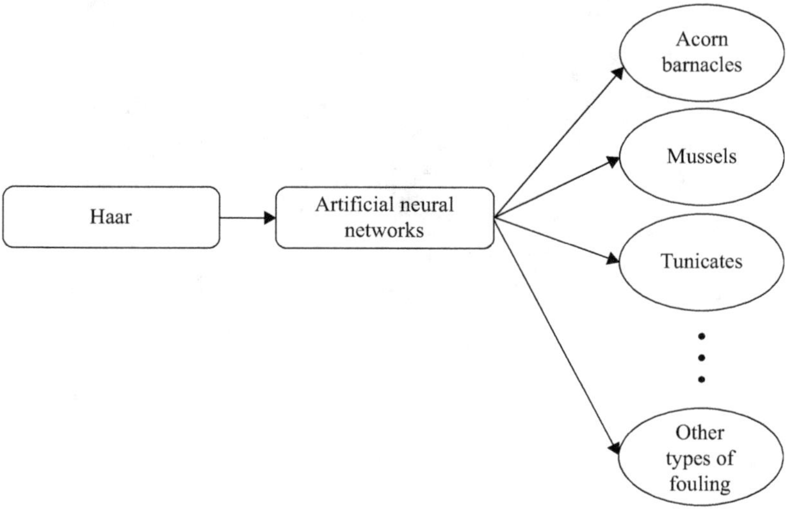

Figure 7.35 Haar cascade with neural networks

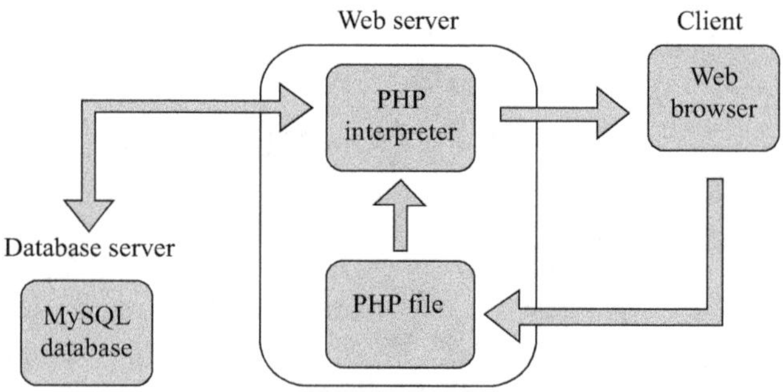

Figure 7.36 Webpage design concept

design for the user to upload images by dragging and dropping into the box provided. Figures 7.37–7.41 show the process when a user uploads images into the database.

Step 1: Access the webpage.

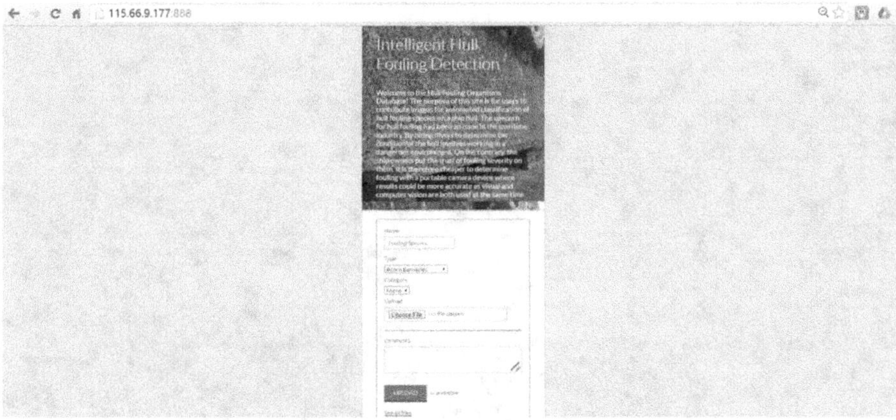

Figure 7.37 Initial webpage design

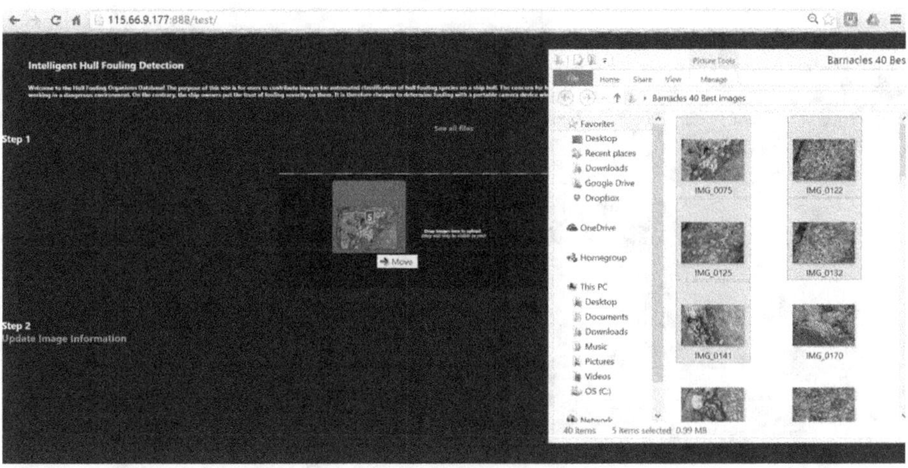

Figure 7.38 PHP webpage Step 2

Step 2: Select images and drop them into the box provided on the webpage. The process will take some time to upload the images. A tick will appear when uploading process is completed. A maximum of five images is allowed to be uploaded at each time. The user will click on "Update Image Information."

Step 3: Information (name, type, category) of fouling has to be inserted manually. After all the relevant information is provided, users will click on "Update All" at the bottom of the page. The image will now be stored in the database.

7.1.3 Testing

The system was fabricated using aluminum with the black anodized coating. Three O-rings were installed in the camera enclosure to ensure watertight capability.

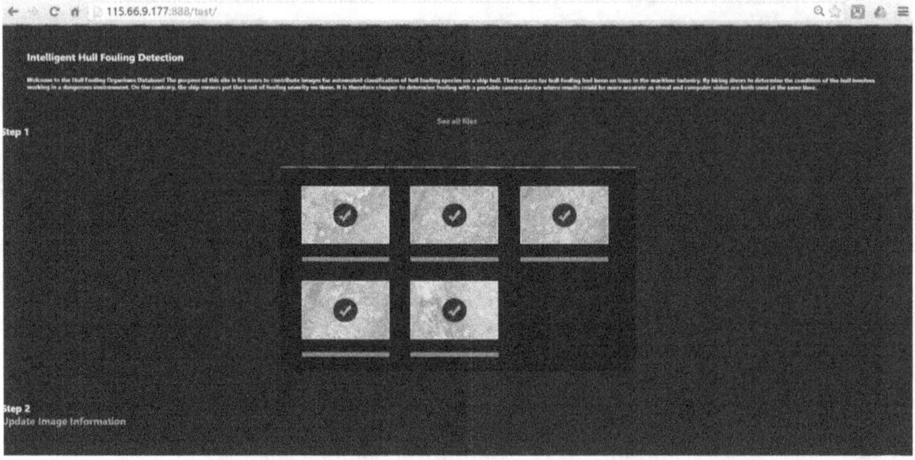

Figure 7.39 PHP webpage

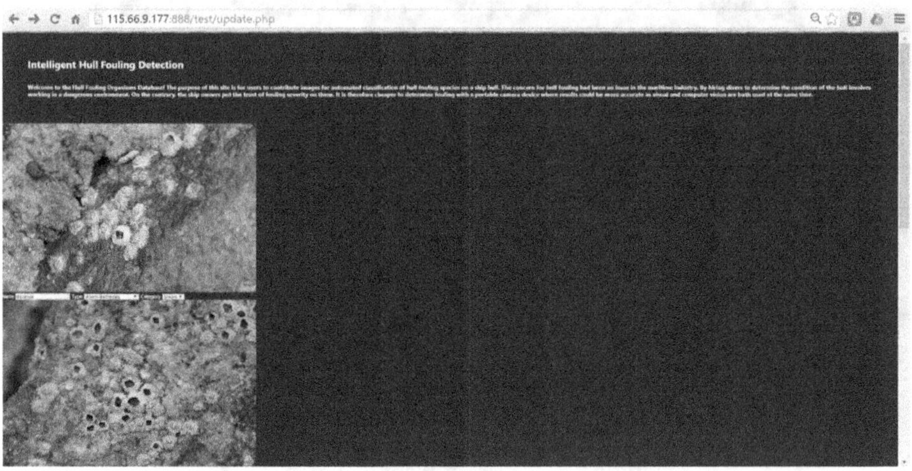

Figure 7.40 PHP webpage for update 1

The rod clamp was replaced with a carabiner for hoisting operation. Actual testing of the mounting on the gunwale of a fishing boat is shown in Figure 7.42. The gunwale width of the fishing boat was 110 mm. Thus, the suction cup was not able to support the camera assembly.

An underwater test was conducted by mounting the clamp on a water polo's goal post as a substitute for the gunwale of 8 mm since the bar of the goalpost is approximately 6 mm thick. The experiment was conducted in a 1.8 m depth swimming pool with the goal post being 0.9 m tall as shown in Figure 7.43.

Minimal air bubbles were observed between the enclosure cover and the enclosure (see Figure 7.44) when the prototype was placed under water. Grease was applied to the surfaces, and the camera enclosure was tested in the pool. The Wi-Fi

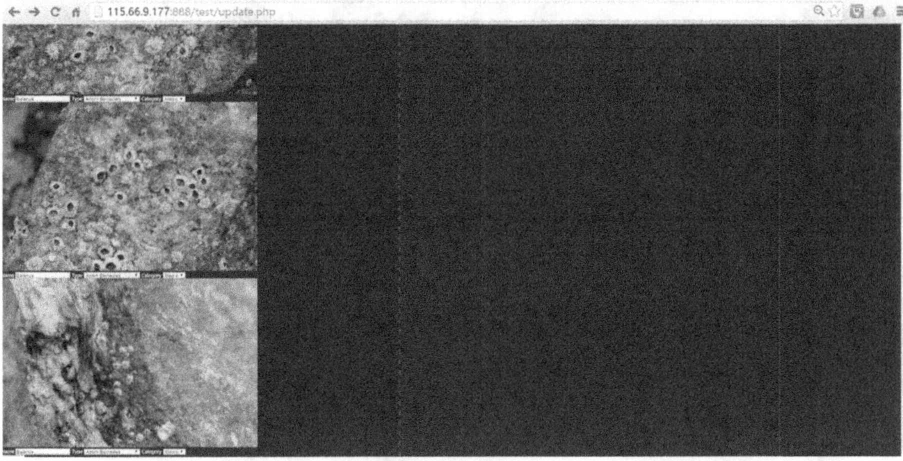

Figure 7.41 PHP webpage for update 2

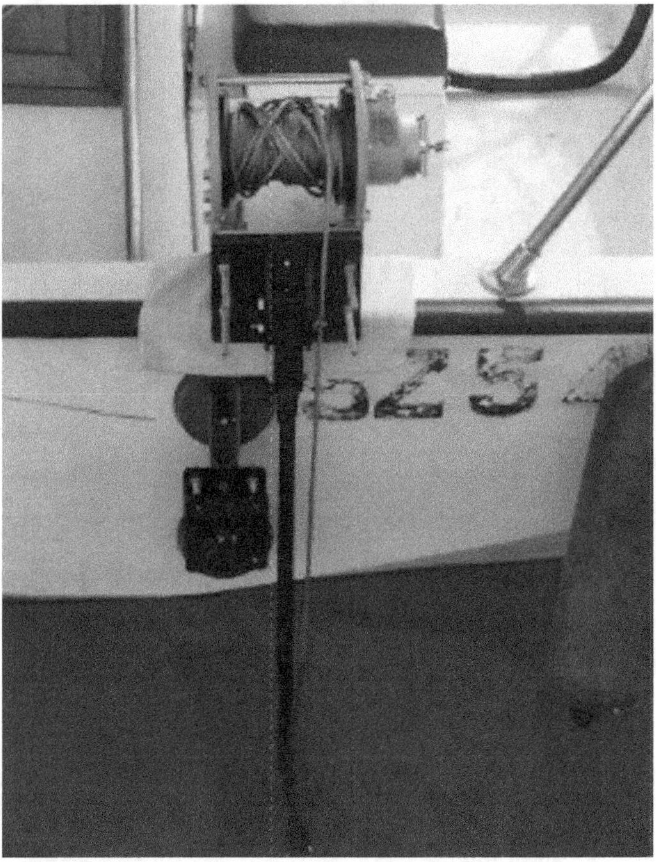

Figure 7.42 Mounting of the system on a fishing boat

Figure 7.43 Experimental setup

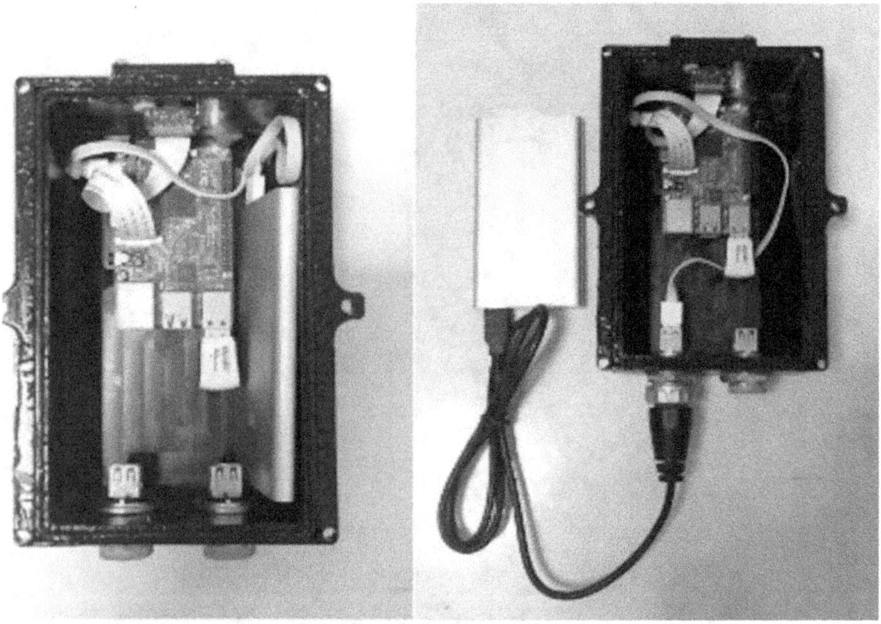

Figure 7.44 Enclosure components layout

connectivity was achieved through a 5 m RG174 placed with one end inside the enclosure and the other end near to the surface router. The Raspberry Pi was powered by a portable charger inside the enclosure. The slow frame rates of the live video can be observed in Figure 7.45.

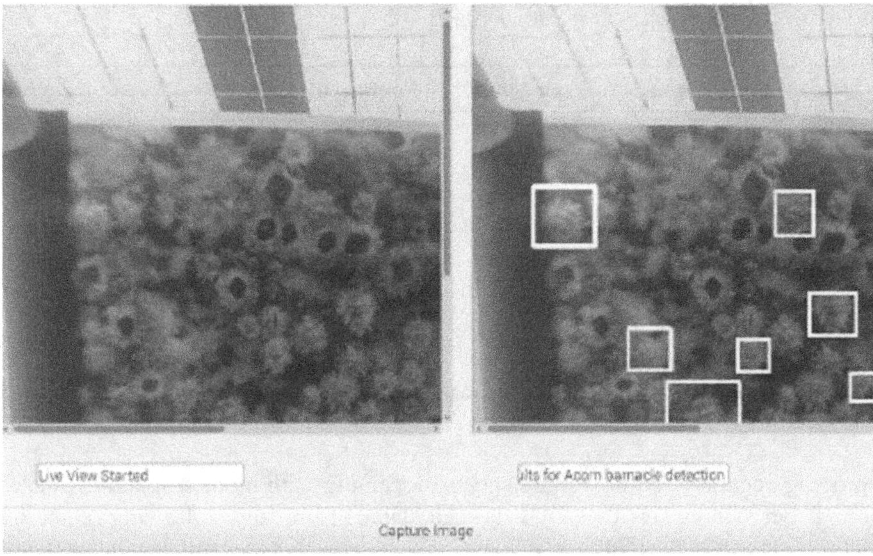

Figure 7.45 Underwater test results

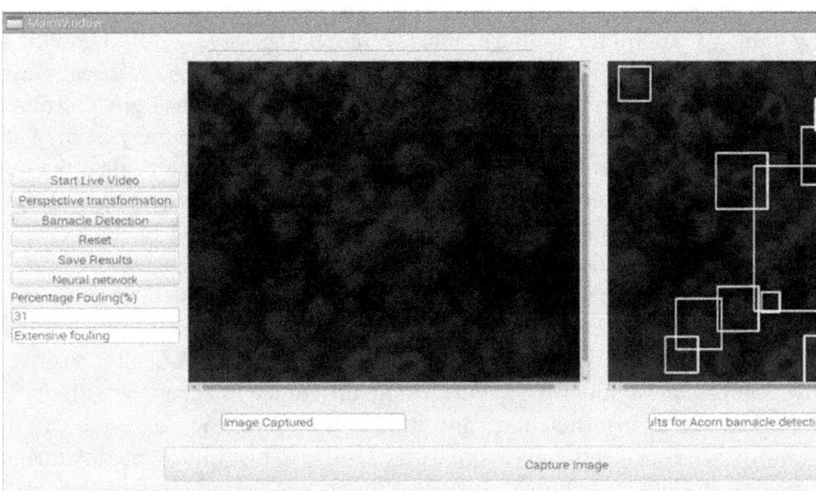

Figure 7.46 Example of test results

Laminated images of barnacles were used to test the algorithm. The algorithm will convert the images to grayscale before comparing the image with the cascade file. Figure 7.46 shows the results obtained from the GUI with a fouling percentage of 17% when tested underwater. The results show that the outcome of the algorithm was considered weak as many of the barnacles seen in the image are not identified.

Table 7.6 Fouling percentage test results

Test	Manual count	Haar count	Fouling by Haar (%)	Accuracy (%)
1	46	31	17	67.39
2	43	15	23	34.88
3	20	11	17	55.00
4	5	3	5	60.00
5	40	24	31	60.00
6	41	20	38	48.78
7	17	9	10	52.94
8	34	15	12	44.12
9	34	15	25	44.11
10	37	14	30	37.84

Improving accuracy requires more stages of training with more cropped barnacles. Fouling percentage in the GUI is calculated from the following equation: $\text{Fouling} = \frac{\text{Area of fouling}}{\text{The total area of the picture}} \times 100\%$. Around ten images filled with barnacles were tested to find out the mean area containing barnacles. Figure 7.46 shows one set of the test result from the GUI. The mean area of fouling was approximately 20.8%. Seven of the results have a fouling rank of 4 (Extensive fouling), 1 has a fouling rank of 2 (Slight fouling), and the remaining gave a fouling rank of 3 (Considerable fouling). The test results are shown in Table 7.6. A direct comparison with other algorithm is unable to perform since the ROI was trained differently and can only be improved if more images are trained. The accuracy of the system can be determined by manual counting of barnacles in an image against the output from the GUI. Another ten tests were performed, and the mean accuracy of the test was found to be around 50.51%, with a standard deviation of 10.44%. False detections were observed due to poor quality of the camera images.

The results showed a success rate of 89.5% where some of the false detection was due to detection of brand logos. It is shown that using Haar-like cascade algorithms can produce a high accuracy of detection rate if large numbers of good-quality positive and negative samples are trained. The considerable difference in accuracy with the algorithm used is due to insufficient quality images used, and each barnacle can differ significantly. We considered only acorn barnacles due to insufficient quality images of other species of fouling organisms. Images of barnacles trained with images from the internet produce poor results. Thus, the need for using a DSLR camera to retrain the algorithm will be needed. The design of the system requires the angle of the detection to be specified and manually adjusted before lowering the assembly underwater to check the fouling density. The distance between the gunwale and the ROI for detection was limited by the poles. The suction cups used in the system was limited to flat surfaces of the ship hull and was unable to accommodate different surfaces.

In summary, continually monitoring the fouling condition on the ship hull will enable timely planned docking schedules to reduce fouling activity. The application of the proposed system would be cost-effective as the vast amount of savings can be achieved not only on fuel costs but also on maintenance costs, as less time is required to check on fouling on the hull surface. Manufacturers of AF paint systems could use such technology to obtain test results. The results obtained from the system would also allow better efficiency in the research of formulating new AF coatings and improved hull forms. With the creation of the fouling database, it will provide more images for Haar cascade algorithm. Thus, the detection accuracy will be improved in a shorter period. Further developments using neural networks can be achieved to identify different organisms.

7.2 Multi-hop microprocessor-based prototype system for remote vibration and image monitoring

The details in this section can be found in the following paper [9]. The increasing need for more energy sources in recent years has driven research for safe and reliable methods for the extraction of oil from below sea water. The dispersed nature of such oil field localities has led to the construction of offshore oil platforms on topographies ranging from shallow water to deep water. While offshore platforms fixed in either locality are subject to dynamic forces due to waves, currents, wind, and earthquakes, offshore platforms fixed in the deep sea are also susceptible to vibrations from the breakage of ice formations accumulated around the support legs at extremely frigid environments. Uncontrolled vibrations caused by the factors above may result in equipment malfunction, damage to the structure, discomfort to the workers on the platform structure, and even stimulate structural failures due to fracture growth over a prolonged period.

Conventionally, offshore structures have been built without due considerations towards the ill effects of ice-induced vibrations, which culminated in their collapse [10]. Hence, new highly expensive structures that were overly rigid were designed. Although there was a notable decrease in vibration, human discomfort and equipment malfunction were still an issue. A recent study about the effects of ice-induced vibrations has yielded various techniques to mitigate such vibrations caused by the ice breaking against the structures. One such example is the implementation of conical legs in place of standard vertical legs [11,12], which transforms the breaking force of the ice from crushing to bending as the ice is lifted off the base. The resultant force exerted on the legs, although was much lower than the original force, was still notable and relatively periodic. Also, it was showed that at high ice speeds, the breaking ice-force could be at the same frequency as the natural frequency of the structure, causing an extreme amplification of the vibration [11,12]. This effect, coupled with the resultant force of current techniques for reducing ice-induced vibrations, may result in severe damages to the structures and equipment.

Despite the limited availability of groundwork on the existing vibration and ice-formation monitoring systems, prominent research has been done on the related subcomponents of the system categorized are, namely, vibration measurement, structural integrity verification, CV, and communication infrastructure. A feasible vibration measurement technique [13] was used. It consists of a spectral analyzer and a set of synchronized tri-axial accelerometers as vibration sensors for acceleration-derived motion and vibration measurement. The vibration sensors were discretely placed on structural elements, such as the plate and the grating, in an orderly arrangement to form a grid with crossing points equally separated around the vibrating machinery.

Flooded member detection, for identifying through-thickness cracks in structures underwater [14], was proposed. An improvised approach exploited the usage of ultrasonic sound technology [14]. The approach utilized a microcontroller-based transmitter, a signal booster, a piezoelectric transducer, an amplifier, a real-time digital signal processing system, and a computer for detecting flooding in hollow sub-sea structures. A successful attempt to benefit from deep water-based CV [15] was used in guiding an autonomous underwater vehicle along a prelaid power cable under the sea. The vision-based guidance enabled the underwater vehicle to locate and follow the cable lines found on the seabed with an average accuracy of over 90% for an input frame rate of 25 fps. Additionally, a ROV [16–18] model can be used to detect cracks in structures underwater. However, it is expensive to build, operate, and maintain.

The usage of multi-hop wireless networks complements the connectivity of the entire system while maintaining performance and supporting extensibility [19,20] was applied. The multi-hop topography is based on the hop-by-hop retransmission scheme, which breaks a single long transmission originating from a particular node towards the server into multiple node-to-node (link) localized packet retransmissions.

In summary, this section aims to propose a working prototype for remote vibration and image monitoring of underwater offshore platform using a combination of open-source ARDUINO® platform, accelerometer, and Raspberry Pi.

7.2.1 System design and architecture

Based on the analysis of similar research works mentioned earlier, a system to satisfy the following requirements is proposed.

- Ability to measure the vibration precisely along the rig legs;
- Possess video-streaming capabilities (for monitoring ice formation along legs);
- Provide an intuitive user interface for viewing vibration intensities, stream videos, and to control the hardware resources (sensors and cameras);
- Automate alerts generation based on inputs from vibration sensors and camera frames.

The system can be broken down into three subsystems such as sensor network, interface, and client terminal. The sensor network, composed of individual nodes, is typically installed underwater along the legs of an offshore platform. As shown in Figure 7.47, each node of the sensor network primarily houses a vibration sensor and a camera module (collectively termed as peripherals) controlled by an onboard node controller. The node controller serves the undermentioned purposes that are to

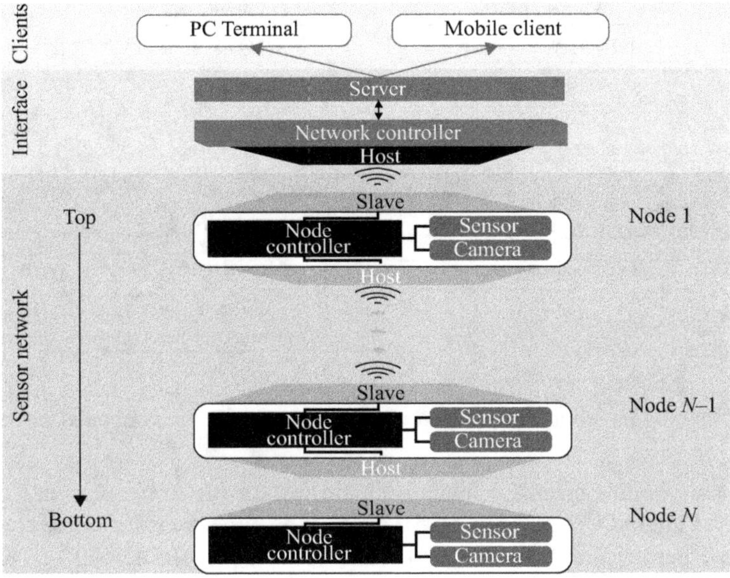

Figure 7.47 High-level architecture of the proposed system

coordinate activities between the peripherals, and to communicate with other nodes via the integrated host and slave wireless modules. The interface subsystem primarily serves as a middleman between the wireless sensor network and the end clients, such as a personal computer. A suitable hardware interconnection scheme to support the system architecture is described. The system is composed of four main subsystems: node, sensor network, interface, and clients.

Each node is a low-powered microprocessor system capable of transmitting real-time vibration data and streaming video from the camera to the neighboring node via the integrated wireless modules. The node comprises four components, namely, node controller, camera, vibration sensor, and slave or host wireless module. The node controller is the "CPU" of the node, coordinating the data input from the connected vibration sensor and camera, transmitting them to the next node, and receiving control instructions from the next node. Raspberry Pi was used as the node controller, due to its high features per cost ratio. Figure 7.48 depicts the interconnection between the Raspberry Pi and other peripherals.

The camera is used exclusively for monitoring ice formation on legs and joints to identify cracks. The Raspberry Pi 5MP camera board was selected due to the cost, and it provides sufficiently clear images at full HD resolutions and can be readily interfaced with the Raspberry Pi. The camera module is connected to the Raspberry Pi using the CSI Interface as depicted in Figure 7.48. Due to the presence of Linux® OS on the Raspberry Pi, despite the 700 MHz clock rate, a high measurement frequency using the GPIO pins is not expected to be feasible, due to the limited memory bandwidth and the low-powered single core ARM processor. Similarly, the Triple Axis Accelerometer, though expensive, requires averaging

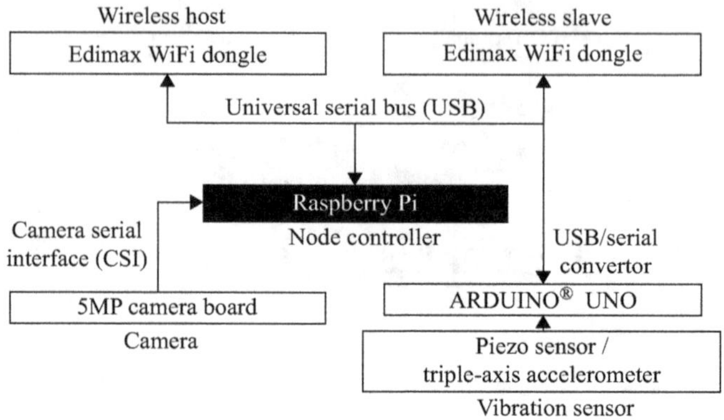

Figure 7.48 Node interconnection schematic for the proposed system

over a fixed window size to achieve accurate results due to the presence of noise. Thus, an ARDUINO® UNO is used for controlling either sensors and averaging the data before passing the value back to the Raspberry Pi. ARDUINO® UNO, being an AVR-based microcontroller, despite the meager clock rate of 16 MHz, can access the low-level hardware (sensors) faster due to the lack of OS overhead (code written gets full access time on CPU). The ARDUINO® UNO is powered using the USB port connected from Raspberry Pi's USB port. It communicates with the latter using an inbuilt Serial to USB converter.

Since each node is an isolated entity, separated the nodes over a relatively long distance may not be economical or feasible to run wiring to connect the nodes. Thus, a Wi-Fi-based wireless communication infrastructure is finalized. In this infrastructure, two Wi-Fi adapters are connected to the Node Controller to act as a client for the upper node and to act as a Host for the lower node. Edimax 150 Mbps Wi-Fi adapters are connected to Raspberry Pi via the USB to serve the purpose above. The Raspberry Pi Model B has only 2 USB ports; the adapters should be connected using a USB hub. There is a need to monitor vibration and ice-formation levels at shallow depths, and the limited wireless range of the Wi-Fi adapters prevents long-range transmissions, a multi-hop transmission scheme is implemented as shown in Figure 7.49.

In this scheme, the lower node is connected to the upper node using the Client-Host Wi-Fi interface method. Thus, each node acts as a host to the lower node and as a client to the upper node. Two types of communication take place between the nodes.

- Downward communication—control signals from the network controller;
- Upward communication—vibration sensor data and video-streaming data from the nodes up to the network controller.

As control signals from a negligible portion compared to video streaming, the transmission scheme is effectively an upwards transmission scheme. This transmission follows a pipelining technique to convey information between the nodes. For example, Node 5 would first send data upwards to Node 4, which would send

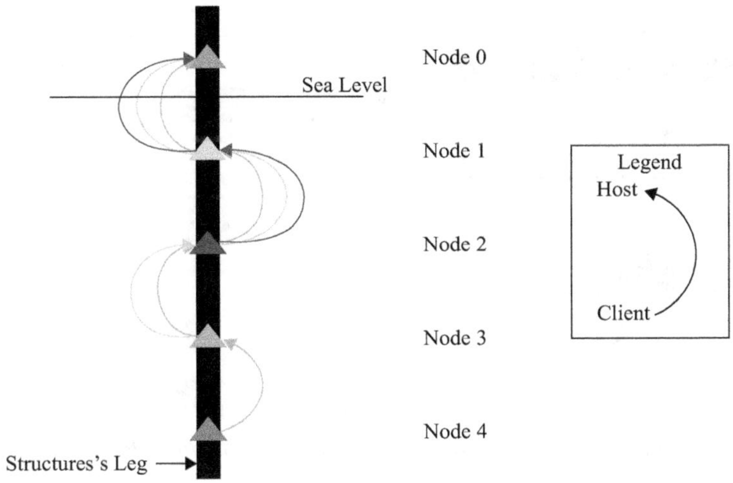

Figure 7.49 Internode multi-hop transmission scheme

that data and the locally generated data upwards to Node 3 in the next time frame, and so on until it reaches Node 0. By adopting this strategy, although the maximum "effective" range is proportional to the number of nodes present along the leg, the transmission speed between the nodes is not constant due to the varying level of unwanted noise affecting the signal-to-noise ratio (SNR). Thus SNR was minimized by choosing a locally optimal separation distance between following pairs of nodes, instead of resorting to a fixed separation distance. Furthermore, this scheme gives rise to an inversed pyramid workload on the nodes, i.e., the transmission workload on a given load is directly proportional to the number of nodes under it. Thus, the maximum achievable bandwidth between the upper nodes poses a bottleneck for the number of nodes pipelined below them. This behavior is illustrated in Figure 7.49 where Node 1 to Node 0 experiences the highest workload.

The interface subsystem forms the core of control and logging system. The primary purpose of this subsystem is to provide a hardware interface between the client PC and wireless sensor network. The interface subsystem is essentially a Raspberry Pi (termed henceforth as an auxiliary controller), connected directly to a single Wi-Fi adapter (as a Host to the Top-most Node) on one end, and a server PC on the other end via a network hub. The server PC serves two vital roles, namely, provide high-level control over the sensor network and perform video processing and logging.

High-level control: The server provides high-level control over the operation of the individual nodes of the wireless sensor network by transcoding control requests from clients to the relevant protocol format and sending them to the specific node via the auxiliary controller. Control operations on individual nodes include the following: turn on/off video streaming, alter video streaming resolution, alter video streaming fps, and turn on/off vibration monitoring.

Video processing and logging: The server is also used to takeover intensive tasks, such as to generate an alert. Image-processing algorithms and vibration-level

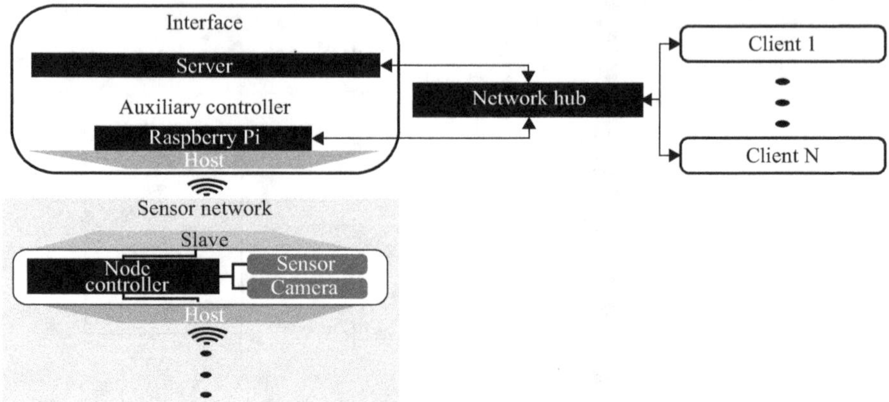

Figure 7.50 Interface subsystem architecture design

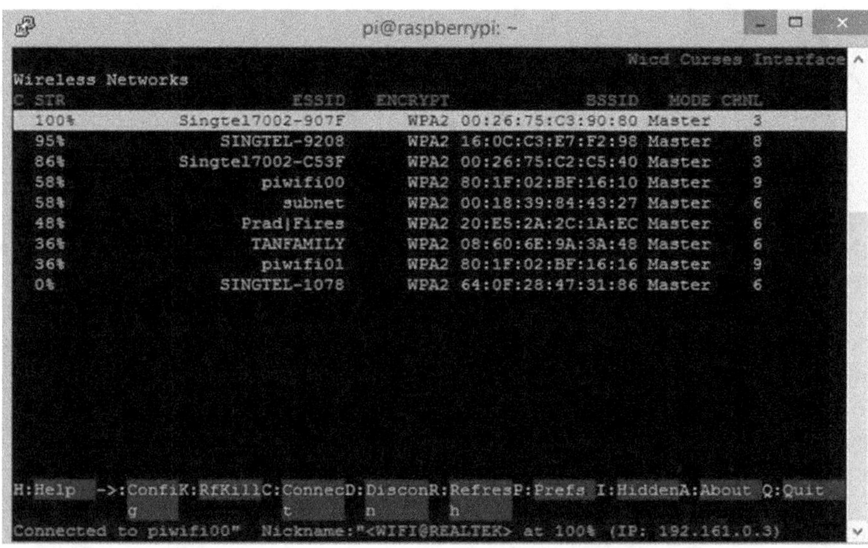

Figure 7.51 Wicd curses and wireless network GUI

monitoring are executed on the server real time, and the results logged, to provide for a centralized data storage system to support multiple clients. The server is mostly a computer which is always on, connected to the auxiliary controller and multiple clients on the same network as illustrated in Figure 7.50.

The client's subsystem refers to the group of devices that serve as end terminals for monitoring and exercising control over the sensor network. The clients which are connected to the same network of the interface server are shown in Figure 7.51. The user interface includes the following: selection of a node to monitor, display detailed vibration graph, video streaming, overall system status of all nodes, and exercise high-level control. The vibration sensor was connected directly to the Raspberry Pi. There was a significant delay in response to vibration. A way to overcome this issue was to use ARDUINO® as a mediator between the vibration sensor and Raspberry Pi. The experiment was conducted using ARDUINO® UNO as a transmitter to Raspberry Pi and a receiver of vibration signals from the vibration sensor. Significant improvement was observed. Through this experiment, it was realized that Raspberry Pi has a better capability than an ARDUINO® UNO. The Raspberry Pi will have a delay in accessing data from the vibration sensor. An ARDUINO® UNO does not have an OS. It will not run any other tasks than reading data from the vibration sensor. Therefore, it is more capable of providing data in a shorter time delay than Raspberry Pi. Thus, it was decided to use ARDUINO® UNO as a vibration transmitter.

Accelerometer framework: It was realized that the SPI interface is much simpler and faster than I2C. Master and slave are linked by three data wires, usually called MISO (Master in, Slave out), MOSI (Master out, Slave in), and M-CLK. It was decided to connect Accelerometer to ARDUINO® using an SPI interface circuit. Accelerometer sends vibration reading to ARDUINO® as a movement for x-, y-, and z-directions in m/s^2 or gravity as selected from the program.

ARDUINO® framework: ARDUINO® was programmed using Visual Studio. It was programmed to handle the process of receiving vibration reading and to transmit reading to the Raspberry Pi using two different treads in a single program. Thread 1 which can be termed as Accelerometer Handler, de-serialize the reading from Accelerometer at a baud rate of 115,200, the reading in x, y, and z is then stored into the respective array of window size 32. Thread 2 that can be termed as Communication Handler. It primarily communicates with Raspberry Pi and checks if data are required to be sent. Once Raspberry Pi is ready to receive the communication, handler in ARDUINO® will average out the vibration data collected in array x, y, and z to avg (x), avg (y), and avg (z). These data are then serialized before sending them to Raspberry Pi.

Raspberry Pi and camera framework: Python programming language was used to write programs in Raspberry Pi. It is programmed to handle the process of receiving vibration reading from ARDUINO® than to send it to the server and to read image frames from Pi Camera then to send it to the server. A single program will have two different threads to handle these receive–send processes. Once the Raspberry Pi boots up Thread 1, which can be termed as ARDUINO®, Handler will send a Ready-To-Receive signal to ARDUINO® and then de-serialize the reading from ARDUINO® x, y, and z and store them into the respective array of chunk size 16; this chunk is then serialized and sent to the server. Thread 2 which can be termed as camera handler captures images every 600 ms and sends it to the server.

Wireless connection from the last node to node 0 (first node) is established after setting up Wi-Fi Client and Host. Vibration data and visual images are transmitted to the server via this connection. It will only provide only one-way connectivity. Hence, the connectivity between two nodes was set up to perform maintenance and update of the node from the server. Each node in the system is a client. The computer on which the GUI runs is the server. Each node was programmed to multi-hop vibration data and image streaming to the server that runs the GUI.

7.2.2 Programming and system implementation

The following steps were carried out to configure one node only (node 0). The SD card image was copied into two other SD cards. Then necessary IP address changes were made to the Host and Client setup files. Due to a limitation in the number of available USB ports in Raspberry Pi, ARDUINO® was connected to Raspberry Pi via a USB Hub. As the accelerometer was connected to ARDUINO®, the program for vibration reading was written using Visual Studio C++ language. The vibration reading was then transmitted to Raspberry Pi by physically connecting ARDUINO® and Raspberry Pi via USB cable.

Basic Setup (Step1): Raspberry Pi OS, Raspbian was installed on the 16 GB SD card. The latest version of Raspbian was downloaded to a Windows PC. Win 32 disk image was downloaded, and the application was unzipped. SD card was formatted, and the Win 32 disc imager was run as administrator. The image was written to the SD card. The SD card was inserted into Raspberry Pi. HDMI cable was connecting between Raspberry Pi and a TV, wireless keyboard and a USB mouse was also connected to establish control. Raspberry Pi was boot up and was successfully able to operate Raspberry Pi in its Raspbian OS.

Configuring Raspberry Pi (Step 2): Configuration required for the first time was set. REAL VNC Viewer was used to set remote access; there are many VNC viewers available. Real VNC Viewer was chosen as it was mostly used with Raspberry Pi. Next, a physical connection between the camera and Raspberry Pi was established by connecting a CSI interface. The setup of Wi-Fi Client (1st Wi-Fi Adaptor) and Dynamic Host Configuration Protocol (DHCP) is shown. A Wi-Fi adaptor was plugged into the bottom USB slot of Raspberry Pi. (Both Wi-Fi adaptors were not plugged into the two USB slots as it would cause a conflict.)

Step 1: The below codes were entered in steps to set up hostapd to provide the wireless host access point service.

```
sudo apt-get install hostapd
wget http://www.daveconroy.com/wp3/wp-content/uploads/2013/07/hostapd.zip
unzip hostapd.zip
sudo mv /usr/sbin/hostapd /usr/sbin/hostapd.bak
sudo mv hostapd /usr/sbin/hostapd.edimax
sudo chmod +x /usr/sbin/hostapd.edimax
sudo ln -sf /usr/sbin/hostapd.edimax /usr/sbin/hostapd
```

Configure Hostapd (Step 2): In order to set sudo scite /etc/hostapd/hostapd. conf was entered to set the Wi-Fi SSID to: RPiNode1.

Configure Hostapd (Step 3): To edit hodtabd file, sudo nano /etc/default/hostapd was entered. The command DAEMON_CONF= /etc/hostapd/hostapd.conf was added to the file and saved.

Configure Interface (Step 4): To set static IP for the Raspberry Pi, sudo nano/ etc/network/interfaces was entered by setting the following in the interfaces file.

Configure Interface (Step 5): To eliminate bug when the Wi-Fi is set to use static IP, the coding in ifplugd file was changed by entering sudo nano /etc/default/ ifplugd.

To install DHCP service (Step 6): sudo apt-get install udhcpd was entered. To enter the setting page to set DHCP, sudo nano /etc/udhcpd.conf was entered. The following two lines indicate the range of IPs that the Wifi hotspot will give to client devices:

To enable DHCPD, /etc/default/udhcpd was entered and to start DHCPD as a service using sudo service udhcpd start.

The setup of Wi-Fi Client (Second Wi-Fi Adaptor) and Wireless transmission of vibration reading and visual streaming between nodes are shown. The power was switched off and the existing Wi-Fi adaptor was unplugged from the bottom USB port. Another Wi-Fi adaptor was plugged into the top USB port. Raspberry Pi was connected to the internet via Ethernet cable. Plugin only one Wi-Fi adaptor into the TOP USB slot of Raspberry Pi. (Two Wi-Fi adaptors into both the USB slots will cause conflict.)

Step 1: Power on and SSH to the respective RPI. sudo apt-get install wicd-curses was entered to install Wicd-curses a user-friendly GUI Wi-Fi setup utility. With Wicd-curses, the setup of Wi-Fi on Raspberry Pi is simple and straightforward. To enable the OS to configure the second Wi-Fi USB port (TOP USB slot of Raspberry Pi), enter sudo wicd-curses. A preferred Wi-Fi Network's SSID was selected from a list of the wireless network found by the Raspberry Pi, as shown in Figure 7.51.

Step 2: Wi-Fi Network was selected by highlighting it as shown in Figure 7.51. The desired IP (e.g., 192.168.2.135), the subnet mask (255.255.255.0), and the router's IP (e.g., 192.168.2.1) was set as a static IP. It was assured that the IP of the Raspberry Pi and the gateway IP has the same index. If the router has 192.168.2.ip, then the Raspberry Pi should have the same except for the last ip digits. Raspberry Pi was connected to the router via Wi-Fi and able to Host Wi-Fi for other devices.

7.2.3 Testing

Wireless connection from the last node to node 0 (first node) is established. Vibration data and visual images are transmitted to the server. It provides single-way connectivity. Bi-directional connectivity between nodes is required to perform maintenance and update any of the nodes from the server. The following commands using sudo, netmask, and gw were used to establish two-way connectivity between nodes. Each node in the system is considered as a client, the computer in which the GUI is the server. Each node was programmed to multi-hop vibration data and image streaming to the server. Server's IP address is programmed in each node for identification. Each node will use two ports in the server, one to transmit vibration data and the other to transmit visual streaming. Therefore, these port numbers were set into each node at the client program file called main.py. For example: image_client = ImageClient('PI_0', '192.161.0.5', 10000) and vibration_client = VibrationClient('PI_0', '192.161.0.5', 10100). The code above shows that the server's IP address was set to "192.161.0.5". As shown above, the port

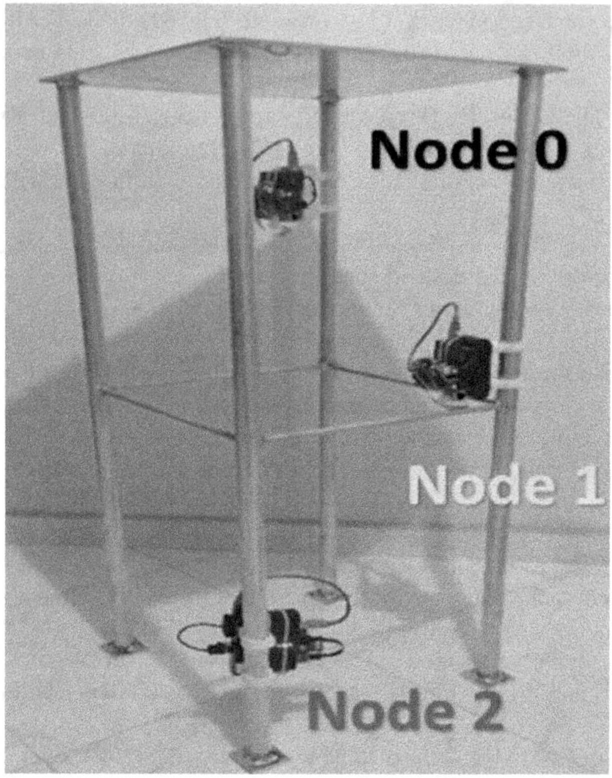

Figure 7.52 Three nodes fixed onto proposed offshore structure model

used to receive the image in the server was set to 10,000 and 10,100 to receive vibration data.

The GUI was programmed together with the server–client program as a single server program file. It will display video streaming and a graph that shows vibration reading in the *x*-, *y*-, and *z*-axes. The highest vibration sensed at any axis at a particular time is displayed in digital value in m/s^2. It can automatically extend the above settings to the nodes connected in future if the system is expended to have more nodes. The SD card's image in PI 0 (first Raspberry Pi) was duplicated for two more Raspberry Pi. The Raspberry Pi was accessed via SSH and instructions. The IP addresses were changed in the different sections of Host and Client files as follows: configure Interface, setting up DHCP (Host), and Wi-Fi Client.

As seen in Figure 7.52, a rig-like model of 1,000 mm height 400 × 400 mm square-shaped structure was constructed to fix the nodes onto its legs. A battery pack, Raspberry Pi, USB hub, ARDUINO®, and an accelerometer were packaged as a node as shown in Figure 7.53. Three such nodes were fixed at various joints of the structure as shown in Figure 7.52.

The structure consists of three nodes. The constant vibration was applied to the structure. The result of vibration was shown in the GUI. The video streaming was

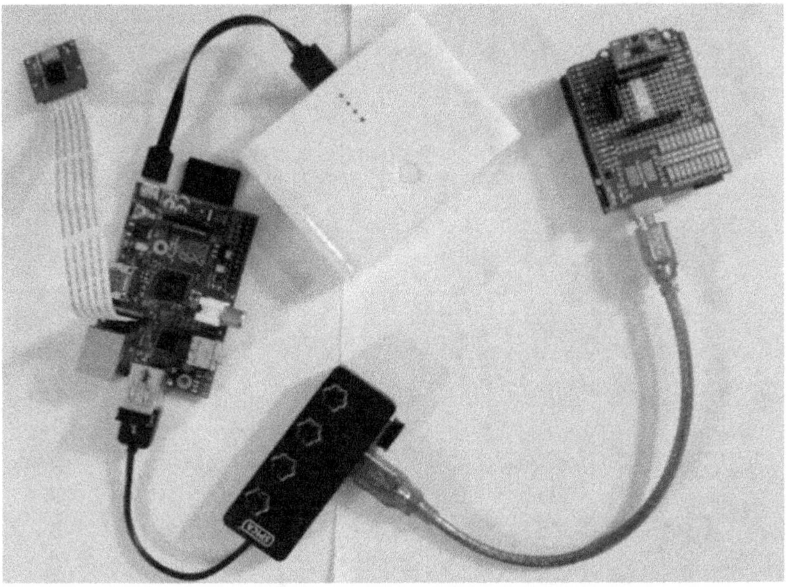

Figure 7.53 Devices and components packed in one node

also accomplished successfully as seen in Figure 7.54. The values of the vibration were verified by Fluke 805 Vibration Meter. During the testing process, it was realized that there was a delay in the system. The GUI did not provide a real-time update when more nodes were connected. With Node 0 and Node 1, the system worked well, but when Node 2 was connected, the video transmission was lagged. It was limited by a low-bandwidth Wi-Fi adaptors. The Wi-Fi adaptors used in this prototype were not capable of transmitting significant data in real time. It was resolved by replacing the Wi-Fi adaptors with a higher-bandwidth Wi-Fi adaptor. It was observed that the cheap battery pack used for each node was draining the power too fast and caused a time lag in the system. As a result, a high-quality battery pack was used. Despite the fact that CV is a powerful technique to identify features such as ice formation along the legs of the rig using input frames from the camera, the environment plays a critical role in the effectiveness of the algorithm.

The suggested architecture supports the application of multiple clients (devices). For example, a standalone application targeted for a windows platform may not work for Android- or Mac-based devices. However, the development of a website-based application was identified to circumvent this issue. Good knowledge of XHTML and JAVA Script was required to develop a web-based application. The proposed GUI can be run in both Windows- and Linux®-based operating system. A vibration monitoring system with image-processing capability will be employed to detect anomalies in vibrations and surface issues. As shown in the proposed system architecture, it concluded that even a low-cost microprocessor-based system was quite efficient in measuring vibration accurately. However, it can be useful in

Figure 7.54 GUI on vibration data and video transmitted by three nodes

measuring the vibration and image processing if it is properly enclosed in a pressure-tolerant enclosure.

As the Wi-Fi works in the high-frequency range, the signals cannot penetrate through the water medium. However, these modules will be replaced with those of low-frequency modules for real-time applications. The image quality of a regular camera will reduce exponentially in a harsh environment in underwater as compared to the lab environment. Proper enclosure design will be performed before actual implementation in underwater.

7.3 Face recognition system

A visual-based surveillance system can be developed using Raspberry Pi. The intelligent system can enhance the modern world of security and surveillance. Local binary patterns (LBP) and Haar cascade classifiers are the two most popular face detection techniques. LBP represents the image as a local representation of matrices with gray-level co-occurrence. The main benefit of implementing LBP is the ability to capture extreme fine details on an image [21]. Haar cascade classifier (see Section 7.1) is a machine-learning approach where the functions of cascades are trained using positive and negative images. The positive images refer to the image of objects and the negative images refer to the images without the object [22].

Face recognition in CV is capable of identifying a person by comparing the facial features detected from the video against the facial database. Two of the most effective detection methods are Eigenfaces and Fisherfaces. Eigenfaces method use principal component analysis (PCA) to represent a gray two-dimensional image with a set of correlated variables represented in the form of the covariance matrix. Eigenvalues and Eigenvectors are calculated and projected over the training samples in a PCA subspace to recognize the facial images. Fisherfaces use linear discriminant analysis (LDA) that determines the combination of features by maximizing the ratio of within-classes from between-classes. The clustering of identical classes and different classes separately are performed in low dimensions [23]. Fisherfaces methods are best in interpolating and extrapolating the lighting situations and better in handling the facial expressions.

7.3.1 System design and architecture

The design of the prototype is shown in Figures 7.55 and 7.56. It consists of the mechanical enclosure with electronics parts. The setup of the Raspberry Pi is similar to the previous section. Hence, it will be omitted in this section. To start the Raspberry Pi, you have to install a few packages. The list of hardware and software used can be seen in Tables 7.7 and 7.8, respectively. The completed assembly of the Raspberry Pi is shown in Figure 7.57.

7.3.2 Programming and system implementation

Python is a widely used high-level, general-purpose, and dynamic programming language. Its design philosophy emphasizes code readability, and its syntax allows

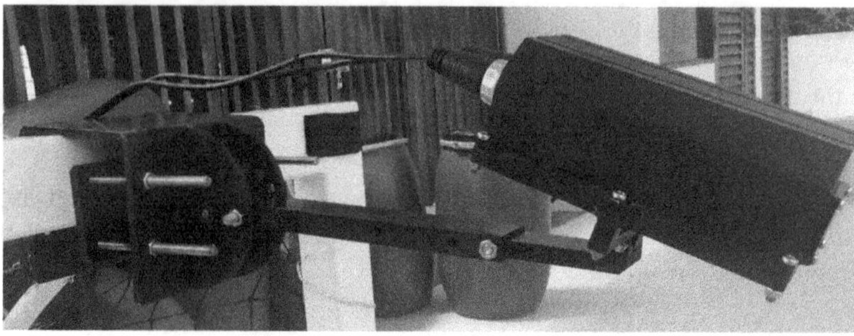

Figure 7.55 Fabricated model

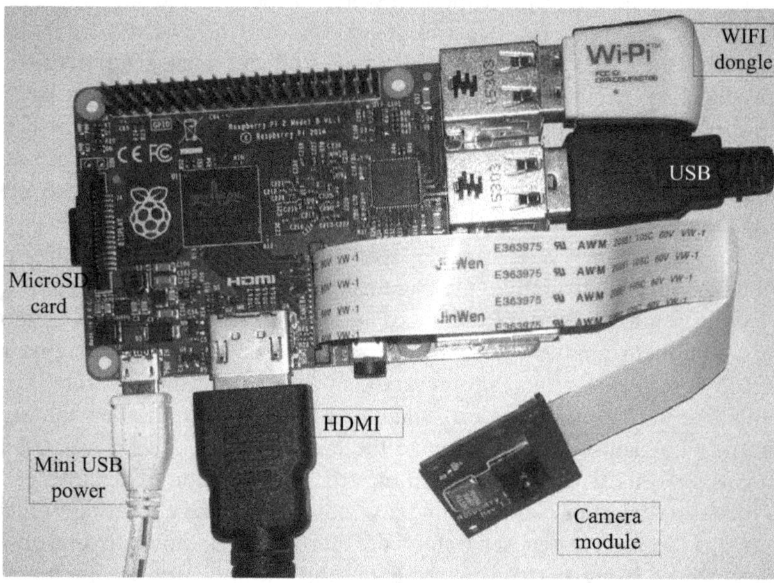

Figure 7.56 Assembled Raspberry Pi

Table 7.7 List of hardware

Hardware	Quantity
Raspberry Pi	1
Pi Camera module	1
WIFI Dongle	1
Micro-USB power cable	1
MicroSD card	1

Table 7.8 List of software

Software	**Purpose**
Raspbian OS	Linux® operating system
Python 2.7	Programing language
TightVNC (Client and server)	SSH remote access
Qt designer	GUI designer
SolidWorks 2011	3D modeling

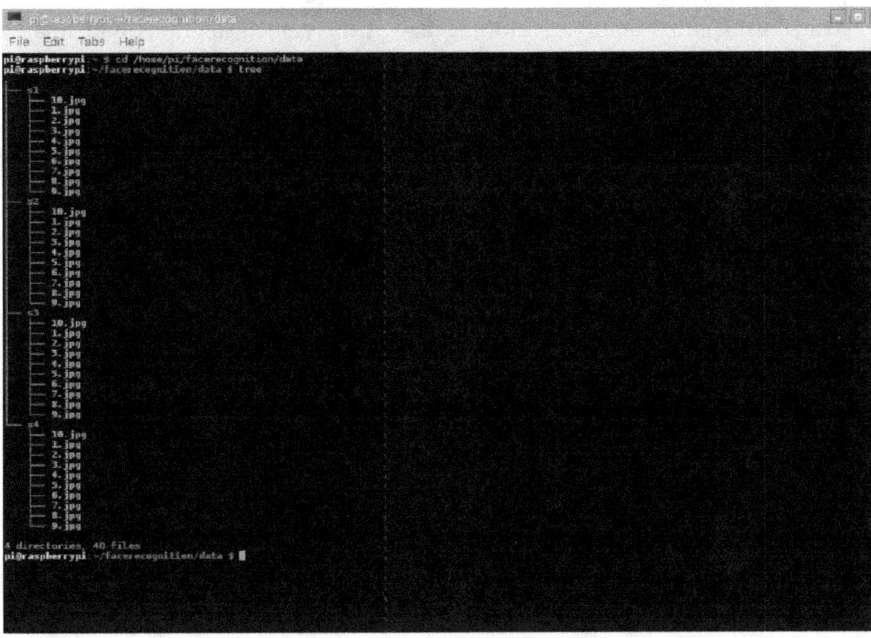

Figure 7.57 Database hierarchy: an example

programmers to express concepts in fewer lines of code that would be possible in languages such as C++ or Java. The language provides specific programs on both a small and large scale. Python supports multiple programming paradigms including object-oriented, imperative, and functional programming or procedural styles. It features a dynamic type system, automatic memory management and has an extensive and comprehensive standard library. Python is chosen as the programming IDE for Raspberry Pi as it has many features, libraries, and also supports the Internet community. There are additional libraries for python for image processing that have to be downloaded separately. As a graphical installer is now available, python packages can be easily installed instead of using the tedious conventional way of using the command prompt. The python packages in the Synaptic Installer are listed as follows:

- **NumPy** extends Python's programming ability to compute fast arrays in multi-dimension. It is essential as image processing is based on manipulating individual pixels. In the picture, the image is converted into arrays of pixels, and each of them has a set of red, blue, green, and alpha values that represents the color.
- **SciPy** stacks onto NumPy as an extension to include many modules such as image-processing modules.
- **Matplotlib** is a library for plotting and displaying histograms and images. Histograms can help to illustrate specific properties of the image such as finding the level of "Red" in the image.
- **Scikit** is also an extension of Python that builds on NumPy and Matplotlib for image processing functions such as color detection and color clustering.
- **OpenCV** can be added to Python to utilize image-processing capabilities. It can also be used for real-time image processing and complements SciPy.
- **IPython** adds on to the Python IDE as a helping tool. It provides easy help, reference, and also running of scripts.
- **"Python imaging-tk"** has to be installed separately using the command prompt as it is missed out in the graphical installer. By typing the following line into the LXTerminal, you will be able to install this file.

```
sudo apt-get install python-imaging-tk
```

The command line tool for capturing still photographs with the Raspberry Pi camera module is as follows.

```
raspistill -o cam.jpg
```

With the camera upside down, images can be rotated 180° to be displayed correctly, and the way to correct for this is to apply both a vertical and a horizontal flip by passing in the -vf and -hf flags:

```
raspistill -vf -hf -o cam2.jpg
```

A Blob is a group of connected pixels in an image that shares some common property (e.g., grayscale value). In a grayscale image, the dark connected regions are blobs, and the goal of blob detection is to identify and mark these regions.

```
# Standard imports
import cv2
import numpy as np;

# Read image
im = cv2.imread("blob.jpg", cv2.IMREAD_GRAYSCALE)

# Set up the detector with default parameters.
detector = cv2.SimpleBlobDetector()

# Detect blobs.
keypoints = detector.detect(im)

# Draw detected blobs as red circles.
# cv2.DRAW_MATCHES_FLAGS_DRAW_RICH_KEYPOINTS ensures the size of the
circle corresponds to the size of blob
im_with_keypoints = cv2.drawKeypoints(im, keypoints, np.array([]), (0,0,255),
cv2.DRAW_MATCHES_FLAGS_DRAW_RICH_KEYPOINTS)

# Show keypoints
cv2.imshow("Keypoints", im_with_keypoints)
cv2.waitKey(0)
```

SimpleBlobDetector, as the name implies, is based on a rather simple algorithm described below. The algorithm is controlled by parameters (shown in bold below) has the following steps. The detailed information can be found from the following website: https://docs.opencv.org/3.3.1/d0/d7a/classcv_1_1SimpleBlobDetector.html

- Thresholding: Convert the source images to several binary images by thresholding the source image with thresholds starting at **minThreshold**. These thresholds are incremented by **thresholdStep** until **maxThreshold**. So, the first threshold is **minThreshold,** the second is **minThreshold** + **thresholdStep,** the third is **minThreshold + 2 x thresholdStep**, and so on.
- Grouping: In each binary image, connected white pixels are grouped as binary blobs.
- Merging: The centers of the binary blobs in the binary images are computed, and blobs located closer than **minDistBetweenBlobs** are merged.
- Center and radius calculation: The centers and radii of the newly merged blobs are computed and returned.

For motion detection based on the spatiotemporal filter, the motion is characterized via the entire three-dimensional (3D) spatio-temporal data volume spanned by the moving person in the image sequence. The advantages are low computational complexity and simple implementation process. Here, the background elimination algorithm is as follows.

```
# Calculate the temporal difference error
  def TDError(self, k, t):
    if t == 10:
      delta = self.reward + self.discount * 0 - self.getQ(k, t, self.state, self.action)
    else:
      delta = self.reward + self.discount * self.getQ(k, t+1, self.nextState,
              self.nextAction) - self.getQ(k, t, self.state, self.action)
    return delta
```

It is a Gaussian mixture-based background/foreground segmentation algorithm. It uses a method to model each background pixel by a mixture of K-Gaussian distributions ($K = 3$ to 5). The weights of the mixture represent the time proportions of those colors stay in the scene. The probable background colors are the ones who stay longer and more static. While coding, we need to create a background object using the function, **cv2.createBackgroundSubtractorMOG ()**. It has some optional parameters such as the length of history, number of Gaussian mixtures, and threshold. Then inside the video loop, use background subtractor.apply() method to get the foreground mask ("OpenCV: Background Subtraction").

```
import numpy as np
import cv2

cap = cv2.VideoCapture('vtest.avi')
fgbg = cv2.createBackgroundSubtractorMOG()
while(1):
et, frame = cap.read()
  fgmask = fgbg.apply(frame)

  cv2.imshow('frame',fgmask)
  k = cv2.waitKey(30) & 0xff
  if k == 27:
    break
cap.release()
cv2.destroyAllWindows()
```

A critical feature of Gaussian mixture-based background/foreground segmentation algorithm is to find an appropriate number of Gaussian distribution for each pixel. It provides better adaptability to varying scenes due to illumination changes. We have to create a background subtractor object. Here, you have an option of selecting whether shadow or not. If detectShadows = True (by default), it detects and marks shadows. However, it decreases the computational speed.

```
import numpy as np
import cv2

cap = cv2.VideoCapture('vtest.avi')

fgbg = cv2.createBackgroundSubtractorMOG2()

while(1):
    ret, frame = cap.read()

    fgmask = fgbg.apply(frame)

    cv2.imshow('frame',fgmask)
    k = cv2.waitKey(30) & 0xff
    if k == 27:
        break

cap.release()
cv2.destroyAllWindows()
```

This algorithm combines statistical background image estimation and per-pixel Bayesian segmentation. It uses frames for background modeling using probabilistic foreground segmentation algorithm that identifies possible foreground objects using Bayesian inference. Several morphological filtering operations such as closing and opening are used to remove unwanted noise. It would be better to apply morphological opening to remove the noises.

```
import numpy as np
import cv2
cap = cv2.VideoCapture('vtest.avi')
kernel = cv2.getStructuringElement(cv2.MORPH_ELLIPSE,(3,3))
fgbg = cv2.createBackgroundSubtractorGMG()
while(1):
    ret, frame = cap.read()
    fgmask = fgbg.apply(frame)
    fgmask = cv2.morphologyEx(fgmask, cv2.MORPH_OPEN, kernel)

    cv2.imshow('frame',fgmask)
    k = cv2.waitKey(30) & 0xff
    if k == 27:
        break
    cap.release()
cv2.destroyAllWindows()
```

As shown in Table 7.9, a comparison study (https://docs.opencv.org/master/db/d5c/tutorial_py_bg_subtraction.html) was performed for different approaches to human motion detection. The Background Subtraction with MOG2 was selected as the method to segment the motion from the background. The Python environment is explained below. The model is mostly static, and lightings are primarily stable over consecutive frames.

Table 7.9 Comparison table for human motion detection

	Accuracy	**Computational time**	**Comments**
BLOB Identification	Moderate	Moderate to high	Works well with images of different scale and pixel ratings but suffers from high computational speed and the accuracy depends on the location and scale of blob
Spatial-temporal filter	Moderate to high	Low to moderate	Works well with the low-resolution scenarios but suffer from noise issues
Background subtraction-			
MOG	Moderate	Moderate	Simple implementation and good performance but not so well with the dynamic background. It can capture multimodel scenario
MOG2	Moderate to high	Moderate	Simple implementation and excellent performance and has better adaptability with varying illumination
GMG	Moderate	Moderate to high	Introduces a problematic foreground segmentation to find the motion, adaptive to the illumination changes but suffers from noise and higher computational requirements

```
# import the necessary packages
# The Package " imutils " is a combination of functions, which make image processing task lighter.
import argparse
import datetime
import imutils
import time
import cv2

# construct the argument parser and parse the arguments
# The "—video " is  used for define a prerecorded video to find the motion.
# The --min-area is  defined to state the minimum size for a region of the image to be considered
ap = argparse.ArgumentParser()
ap.add_argument("-v", "--video", help="path to the video file")
ap.add_argument("-a", "--min-area", type=int, default=500, help="minimum area size")
args = vars(ap.parse_args())

# if the video argument is None, then we are reading from webcam
if args.get("video", None) is None:
        camera = cv2.VideoCapture(0)
        time.sleep(0.25)

# otherwise, we are reading from a video file
else:
        camera = cv2.VideoCapture(args["video"])

# initialize the first frame in the video stream
firstFrame = None
# loop over the frames of the video
while True:
# grab the current frame and initialize the occupied/unoccupied
# text
(grabbed, frame) = camera.read()
text = "Unoccupied"
        # if the frame could not be grabbed, then we have reached the end
        # of the video
        if not grabbed:
                break
        # resize the frame, convert it to grayscale, and blur it
        frame = imutils.resize(frame, width=500)
        gray = cv2.cvtColor(frame, cv2.COLOR_BGR2GRAY)
        gray = cv2.GaussianBlur(gray, (21, 21), 0)
        # if the first frame is None, initialize it
        if firstFrame is None:
                firstFrame = gray
                continue
```

A call to camera.read() returns a 2-tuple. The first value of the tuple is grabbed, indicating whether the frame was successfully read from the buffer. The second value of the tuple is the frame itself. On the other hand, "frame = imutils.resize (frame, width=500)" will resize the image to have a width of 500 pixels. The "gray = cv2.cvtColor(frame, cv2.COLOR_BGR2GRAY) is to convert the image to grayscale and do the Gaussian Blurring to smooth the images as it has no impact on the motion detection algorithm. The Gaussian smoothing is used to average pixel intensities across an 11×11 region. It helps to smooth out high-frequency noise that could affect the detection algorithm.

```
# compute the absolute difference between the current frame and
        # first frame
        frameDelta = cv2.absdiff(firstFrame, gray)
        thresh = cv2.threshold(frameDelta, 25, 255, cv2.THRESH_BINARY)[1]

        # dilate the thresholded image to fill in holes, then find contours
        # on thresholded image
        thresh = cv2.dilate(thresh, None, iterations=2)
        (cnts, _) = cv2.findContours(thresh.copy(), cv2.RETR_EXTERNAL,
                cv2.CHAIN_APPROX_SIMPLE)

        # loop over the contours
        for c in cnts:
                # if the contour is too small, ignore it
                if cv2.contourArea(c) < args["min_area"]:
                        continue

                # compute the bounding box for the contour, draw it on the frame,
                # and update the text
                (x, y, w, h) = cv2.boundingRect(c)
                cv2.rectangle(frame, (x, y), (x + w, y + h), (0, 255, 0), 2)
                text = "Occupied"
```

Computing the difference between two frames is a simple subtraction, where we take the absolute value of their corresponding pixel intensity differences: *delta* = */background_model* – *current_frame/*. Not that the background of the image (see https://docs.opencv.org/master/db/d5c/tutorial_py_bg_subtraction.html) is apparently black. However, regions that contain motion are much lighter. It implies that larger frame deltas indicate that motion is taking place in the image. We will threshold the frameDelta to reveal regions of the image that have significant changes in pixel intensity values. If the delta is less than 25, we discard the pixel and set it to black (i.e., background). If the delta is higher than 25, we will set it to white (i.e., foreground). An example of our thresholded delta image can be seen below. The background of the image is black, whereas the foreground (and where the motion is taking place) is white. Given this threshold image, it is simple to apply contour detection to find the outlines of these white regions. If the contour area is more substantial than our supplied "--min-area," a bounding box surrounding the foreground and motion region on is drawn.

```
# draw the text and timestamp on the frame
        cv2.putText(frame, "Room Status: (de Oliveira et al.)".format(text), (10, 20),
                cv2.FONT_HERSHEY_SIMPLEX, 0.5, (0, 0, 255), 2)
        cv2.putText(frame, datetime.datetime.now().strftime("%A %d %B %Y %I:%M:%S%p"),
                (10, frame.shape[0] - 10), cv2.FONT_HERSHEY_SIMPLEX, 0.35, (0, 0, 255),
1)

        # show the frame and record if the user presses a key
        cv2.imshow("Security Feed", frame)
        cv2.imshow("Thresh", thresh)
        cv2.imshow("Frame Delta", frameDelta)
        key = cv2.waitKey(1) & 0xFF

        # if the `q` key is pressed, break from the lop
        if key == ord("q"):
                break

# cleanup the camera and close any open windows
camera.release()
cv2.destroyAllWindows()
```

The remainder of codes simply combines all the previous commands. We draw the room status on the image in the top-left corner, followed by a timestamp (like "real" security footage) on the bottom-left. A texture descriptor that is called LBPs was used. Unlike the texture features that can be computed by the global representation of texture-based gray level co-occurrence matrix, LBPs will compute a local representation of textures. This local representation is constructed by comparing each pixel with its surrounding neighbor pixel. The first step in constructing the LBP texture descriptor is to convert the image to grayscale. For each pixel in the grayscale image, we select a neighborhood of size r surrounding the center pixel. The LBP value is then calculated for this center pixel and stored in the output 2D array with the same width and height as the input image. A primary benefit of LBP implementation is that we can capture extremely fine-grained details in the image. However, the ability to capture details at such a small scale is one of the drawbacks of the algorithm. For example, we cannot capture details at varying scales, only the fixed 3×3 scale. The number of uniform prototypes in the LBP is entirely dependent on the number of points labeled as p. As the value of p increases, the dimensionality of the histogram increases. However, for a given number of point p in the LBP, there are $p + 1$ uniform patterns. The final dimensionality of the histogram is $p + 2$, where the added entry tabulates all nonuniform patterns. The uniform LBP patterns are important as they will add an extra level of rotation and grayscale invariance. Hence, they are commonly used when extracting LBP feature vectors from images.

```
# import the necessary packages
from skimage import feature
import numpy as np

class LocalBinaryPatterns:
	def __init__(self, numPoints, radius):
		# store the number of points and radius
		self.numPoints = numPoints
		self.radius = radius

	def describe(self, image, eps=1e-7):
		# compute the Local Binary Pattern representation
		# of the image, and then use the LBP representation
		# to build the histogram of patterns
		lbp = feature.local_binary_pattern(image, self.numPoints,
			self.radius, method="uniform")
		(hist, _) = np.histogram(lbp.ravel(),
			bins=np.arange(0, self.numPoints + 3),
			range=(0, self.numPoints + 2))

		# normalize the histogram
		hist = hist.astype("float")
		hist /= (hist.sum() + eps)

		# return the histogram of Local Binary Patterns
		return hist
```

The class LocalBinaryPatterns defines the constructor for LBPs and takes the required parameter that is the radius of the pattern surrounding the central pixel, along with the number of points along the outer radius. The LBP variable returned by the local_binary_patterns function is not directly usable as a feature vector. Instead, LBP is a 2D array with the same width and height as our input image. Each value inside LBP ranges from [0, numPoints + 2] to a value for each of the possible numPoints + 1 possible rotation invariant prototypes. Hence, to construct the actual feature vector, we need to make a call to np.histogram that counts the number of times each LBP prototypes appears. The returned histogram is *numPoints* + *2*-dimensional will be an integer count for each prototype. We then take this histogram to normalize it such that it gives a total of 1 before returning to the calling function.

```
# construct the argument parse and parse the arguments
ap = argparse.ArgumentParser()
ap.add_argument("-t", "--training", required=True,
	help="path to the training images")
ap.add_argument("-e", "--testing", required=True,
	help="path to the tesitng images")
args = vars(ap.parse_args())

# initialize the local binary patterns descriptor along with
# the data and label lists
desc = LocalBinaryPatterns(24, 8)
data = []
labels = []
```

Two switches are used for parsing our command line argument. The following set of codes is used to extract LBP features from the training images.

```
# loop over the training images
for imagePath in paths.list_images(args["training"]):
        # load the image, convert it to grayscale, and describe it
        image = cv2.imread(imagePath)
        gray = cv2.cvtColor(image, cv2.COLOR_BGR2GRAY)
        hist = desc.describe(gray)

        # extract the label from the image path, then update the
        # label and data lists
        labels.append(imagePath.split("/")[-2])
        data.append(hist)

# train a Linear SVM on the data
model = LinearSVC(C=100.0, random_state=42)
model.fit(data, labels)

# loop over the testing images
for imagePath in paths.list_images(args["testing"]):
        # load the image, convert it to grayscale, describe it,
        # and classify it
        image = cv2.imread(imagePath)
        gray = cv2.cvtColor(image, cv2.COLOR_BGR2GRAY)
        hist = desc.describe(gray)
        prediction = model.predict(hist)[0]

        # display the image and the prediction
        cv2.putText(image, prediction, (10, 30), cv2.FONT_HERSHEY_SIMPLEX,
                1.0, (0, 0, 255), 3)
        cv2.imshow("Image", image)
        cv2.waitKey(0)
```

Finally, we can compute a histogram. Haar-like features (as seen in Section 7.1) in two dimensions consist of adjacent light and dark rectangles. The typical features used for recognition are used to detect and recognize objects for real-time facial detection. It is a machine-learning-based approach where a cascade function is trained from many positive and negative images. It is used to detect objects in other images. The presence of a Haar-like feature is determined by adjacent rectangular regions at a particular location within the detection window. The pixel intensities in these regions are summed, and the differences between these regions are calculated. It is then used to place different subsections of an image into different categories. If the difference exceeds a threshold, then the feature is present in the image. To compute the existence of several hundred Haar-like features at different scales and locations in a single image, an integral image technique has been used. Each pixel is assigned as the sum of all pixels above it and to its left. An extension to their approach has also been proposed that analyzes a set of rotated Haar-like features. This set enriches the simple features and can also be computed efficiently. As a result, an average of 10% lower false alarm rate at a given hit rate is achieved. To train our classifier, we need samples of many images that show the object we want to detect (positive sample) and even

Table 7.10 Comparison of face detection methods

Method	Accuracy	Computational time	Comments
Local binary patterns	Moderate–high	High	Higher accuracy rate to detect the very detailed information and is affected by the noise issues. Because of this, its computational requirements are high too.
Haar cascade	High	Moderate	High level of accuracy in finding the facial detection with moderate computational requirements.

more images without the object (negative sample). The number of images depends on a variety of factors, including the quality of the images, the object you want to recognize, the method (see Table 7.10) to generate the samples and the computational power.

Instead of applying all the 6,000 features on a window, group the features into different stages of classifiers and apply one by one. If a window fails the first stage, discard it. We do not consider remaining features on it. If it passes, apply the second stage of features and continue the process. The OpenCV already contains many pretrained classifiers for face, eyes, and smile. Those XML files are stored in opencv/data/haarcascades/ folder. The required XML classifiers can be loaded as follows.

```
face_cascade = cv2.CascadeClassifier('haarcascade_frontalface_default.xml')
eye_cascade = cv2.CascadeClassifier('haarcascade_eye.xml')
```

Then, we need to load input image in the grayscale mode as follows.

```
img = cv2.imread('xfiles4.jpg')
gray = cv2.cvtColor(img, cv2.COLOR_BGR2GRAY)
```

Cv2.CascadeClassifier.detectMultiScale() is used to find faces or eyes. It is defined as follows.

```
cv2.CascadeClassifier.detectMultiScale(image[, scaleFactor[, minNeighbors[, flags[,
minSize[, maxSize]]]]])
```

The parameters used are as follows.

- **Image**: Matrix of the type CV_8U containing an image where objects are detected.
- **scaleFactor**: How much the image size is reduced at each image scale? This scale factor is used to create scale pyramid. With the scale factor is 1.03, it means that using a small step for resizing, it reduces the size by 3%.
- **minNeighbors** : How many neighbors each candidate rectangle should have to retain it? This parameter will affect the quality of the detected faces. A higher value results in fewer detections with higher quality.
- **Flags**: Parameter with the same meaning for an old cascade as in the function cvHaarDetectObjects. It is not used for a new cascade.
- **minSize** : Minimum possible object size. Objects smaller will be ignored.
- **maxSize** : Maximum possible object size. Objects larger will be ignored.

If faces are found, it returns the positions of the detected faces as Rect(x,y,w,h).

```
import numpy as np
import cv2
from matplotlib import pyplot as plt

face_cascade = cv2.CascadeClassifier('haarcascade_frontalface_default.xml')
eye_cascade = cv2.CascadeClassifier('haarcascade_eye.xml')

img = cv2.imread('sajin.jpg')
gray = cv2.cvtColor(img, cv2.COLOR_BGR2GRAY)

faces = face_cascade.detectMultiScale(gray, 1.3, 5)


for (x,y,w,h) in faces:
   cv2.rectangle(img,(x,y),(x+w,y+h),(255,0,0),2)
   roi_gray = gray[y:y+h, x:x+w]
   roi_color = img[y:y+h, x:x+w]
   eyes = eye_cascade.detectMultiScale(roi_gray)
   for (ex,ey,ew,eh) in eyes:
      cv2.rectangle(roi_color,(ex,ey),(ex+ew,ey+eh),(0,255,0),2)

cv2.imshow('img',img)
cv2.waitKey(0)
cv2.destroyAllWindows()
```

The AT&T Face database is sometimes known as ORL Database of Faces. It contains ten different images of 40 distinct subjects. For some subjects, the images were taken at different times, varying the lighting, facial expressions (open/closed eyes, smiling/not smiling) and facial details (glasses/no glasses). All the images were taken against a dark homogeneous background with the subjects in an upright and frontal position (with tolerance for some side movement). The face images are stored in a folder hierarchy similar to the AT&T face database as <database name>/<subject name>/<filename >.<ext> as shown in Figure 7.57.

The following function can be used to read the images of each subfolder in a given directory. Each directory is given a unique (integer) label and stores the folder name. The function returns the images and the corresponding classes.

```
def read_images(path, sz=None): c=0
X,y = [], []
for dirname , dirnames , filenames in os.walk(path):
for subdirname in dirnames:
subject_path = os.path.join(dirname , subdirname) for filename in os.listdir(subject_path):
try:
im = Image.open(os.path.join(subject_path , filename)) im = im.convert("L")
# resize to given size (if given)
if (sz is not None):
im = im.resize(sz, Image.ANTIALIAS) X.append(np.asarray(im, dtype=np.uint8))
y.append(c)
except IOError:
print "I/O error((0)): (1)".format(errno, strerror)
except:
print "Unexpected error:", sys.exc_info()[0] raise
c = c+1 return [X,y]
```

The problem with the image representation is given a high dimensionality. Two-dimensional p $\times$ q grayscale images span a pq-dimensional vector space, so an image with 100 $\times$ 100 pixels lies in a 10,000-dimensional image space. It contains too much information that may not be useful. As a result, the components that account for most of the information are taken into account. Hence, a PCA can be proposed. The idea is that a high-dimensional dataset is often described by correlated variables. There exist a few meaningful dimensions for the most dominant information. The PCA method finds the directions with the most significant variance in the data called principal components. Eigenfaces method can be used. The following code is defined to reshape the multidimensional data.

```
def asRowMatrix(X): if len(X) == 0:
return np.array([])
mat = np.empty((0, X[0].size), dtype=X[0].dtype) for row in X:
mat = np.vstack((mat, np.asarray(row).reshape(1,-1))) return mat
def asColumnMatrix(X): if len(X) == 0:
return np.array([])
mat = np.empty((X[0].size, 0), dtype=X[0].dtype) for col in X:
mat = np.hstack((mat, np.asarray(col).reshape(-1,1))) return mat
```

The following set of codes will implement the PCA. It also implements the inner-product PCA formulation that occurs if there are more dimensions than samples.

```
def pca(X, y, num_components=0): [n,d] = X.shape
if (num_components <= 0) or num_components = n
mu = X.mean(axis=0) X = X - mu
if n>d:
C = np.dot(X.T,X)
[eigenvalues ,eigenvectors] = else:
C = np.dot(X,X.T)
[eigenvalues ,eigenvectors] = eigenvectors = np.dot(X.T,eigenvectors) for i in xrange(n):
eigenvectors[:,i] = eigenvectors[:,i]/np.linalg.norm(eigenvectors[:,i])
(num_components >n):
np.linalg.eigh(C)
np.linalg.eigh(C)
# or simply perform an economy size decomposition
# eigenvectors , eigenvalues , variance = np.linalg.svd(X.T,
# sort eigenvectors descending by their eigenvalue
idx = np.argsort(-eigenvalues)
eigenvalues = eigenvalues[idx]
eigenvectors = eigenvectors[:,idx]
# select only num_components
eigenvalues = eigenvalues[0:num_components].copy() eigenvectors =
eigenvectors[:,0:num_components].copy() return [eigenvalues , eigenvectors , mu]
full_matrices=False)
```

The data are arranged as shown.

```
def project(W, X, mu=None):
if mu is None:
return np.dot(X,W)
return np.dot(X - mu, W)

def reconstruct(W, Y, mu=None):
if mu is None:
return np.dot(Y,W.T)
return np.dot(Y, W.T) + mu
```

The face images are read followed by performing the PCA.

```
import matplotlib.cm as cm
# turn the first (at most) 16 eigenvectors into grayscale # images (note: eigenvectors are
stored by column!)
E = []
for i in xrange(min(len(X), 16)):
e = W[:,i].reshape(X[0].shape)
E.append(normalize(e,0,255))
# plot them and store the plot to "python_eigenfaces.pdf"
subplot(title="Eigenfaces AT&T Facedatabase", images=E, rows=4, cols=4, sptitle="
Eigenface", colormap=cm.jet, filename="python_pca_eigenfaces.pdf")
```

To see how the grayscale values are distributed with the specific Eigenfaces, few color maps are used. Eigenfaces do not only encode facial features but also illuminate the images as shown in Figure 7.58.

The face form can be reconstructed from a lower-dimensional approximation as shown.

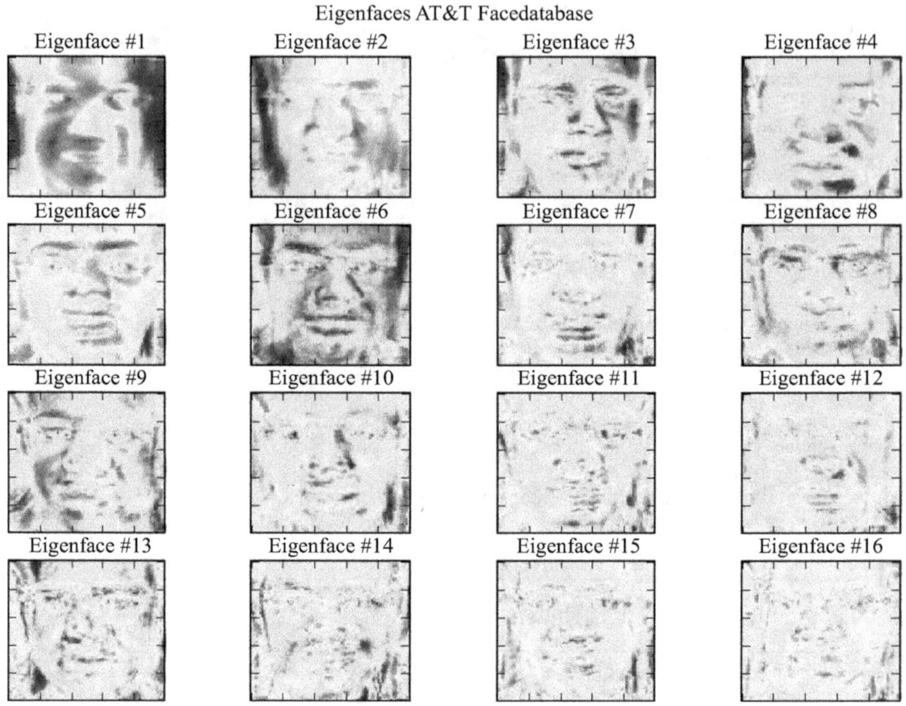

Figure 7.58 Examples of how grayscale values distributed with specific Eigenfaces

```
from tinyfacerec.subspace import project , reconstruct
# reconstruction steps
steps=[i for i in xrange(10, min(len(X), 320), 20)] E = []
for i in xrange(min(len(steps), 16)):
numEvs = steps[i]
import matplotlib.cm as cm
# turn the first (at most) 16 eigenvectors into grayscale # images (note: eigenvectors are
stored by column!)
E = []
for i in xrange(min(len(X), 16)):
e = W[:,i].reshape(X[0].shape)
E.append(normalize(e,0,255))
# plot them and store the plot to "python_eigenfaces.pdf"
subplot(title="Eigenfaces AT&T Facedatabase", images=E, rows=4, cols=4, sptitle="
Eigenface", colormap=cm.jet, filename="python_pca_eigenfaces.pdf")
P = project(W[:,0:numEvs], X[0].reshape(1,-1), mu) R = reconstruct(W[:,0:numEvs], P, mu)
# reshape and append to plots
R = R.reshape(X[0].shape) E.append(normalize(R,0,255))
# plot them and store the plot to "python_reconstruction.pdf"
subplot(title="Reconstruction AT&T Facedatabase", images=E, rows=4, cols=4, sptitle="
Eigenvectors", sptitles=steps, colormap=cm.gray, filename="
python_pca_reconstruction.pdf")
```

Around ten Eigenvectors are apparently not sufficient for a proper image reconstruction; 50 Eigenvectors can be sufficient to encode critical facial features. You will get an accurate reconstruction with approximately three hundreds Eigenvectors for the AT&T Facedatabase as seen in one of the following databases (see all databases in http://www.face-rec.org/databases/). There is the rule of thumbs how many Eigenfaces to choose for successful face recognition. However, it depends on the input data used.

The LDA was proposed by the great statistician Sir R. A. Fisher who successfully used it for classifying flowers using multiple measurements in taxonomic problems. The PCA finds a linear combination of features that maximize the total variance in data. It is a powerful way to represent data. However, it does not consider any classes, and discriminative information lost if components are removed. The components identified by a PCA do not necessarily contain discriminative information. As a result, the projected samples are smeared together, and classification becomes difficult. To find the combination of features that separate them into different classes, LDA maximizes the ratio of between-classes to within-classes scatter. The idea is simple: same classes should cluster tightly together, while different classes are grouped as far away as possible. The Fisherfaces in Python language can be written as follows.

```
def lda(X, y, num_components=0): y = np.asarray(y)
[n,d] = X.shape
c = np.unique(y)
if (num_components <= 0) or (num_component >(len(c)-1)): num_components = (len(c)-
1)
meanTotal = X.mean(axis=0)
Sw = np.zeros((d, d), dtype=np.float32) Sb = np.zeros((d, d), dtype=np.float32) for i in c:
Xi = X[np.where(y==i)[0],:]
meanClass = Xi.mean(axis=0)
Sw = Sw + np.dot((Xi-meanClass).T, (Xi-meanClass))
Sb = Sb + n * np.dot((meanClass - meanTotal).T, (meanClass - meanTotal))
eigenvalues , eigenvectors = np.linalg.eig(np.linalg.inv(Sw)*Sb)
idx = np.argsort(-eigenvalues.real)
eigenvalues , eigenvectors = eigenvalues[idx], eigenvectors[:,idx]
eigenvalues = np.array(eigenvalues[0:num_components].real, dtype=np.float32, copy=
True)
eigenvectors = np.array(eigenvectors[0:,0:num_components].real, dtype=np.float32,
copy=True)
return [eigenvalues , eigenvectors]
```

Using the code above, the functions to perform a PCA and LDA are now defined. We can implement the Fisherfaces.

```
def fisherfaces(X,y,num_components=0): y = np.asarray(y)
[n,d] = X.shape
c = len(np.unique(y)) [eigenvalues_pca , eigenvectors_pca , [eigenvalues_lda ,
eigenvectors_lda]
mu_pca] = pca(X, y, (n-c))
= lda(project(eigenvectors_pca , X,
mu_pca), y,
num_components)
eigenvectors = np.dot(eigenvectors_pca ,eigenvectors_lda) return [eigenvalues_lda ,
eigenvectors , mu_pca]
```

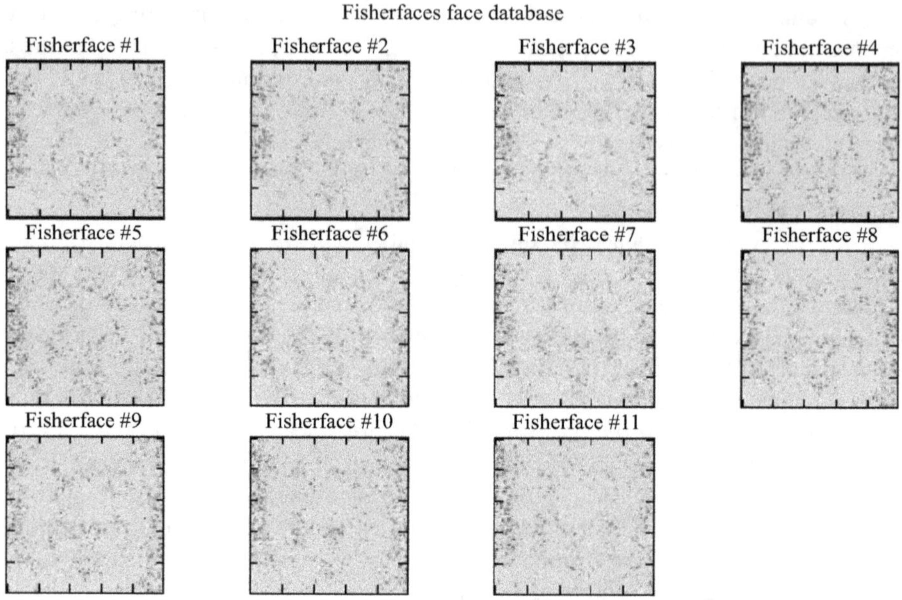

Figure 7.59 Example of how grayscale values are distributed with specific Fisherfaces

The Fisherfaces method learns a class-specific transformation matrix. It does not capture illumination as naturally as the Eigenfaces method as seen in Figure 7.59. The discriminant analysis determines the facial features to discriminate between the persons. The performance of the Fisherfaces depends on the input data.

```
import sys
# append tinyfacerec to module search path
sys.path.append("..")
# import numpy and matplotlib colormaps
import numpy as np
# import tinyfacerec modules
from tinyfacerec.subspace import fisherfaces
from tinyfacerec.util import normalize , asRowMatrix , read_images
from tinyfacerec.visual import subplot
# read images
[X,y] = read_images("/home/pi/facerecognition/data ") # perform a full pca
[D, W, mu] = fisherfaces(asRowMatrix(X), y)
#import colormaps
import matplotlib.cm as cm
# turn the first (at most) 16 eigenvectors into grayscale
# images (note: eigenvectors are stored by column!)
E = []
for i in xrange(min(W.shape[1], 16)):
e = W[:,i].reshape(X[0].shape)
E.append(normalize(e,0,255))
# plot them and store the plot to "python_fisherfaces_fisherfaces.pdf"
subplot(title="Fisherfaces AT&T Facedatabase", images=E, rows=4, cols=4, sptitle="
Fisherface", colormap=cm.jet, filename="python_fisherfaces_fisherfaces.pdf")
```

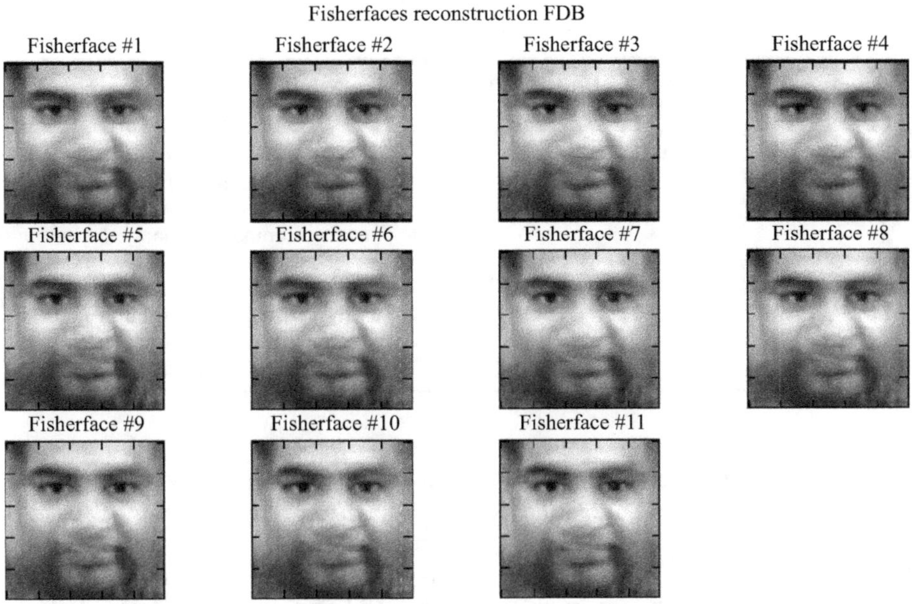

Figure 7.60 Fisherfaces face reconstruction

The Fisherfaces allow a reconstruction of the projected image, just like the Eigenfaces. However, since we can only identify the features to distinguish between subjects, you cannot expect an excellent approximation of the original image as shown in Figure 7.60.

```
from tinyfacerec.subspace import project , reconstruct
E = []
for i in xrange(min(W.shape[1], 16)):
e = W[:,i].reshape(-1,1)
P = project(e, X[0].reshape(1,-1), mu) R = reconstruct(e, P, mu)
# reshape and append to plots
R = R.reshape(X[0].shape) E.append(normalize(R,0,255))
# plot them and store the plot to "python_reconstruction.pdf"
subplot(title="Fisherfaces Reconstruction Yale FDB", images=E, rows=4, cols=4, sptitle
="Fisherface", colormap=cm.gray, filename="python_fisherfaces_reconstruction.pdf")
```

Table 7.11 explains and compares the different methods of the face recognition [23,24].

7.3.3 GUI using PyQt

GUI is a software that works at the point of contact (interface) between a computer and its user. It employs few graphic elements (dialog boxes, icons, menus, and scroll bars) instead of text characters to allow the user to interact with the computer. The GUI elements are usually accessed through a pointing device such as a mouse,

Table 7.11 Comparison table on face recognition methods

Method	Accuracy	Computational time	Comments
Eigenfaces	Moderate–high	Moderate	Higher accuracy rate to reconstruct the file form projected image. The accuracy of the Eigenfaces suffers from illumination changes.
Fisherfaces	High	Moderate	The Fisherfaces method appears to be the best at extrapolating and interpolating over variation in lighting. Found to be better at handling expressions also.

pen, or stylus. All programs running under a GUI use a consistent set of graphical elements. Pioneered by Xerox and developed by Apple computers, GUI is now employed by all modern operating systems and application programs. We are using PyQt as GUI developer (https://www.qt.io/). PyQt is a GUI widgets toolkit. It is a Python interface for Qt, one of the most powerful and accessible cross-platform GUI library. PyQt is a blend of Python programming language and the Qt library. To install the PyQT and the Qt designer, run the following commands.

```
apt-get install python-pyqt5 pyqt5-dev-tools qt5-designer
```

The Qt designer makes the Designing GUI much easier. It is easy to drag and drop the widgets and other functions to the window. The steps to get GUI are explained below. First, open a new window as shown.

```
Open QT designer > select Widgets> click on Create
```

It will create a window as shown below. The widgets that we need can be inserted into the designer window as shown in Figure 7.61.

The Property on the left side of the window can be used to specify the Widgets properties and the object name in the main program (see Figures 7.62 and 7.63).

Once the GUI is fixed with the widgets and the buttons, a preview can be performed.

It will show the preview of the GUI as shown in Figure 7.64.

After GUI design is completed, save the file as ".ui file. Then go to terminal and convert the file to python file by using the following command.

```
pyuic –x <filename.ui> -o <filename.py>"
```

The new file is the python script for the GUI. Moreover, we can use the object names in the code to work on GUI as shown in Figure 7.65.

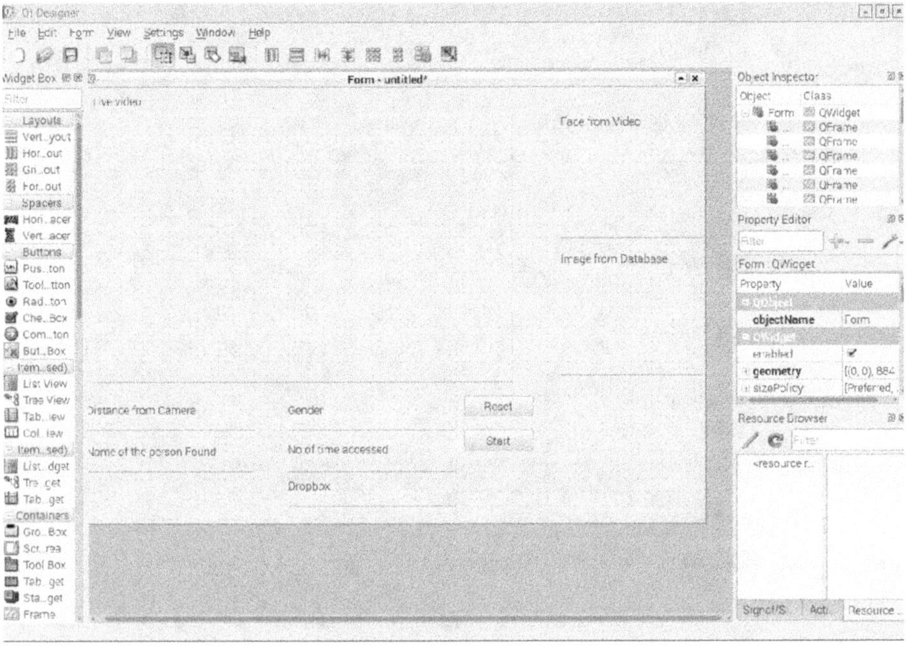

Figure 7.61 Qt designer window

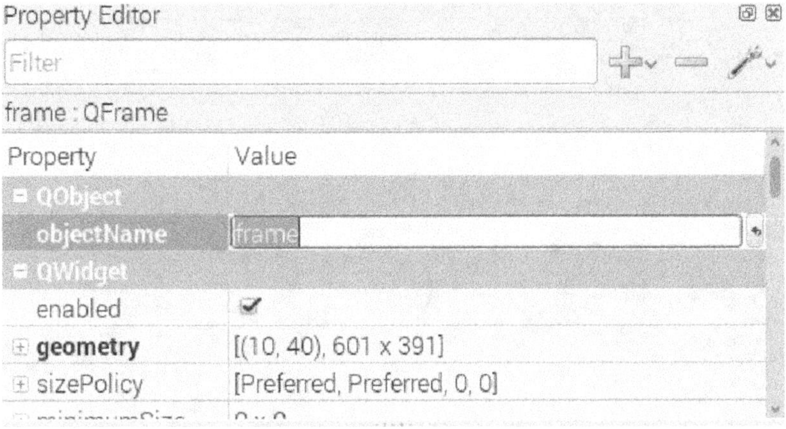

Figure 7.62 Qt designer Property Editor

7.3.4 Testing

The system mainly incorporates image-processing techniques and the CV together with a surveillance system. The capabilities of the system are as follows.

- Monitor the premises closely;
- Provide captured photo with time and date stamp;

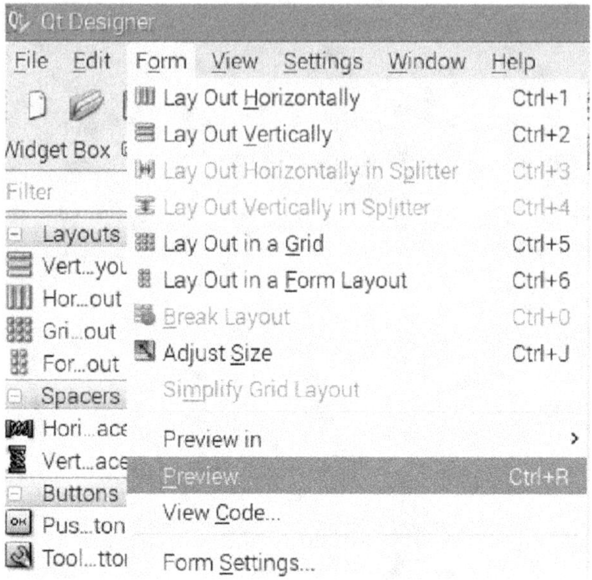

Figure 7.63 Menu for preview

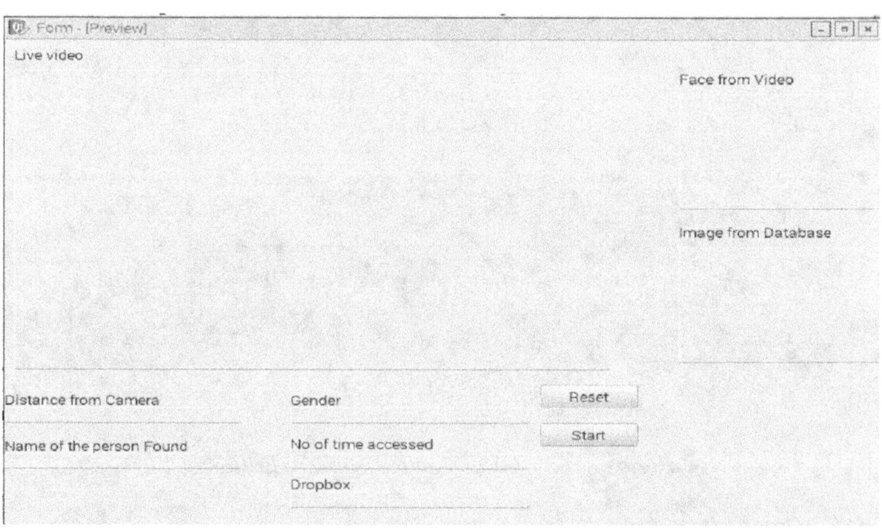

Figure 7.64 Preview window

Figure 7.65 Python script generated

- Compare the captured photo with the database to find the person;
- Give information like distance from camera and number of times entering the room;
- Integrate with Dropbox.

Raspberry Pi and Camera Module act as the backbone of the system for processing. Python is installed in the Raspbian (a Linux®-based GUI). The OpenCV is used as the key module for CV and Python Imaging Library modules. The system uses SSH that is an encrypted protocol that allows remote login to Host from different clients connected to the same Network over WIFI as shown in Figure 7.66.

The Raspberry Pi acts as the host that enables the network to interact with the Raspbian. The Wi-Fi modem acts as the server for the system that provides the Internet access to the host and clients. It acts as a bridge for the information exchange. The client will be useful for SSH as the information from the Raspberry Pi can be accessed from their platforms of comfort such as Android, IOS, OSX, Windows, and Linux® (see Figure 7.67).

The CV-based algorithms are scripted in the Python as shown in the following flowchart in Figure 7.68.

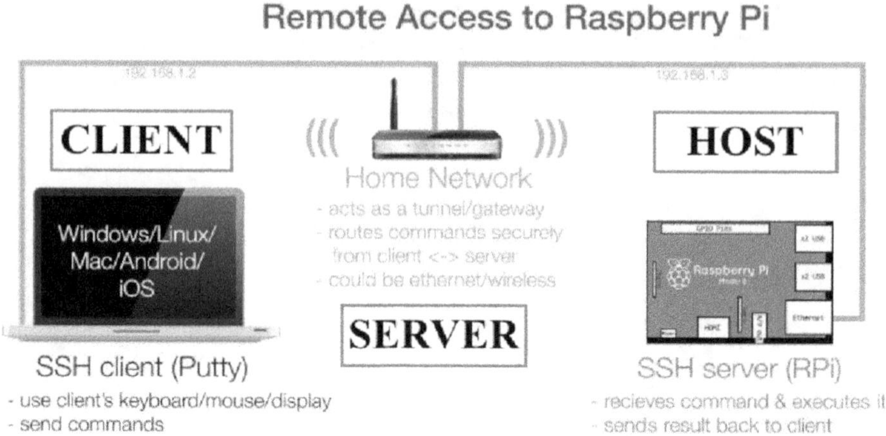

Figure 7.66 SSH network overview

Figure 7.67 SSH Clients accessing Raspberry Pi

Distance calculation from camera to an object is a well-studied problem in CVs. There are a few simple and straightforward methods such as triangle similarity method that give very sophisticated methods. In this section, the triangle similarity method was used. The triangular similarity method used an algebraic approach to find the distance between the camera and the object. For example, the Object with a width W placed at D distance from the camera, a picture is taken, and the apparent width is measured in pixels. With these values, the focal length F of the camera can be calculated.

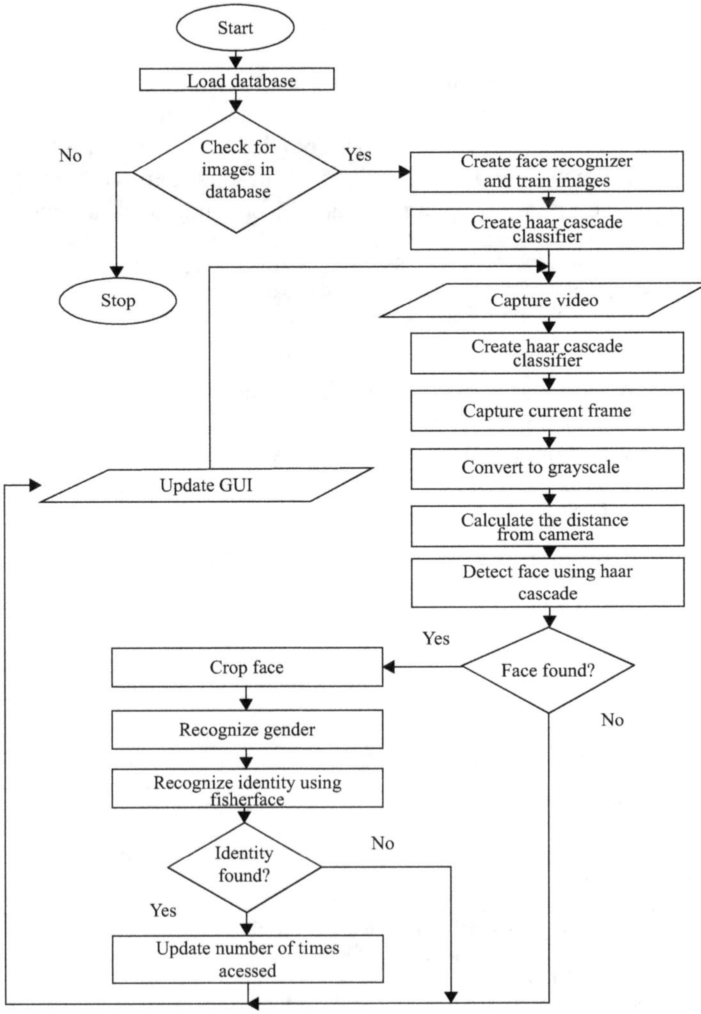

Figure 7.68 Flowchart of the proposed system

```
# import the necessary packages
import numpy as np
import cv2

def find_marker(image):

        gray = cv2.cvtColor(image, cv2.COLOR_BGR2GRAY)
        gray = cv2.GaussianBlur(gray, (5, 5), 0)
        edged = cv2.Canny(gray, 35, 125)

        (cnts, _) = cv2.findContours(edged.copy(), cv2.RETR_LIST,
cv2.CHAIN_APPROX_SIMPLE)
        c = max(cnts, key = cv2.contourArea)

        return cv2.minAreaRect(c)
```

The find_marker is defined to accept single argument image to find the object. The Markers used cv2.findContours function and the Counter that have the maximum selected. With this counter that corresponds to the marker, a bounding box with the *x*, *y* coordinates, width and height of the box were computed. The following codes will compute the distance to an object using the triangular similarity.

```
def distance_to_camera(knownWidth, focalLength, perWidth):
    # compute and return the distance from the maker to the camera
    return (knownWidth * focalLength) / perWidth
```

This function takes a known width of the marker, a computed focalLength, and perceived width of an object in an image (measured in pixels), and applies the triangle similarity to compute the actual distance to the object.

```
KNOWN_DISTANCE = 12.0

# initialize the known object width, which in this case, the piece of
# paper is 11 inches wide
KNOWN_WIDTH = 12.0

# initialize the list of images that we'll be using
IMAGE_PATHS = ["images/2ft.png", "images/3ft.png", "images/4ft.png"]

# load the furst image that contains an object that is KNOWN TO BE 2 feet
# from our camera, then find the paper marker in the image, and initialize
# the focal length
image = cv2.imread(IMAGE_PATHS[0])
marker = find_marker(image)
focalLength = (marker[1][0] * KNOWN_DISTANCE) / KNOWN_WIDTH
# loop over the images
for imagePath in IMAGE_PATHS:
    # load the image, find the marker in the image, then compute the
    # distance to the marker from the camera
    image = cv2.imread(imagePath)
    marker = find_marker(image)
    inches = distance_to_camera(KNOWN_WIDTH, focalLength, marker[1][0])

    # draw a bounding box around the image and display it
    box = np.int0(cv2.cv.BoxPoints(marker))
    cv2.drawContours(image, [box], -1, (0, 255, 0), 2)
    cv2.putText(image, "%.2fft" % (inches / 12),
        (image.shape[1] - 200, image.shape[0] - 20),
cv2.FONT_HERSHEY_SIMPLEX,
        2.0, (0, 255, 0), 3)
    cv2.imshow("image", image)
    cv2.waitKey(0)
```

Gender identification from a facial image is one of the primary tasks in face recognition algorithms. The human face has many visual elements that provide one or more information sources for gender identification. It is easy to identify the gender of a person from the facial features using the database as seen in Figure 7.69. However, for the CV, the situation is different. PCA for face images is

Figure 7.69 Database for gender detection training

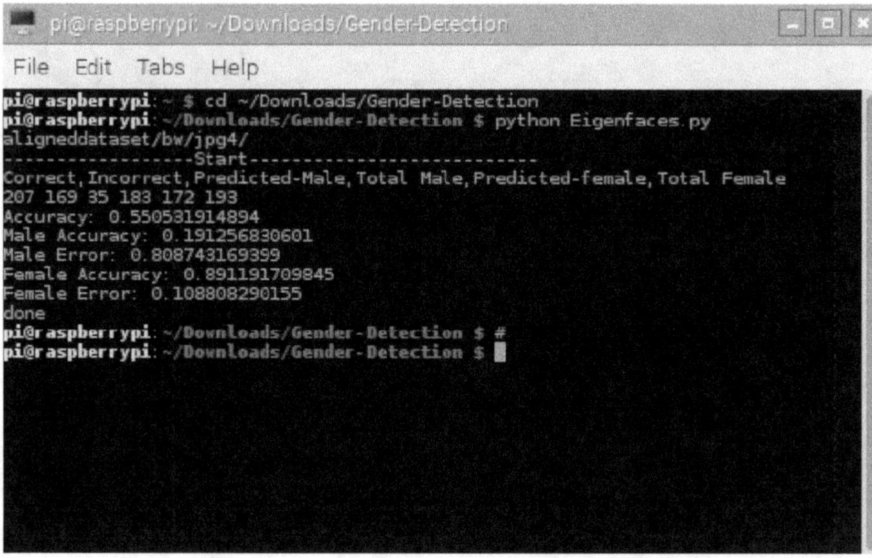

Figure 7.70 Gender detection training results

a well-established technique for dimension reduction. The first step is to apply PCA to reduce the dimensionality and then using the Eigenmatrix to find Eigenfaces. For each image in test dataset, the Eigenfaces Matrix was computed to classify the image. The results can be seen in Figure 7.70.

```
from numpy import array, dot, mean, std, empty, argsort ,size ,shape ,transpose
from numpy.linalg import eigh, solve
from numpy.random import randn
from matplotlib.pyplot import subplots, show
import Image
import numpy as np
import os

def pca(data,k,m):
   """
       Principal component analysis using eigenvalues
       note: this mean-centers and autoscales the data (in-place)
   """
   mean_mat =data-m
   C = np.dot(mean_mat.transpose(),mean_mat)

#Eigen values and V is eigenvectors
  E, V = eigh(C)
#key will have indices of array
   key = argsort(E)[::-1][:k]

   E, V = E[key], V[:, key]
#eigen matrix ka transpose bhj rha hu i.e. k*48
   V=V.T
   #print "Dim of eigen matrix",V.shape,data.shape
```

```
# U is projection matrix
 U = np.dot(V,data.transpose())
   return U,V
#----------------Start of program---------------
dirname="aligneddataset/bw/jpg4/"
print dirname
count=0

#for color pics change it to 140,140
size=48,48

for filename in os.listdir(dirname):
        count=count+1
        image_file=Image.open(dirname+filename)
#initializing size of the image
size=image_file.size

#total no. of images
total=count

#trainging data is half the size of total no. of images
data=np.zeros((total//2,size[0]*size[1]))

count=0
trainglabel=[]                                          #training label
testlabel=[]                                            #testing label
trainf=[]                                                       #training filename
testf=[]                                                #training filename,used
later for testing each test image
j=0
for filename in os.listdir(dirname):

        if count < total//2:
                trainglabel.append(filename[7])
                trainf.append(filename)
                #print filename[7]
                image_file=Image.open(dirname+filename)
                temp = array(image_file.convert('1'))
                A = np.asarray(temp).reshape(-1)
                #print temp.shape,A.shape
                data[count,:]=A

        else:
                testf.append(filename)
                testlabel.append(filename[7])
        count=count+1
```

```
#print label
print "------------------Start--------------------------"
m=mean(data, 0)
#print "Dim of mean ",m.shape
k=200                           #751/10=75
projmatrix,eigenmatrix = pca(data,k,m)

#saving eigen values for backup
np.save("eigmat1.txt",eigenmatrix)
#print projmatrix.shape
#print "Eigen Matrix",eigenmatrix.shape

#Eigen Matrix of size==k X size[0]*size[1]
eigenfaces=eigenmatrix*m                                #eigenfaces=eigenmatrix*mean
of data

wttraining=np.dot(eigenfaces,data.transpose()).transpose()              #sizze of weight
for trainging data == SIZE(training X k)

#print wttraining.shape
#print data.shape[0],len(trainglabel),len(trainf),len(testlabel),len(testf)

male=0
female=0
correct=0
incorrect=0

for j in range(len(testf)):
        testimg=Image.open(dirname+testf[j])
        testimg=np.asarray(testimg).reshape(-1)
        #print eigenfaces.shape,testimg.transpose().shape
        wttest=np.dot(eigenfaces,testimg.transpose()).transpose()                #size of
weight for teseting data==SIZE(1 X K)
        #print wttest.shape
```

```
        cur=1000000000
        mini=1000000000
        maxi=0
        ind=0

        for i in range(len(data)):
                cur=np.linalg.norm(wttraining[i,:]-wttest[:])                    #Dist of
ith-EigenFace with current test data
                if mini>cur:
                        mini=cur
                        ind=i
        if trainglabel[ind]==testlabel[j]:
        #ind is the Least distance,i.e.predicted value

                correct=correct+1

                if testlabel[j]=='m':
                        male=male+1
                else:
                        female=female+1

        else:
                incorrect=incorrect+1

print "Correct,Incorrect,Predicted-Male,Total Male,Predicted-female,Total Female"
print correct,incorrect,male,testlabel.count('m'),female,testlabel.count('f')
acc=correct*1.0/(incorrect+correct)
print "Accuracy:",acc
print "Male Accuracy:",(male*1.0)/testlabel.count('m')
print "Male Error:",1-(male*1.0)/testlabel.count('m')
print "Female Accuracy:",(female*1.0)/testlabel.count('f')
print "Female Error:",1-(female*1.0)/testlabel.count('f')
print "done"
```

The integration with Dropbox allows the images to upload to Dropbox. It will give great flexibility to access the images from anywhere. The integration steps are as follows. First, the Dropbox API app was created. Click on "Create your app" as shown in Figures 7.71 and 7.72.

Once the app is created, take down the "App Key" and the following information as shown in Figure 7.73.

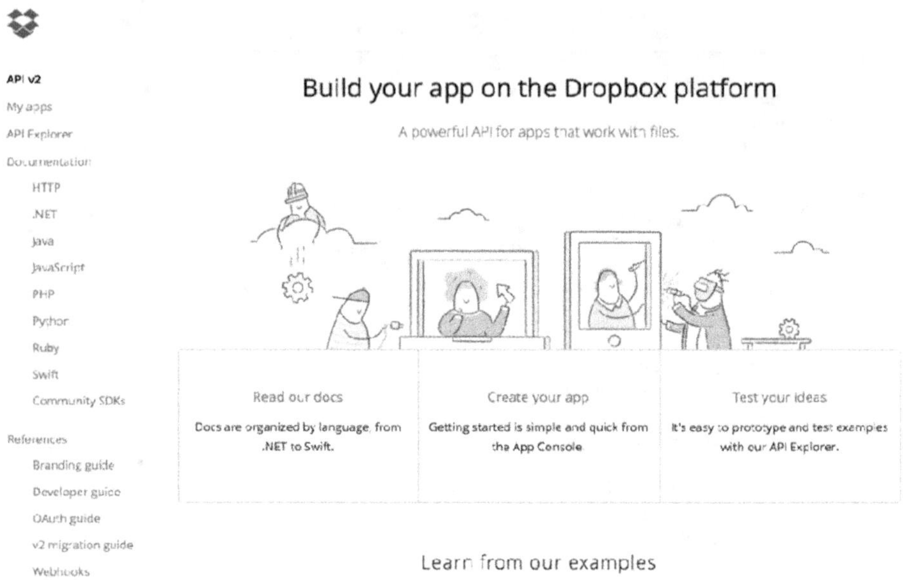

Figure 7.71 Create API app for Dropbox 1

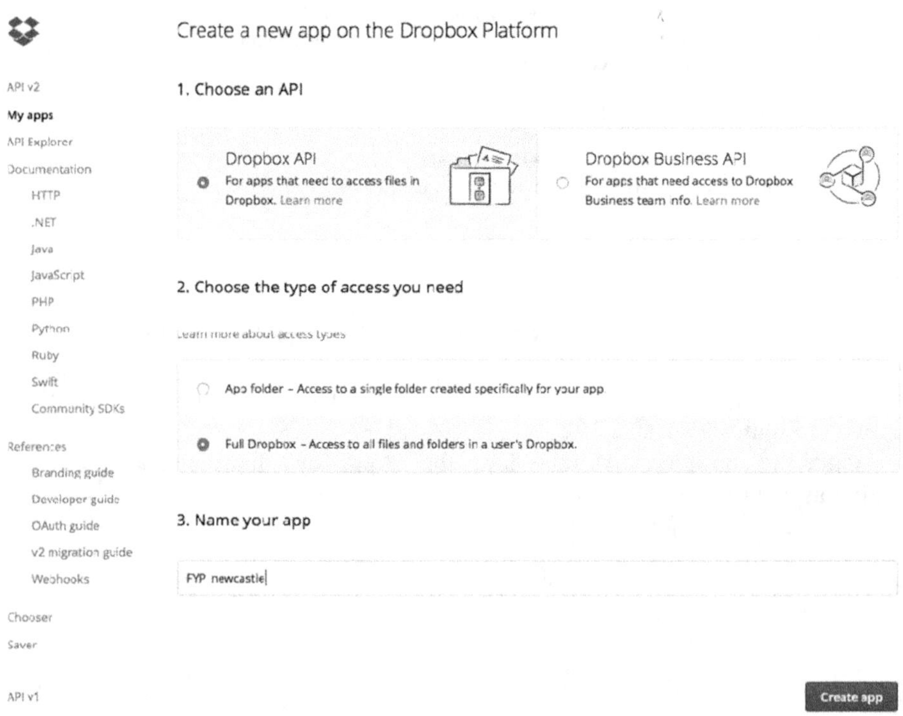

Figure 7.72 Create API app for Dropbox 2

```
# import the necessary packages
from pyimagesearch.tempimage import TempImage
from dropbox.client import DropboxOAuth2FlowNoRedirect
from dropbox.client import DropboxClient
from picamera.array import PiRGBArray
from picamera import PiCamera
import argparse
import warnings
import datetime
import imutils
import json
import time
import cv2

# construct the argument parser and parse the arguments
ap = argparse.ArgumentParser()
ap.add_argument("-c", "--conf", required=True,
        help="path to the JSON configuration file")
args = vars(ap.parse_args())

# filter warnings, load the configuration and initialize the Dropbox
# client
warnings.filterwarnings("ignore")
conf = json.load(open(args["conf"]))
client = None

if conf["use_dropbox"]:
        # connect to dropbox and start the session authorization process
        flow = DropboxOAuth2FlowNoRedirect(conf["dropbox_key"],
conf["dropbox_secret"])
        print "[INFO] Authorize this application: ".format(flow.start())
        authCode =
"2gfc5tlVceAAAAAAAAAAPIBfaYiI582zXosBfU7KVHnwQAlW8AFnc1AapyJKh
Szg"

        # finish the authorization and grab the Dropbox client
        (accessToken, userID) = flow.finish(authCode)
        client = DropboxClient(accessToken)
        print "[SUCCESS] dropbox account linked"
```

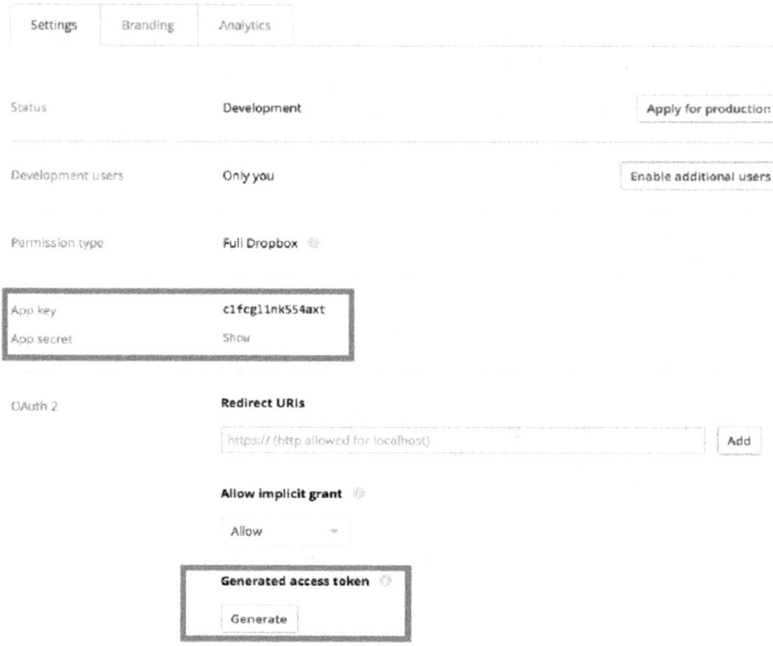

Figure 7.73 Dropbox app API settings

JSON Configuration file is as follows.

```
{
        "show_video": true,
        "use_dropbox": true,
        "dropbox_key": "c1fcgl1nk554axt",
        "dropbox_secret": "xtbjaft3rqfkt10",
        "dropbox_base_path":
"https://www.dropbox.com/developers/apps/info/c1fcgl1nk554axt",
        "min_upload_seconds": 3.0,
        "min_motion_frames": 8,
        "camera_warmup_time": 2.5,
        "delta_thresh": 5,
        "resolution": [640, 480],
        "fps": 16,
        "min_area": 5000
}
```

The experiments were conducted on the proposed system as shown in Figure 7.74.

The AT&T database model was used in the experiments. The database contains 25 images of the person detected. All images were converted to grayscale, cropped to 100×100 and aligned to match the eyelevel using the python program "train.py". The "train.py" was coded to capture the images using camera, detect

Figure 7.74 Test setup photos of the completed prototype

the face using Haar cascade classifier, convert them into gray-scaled, crop them to align the eye level, and save the images in the dataset folder as shown in Figure 7.75.

The main program is saved in apps/videofacerec. To run the program, run the python file videofacerec.py by python videofacerec.py –t database location. Once the Start button is pressed in the GUI window in Figure 7.76, the system will start the live video from the camera and begin to detect the faces.

The software was tested with a database of multiple facial images of different people in a wide variety of lighting situations. The face detection using Haar Cascade has a good accuracy up to a distance of 5 m under good lighting and up to 3 m in a low-light situation. Once the distance exceeds the limit, the accuracy decreases. The face recognition depends on lighting and the number of images in the database. The accuracy of the results increases with more images. The initial test was done using 25 images in the database with poor recognition rate. To overcome the situation, the system with 200 facial images taken in different lightings was used. The classifier was trained with 750 images of different gender with color images (RGB) and grayscale (B/W) images. The color images were sized to 140×140 and the B/W to 64×48. The computational requirements for gender detection are high. The Raspberry Pi showed severe lag when tested with the face recognition and the gender recognition. The GPU performance of 900 MHz and 1 GB RAM was not enough for the program. The image resolutions for the video were reduced. The system was tested under different lighting conditions and different background scenarios. Each test was repeated for ten times, and the results for each feature are tabulated as shown in Figure 7.77.

The performance of the system increases with the number of images in the database. However, the size of the database will affect the hardware performance

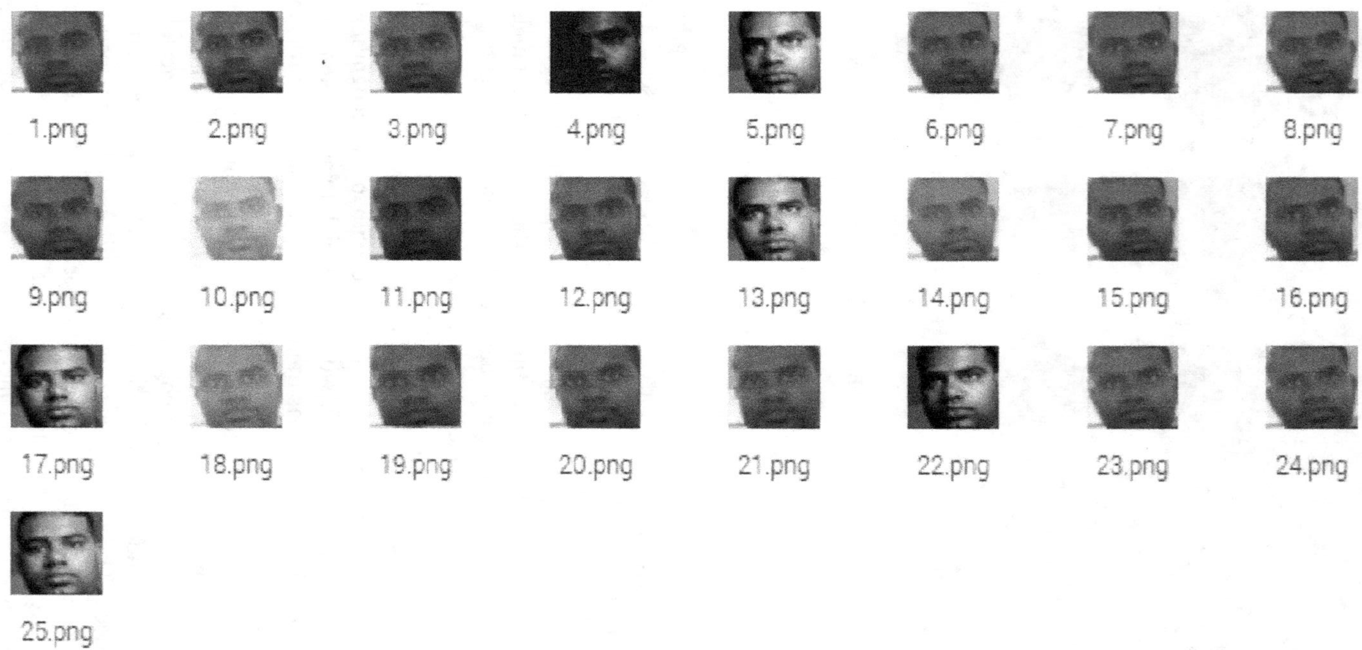

Figure 7.75 Example from facial database

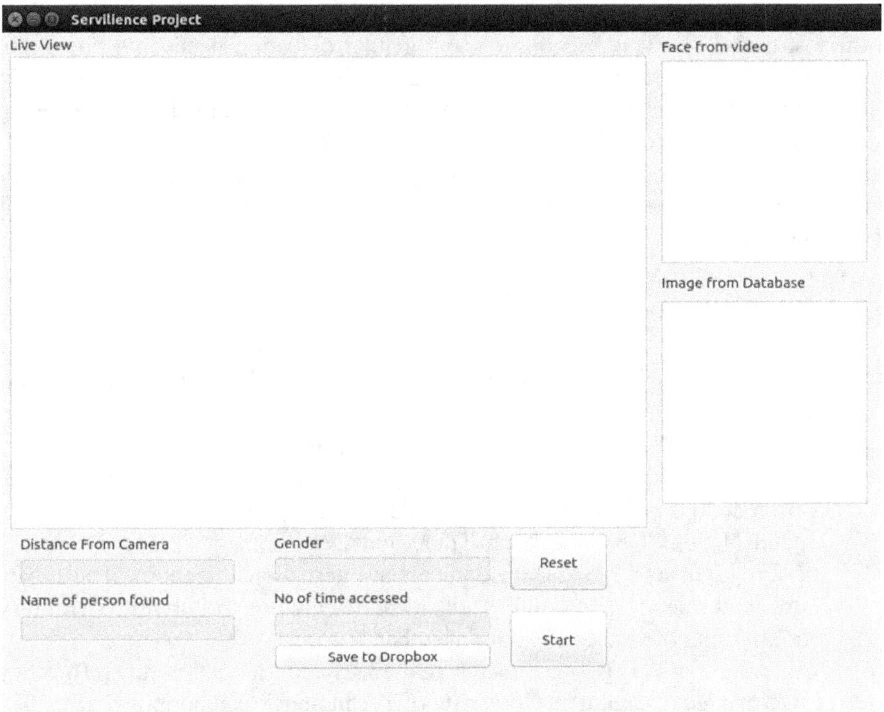

Figure 7.76 GUI application

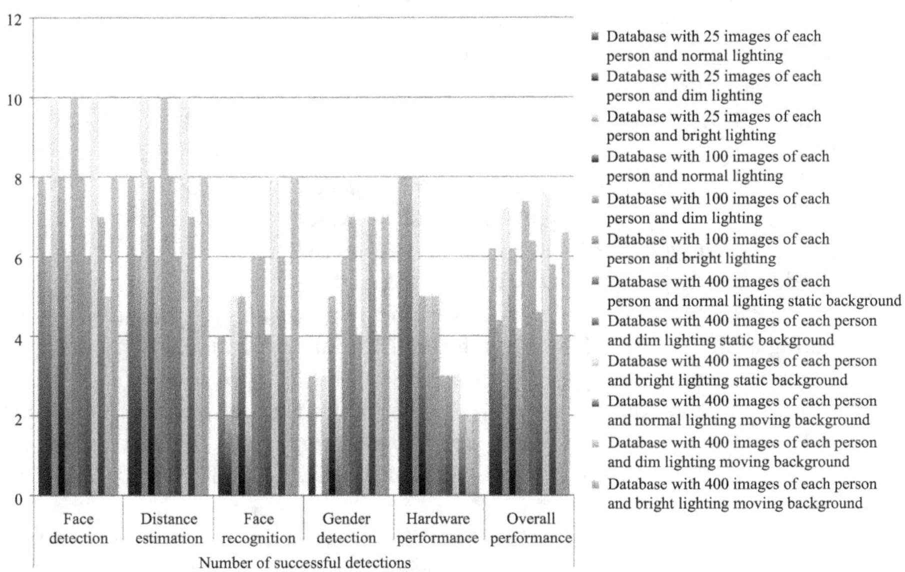

Figure 7.77 Comparison bar chart

and start to show latency. Face detection and distance estimation show good results in the test trials. The face recognition and gender detection need high processing power. The gender detection and identity recognition accuracy can be improved by training the classifiers with more images. The system intended to enhance the safety and surveillance of the maritime industry for human motion detection and identification. From the experiments, the system is capable of handling different environmental situations to detect the motion and estimate the distance in a variety of lighting conditions.

References

[1] Guo, J.Y., Chin, C.S., Maode, Ma, and Clare, A.S. (2016). Interactive vision-based intelligent system for active macfouling and microfouling detection on hull. OCEANS 16 MTS/IEEE, Shanghai, China: IEEE.

[2] Schultz, M. (2007). Effects of coating roughness and biofouling on ship resistance and powering. *Biofouling*, 23(5), 331–341.

[3] United Nations Conference on Trade and Development (2015). Review of maritime transport 2015: Developments in International Seaborne Trade. New York and Geneva, Available at: http://unctad.org/en/PublicationsLibrary/rmt2015_en.pdf. Accessed: 25 November 2015.

[4] Wigforss, J. (2012). Benchmarks and measures for better fuel efficiency. *Undergraduate*, Chalmers University of Technology, Goteborg, Sweden, 2012. Available at: http://publications.lib.chalmers.se/records/fulltext/159671.pdf

[5] Floerl, O., Inglis, G., and Hayden, B. (2005). A risk-based predictive tool to prevent accidental introductions of nonindigenous marine species. *Environmental Management*, 35(6), 765–778.

[6] Millett, J. and Anderson, C.D. (1997). Fighting fast ferry fouling. Fast '97. Conference Papers. Australia, 1, 493–495.

[7] Ismail, S.B., Salleh, Z., Yusop, M.Y.M., and Fakhruradzi, F.H. (2013). Monitoring of barnacle growth on the underwater hull of an FRP boat using image processing. *Procedia Computer Science*, 23, 146–151.

[8] Goh, J.Y. (2015). Intelligent surveillance system for hull fouling. *Undergraduate*, University of Newcastle upon Tyne, UK.

[9] Krishnamoorthy, P., Chin, C.S., Gao, Z., and Lin, W. (2015). A multi-hop microprocessor based prototype system for remote vibration and image monitoring of underwater offshore platform. IEEE 7th International Conference on Cybernetics and Intelligent Systems (CIS) and RAM, pp. 268–275.

[10] Qianjin, Y. and Xiangjun, B. (2000). Ice-induced jacket structure vibrations in Bohai sea. *Journal of Cold Regions Engineering*, 14(2), June. Available at: https://ascelibrary.org/doi/pdf/10.1061/%28ASCE%290887-381X%282000%2914%3A2%2881%29

[11] Wang, S-Q and Li, N. (2013). Semi-active vibration control for offshore platforms based on LQG method. *Journal of Marine Science and Technology*, 21(5), 562–568.

[12] Villanyeva, H., Antonio, Molina, V., and Oscar, L. (1996). Vibration measurement in offshore structure, 11th World Conference on Earthquake Engineering, Paper no. 1667, 1–6, ACAPULCO, MEXICO.

[13] Mijarez, R., Gaydecki, P., and Burdekin, M. (2003). Continuous monitoring chirp encoded ultrasound sensor for flood detection of oil rig leg cross beams. Proceedings Sensors and Their Applications XII, IOP Publishing Ltd, pp. 545–550.

[14] Ortiz, A., Simó, M., and Oliver, G. (2002). A vision system for an underwater cable tracker. *Machine Vision and Applications*, 13(3), 129–140.

[15] Sheikholeslami, A., Ghaderi, M., Pishro-Nik, H., and Goeckel, D. (2014). Jamming-aware minimum energy routing in wireless networks. 2014 IEEE International Conference on Communications (ICC), pp. 2313–2318.

[16] Chin, C., Lau, M., Low, E., and Seet, G. (2008). Robust and decoupled cascaded control system of underwater robotic vehicle for stabilization and pipeline tracking. Proceedings of the Institution of Mechanical Engineers, *Part I: Journal of Systems and Control Engineering*, 222(4), 261–278.

[17] Chin, C.S., Lau, M.W.S., and Low, E. (2011). Supervisory cascaded controller design: Experiment test on a remotely operated vehicle. Proceedings of the Institution of Mechanical Engineers, *Part C: Journal of Mechanical Engineering Science*, 225(3), 584–603.

[18] Chin, C. and Lau, M.W.S. (2012). Modeling and testing of hydrodynamic damping model for a complex-shaped remotely-operated vehicle for control. *Journal of Marine Science and Application*, 11(2), 150–163.

[19] Sadat, A. and Karmakar, G. (2010). Optimal reliable and energy-aware intercluster communication in wireless sensor networks, 16th Asia-Pacific Conference on Communications, 2010, pp. 194–199.

[20] Qiu, F. and Xue, Y. (2014). Robust joint congestion control and scheduling for time-varying multi-hop wireless networks with feedback delay. *IEEE Transactions on Wireless Communications*, 13(9), 5211–5222.

[21] Ojala, T., Pietikäinen, M., and Mäenpää, T. (2002). Multiresolution gray-scale and rotation invariant texture classification with local binary patterns. *IEEE Transactions on Pattern Analysis and Machine Intelligence*, 24(7), 971–987.

[22] Viola, P. and Jones, M. (2001). Computer vision and pattern recognition, 2001. CVPR 2001. Proceedings of the 2001 IEEE Computer Society Conference on IEEE, Salt Lake City, Utah.

[23] Belhumeur, P.N., Hespanha, J.P., and Kriegman, D.J. (1997). Eigenfaces vs. fisherfaces: Recognition using class specific linear projection. *IEEE Transactions on Pattern Analysis and Machine Intelligence*, 19(7), 711–720.

[24] Jaiswal, S. (2011). Comparison between face recognition algorithm – eigenfaces, fisherfaces and elastic bunch graph matching. *Journal of Global Research in Computer Science*, 2(7), 187–193.

Index

AA rechargeable batteries 294
acceleration in heave direction 85
accelerometer
  bias 46
  framework 393
actual dynamic model 271
actual magnetizing current 270
actuator(s) 123
  driver design 39–40
  driver layer 30
ADC: *see* analog-to-digital converter
adhesion 328–9
ad-hoc process 171
AF systems: *see* anti-fouling systems
air gap reluctance 269
algebraic Riccati equation (ARE) 171
amplifier circuit 240
  connection circuit diagram 57
  design 238
analog
  circuitry 126
  in xPC-Target 151–8
analog-to-digital converter (ADC) 236–7, 240
Analog Input block 158
analogWrite() function 321–2
angular acceleration 90
angular speeds 186–7
angular velocity 89
ANN identification: *see* artificial neural networks identification
ANSYS-CFX nonlinear damping coefficients 191
anti-fouling systems (AF systems) 346
Apache Web Server 359, 377
App Inventor Blocks Editor 315–16
AR9331 processor 303
ARDUINO® embedded system design 294
  sloth bear-inspired pole-climbing robot 327–44
  smart control of marine-tracked vehicle for surveillance 313–27
  see also PIC embedded system design
ARDUINO® 2, 10
  example of embedded systems using 3–5
  fouling image recognition system setup 7
  framework 393
  motor shield 320, 325
  sketches 320–1, 325
  software IDE 303, 318
  system architecture design of low-cost ROV 6
ARDUINO® Uno 314, 390, 393
  board 325
  electrical schematic of ARDUINO® Uno Rev3 334
ARDUINO® Yun™ board 5, 294, 305–6
ARE: *see* algebraic Riccati equation
artificial neural networks identification (ANN identification) 377–81
ASCII commands 50
ASCII decode V2 setup 156
assembled Raspberry Pi 400
assembly process 127–8
AT&T database model 434–5
AT&T Face database 413–14
Athena-II/Hercules-II SBC Linux® 2.6.23 20

Athena II PC104
computer 125
as target PC 139–42
Athena II–Prometheus CPU card 125
Athena II–Prometheus panel I/O card 125
Athena III PC104, embedded system example using 2–3
Athena PC104 platform 56
Atheros AR9331 5
ATmega328 microcontroller 325
ATmega32U4 microcontroller 5, 294, 306–7
AutoCAD™ 12
Automatic Identification System 345
autonomous underwater vehicle 178–9
AUV Docking Hoop
control system panel 64–5
function modules for 64
GUI 63
GUI control system panel 60

Barbalat's lemma 283
barnacles images 372, 376
Baseboard RS232 Send/Receive
basic setup for 155
board and FIFO setup for 155
FIFO OUT block diagram 154
Basic Input/Output System (BIOS) 135
bg.txt file 370
bi-directional connectivity 395
biomimetic adhesives 243–4
bio-mimicry development 327
BIOS: *see* Basic Input/Output System
blob 402–3
Block Editor Interface 316
Block Viewer 316
Bluetooth 293, 314
body-fixed angular velocities 181
body-fixed coordinate 102–3
body-fixed reference frame 71
body inertia 91
body temperature 235–6
boundary value problem (BVP) 100
"bridge" library 303, 306
brushless electric motors 293
bulk image converter 373
Butterworth high-pass filter 46
BVP: *see* boundary value problem

C++ 401–2
CAD: *see* computer-aided design
CADI 235
camera
capture system process 53
driver 50
system hardware architecture 55
Camera Module 423
cascade classifier structure 369
cascade.xml 368
cat command 19
C/C++ compiler 9
CCW: *see* counterclockwise
cd command 19
"cd/mnt/sda1" command line 305
CFD-ANSYS CFX method 181
CFD method: *see* computational fluid dynamics method
children processes 33
chromosome 105–6
class LocalBinaryPatterns 410
CLI: *see* command line interval
client–server model 54
client's subsystem 392
client terminal 388
climbing robot 243, 248, 259
clockwise (CW) 132
closed-loop system 168
command line 305
command line interval (CLI) 304
communication
ports 11
types 390
competitive markets 1
computational effective triangular membership functions 108
computational fluid dynamics method (CFD method) 69

computer-aided design (CAD) 79–80
  model 93
  software 9, 12
computer vision (CV) 346
constant heading velocity 89
contemporary systems 1
controller design 199
  SMC law 200
  SMC system using xPC-Target block diagrams 201
  time response of controllers for certain angle 202
ControlProcess 30
control process design 39
control system/techniques 9
  depth sensor unit 56–9
  design 12, 102
  DVL sensor unit 47–50
  flowchart 31
  image video unit 50–6
  inertial measurement unit 42–7
  PID 110, 112–16
  proposed fuzzy-based genetic algorithm for SMC 105–13
  ROV dynamics 103
  sliding-mode control 104–5
  software layers communication in 29
  testing of components for 42
conventional conveyor vehicles 261
conventional SMC 110
  fuzzy logic controller 112
  proposed FGA for SMC (FGA_SMC) 113–16
convergence 75
conveyor system 261
coordinate systems 181
coordinator
  MCU 240–1
  set design 240
Coriolis and centripetal matrix 71–2, 79
count 32
counterclockwise (CCW) 132
counts per revolution (CPR) 160
CPR: *see* counts per revolution
C program 214
  codes 217
  development process uses Linux® and 26
  files on main and subprograms 27
  software 10
CRC: *see* cyclic redundancy check
CV-based algorithms 423
CV: *see* computer vision
"cvtColour" script 371
CW: *see* clockwise
cyclic redundancy check (CRC) 240

DAC program flowchart 40
damping
  coefficients 78, 85, 99–100, 189
  function 74
DAQ: *see* data acquisition
data acquisition (DAQ) 2, 39
database
  creating tables in 365
  creation 363
  fields in 365
  layout 365
  using PHPMyAdmin 364
  PHP template for uploading images to 366
DDM: *see* direct drive mode
dead-reckoning filter
  flowchart 48
  method 46
  velocity and position signal after 49
debugging 25
decision layer 30
deep water-based CV 388
degrees of freedoms (DOFs) 69
depth
  sensor 35, 56–9, 84
  validation 84
desktop PC as Target PC, xPC-Target real-time kernel creation using 136–9
device file 51
DGM-202 gauss meter 268

DHCP: *see* Dynamic Host Configuration Protocol
Diamond Systems ATHENA III PC104 2
Diamond Systems ATHENA PC104 28
digital I/O in xPC-Target 151–8
digital single-lens reflex camera (DSLR camera) 368
digital speed control of servo DC motor 175
  block diagram for digital PID control of DC motor 176
  rapid control systems prototyping using hardware-in-the-loop 177
  response time of digital PID control 176
digitalWrite() function 321
DIP: *see* dual inline package
DIPIC-SN754410 connection to H-bridge 130
direct drive mode (DDM) 314
direction movements 321
directories 17
DOFs: *see* degrees of freedoms
Doppler velocity log (DVL) 3, 84
  DVL-aided INS system 37
  NavQuest 600 MicroSensor 48–9
  sensor 35, 47–50
Dormand–Prince solver 92–3
downward communication 391
drag force as function of velocity in surge, sway, and heave 78
Dropbox API app 431
Druck pressure sensor 3
DSLR camera: *see* digital single-lens reflex camera
dual inline package (DIP) 130
DVL: *see* Doppler velocity log
Dynamic Host Configuration Protocol (DHCP) 394
Dynamics modeling of remotely operated underwater vehicle 71
  hydrodynamic-added mass model 79–84
  hydrodynamic damping model 73–9
  hydrodynamic forces and moments 72

earth-fixed coordinate error 104
earth-fixed current velocity vector 97–8
Easyphp 361–2
EDA: *see* electronic design automation
Eigenfaces 418
EKF: *see* extended Kalman filter
electro-adhesion 244
electromagnetic adhesive 243
electromagnetic system 261
electromagnet suspension system (EMS) 261
  required specification and design constraints of 267
electronic circuit board 256–7
electronic design automation (EDA) 325
electrostatic adhesion 243–4
elites selection in GA 106
embedded Linux® system 30
embedded microprocessor 122
embedded system 1–2, 121
  building Linux® for 20
  embedded control system 2
  example using ARDUINO® 3–5
  example using Athena III PC104 2–3
  example using PIC 7–8
  example using Raspberry PI™ 5–7
  motivations 8–10
  systematic design approach for prototyping 10–14
embedded underwater vehicle system
  control systems design 102–16
  dynamics modeling of remotely operated underwater vehicle 71–84
  launch and recovery process model 101–2
  remotely operated underwater vehicle 69–71

remotely operated underwater vehicle sea trial 117–19
simulating external disturbance for remotely operated underwater vehicle model 97–101
simulation of remotely operated underwater vehicle model 92–7
validation of experimental results 84–92
EMS: *see* electromagnet suspension system
encoder block diagrams 162
environmental disturbances 97
Eric4 Python IDE 63
Ethernet 122, 180
Euler's transformation 95, 98, 181
exit command 19
extended Kalman filter (EKF) 37–9
External disturbance simulation for remotely operated underwater vehicle model 97–101

fabricated model 400
face detection
and distance estimation 438
methods 412
face recognition system 399
assembled Raspberry Pi 400
comparison bar chart 437
create API app for Dropbox 1 432
database for gender detection training 427
database hierarchy 401
Dropbox API app 431
Dropbox app API settings 434
example from facial database 436
fabricated model 400
face detection and distance estimation 438
facial image 426
flowchart of proposed system 425
gender detection training results 428
GUI application 437
GUI using PyQt 419–23
hardware 400
programming and system implementation 399, 401–19
software 401
SSH Clients accessing Raspberry Pi 424
SSH network overview 424
system design and architecture 399
testing 421
test setup photos of completed prototype 435
facial database 436
Fedora 20
Fedora 11 live OS 21
ferrite or alnico magnets 243
FGA_SMC: *see* fuzzy-based genetic algorithm on SMC
FIFO: *see* first-in-first-out
file
command 19
permissions 19
system 18
FileZilla
GUI 360
input SSH information for 360
first-in-first-out (FIFO) 35
FIFO read hdrs setup 156
pipe model 37
first-order plus dead-time model 165
Fisherfaces 418
5 V supply 132
FLC: *see* fuzzy logic controller
flooded member detection 388
fluid domain 74
fluid velocities 103
Fluke 805 Vibration Meter 397
fouling 386
image recognition system setup 7
organisms 345, 366
rank 346
fouling detection system 345
alignment of system 353
ANN identification 377–81
database creation 363–6

first-boot procedures 353–5
Haar cascade object recognition 367–71
hardware components 348
installing MySQL and Easyphp and PhpMyAdmin on computer 361–2
installing Raspberry Pi cam 355–8
Maxpull GM-1-SI winch 349–50
merging vector files and training cascade 371–7
OpenCV 366–7
PHP template for uploading images to database 366
programming and system implementation 367
prototype design orientation 2 350–2
remotely controlling Raspberry Pi terminal 359
remotely copy files to and from Raspberry Pi 359
setting up database and PHP webpage 359–61
setting up Wi-Fi to enable remote control for Raspberry Pi 358–9
software components 347
system design and architecture 347
testing 381–7
voyage 345–6
Wi-Fi adapter 351
see also multi-hop microprocessor-based prototype system

Four-Legged Multipurpose Rope Climbing Robot 327
Fritzing software 325
physical electrical layout of marine vehicle by 319
schematic circuit diagram for marine vehicle by 319
fusion sensor data 40
fuzzy-based crossover and mutation in GA 106
fuzzy-based evolutionary process 114
fuzzy-based genetic algorithm for SMC 105–10
final remotely operated underwater vehicle 113
hydrodynamic damping and added mass coefficients 113
position response 111
velocity response 112
visualization of wave effect 112
fuzzy-based genetic algorithm on SMC (FGA_SMC) 107, 113–16
fuzzy inference mechanism 107–8
fuzzy logic controller (FLC) 108, 112

GA: *see* genetic algorithm
Gaussian mixture-based background/foreground segmentation algorithm 404
GCC compiler 26
GDB: *see* GNU Debugger
gecko pad 243, 252
gender detection training
database for 427
results 428
genetic algorithm (GA) 105
GNC system: *see* guidance, navigation, and control system
GNU Debugger (GDB) 25–6
GNU/Linux® system 17
GNU software 17
graphical-simulation and control system design tool 123–4
graphical programming tool 121
graphical Simulink® models 124
graphical user interface (GUI) 25, 59
application 437
control panel layout 62
FileZilla 360
function modules for AUV Docking Hoop GUI 64
menu for preview 422
multithreading techniques based on Python language 65
preview window 422
using PyQt 419–21

Python script generated 423
Qt user-interface and application flowchart 63
telepresence systems 59–60
thruster status 67
velocity error in Y-direction 61
on vibration data and video transmission 398
video image UDP thread program flowchart 66
graphic elements 419–20
grayscale script, image converting to 377–8
gripping mechanism 328–30
ground 132
GTK+, 50
GUI: *see* graphical user interface
guidance, navigation, and control system (GNC system) 69
guideway 284

H-bridge circuit 128, 214
H-Infinity control 274
H-Infinity controller 284–5
Haar/Haar-like cascade
algorithm 367, 386
Haar-like feature 411–12
with neural networks 380
object recognition 367–71
"Haar training Utility", 371
Hall Effect sensor 252
Hand-bot 327
hardware interfacings 124
12A8 PWM driver connection to DC motor 131
Athena CPU card with location of jumper setting 126
Athena II–Prometheus CPU card 125
Athena II–Prometheus panel I/O card 125
DIPIC-SN754410 connection to H-bridge 130
H-bridge circuit 129
image of HIL setup for DC motor control 132
J13 Jumper block setting 127
Pandora enclosure 128
pins location on 12A8 PWM driver 133
PWM output signals 130
for simulations testing 123–4
hardware-in-the-loop simulation (HILS) 121
solution on underwater robotic vehicle 123
hardware-in-the-loop testing 176–7, 234
and desktop PC-NI PCI card 135
real-time kernel creation using Athena II-PC104 as target PC 139–42
real-time kernel creation using desktop PC as Target PC 136–9
real-time Window Target and NI PCI card 135
using xPC-Target 134
HC-06 TTL 321
HDMI port: *see* high-definition multimedia interface port
heave
model identification 84–8
motion 84
tests 117
velocity response 85
high-definition multimedia interface port (HDMI port) 1
high-quality DSLR camera 368, 371
high-speed trains 260
HILS: *see* hardware-in-the-loop simulation
horizontal subsystem model 188–98
host-target communication 180
HSST 261
human motion detection 406
hybrid electromagnetic levitation system 286
hybrid EMS system 262, 272
hydrodynamic-added mass model 79–84

hydrodynamic coefficients 178, 183, 198
hydrodynamic damping
  forces 75–6
  model 73–9
hydrodynamic forces and moments 72, 182
hydrodynamic parameters 196, 206
HyperTerminal 152
  results 153
hypertext pre-processor (PHP) 377
  scripts 241
  step 2 381
  template for uploading images to database 366
  for update 383
  webpage 382
hyperthermia 236
hysteresis 1323

ice-induced vibrations 387
ICs: *see* integrated circuits
IDE: *see* integrated development environment
ideal current command 282
IF-ELSE function 33
image
  acquisition program flowchart 53
  cropping using Objectmarker 379
  database 363
  image-processing methods 346, 391–2
  info file generation 379
  multi-hop microprocessor-based prototype system 387–99
  positive sample creation 380
  video unit 50–6
IMU: *see* inertia(l) measurement unit
in-circuit-debugging 214–15
incremental encoder model 159–64
  channel A and B outputs from xPC-Target 160
  channel A of encoder block diagrams 163
  configurations for count up for 60 s and 1 s 163
  encoder block diagrams 162
  lookup table editor 163
  overall simulation diagram for motor with encoder 161
  planetary gear reduction of Transmotec DC motor 161
  speed *vs*. voltage curve 164
inertial coefficients 189
inertia(l) measurement unit (IMU) 3, 42–7, 84
inertial navigation system (INS) 37
  integrated INS structure 37
info command 19
infrared sensor (IR sensor) 158–9, 214, 252, 335
inhibit line 132
initial variable values 74–5
input/output (I/O) 30
  boards 9, 135
  management 30
input voltage 187
INS: *see* inertial navigation system
integral control action 168
integrated circuits (ICs) 325
integrated development environment (IDE) 2
integrated INS structure 37
intelligent climbing robot climbing table leg 260
intelligent surveillance system 346
intelligent vacuum robot system design 212
  Add Files in Project Pane Window 225
  build program 218
  components using on target board 213
  connecting to PICkit2 218
  C Program codes 217
  device dialog selection 220
  editing source file-main. c 226
  flowchart of control architecture 215

Language Toolsuite dialog selection 223
mouse over toolbar icon 220
MPLAB-IDE v8. 30 Desktop 219
MPLAB PICkit2 216
new project dialog 222
Open Project dialog selection 224
using PIC 227–34
using PICkit2 215
program device 219
programming and system implementation 216–34
Project Pane Window 224
setting language tool location dialog 222
simulator mode selection 221
system design and architecture 212–16
target board schematic diagram 213
target board with components mounted 212
testing 234–5
User Code Section 226–7
see also PIC embedded system design
interface subsystem architecture design 392
interfacing circuit 123
internode multi-hop transmission scheme 391
interrupt service routines (ISRs) 25
I/O: *see* input/output
IPython 402
IR sensor: *see* infrared sensor
ISRs: *see* interrupt service routines

J13 Jumper block setting 127
Java 401–2
JAVA Script 397

Kalman filter(ing) 33, 37, 59–60
kernel source 17, 19, 22
kinematics equation: *see* Euler's transformation

L293 quadruple high-current half-H driver 255–6
L298
Dual H-Bridge integrated chip 263
integrated circuit 255
Lagrange equations 181
Laplace transforms 164
laser sensors 252
launch and recovery process model 101–2
LBP: *see* local binary patterns
LDA: *see* linear discriminant analysis
least-square method 87
LED 252
levitation force 269
linear coefficients 74
linear controller 199
linear damping coefficients 79, 194–5
linear discriminant analysis (LDA) 399
linearized model 37
linear quadratic regulator (LQR) 171
block diagram of LQR control of DC motor 172
block diagram of switching control of DC motor 175
response time of switching control of dc motor under disturbance 175
response to switching control of DC motor 174
sinusoidal disturbance into system 174
switching control of DC motor 173
time response of LQR control of DC motor 173
linear synchronous motor (LSM) 262–3
design of LSM using contactless sensing 264
simplified schematics of LSM system 265
linear wave model 98
links 17

Linux® 5
  kernel 17
  Linux®-based embedded systems 9
  Linux®-based GUI 423
  OS 51
  process-based threads 26
Linux®-based embedded system design
  actuator driver design 39–40
  building Linux® for embedded systems 20–5
  control process design 39
  development process for building OS layer 30
  example of packaged Linux® distributions from Ubuntu 16.04.2 LTS 18
  GUI 59–67
  Kalman filter 59
  Linux® commands 19
  Linux® operating system 17
  main process design 30–3
  network communication thread design 40–2
  program layouts in Linux® 25–7
  remotely operated underwater vehicle design 28
  sensor fusion thread design 37–9
  sensor process design 33–7
  software layers communication in control system 29
  standard root partition 18–19
  system design and architecture 27
  testing of components for control systems 42–59
  see also ARDUINO® embedded system design; Raspberry Pi-embedded system design
linuxuvc 51, 64
lithium iron phosphate battery ($LiFePO_4$) 11
local binary patterns (LBP) 399, 410
lock 32
"locking system", 32
loop() function 320
loosely integrated IMU/DVL navigation method 59
lower frequency coefficient 44
low-order method results of sphere 81
LQR: *see* linear quadratic regulator
ls command 19
LSM: *see* linear synchronous motor
lumped uncertainty vector 271
Lyapunov function candidate 283

Mabuchi motor FF-130RA-18100 251
Mabuchi motor FF-130SH-14230 251
macrofouling organisms 346
MagLev in China 261
magnetic conveyor system design 260
  air core linear drive commutation sequence 267
  comparisons of controllers' closed-loop time response 286
  controllers' responses *vs.* compared to sliding-mode controller 285
  design of LSM using contactless sensing 264
  estimated thrust force *vs.* input current 271
  experimental setup of magnetic levitation system 274
  front view of magnetic conveyance system 263
  hybrid EMS system 272
  magnetic flux path circuit 268
  magnetic polarity 264
  nonlinear EMS system model 272
  open-loop time response of air gap 273
  oscilloscope signal output for sliding-mode control system 284
  overall oscilloscope output for open-loop system 273
  physical arrangements of electromagnet with track 268
  programming and system implementation 274–81
  programming flowchart for air-core linear drive 266

required specification and design constraints of EMS 267
simplified electrical schematic for hybrid EMS system 263
simplified schematics of LSM system 265
sliding-mode closed-loop time response of air gap 285
sliding-mode controller for hybrid EMS system 283
system design and architecture 261–74
testing 284–6
3D view of magnetic conveyance system 262
truth table of logic gate block 266
see also PIC embedded system design
magnetic levitation 262, 268
experimental setup 274
force 261, 269
systems 261
magnetic sensors 252
main control mode 64
main process design 30–3
flowchart 34
man command 19
Mandriva 17
marine robotic vehicle with uncertainties using xPC-Target system 177
actual motion sequence of underwater robotic vehicle 204
controller design 199–202
hydrodynamic parameters 206
implementation and testing 202
PC demo interface program for magnetic compass 180
rapid control system prototyping process flow 179
steady-state thruster's dynamics 183–8
switching signal of underwater robotic vehicle 205
system design and architecture 178
TCP/IP network 206–7
test-bed underwater robotic vehicle with coordinate systems 178
time responses of underwater robotic vehicle 205
underwater robotic vehicle 188–98
underwater robotic vehicle dynamic model 181–3
xPC-Target block diagram for obstacle avoidance 203
marine-tracked vehicle 326
Marine Track Vehicle by Bluetooth Protocol 314
marine vehicles 69
marine vessel control system 69
mass
coefficients 84
inertia matrix 71
Master Boot Record (MBR) 141
Master in, Slave out (MISO) 393
Master out, Slave in (MOSI) 393
MATLAB® 9
GUIDE development tool 206
MATLAB®/Simulink® environment 92–3, 121–2, 134
routine 100
script 84
Matplotlib 402
Maxpull GM-1-SI winch 349–50
MBR: *see* Master Boot Record
M-CLK 393
MCU: *see* PICmicro microcontroller
mean square error (MSE) 100
mechatronics engineer 1
merging vector files 371–7
mesh settings 76
metal-oxide semiconductor field-effect transistor (MOSFET) 299, 303
Meteorological Service Singapore 117
microcontroller 10

microprocessor 294
  microprocessor-based mechatronics application 121
  technologies 1
micropump system 293
microwave liver ablation 235
MISO: *see* Master in, Slave out
MIT App Inventor 314
  Block Editor Interface 317
  designer GUI 316
  interface 315
MiWi wireless temperature-monitoring system 236
MiWi wireless transceiver module 239
MLX in Japan 261
Moore–Penrose pseudoinverse 87, 104
morphological filtering operations 405
MOSFET: *see* metal-oxide semiconductor field-effect transistor
MOSI: *see* Master out, Slave in
motion tracker 46
motor controller 255
motor syringe system 293–4
MPLAB IDE 211–12, 253
MPLAB SIM simulator 218
MRF24J40MA transceiver, network coverage range of 241
MSE: *see* mean square error
MTi-30 AHRS connection specifications 43
MTI 10 altitude sensor 35
multi-hop
  topography 388
  wireless networks 388
multi-hop microprocessor-based prototype system 387
  availability of groundwork 387–8
  high-level architecture 389
  interface subsystem architecture design 392
  internode multi-hop transmission scheme 391
  node interconnection schematic 390
  programming and system implementation 394–5
  system design and architecture 388
  testing 395–9
  user interface 392–3
  wireless connection 393–4
  see also fouling detection system
multiple tasks approach 25
multiplier 58
MULTISURF™ 80–2
MySQL database 241
  installing on computer 361–2
  installing on Raspberry Pi 362

named pipes 18
NavQuest 600 47
NavQuest DVL 3
n-channel MOSFETs 302
neodymium magnet 243, 246
network
  communication thread design 40–2
  connection 64
  socket 40
neural network (NN) 100
Newton's Second Law of Motion 181
NN: *see* neural network
noncontact temperature-measuring device 235
nonlinear EMS system model 272
normothermia 236
numerical modeling 9
  and simulation techniques 69
NumPy 402

Objectmarker, images cropping using 379
ODE45 variable-type solver 281
offshore structure model 396
online database 236, 241
OpenCV 366–7, 402
  OpenCV 3. 0 source code 366–7
open-loop
  speed regulation block diagram 166
  transfer function 165, 187

open-source
  platform 10, 27
  software 371
OpenWrt wireless 5
  stack 294
OpenWrt-Yun
  Linux® distribution 294
  system 304
operating system (OS) 17
  layer 30
opkg update command 305
optical incremental encoders 159–60
optimal state-feedback design: *see* linear quadratic regulator (LQR)
ORK() function 33
ORL Database of Faces 413–14
OS: *see* operating system

Pandora enclosure 127–8
passwd command 19
patient MCU 240–1
patient set design 238–40
PC104 computer 134, 180
PCA: *see* principal component analysis
PCBs: *see* printed circuit boards
PCDemo Interface 4. 01 152
PC demo interface program
  for compass 153
  for magnetic compass 180
PDCR1830 Pressure Sensor 56
per-pixel Bayesian segmentation 405
peristatic pumps 293
Perl programming languages 60
permanent magnets 243
PHP: *see* hypertext pre-processor
PHPm 359
PhpMyAdmin installing on computer 361–3
PIC-Microchip™ 9–10
PIC18F4520 253
  PIC18F4520 MCU 258, 262
PIC embedded system design 211
  example 7–8
  MPLAB IDE 211–12
  remote temperature-sensing system design for patients 235–42
  see also ARDUINO® embedded system design
PICkit2 debugger 218
PICkit™ 2 Development Programmer/Debugger 7, 214
PICmicro microcontroller (MCU) 211, 214, 258
PIC™ microcontroller 2
PID: *see* proportional-integral-derivative
pipe(s) 35
pixel 408
planetary gear reduction of Transmotec DC motor 161
plant dynamics 1
platform layer 30
pole-climbing robot, tasks of 331
pole assembly connection 352
Pololu Zumo chassis kit 318
polyvinyl chloride (PVC) 291
  holders for mounting of camera housing 298
power MOSFETs 302
predictive methods 183
pressure sensor 59
Primex 235
principal component analysis (PCA) 399
printed circuit boards (PCBs) 325
process-based user thread 25
producer-and-consumer concept 32
producer–consumer model with FIFO PIPE 35
Pro/ENGINEER™ 12
program layouts in Linux® 25–7
programming languages 60
Project Pane Window 225
propellers 294
  speed 187, 199
  thrust and torque model 184
proportional-integral-derivative (PID) 110
  controller 274, 282

control of DC motor 167
conventional SMC 110, 112–16
speed control of servo DC motor 168–9
prototype/prototyping
designs 245, 348–9
systematic design approach for prototyping embedded systems 10–14
temperature-monitoring system 238
proximity sensors 252
pulse-echo ultrasound 235
pulse-width modulation method (PWM) 129–30
PVC: *see* polyvinyl chloride
pwd command 19
PWM: *see* pulse-width modulation method
PyQt: *see* Qt library with Python programming language
Python 399
program flowchart 367–8
programming languages 60, 393
Python-based IDE 2, 27
Python-Qt open-source codes 117
Python imaging-tk 402

Qt library with Python programming language (PyQt) 60, 420
Qt user-interface and application flowchart 63
"quarter decay" closed-loop response 168–9

rapid control system prototyping
using hardware-in-the-loop 177
process flow 179
rare-earth magnet 243
Raspberry Pi-embedded system design
face recognition system 399–438
fouling detection system 345–87
multi-hop microprocessor-based prototype system 387–99
see also ARDUINO® embedded system design; Linux®-based embedded system design; xPC-Target embedded system design
Raspberry Pi MicroSD slot 354
Raspberry PI™ microcontroller 2, 10
configuration for fouling image recognition system 7
example of embedded system using 5–7
Raspbian 394
RDE: *see* Riccati Differential Equation
real-time embedded OS Linux® environment 25
real-time operating system (RTOS) 25
real-time Window Target and NI PCI card 135
Real-Time Workshop 9, 124
real-time xPC-Target spy screen 157
REAL VNC Viewer 394
rectal measurement of body temperature 236
recursive least square (RLS) 87
RedHat 17
region of interest (ROI) 371
remote debugger 25
remotely operated underwater vehicle: *see* remotely operated vehicles (ROVs)
remotely operated vehicles (ROVs) 2–3, 69–84, 291
circuit connection 300
command line for SD card and video streaming 305
completed ROV prototype 292
components 5
connection to Yun via SSH 304
design 28
detect widget for control mode and status of 66
experimental test results 307

gluing PVC frame onto holes of Tupperware with hot glue gun 296
heave response between simulated and experiment results 96
hole saw drilling hole in Tupperware 295
horizontal PVC frame assembling with Tupperware 296
maneuvering forward 309
model 37, 92
moving in yaw direction 309
nonlinear remotely operated underwater vehicle model 93
OKPG update and installation of drivers 305
opening terminal application on Ubuntu Linux® 304
plan and side view 299
plan view of prototype 292
program codes 310–3
programming and system implementation 303–7
PVC holders for mounting of camera housing 298
rear view of prototype 293
ROV prototype with electronics and thrusters 298
sea trial 117–19
sensor readout of 64
simulating external disturbance for 97–101
skip argument on CLI 305
system architecture design for 4
system architecture design of low-cost ROV 6
system design and architecture 291–303
tested underwater 308
testing 307–13
thrust inputs model in Simulink® 94
Tupperware exterior with PVC fittings 295
Ubuntu operating system on VMware player 303
user interface 96
V-realm builder screenshot and virtual reality world 95
wire connections 301–2
yaw response between simulated and experiment results 97
see also ARDUINO® embedded system design
remote temperature-sensing system design for patients 235
amplifier circuit design 238
coordinator set design approach 237
programming and system implementation 240–1
prototype temperature-monitoring system 238
system design and architecture 236–40
testing 241–2
time series plot 242
remote underwater dual cameras video image acquisition system 56
remote vibration, multi-hop microprocessor-based prototype system for 387–99
repulsive approach 262
Rexx programming languages 60
Reynolds number 74, 189
Riccati Differential Equation (RDE) 171
rigid-body remotely operated underwater vehicle mass inertia 73
rigid body underwater robotic vehicle dynamic equation 182
RLS: *see* recursive least square
Robo-Sloth 327
robot, front view of 257
robotics vehicle control system 28
robust controller 200
robust control method 282

robust SMC incorporating evolutionary fuzzy-based scheme 102–3
ROI: *see* region of interest
rotation 329
router 294
ROVs: *see* remotely operated vehicles
RS232 connection 122
RS232 in xPC-Target 151
  ASCII decode V2 setup 156
  basic setup for Baseboard RS232 Send/Receive 155
  board and FIFO setup for Baseboard RS232 Send/Receive 155
  connection to PC 152
  FIFO read hdrs setup 156
  host scope viewer 157
  hyper terminal results 153
  PC demo interface program for compass 153
  pin assignment of HMR3000Pin assignment of HMR3000 152
  real-time xPC-Target spy screen 157
  SHARP GP2D120 connection to Athena II PC104 158
  xPC-Target block diagram for RS232 data acquisition 154
RTOS: *see* real-time operating system

samarium-cobalt magnet 243
SARS: *see* severe acute respiratory syndrome
sat(.) function 283
SBC: *see* single-board computer
Scikit 402
SciPy 402
scripting languages 359
SD card image 394
SDK: *see* software development kit
Secure Shell (SSH) 5, 304
  Clients accessing Raspberry Pi 424
  network overview 424
semaphores 32
sensitivity of PCR1830 pressure sensor 58
sensor 121–2
  driver layer 30
  fusion thread design 37–9
  network 388
  process design 33–7
  readout display 66
  readout of remotely operated underwater vehicle 64
  sensitivity 57
SensorProcess 30
server 392
servo DC motor 164–8
  PID speed control 168–9
  sliding-model speed control 169–70
setup() function 320
severe acute respiratory syndrome (SARS) 235
sgn(.)sigmoid function 283
share memory 32
SHARP GP2D120 connection to Athena II PC104 158
sharp IR sensor 256
Sharp sensor 2Y0A41SK 252
Shear-stress-transport (SST) 74
shell 19–20
signal-conditioning functions 128
signal-to-noise ratio (SNR) 391
SimpleBlobDetector 403
Simulink® 9, 123–4
  block diagrams 142, 154
single-board computer (SBC) 2
single-ended command signals 132
single frame data file size 56
singleton membership functions 108
sinusoidal control command 89
sliding
  control law 199
  sliding-model speed control of servo DC motor 169–70
  surface 282

sliding-mode control 104–5
  of DC motor 167
  law 282
sliding-mode controller (SMC) 101, 261, 282
  for hybrid EMS system 283
  oscilloscope signal output 284
  system using xPC-Target block diagrams 201
sloth bear-inspired pole-climbing robot 327, 333
  control unit overview 329
  electrical schematic of ARDUINO® Uno Rev3 334
  entire body mechanism 328
  entire gripper 329
  flowchart of control logic for climbing 336
  flow of servo control logic 343
  gripper and rotation of motor 333
  gripper showing rotation of motor and movement 332
  gripping motion of robot 344
  horizontal climbing with rotation of arms 332
  links with motor in place 328
  picture of gripper 332
  program codes 337–43
  programming and system implementation 335–43
  rotational links 328
  rotational links with motor 329
  servo controller board 335
  side view of robot moving horizontally and vertically 330
  system design and architecture 327–35
  tasks of pole-climbing robot 331
  testing 343–4
smart control of marine-tracked vehicle for surveillance 313
  ARDUINO® motor shield 320
  ARDUINO® sketches 320–1
  ARDUINO® Uno 314
  block diagram of overall system 315
  initial testing of marine-tracked vehicle 318
  MIT App Inventor interface 315, 317
  physical electrical layout of marine vehicle 319
  programming and system implementation 320–5
  schematic circuit diagram for marine vehicle 319
  system design and architecture 314–20
  testing 325–7
SMC: *see* sliding-mode controller
SNAME: *see* Society of Naval Architects and Marine Engineers
SNR: *see* signal-to-noise ratio
Society of Naval Architects and Marine Engineers (SNAME) 70
socket(s) 17
  server–client program flowchart 41
software 12
  layers communication in control system 29
software development kit (SDK) 20
SolidWorks™ 12
special files 17
special magnets 243
speed of individual motor 321
SSH: *see* Secure Shell
SST: *see* Shear-stress-transport
standard disposable tape 242
standard medical temperature probe 242
standard root partition 18–19
STAR CCM+, 77
steady-state
  error 119, 173
  heading velocity 89
  thruster's dynamics 183–8
streamlined semi-empirical model 206
switch control of docking hoop 64
switching control of DC motor 173, 175

Tamiya 70097 Twin-Motor Gearbox Kit 251
tank tracks with gecko pad 256
target PC, xPC-Target real-time kernel creation using desktop PC as 136–9
Taylor series 37
Tcl programming languages 60
TCP: *see* transmission control protocol
telepresence systems 59–60
temperature
  sensor 237, 240, 260
  temperature-monitoring system 236
TempTrak 235
test-bed underwater robotic vehicle 179
  with coordinate systems 178
texture-based gray level co-occurrence matrix 409
thermal imaging camera 235
threads 25
three-dimension (3D) 403
  spatio-temporal data volume 403–4
  visualization of vehicle's motion 123
thrust allocation 93
thruster(s) 183, 292
  assemblies 296–7
  readout 64
  status 67
  tank experiment measuring thrust and voltage input 184
"TightVNC Viewer", 351
time response
  of controllers for certain angle 202
  of LQR control of DC motor 173
  of PID control of DC motor 167
  of SMC of DC motor 168
time-sharing scheduling algorithm 37
time-varying control law 171
tissue temperature 235
torque sensor 185
training cascade 371–7
transformation 182
transistor transistor logic (TTL) 321
transmission control protocol (TCP) 17
transmission scheme 390–1
Transmotec DC motor 160–1
transportation of cargo 345
TransRapid in Germany 261
traveling mechanism 329
trial-and-error method 121
triangular similarity method 424
Triple Axis Accelerometer 389
TTL: *see* transistor transistor logic
tuning coefficients 106
Tupperware 297
  exterior with PVC fittings 295
  gluing PVC frame onto holes of Tupperware with hot glue gun 296
  hole saw drilling hole in 295
  horizontal PVC frame assembling with 296
turbulence properties 74
12A8 PWM driver connection to DC motor 131
12A8 PWM servo drive 130
204:1 gear ratio 251
two-wheel drive vacuum robot using PIC18F4520 8
tympanic probe 242

Ubuntu 20
  opening terminal application on Ubuntu Linux® 304
  operating software 2
  operating system on VMware player 303
  Ubuntu 16.04.2 LTS 18
UDP: *see* User Datagram Protocol
ultrasonic sensors 214, 252
underwater robotic systems development 177
underwater robotic vehicles 69, 177, 188, 199
  actual motion sequence 204

ANSYS-CFX nonlinear damping coefficients for 191
control 178
dynamic model 181–3
hydrodynamic parameters 196
measured values for test-bed underwater robotic vehicle 195
meshing for flow domain around 194
nonlinear damping force in X, Y, and moment 192
Pro/ENGINEER software to compute moment of inertia 189
Reynolds number 193
rig for drag testing in X-direction 190
switching signal 205
time responses 205
xPC-Target in horizontal plane used in water-tank test 197
underwater test 382, 385
universal serial bus port (USB port) 1
upward communication 391
USB Logitech Webcams C170 cameras 50, 55, 294
USB port: *see* universal serial bus port
USB Video Class Linux® device driver (UVC) 55
USB webcam 5
driver module 51
user-friendly graphical interface 123
User Datagram Protocol (UDP) 40
user interface 392–3
layer 30
UTM in Korea 261
UVC: *see* USB Video Class Linux® device driver

V4L2: *see* Video4Linux2
vacuum cleaning robot 7
vacuum robot 234
vacuum suction 243–4
validation of experimental results 84
heave model identification 84–8
yaw model identification 88–92
van der Waals forces 243
vehicle body-fixed frame 98
vibration
measurement technique 388
monitoring system 397
sensors 388
vibration-level monitoring 391–2
video
display 64
image acquisition programs 54–6
image capture program 51
image UDP thread program flowchart 66
processing and logging 391–2
streaming 396–7
streams 40
Video4Linux2 (V4L2) 50
Virtual Reality Modeling Language file format 93
vision-based guidance 388
visual-based surveillance system 399
Visual Studio C++ language 394
VMware player 304

wait queue 32
waking 32
wall-climbing robot system design 243
comparison between different types of adhesion 244
comparison of sensors 253
electronic circuit board 256
flowchart for climbing robot 253
free body diagram of climbing robot front body 249
free body diagram of robot transiting from horizontal to vertical surface 250
front view of robot 257
Gecko pad adhesion test results 252
intelligent climbing robot climbing table leg 260

isometric view of prototype design 245–6
Mabuchi motor specifications 251
nanotubes 244
overview of climbing robot 259
PIC program codes 254–5
programming and system implementation 253–9
sensor calibration test setup and output voltage waveform 258
sensor characteristics 258
specifications of back body and front view 247
system design and architecture 245–53
tank tracks with gecko pad 256
testing 259–60
transition from horizontal to vertical surface 249
see also PIC embedded system design
WAMIT™ solver 80, 82
water jet thruster 293
water-tank test, xPC-Target in horizontal plane used in 197
web-based wireless remote temperature-monitoring device 242
webpage
design concept 380
initial webpage design 381
Wi-Fi 242, 399
adapters 391, 394
interface of AR9331 306
Wicd curses 392, 395
wide-area network 207
wind-tunnel or towing-tank model tests 183
Window 98/XP/Vista/7 121–2
Window Operating System 121–2
Windows-based embedded systems 9
Windows® programming interface 10, 214
wired cables 293
wireless local area network (WLAN) 5
wireless network
GUI 392
protocols 242
wireless remote temperature-monitoring system 236, 242
wireless temperature sensor system 236
WLAN: *see* wireless local area network

XHTML 397
xPC Target 9
using Athena II-PC104 as target PC 139–42
block diagram 187, 203
and desktop PC-NI PCI card 135
using desktop PC as Target PC 136–9
in horizontal plane used in water-tank test 197
real-time kernel creation 136
Simulink® block diagrams creation 142–51
xPC-Target embedded system design
creating xPC-Target Simulink® block diagrams 142–51
digital speed control of servo DC motor 175–7
example of HILS solution on underwater robotic vehicle 123
hardware-in-the-loop testing using xPC-Target 134–42
hardware interfacings 124–34
hardware interfacings for simulations testing 123–4
identification of servo DC motor 164–8
incremental encoder model 159–64
infrared sensor model 158–9
LQR 171–4
marine robotic vehicle with uncertainties using xPC-Target system 177–207

PID speed control of servo DC motor 168–9
using RS232, analog, and digital I/O in xPC-Target 151–8
sliding-model speed control of servo DC motor 169–70
see also ARDUINO® embedded system design; Linux®-based embedded system design; Raspberry Pi-embedded system design
Xsens MTi-30 AHRS sensor 42

yaw
model identification 88–92
motion 84
tests 117
Yun Buddy
app 5, 306, 309
control interface 307

ZENA wireless network analyzer 239
ZigBee 242

www.ingramcontent.com/pod-product-compliance
Lightning Source LLC
LaVergne TN
LVHW011859200926
848484LV00002B/488

9781785613227